Integral operators in spaces
of summable functions

Monographs and textbooks on mechanics of solids and fluids

editor-in-chief: G. Æ. Oravas

Mechanics analysis

editor: V. J. Mizel

Integral operators in spaces of summable functions

M. A. Krasnoselskii, P. P. Zabreiko
E. I. Pustylnik, P. E. Sbolevskii

Translated by

T. Ando

Hokkaido University
Sapporo, Japan

NOORDHOFF INTERNATIONAL PUBLISHING – LEYDEN

A division of A. W. Sijthoff International Publishing Company B.V.

Leyden, The Netherlands

ISBN: 90 286 0294 1

Printed in The Netherlands by Kon. Drukkerij van de Garde B.V., Zaltbommel

Contents

Preface

The investigation of many mathematical problems is significantly simplified if it is possible to reduce them to equations involving continuous or completely continuous operators in function spaces. In particular, this is true for non-linear boundary value problems and for integro-differential and integral equations.

To effect a transformation to equations with continuous or completely continuous operators, it is usually necessary to reduce the original problem to one involving integral equations. Here, negative and fractional powers of those unbounded differential operators which constitute 'principal parts' of the original problem, are used in an essential way. Next there is chosen or constructed a function space in which the corresponding integral operator possesses sufficiently good properties. Once such a space is found, the original problem can often be analyzed by applying general theorems (Fredholm theorems in the study of linear equations, fixed point principles in the study of non-linear equations, methods of the theory of cones in the study of positive solutions, etc.).

In other words, the investigation of many problems is effectively divided into three independent parts: transformation to an integral equation, investigation of the corresponding integral expression as an operator acting in function spaces, and, finally, application of general methods of functional analysis to the investigation of the linear and non-linear equations.

This book concerns the second of the above divisions. Here, linear operators acting in spaces L_α of summable functions are studied and criteria are established for continuity, regularity, compactness of linear operators and their fractional powers. Such properties are considered also for non-linear operators, as is the problem of differentiability of non-linear operators and so on. We emphasize that properties of integral operators are not only important in the study of integral equations, but that such operators and their fractional powers also play an important role in the study of convergence of Fourier series, in the investigation of approximation methods, in the consideration of various evolution problems, in the theory of oscillation, and so forth.

In the investigation of concrete integral operators, the possibility of considering them in various spaces L_α is often important. In this connection operators are here studied simultaneously in the whole family of spaces L_α. For linear operators such a study permits one to make wide use of interpolation theorems, which had their origin in the famous works of M. Riesz. It also permits one to formulate in a simple manner theorems on general properties of fractional powers of linear operators. Since we wish to describe general properties of both linear and non-linear operators with respect to the family of spaces L_α, we consistently utilize the so-called L-characteristics of such operators.

The book consists of six chapters.

In the first chapter general properties of linear operators are studied. Here the Riesz and Marcinkiewicz interpolation theorems play a central role. In addition compactness properties of linear operators are discussed in detail.

In the second chapter it is shown that a linear integral operator possesses many supplementary properties in comparison with abstract linear operators. There are presented various sufficient conditions for continuity and compactness of an integral operator with a kernel $K(s, t)$, in terms of properties of the kernel. In particular, important theorems of L. V. Kantorovic are presented and some modifications of them are analyzed. Also the interpolation theorem of Stein-Weiss is presented; this leads to a simple proof of the limit theorem of S. L. Sobolev and V. P. Ilin on operators of potential type.

The third chapter concerns fractional powers of selfadjoint operators. Here we establish theorems on splitting of selfadjoint positive definite operators A, on properties of fractional powers of A, on inequalities connected with fractional powers, and so on.

In the fourth chapter, fractional powers of operators acting in Banach spaces are studied. Here we discuss from which spaces to which other spaces fractional powers of positive type operators act. As an application, elliptic differential operators are considered.

The fifth chapter concerns non-linear operators. Here, known criteria on continuity and complete continuity of non-linear integral operators are systematized and new criteria are given. The problem of differentiability of non-linear operators at distinguished points in spaces L_α is subjected to special analysis.

In the final (sixth) chapter are indicated the simplest applications of the general theorems established in the principal parts of the book. Here we describe general ways of applying theorems on continuity and compactness of integral operators in order to do the following: prove solvability of various equations; estimate the number of solutions; justify convergence of some approximation methods; investigate bifurcation values of parameters; and so on. Likewise, applications of fractional powers of operators to the analysis of convergence of Fourier series and to Fourier's method of solution of boundary value problems are presented. Finally, translation operators along the trajectory of differential equations with unbounded operators in Banach spaces are studied; fixed points of such operators determine, for instance, periodic solutions of parabolic equations.

The book can be understood by readers, acquainted with only the fundamental notions of functional analysis given in a university program (for instance, familiarity with the principal chapters of the monograph of L. A. Ljusternik and V. I. Sobolev is sufficient). That part of the theory of unbounded selfadjoint operators which is used in the book is written out in detail. The authors also considered it appropriate to present, with complete proofs, some theorems on generators of semi-groups of operators.

The authors express their hearty thanks to B. S. Mityagin and Y. B. Rutickii, who read the manuscript in its original form. Their many suggestions helped to improve the book. The authors also thank E. M. Semenov, S. G. Krein, and S. D. Eidelman for numerous suggestions.

1

Linear operators in L_α spaces

§ 1 The spaces L_α

1.1 *Description of the spaces*[1]

In the sequel some basic information is presented concerning certain spaces of real or complex functions $x(s)$ defined on a set Ω of finite Lebesgue measure in a finite dimensional space. It is also possible to take Ω to be an arbitrary set with a finite continuous measure[2].

We let $S = S(\Omega)$ denote, as usual, the space of all almost finite measurable functions $x(s)$ with metric $\rho(x, y) = \|x - y\|$, where[3]

$$\|x\| = \inf_{a>0}\{a + \text{mes}\{s; |x(s)| \geqq a\}\}. \tag{1.1}$$

Convergence in the space S coincides with the usual notion of convergence in measure. At the same time, almost everywhere convergence in S is also important. Almost everywhere convergence of a sequence $x_n(s)$ implies its convergence in measure. The converse is not true, but it is possible to choose an almost everywhere convergent subsequence from each sequence which is convergent in measure.

---

[1] Properties of the spaces L_α, which are given in this paragraph and used throughout this book, are presented in full detail in the book of I. P. Natanson [1] (see also L. V. Kantorovic and G. P. Akilov [1] and S. L. Sobolev [2]).

[2] A measure is called continuous, if each measurable set $D \subset \Omega$ can be split into two parts of equal measure.

[3] $\{t: |x(t)| \geqq a\}$ denotes the set of those points t, at which $|x(t)| \geqq a$.

Let α be a positive real number. Then $L_\alpha = L_\alpha(\Omega)$ denotes the set of measurable functions $x(s)$, for which the integral $\int_\Omega |x(s)|^{1/\alpha} ds$ is finite. The number

$$\|x\|_\alpha = \{\int_\Omega |x(s)|^{1/\alpha} ds\}^\alpha \tag{1.2}$$

is called the *norm* of an element $x \in L_\alpha$.

When $\alpha \leqq 1$, the space L_α becomes a Banach space (with norm (1.2)). When $\alpha > 1$, the space L_α is not normable; however it becomes a complete metric space with metric $\rho(x, y) = (\|x - y\|_\alpha)^{1/\alpha}$. Convergence in L_α is usually called *convergence in mean with exponent* $1/\alpha$.

Together with the spaces L_α there is usually considered the space M of essentially bounded functions $x(t)$ with norm

$$\|x\|_0 = \operatorname*{ess\,sup}_{s \in \Omega} |x(s)|. \tag{1.3}$$

This space is also denoted by L_0, because

$$\lim_{\alpha \to 0} \|x\|_\alpha = \|x\|_0 \qquad (x \in L_0 = M). \tag{1.4}$$

In contrast to the spaces L_α $(0 < \alpha < \infty)$, the space L_0 is non-separable. The space C of uniformly continuous functions on Ω is the most important subspace of the space L_0. The space C is separable if the set Ω is bounded.

In the sequel it is necessary to examine different spaces L_α simultaneously. In such cases Hölder's inequality:

$$\|x \cdot y\|_{\alpha+\beta} \leqq \|x\|_\alpha \cdot \|y\|_\beta \qquad (x \in L_\alpha,\ y \in L_\beta), \tag{1.5}$$

valid for any $\alpha, \beta \in [0, \infty)$, plays a basic role. From Hölder's inequality it follows, in particular, that the function $\lambda(\alpha) = \|x\|_\alpha$ for any fixed element $x(s)$ is continuous and logarithmically convex. This means that for any $\tau \in (0, 1)$ the following inequality is fulfilled:

$$\|x\|_{\alpha(\tau)} \leqq \|x\|_{\alpha_0}^{1-\tau} \|x\|_{\alpha_1}^{\tau}, \tag{1.6}$$

where

$$\alpha(\tau) = (1 - \tau)\alpha_0 + \tau\alpha_1.$$

The inequality (1.5) implies that for $\alpha_1 < \alpha_2$ the space L_{α_1} is imbedded in the space L_{α_2}, i.e. $L_{\alpha_1} \subset L_{\alpha_2}$ and

$$\|x\|_{\alpha_2} \leqq k \|x\|_{\alpha_1} \quad (x \in L_{\alpha_1};\ k = (\text{mes } \Omega)^{\alpha_2 - \alpha_1}). \tag{1.7}$$

1.2 *Criteria for compactness*

Let D be a measurable subset of Ω. Denote by $\boldsymbol{P}_D$ the linear operator, defined by the relation

$$\boldsymbol{P}_D x(s) = \begin{cases} x(s), & \text{if } s \in D \\ 0, & \text{if } s \notin D. \end{cases} \tag{1.8}$$

It is clear that $\boldsymbol{P}_D^2 = \boldsymbol{P}_D$ and[1] $\|\boldsymbol{P}_D\|_{\alpha \to \alpha} = 1$, if $\text{mes}(D) \neq 0$.

The norm in the spaces L_α for $\alpha > 0$ possesses the property of *absolute continuity*. This means that for each fixed function $x(s) \in L_\alpha$

$$\lim_{\text{mes } D \to 0} \|\boldsymbol{P}_D x\|_\alpha = 0.$$

A family $\mathfrak{M}$ of functions in L_α is said to have *equi-absolutely continuous norms*, if for any $\varepsilon > 0$ there can be found a $\delta > 0$ such that $\text{mes}(D) < \delta$ and $x(s) \in \mathfrak{M}$ imply the inequality $\|\boldsymbol{P}_D x(s)\|_\alpha < \varepsilon$.

It is clear that convergence of a sequence of functions $x_n(s)$ to $x_0(s)$ in the norm of L_α, implies convergence of this sequence in measure (but not almost everywhere!). The converse is not true. What is valid for the spaces L_α $(0 < \alpha < \infty)$ is that a sequence $x_n(s)$, convergent to a function $x_0(s)$ in measure, converges in the norm of L_α $(0 < \alpha < \infty)$ if and only if the functions $x_n(s)$ have equi-absolutely continuous norms.

This assertion immediately implies the following simple criterion for compactness in the spaces L_α $(0 < \alpha < \infty)$.

LEMMA 1.1: *A family $\mathfrak{M}$ of functions $x(s) \in L_\alpha$ $(0 < \alpha < \infty)$ is compact if and only if it is compact in measure and $\mathfrak{M}$ has equi-absolutely continuous norms.*

---

[1] $\|\boldsymbol{A}\|_{\alpha \to \beta}$ denotes norm of $\boldsymbol{A}$ as an operator acting from L_α to L_β (see 2.1°).

In the study of concrete families of functions, a direct verification of compactness in measure is often difficult to achieve. Hence the following scheme of arguments is usually applied. First it is shown that the set $\mathfrak{M}$ of functions in L_α which is in question is compact in some space L_{α_0}, where $\alpha_0 > \alpha$. It then follows that $\mathfrak{M}$ is compact in measure.

To check equi-absolute continuity of norms of the functions in a set $\mathfrak{M}$, it is sometimes convenient to use the following criterion of Vallée-Poussin:

The norms of functions $x(s)$ in a set $\mathfrak{M}$ are equi-absolutely continuous, if and only if there exists an even continuous function $\Phi(u)$, satisfying the condition

$$\lim_{u\to\infty} \frac{\Phi(u)}{u} = \infty,$$

for which

$$\|\Phi[x(s)]\|_\alpha \leqq A \qquad (x \in \mathfrak{M}).$$

Now examine the space L_0. The norm in this space does not possess the property of absolute continuity. In fact, for each non-zero function $x(s) \in L_0$, it is possible to find sets $D_n \subset \Omega$ such that

$$\operatorname{mes}(D_n) \to 0 \ \text{ but } \ \|\boldsymbol{P}_{D_n}x\|_0 = \|x\|_0 \qquad (n = 1, 2, \ldots).$$

LEMMA 1.2: *A bounded set $\mathfrak{M}$ of functions in L_0 is compact if and only if for any $\varepsilon > 0$ there exists a partition of the set Ω into a finite number of subsets $\Omega_1, \ldots, \Omega_n$,*

$$\Omega = \Omega_1 \cup \Omega_2 \cup \ldots \cup \Omega_n$$

such that for any function $x(s) \in \mathfrak{M}$

$$\operatorname*{ess\,sup}_{s', s'' \in \Omega_i} |x(s') - x(s'')| < \varepsilon \qquad (i = 1, 2, \ldots, n).$$

This condition for compactness can be modified, if $\mathfrak{M}$ lies in some separable subspace of L_0. For instance, if $\mathfrak{M} \subset C$ (C being the space of uniformly continuous functions) and Ω is a bounded set, then Arzela's criterion for compactness is valid: a set $\mathfrak{M} \subset C$ is compact, if and only if it is bounded in norm and the functions $x(s)$ in $\mathfrak{M}$ are equi-uniformly continuous.

The criteria for compactness, presented in this section, will later be used to prove the compactness of some linear and non-linear operators. Lemma 1.1 is systematically used for this purpose in M. A. Krasnoselskii [1], M. A. Krasnoselskii and Y. B. Rutickii [4, 5] and elsewhere; Lemma 1.2 is given in N. Dunford and J. T. Schwartz [1]. Criteria for compactness, due to A. N. Kolmogorov [1] and F. Riesz [1] have also been applied by a number of authors. Finally, we note that in the study of concrete operators it is often convenient to use the definition of compactness of sets directly, not relying on any criterion for compactness.

1.3 *Continuous linear functionals and weak convergence*

A functional $f(x)$, defined on a linear space E, is called *linear* and *continuous*, if

$$f(\alpha_1 x_1 + \alpha_2 x_2) = \alpha_1 f(x_1) + \alpha_2 f(x_2)$$

and if $f(x_n) \to f(x_0)$ follows from $\|x_n - x_0\| \to 0$.

There does not exists any non-zero continuous linear functional on the spaces L_α, where $\alpha > 1$. In the sequel, we will describe the continuous linear functionals on those spaces L_α, where $\alpha \in (0, 1]$.

For continuous linear functionals $f(x)$, defined on a Banach space E (in particular, on a space L_α, $\alpha \in [0, 1]$), we introduce the notion of the *norm* of a functional:

$$\|f\| = \sup_{\|x\| \leqq 1} |f(x)|.$$

The totality of continuous linear functionals on a Banach space E also constitutes a Banach space. This space is denoted by E^*; it is called the *dual* space (conjugate, adjoint) of E.

The expression

$$(x, y) = \int_\Omega x(s)\overline{y(s)}\,\mathrm{d}s \tag{1.9}$$

is called the *scalar product* of the functions $x(s)$, $y(s)$. Hölder's inequality implies that $|(x, y)| < \infty$ if $x \in L_\alpha$, $y \in L_{1-\alpha}$ $(0 \leqq \alpha \leqq 1)$. It is of course possible for the scalar product to be finite in other cases.

Let $0 \leqq \alpha \leqq 1$. The following fact plays an important role; $a(t) \in L_{1-\alpha}$, if the inequality $|(x, a)| < \infty$ is fulfilled for all functions $x(s) \in L_\alpha$. Furthermore

$$\|a\|_{1-\alpha} = \sup_{\|x\|_\alpha \leqq 1} |(x, a)|. \tag{1.10}$$

As it turns out, each continuous linear functional $f(x)$, defined on L_α $(0 < \alpha \leqq 1)$, can be written in the form

$$f(x) = (x, a),$$

where $a \in L_{1-\alpha}$, and $\|f\| = \|a\|_{1-\alpha}$. In this connection the space $L_{1-\alpha}$ is called the dual of L_α.

The expression $f(x) = (x, a)$ for $a \in L_1$ becomes a continuous linear functional on L_0; however there exist other continuous linear functionals on L_0; L_1 can be considered as a subspace of the space L_0^* dual to L_0.

A space E is called *reflexive*, if [1] $(E^*)^* = E$. The above observations show that the spaces L_α for $0 < \alpha < 1$ are reflexive, but those with $\alpha = 0,1$ are not reflexive. The space $L_{\frac{1}{2}}$ is Hilbertian.

A sequence $x_n \in E$ is called *weakly convergent*, if, for every $f \in E^*$, the sequence $f(x_n)$ converges. A sequence $x_n \in E$ converges weakly to an element x_0, if for every $f \in E^*$, the sequence $f(x_n)$ converges to $f(x_0)$. If each weakly convergent sequence converges weakly to some limit, the space E is called *weakly complete*. A space E is called *weakly compact*, if it is possible to choose from each bounded sequence a weakly convergent subsequence.

Each reflexive space is weakly complete and weakly compact. Hence the spaces L_α for $\alpha \in (0, 1)$ are weakly complete and weakly compact. The space L_1 is weakly complete but does not possess the property of weak compactness.

LEMMA 1.3:[2] *A set* $\mathfrak{M}$ *of functions in* L_1 *is weakly compact if and only if the norms of the functions in* $\mathfrak{M}$ *are equi-absolutely continuous.*

---

[1] The following relation is understood in the sense that each functional $\varphi \in E^{**}$ has the form $\varphi(f) = f(x_0)$, where x_0 is some element of E.

[2] I. P. Natanson [1], N. Dunford and J. T. Schwartz [1].

In the sequel we will consider a special notion of weak convergence for the space L_0, one which is distinct from what is defined above.

Let F be a linear subspace of the space E^*. A sequence $x_n \in E$ is called *F-weakly convergent*, if, for each $f \in F$, the sequence $f(x_n)$ converges. The notions of F-weak completeness and F-weak compactness are introduced in the obvious way. If F is separable the space E is F-weakly compact.

In the space L_0 we will always mean by 'weak convergence' the weak convergence, defined by the subspace $F = L_1$. Since L_1 is separable, the space L_0 is weakly compact with respect to this convergence. It can be shown that L_0 also possesses the property of weak completeness with respect to this convergence.

In conclusion we note that weak convergence does not imply convergence in measure, and convergence in measure does not imply weak convergence. But if a sequence x_n converges weakly and converges in measure, then the weak limit coincides with the limit in measure. Sometimes it is convenient to use the following remark: *if a sequence x_n converges weakly to x_0 and is compact in measure, then it converges to x_0 in measure.*

1.4 *Semi-ordering in the spaces S and L_α*

It is convenient to consider the spaces S and L_α $(0 \leqq \alpha < \infty)$ in their natural semi-ordering: a function $x(s)$ is smaller than a function $y(s)$, if $x(s) \leqq y(s)$ for almost all $s \in \Omega$. This semi-ordering possesses the following properties:

1°) $x \leqq y$, $y \leqq z$ imply $x \leqq z$.
2°) $x \leqq y$, $y \leqq x$ imply $x = y$.
3°) $x \leqq y$ implies that $x + z \leqq y + z$ for any z.
4°) $x \leqq y$, $\lambda \geqq 0$ implies $\lambda x \leqq \lambda y$.
5°) If $x_n(s) \leqq y_n(s)$ $(n = 1, 2, \dots)$ and the sequences $x_n(s)$ and $y_n(s)$ converge in measure (or in the norm of L_α) respectively to $x^*(s)$ and $y^*(s)$ as $n \to \infty$, then $x^*(s) \leqq y^*(s)$. In other words, inequality is preserved under the limit operation.

In a semi-ordered space (in particular, in S or L_α) there are defined the notions of supremum and infimum of a set $\mathfrak{M}$. An element z_1 (z_2) is called the *supremum* (*infimum*) of a set $\mathfrak{M}$ and is denoted by $\sup \mathfrak{M}$ $(\inf \mathfrak{M})$, if $x \leqq z_1$ $(x \geqq z_2)$ for each $x \in \mathfrak{M}$ and if $x \leqq u$ $(x \geqq u)$ being valid for all elements $x \in \mathfrak{M}$, implies $z_1 \leqq u$ $(u \geqq z_2)$.

Each set in S or L_α which is bounded from above (below), has a supremum (infimum). If the set $\mathfrak{M}$ contains not more than a countable number of elements then the following formula is valid:

$$z_1(s) = \sup \mathfrak{M} = \sup_{x \in \mathfrak{M}} x(s).$$

In the general case the relation takes the form:

$$z_1(s) = \sup_n \{x_n(s)\}$$

where $\{x_n(s)\}$ is any countable set of functions of $\mathfrak{M}$, dense[1] in $\mathfrak{M}$. Similar formulas are valid for the infimum.

Let us present several properties of the spaces L_α, connected with their semi-ordering.

1°) Each element $x \in L_\alpha$ is uniquely represented in the form:

$$x = x_+ - x_- \qquad (x_+, x_- \geqq 0),$$

where

$$x_+ = \sup(x, 0), \quad x_- = \sup(-x, 0).$$

Here

$$\|x_+\|_\alpha + \|x_-\|_\alpha \leqq 2\|x\|_\alpha.$$

2°) $0 \leqq x \leqq y$ implies $\|x\|_\alpha \leqq \|y\|_\alpha$.

3°) In the space L_1 the norm $\|x\|_1$ possesses the property

$$\|x + y\|_1 = \|x\|_1 + \|y\|_1 \qquad (x, y \geqq 0).$$

4°) In the space L_0 the norm $\|x\|_0$ possesses the property

$$\|\sup\{x_1, x_2\}\|_0 = \max\{\|x_1\|_0, \|x_2\|_0\} \qquad (x_1, x_2 \geqq 0).$$

5°) Let $\alpha \in (0, \infty)$. Then each monotone sequence x_n $(n = 1, 2, \ldots)$ which is bounded in norm converges to some element x^*.

---

[1] For $\mathfrak{M} \subset L_0$ this refers to denseness in the S metric; the latter property actually suffices for $\mathfrak{M} \subset L_\alpha$ $0 \leqq \alpha < \infty$ [Ed.].

For $\alpha = 0$ this assertion is not true: the sequence will merely converge to some element almost everywhere.

1.5 *Projections and bases of Haar type*

In the constructions of this section the condition $\alpha \leqq 1$ is essential. We are interested in special bases in the spaces L_α. Their construction is connected with certain finite dimensional projections, acting in the spaces L_α (averaging operators).

Suppose that the set Ω is partitioned into a number of disjoint parts:

$$\Omega = \Omega_1 \cup \Omega_2 \cup \ldots \cup \Omega_q. \tag{1.11}$$

Define the operator $\boldsymbol{P}$ by the relation

$$\boldsymbol{P}x(s) = \int_\Omega K(s, \sigma)x(\sigma)\,\mathrm{d}\sigma \tag{1.12}$$

where

$$K(s, \sigma) = \begin{cases} \dfrac{1}{\operatorname{mes} \Omega_i}, & \text{if } s, \sigma \in \Omega_i \quad (i = 1, 2, \ldots, q) \\ 0, & \text{for other } s, \sigma \in \Omega. \end{cases}$$

In other words

$$\boldsymbol{P}x(s) = \frac{1}{\operatorname{mes} \Omega_i} \int_{\Omega_i} x(\sigma)\,\mathrm{d}\sigma \qquad (s \in \Omega_i,\ i = 1, 2, \ldots, q). \tag{1.13}$$

It is easy to see that the operator $\boldsymbol{P}$ is defined on all summable functions. Its values belong to a finite dimensional space—the linear span of the totality of characteristic functions of sets $\Omega_1, \ldots, \Omega_q$. Thus the operator (1.12) acts in each space $L_\alpha(\Omega)$ $(0 \leqq \alpha \leqq 1)$.

First let $\alpha > 0$. Then for each function $x(s) \in L_\alpha$ we have, by Hölder's inequality,

$$\int_\Omega |\boldsymbol{P}x(s)|^{1/\alpha}\mathrm{d}s = \sum_{i=1}^{q} (\operatorname{mes} \Omega_i)^{1-1/\alpha} \left|\int_{\Omega_i} x(\sigma)\,\mathrm{d}\sigma\right|^{1/\alpha}$$

$$\leqq \sum_{i=1}^{q} (\operatorname{mes} \Omega_i)^{1-1/\alpha} \int_{\Omega_i} |x(\sigma)|^{1/\alpha}\mathrm{d}\sigma \cdot (\operatorname{mes} \Omega_i)^{(1/\alpha)-1} = \int_\Omega |x(\sigma)|^{1/\alpha}\mathrm{d}\sigma$$

i.e.

$$\|Px\|_\alpha \leqq \|x\|_\alpha \quad (x \in L_\alpha). \tag{1.14}$$

In the case $\alpha = 0$, inequality (1.14) follows in trivial way from (1.13). Consequently, the operator (1.12) is continuous in each space $L_\alpha(\Omega)$ $(0 \leqq \alpha \leqq 1)$.

The relation $P^2 = P$ is easily verified. Consequently, P becomes an operator projecting L_α onto a finite dimensional space.

Consider a sequence

$$\Omega = \Omega_1^{(n)} \cup \ldots \cup \Omega_{q(n)}^{(n)} \quad (n = 1, 2, \ldots) \tag{1.15}$$

of partitions of the type (1.11). Define for each of these partitions an operator P_n by using formula (1.12). The sequence P_n $(n = 1, 2, \ldots)$ is called *regular*, if for each measurable set $F \subset \Omega$ and any $\varepsilon > 0$ there can be taken sets $\Omega_{i_1}^{(n)}, \ldots, \Omega_{i_k}^{(n)}$ in the partition (1.15) such that

$$|\mathrm{mes}\,[(\Omega_{i_1}^{(n)} \cup \ldots \cup \Omega_{i_k}^{(n)}) \cap F] - \mathrm{mes}(F)| < \varepsilon \tag{1.16}$$

and

$$|\mathrm{mes}\,[(\Omega_{i_1}^{(n)} \cup \ldots \cup \Omega_{i_k}^{(n)}) \cap F] - \mathrm{mes}(\Omega_{i_1}^{(n)} \cup \ldots \cup \Omega_{i_k}^{(n)})| < \varepsilon. \tag{1.17}$$

It is sufficient for regularity of a sequence P_n, for instance, that the maximum diameter of the sets $\Omega_i^{(n)}$ in the partition (1.15) converges to zero as $n \to \infty$.

LEMMA 1.4: *Let P_n $(n = 1, 2, \ldots)$ be a regular sequence of projections* (1.12). *Then the sequence P_n converges strongly in each space L_α* $(\alpha \in (0, 1])$ *to the identity operator:*

$$\lim_{n \to \infty} \|P_n x(s) - x(s)\|_\alpha = 0 \quad (x \in L_\alpha). \tag{1.18}$$

PROOF: First let $x(s)$ be a characteristic function of a measurable set $F \subset \Omega$. Since the sequence P_n is regular, it is possible to define a sequence of characteristic functions $x_n(s)$ $(n = n_0, n_0 + 1, n_0 + 2, \ldots)$ of sets $\Omega_{i_1}^{(n)} \cup \ldots \cup \Omega_{i_k}^{(n)}$, satisfying the conditions (1.16) and (1.17). It is clear that

$$P_n x_n(s) \equiv x_n(s) \quad (n = n_0, n_0 + 1, \ldots)$$

and

$$\|x_n(s) - x(s)\|_\alpha \leqq 2^\alpha \varepsilon^\alpha \qquad (n = n_0, n_0 + 1, \dots).$$

Thus (1.14) implies that for $n \geqq n_0$

$$\|\boldsymbol{P}_n x(s) - x(s)\|_\alpha$$

$$\leqq \|\boldsymbol{P}_n[x(s) - x_n(s)]\|_\alpha + \|\boldsymbol{P}_n x_n(s) - x_n(s)\|_\alpha$$

$$+ \|x(s) - x_n(s)\|_\alpha \leqq 2\|x(s) - x_n(s)\|_\alpha \leqq 2^{1+\alpha}\varepsilon^\alpha.$$

Consequently, relation (1.18) is fulfilled for characteristic functions.

Linearity of the operators $\boldsymbol{P}_n$ implies that they converge strongly to the identity on the linear span of the set of characteristic functions. However, the linear span of the set of characteristic functions is dense in each $L_\alpha(\Omega)$ $(0 < \alpha \leqq 1)$, and the norms of the operators $\boldsymbol{P}_n$ are uniformly bounded. Hence the operators $\boldsymbol{P}_n$ converge strongly to the identity on the whole of L_α. The lemma has been proved.

In the case of the space $L_0(\Omega)$ the assertion of Lemma 1.4 is not true, because the space $L_0(\Omega)$ is non-separable, while the ranges of all the operators $\boldsymbol{P}_n$ lie in some separable subspace of $L_0(\Omega)$.

Denote by $L_0(\Omega; \{\boldsymbol{P}_n\})$ the set of those functions $x(s) \in L_0(\Omega)$ for which the sequence $\boldsymbol{P}_n x$ converges to x in the metric of the space $L_0(\Omega)$. It is easy to see that the finite dimensional subspaces $\boldsymbol{P}_n L_0(\Omega)$ $(n = 1, 2, \dots)$ may not be included in $L_0(\Omega; \{\boldsymbol{P}_n\})$.

It is interesting to note that convergence of $\boldsymbol{P}_n x_0$ to some element $y_0 \in L_0$ implies the relations $y_0 = x_0$ and $x_0 \in L_0(\Omega; \{\boldsymbol{P}_n\})$.[1] Suppose the contrary: $x_0 \notin L_0(\Omega; \{\boldsymbol{P}_n\})$, $\|y_0 - \boldsymbol{P}_n x_0\|_0 \to 0$. Then the trivial inequality $\|y_0 - \boldsymbol{P}_n y_0\|_0 \leqq \|y_0 - \boldsymbol{P}_n x_0\|_0 + \|\boldsymbol{P}_n(y_0 - \boldsymbol{P}_n x_0)\|_0 \leqq 2\|y_0 - \boldsymbol{P}_n x_0\|_0$ implies that $y_0 \in L_0(\Omega; \{\boldsymbol{P}_n\})$. Then $z_0 = x_0 - y_0 \notin L_0(\Omega; \{\boldsymbol{P}_n\})$ and the sequence of functions $\boldsymbol{P}_n z_0$ converges uniformly to zero. It can be assumed without loss of generality that the function $z_0(s)$ takes positive values on a set of non-zero measure; let F be a set of non-zero measure such that

$$\alpha_0 \leqq z_0(s) \leqq \|z_0\|_0 \qquad (s \in F),$$

[1] This also follows directly from Lemma 1.4 since $\boldsymbol{P}_n x_0 \to x_0$ in L_α, for $\alpha > 0$. (Ed.)

where α_0 is some positive number. Since the sequence $\boldsymbol{P}_n$ is regular, it is possible to find n_0 such that for each $n \geqq n_0$ there can be found sets $\Omega_{i_0}^{(n)}$ in the partition (1.15) satisfying

$$\operatorname{mes}(\Omega_{i_0}^{(n)} \cap F) \geqq \frac{2\|z_0\|_0 + \alpha_0}{2(\|z_0\|_0 + \alpha_0)} \operatorname{mes}(\Omega_{i_0}^{(n)}).$$

Then for $s \in \Omega_{i_0}^{(n)}$ the following inequality will be fulfilled:

$$\boldsymbol{P}_n z_0(s) = \frac{1}{\operatorname{mes}(\Omega_{i_0}^{(n)})} \int_{\Omega^{(n)}i_0} z_0(\sigma)\,\mathrm{d}\sigma$$

$$\geqq \frac{1}{\operatorname{mes}\,\Omega_{i_0}^{(n)}} \left(\int_{\Omega^{(n)}i_0 \cap F} z_0(\sigma)\,\mathrm{d}\sigma - \int_{\Omega^{(n)}i_0 - F} |z_0(\sigma)|\,\mathrm{d}\sigma \right)$$

$$\geqq \alpha_0 \frac{\operatorname{mes}(\Omega_{i_0}^{(n)} \cap F)}{\operatorname{mes}(\Omega_{i_0}^{(n)})} - \|z_0\|_0 \frac{\operatorname{mes}\,\Omega_{i_0}^{(n)} - \operatorname{mes}(\Omega_{i_0}^{(n)} \cap F)}{\operatorname{mes}(\Omega_{i_0}^{(n)})} \geqq \frac{\alpha_0}{2}.$$

This means that the sequence $\boldsymbol{P}_n z_0$ does not converge to zero. We arrived at a contradiction.

LEMMA 1.5: *Let E be an arbitrary separable subspace of the space $L_0(\Omega)$. Then there exists a regular sequence of operators $\boldsymbol{P}_n$ which is strongly convergent to the identity operator in E:*

$$\lim_{n\to\infty} \|\boldsymbol{P}_n x(s) - x(s)\|_0 = 0 \qquad (x \in E). \tag{1.19}$$

PROOF: Let a sequence $x_1, x_2, \ldots$ be dense in the ball $\|x\|_0 \leqq 1$ of the space E. Denote by $D_{k,i}^{(n)}$ the set of points t at which

$$i/n \leqq x_k(s) < (i+1)/n \qquad (i = -n, -n+1, \ldots, 0, 1, \ldots, n).$$

Take as the n-th partition (1.15) the family $\Omega_1^{(n)}, \ldots, \Omega_{q(n)}^{(n)}$ of disjoint sets such that each set $D_{k,i}^{(n)}$ $(k = 1, \ldots, n)$ can be represented in a form of a union of some sets in this family. It is clear that with this choice of the sets $\Omega_i^{(n)}$ $(i = 1, \ldots, q(n);\ n = 1, 2, \ldots)$ the sequence $\boldsymbol{P}_n x_m$ converges to x_m

as $n \to \infty$ for any m. As is easily seen, the sets $\Omega_i^{(n)}$ can be chosen to have such small measure that the sequence $\boldsymbol{P}_n$ becomes regular.

Now let $x(s)$ be an arbitrary function in E. Then for any $\varepsilon > 0$ there is a function $x_m(s)$, such that $\|x - cx_m\|_0 < \varepsilon/2$, where $c = \|x\|_0$. Hence for sufficiently large n the following inequality will be fulfilled:

$$\|\boldsymbol{P}_n x - x\|_0$$

$$\leqq \|\boldsymbol{P}_n(x - cx_m)\|_0 + c\|\boldsymbol{P}_n x_m - x_m\|_0 + \|x - cx_m\|_0 < \varepsilon.$$

The lemma has been proved.

In particular, for the inclusion $C(\Omega) \subset L_0(\Omega; \{\boldsymbol{P}_n\})$ it is sufficient that the maximum diameter of the sets $\Omega_i^{(n)}$ in the partition (1.15) converges to zero as $n \to \infty$.

The assertions of Lemmas 1.4 and 1.5 are equivalent to an assertion on expansion of functions $x(s)$ in the form

$$x(s) = \boldsymbol{P}_1 x(s) + (\boldsymbol{P}_2 - \boldsymbol{P}_1)x(s) + \ldots + (\boldsymbol{P}_{n+1} - \boldsymbol{P}_n)x(s) + \ldots \qquad (1.20)$$

Naturally this raises the problem of whether the series (1.20) is an expansion with respect to some basis.

In order that the series (1.20) be an expansion with respect to a basis, it is necessary that each of operators $\boldsymbol{P}_{n+1} - \boldsymbol{P}_n$ $(n = 1, 2, \ldots)$ be a projection on a one-dimensional subspace. For this it is, in turn, sufficient that the $(n + 1)$-st partition be obtained from the n-th partition (1.15) by replacing one of the sets $\Omega_i^{(n)}$ by two disjoint subsets whose union is equal to $\Omega_i^{(n)}$.

Consequently, the constructions of the preceding section reveal the possibility of constructing various systems of functions which are bases simultaneously in all the spaces L_α $(0 < \alpha \leqq 1)$ as well as in some space $E \subset L_0$ containing the space C. In particular, with a suitable choice of partitions (1.15) we arrive at the well known Haar basis [1] (see also M. A. Krasnoselskii and Y. B. Rutickii [5]).

1.6 *Operators in the spaces L_α*

The study of integral, integro-differential and other types of equations is usually simplified, if they can be considered as operator equations in function spaces. A successful choice of spaces guarantees 'good' properties of the operators in the equations: continuity, complete continuity, differentiability, and so on.

In the study of operators in the spaces L_α it is natural to seek those α, β for which these operators possess needed properties. The notion of L-characteristic of a given operator then becomes a convenient method for the description and study of properties of operators in the spaces L_α for various α.

Let A be some concrete operator, for instance, an integral operator

$$Ax(t) = \int_\Omega K[t, s, x(s)]\,\mathrm{d}s,$$

acting from some space of functions, defined on a set Ω, to a space of functions defined, generally speaking, on another set Ω^*. The set $L(A; \text{def.})$ (Fig. 1.1) of all points $\{\alpha, \beta\}$ in the quadrant $\alpha, \beta \geqq 0$ such that the operator A acts from L_α to L_β, is called the *L-characteristic* of the operator A. It possesses the important property of *extrapolation:* if $\{\alpha_0, \beta_0\} \in L(A; \text{def.})$ then

$$\{\alpha, \beta\} \in L(A; \text{def.}) \quad \text{for} \quad \alpha \leqq \alpha_0,\ \beta \geqq \beta_0.$$

In fact, if A acts from L_{α_0} to L_{β_0}, then it will also act from the smaller space L_α ($\alpha \leqq \alpha_0$) to the larger space L_β ($\beta \geqq \beta_0$).

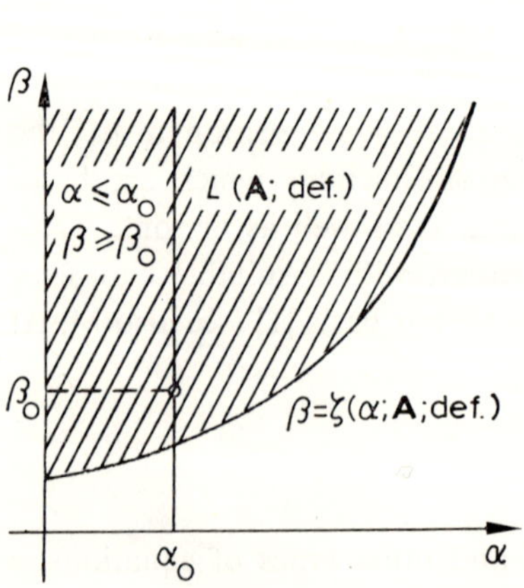

Figure 1.1

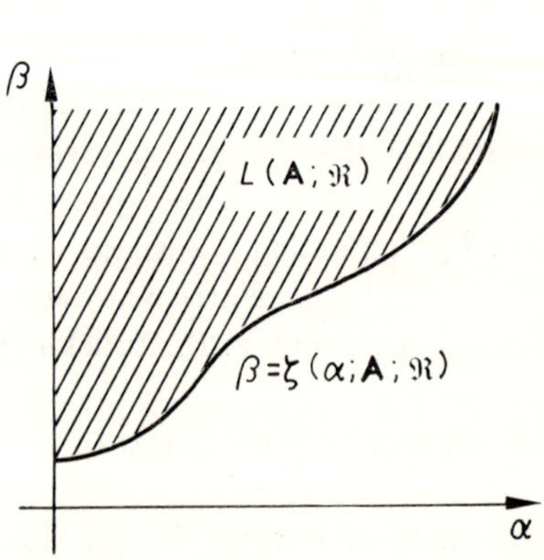

Figure 1.2

The property of extrapolation implies that the function

$$\xi(\alpha) = \xi(\alpha; \boldsymbol{A}; \text{def.}) = \inf_{\{\alpha, \beta\} \in L(\boldsymbol{A}; \text{def.})} \beta \tag{1.21}$$

is defined either on some semi-interval $[0, \alpha_0)$ or on some interval $[0, \alpha_0]$ and is non-decreasing. It is clear that

$$\{\{\alpha, \beta\}; \beta > \xi(\alpha)\} \subset L(\boldsymbol{A}; \text{def.}) \subset \{\{\alpha, \beta\} : \beta \geqq \xi(\alpha)\}. \tag{1.22}$$

The set $L(\boldsymbol{A}; \text{def.})$ possesses some simple properties resulting immediately from the definition.

1°) For the L-characteristics of any two operators $\boldsymbol{A}$ and $\boldsymbol{B}$ the following inclusion is valid:

$$L(\boldsymbol{A} + \boldsymbol{B}; \text{def.}) \supset L(\boldsymbol{A}; \text{def.}) \cap L(\boldsymbol{B}; \text{def.})$$

and so is the inequality:

$$\xi(\alpha; \boldsymbol{A} + \boldsymbol{B}; \text{def.})$$
$$\leqq \max\{\xi(\alpha; \boldsymbol{A}; \text{def.}), \xi(\alpha; \boldsymbol{B}; \text{def.})\}.$$

If $\xi(\alpha; \boldsymbol{A}) \neq \xi(\alpha; \boldsymbol{B})$ for all α, then

$$L(\boldsymbol{A} + \boldsymbol{B}; \text{def.}) = L(\boldsymbol{A}; \text{def.}) \cap L(\boldsymbol{B}; \text{def.}).$$

2°) The following inequality is valid:

$$\xi(\alpha; \boldsymbol{C}; \text{def.}) \leqq \xi(\alpha; \boldsymbol{A}; \text{def.}) + \xi(\alpha; \boldsymbol{B}; \text{def.}),$$

where $\boldsymbol{C}$ denotes the operator, defined by the relation $\boldsymbol{C}x = \boldsymbol{A}x \cdot \boldsymbol{B}x$.

3°) Let $\boldsymbol{A}$ and $\boldsymbol{B}$ be two operators. The L-characteristic $L(\boldsymbol{AB}; \text{def.})$ contains the set of points $\{\alpha, \beta\}$, for which there exists a number γ, such that

$$\{\alpha, \gamma\} \in L(\boldsymbol{B}; \text{def.}), \ \{\gamma, \beta\} \in L(\boldsymbol{A}; \text{def.}).$$

Hence (if $\xi(\alpha; \boldsymbol{A}; \text{def.})$ is continuous from the right)

$$\xi(\alpha; \boldsymbol{AB}; \text{def.}) \leqq \xi[\xi(\alpha; \boldsymbol{B}; \text{def.}); \boldsymbol{A}, \text{def.}].$$

It is natural to examine subsets of the L-characteristic $L(\boldsymbol{A}; \text{def.})$, consisting of points $\{\alpha, \beta\}$ such that the operator $\boldsymbol{A}$ acts from L_α to L_β and possesses some supplementary property $\mathfrak{R}$: continuity, compactness etc. These subsets will be denoted by $L(\boldsymbol{A}; \mathfrak{R})$ (Fig. 1.2) and will also be called *L-characteristics* of the operator $\boldsymbol{A}$. For most of the properties, studied in the sequel, the sets $L(\boldsymbol{A}; \mathfrak{R})$ will possess the property of extrapolation. As with the L-characteristic $L(\boldsymbol{A}; \text{def.})$, the sets $L(\boldsymbol{A}; \mathfrak{R})$ in such cases are defined in essence by monotone functions

$$\xi(\alpha; \boldsymbol{A}; \mathfrak{R}) = \inf_{\{\alpha, \beta\} \in L(\boldsymbol{A}, \mathfrak{R})} \beta.$$

It is clear that $\xi(\alpha; \boldsymbol{A}; \mathfrak{R}) \geqq \xi(\alpha; \boldsymbol{A}; \text{def.})$. We will often study the L-characteristics $L(\boldsymbol{A}; \text{cont.})$ and $L(\boldsymbol{A}; \text{comp.})$, corresponding to the properties of continuity and compactness of the operator $\boldsymbol{A}$.

Consider, as a simple example, an operator $\boldsymbol{A}$ defined by the relation

$$\boldsymbol{A}x(s) = a(s)x(s),$$

where $a(s)$ is a measurable function. Assume that $a(s) \in L_\gamma$. Then (1.5) implies that $\boldsymbol{A}x(s)$ is a function in $L_{\gamma+\alpha}$, if $x(s) \in L_\alpha$, and

$$\|\boldsymbol{A}x(s)\|_{\gamma+\alpha} \leqq \|a(s)\|_\gamma \cdot \|x(s)\|_\alpha.$$

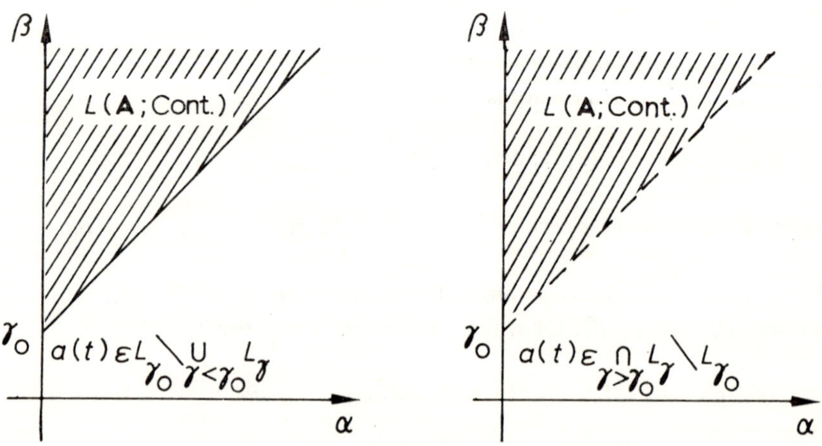

Figure 1.3

In other words, the L-characteristic $L(A; \text{cont.})$ contains the set of points $\{\alpha, \beta\}$, for which $\beta \geqq \gamma + \alpha$.

The L-characteristic $L(A; \text{cont.})$ of this operator can be determined completely. Let γ_0 be such that $a(s) \in L_\gamma$ for $\gamma > \gamma_0$, but $a(s) \notin L_\gamma$ for $\gamma < \gamma_0$. It is not difficult to see that for the case $a(s) \in L_{\gamma_0}$ the set $L(A) = L(A; \text{cont.})$ coincides with the set of points $\{\alpha, \beta\}$, for which $\beta \geqq \gamma_0 + \alpha$, while in the case $a(s) \notin L_{\gamma_0}$, $L(A)$ coincides with the set of points $\{\alpha, \beta\}$, for which $\beta > \gamma_0 + \alpha$ (Fig. 1.3).

The notion of L-characteristics has been systematically applied by many authors to yield a geometric description of interpolation theorems for linear operators. A number of assertions on general properties of L-characteristics are mentioned in the paper of P. P. Zabreiko and M. A. Krasnoselskii [1].

§ 2 Continuous linear operators

2.1 *Linear operators*[1]

In this section we recall certain fundamental definitions concerning linear operators, acting from one space E_1 to another space E_2. An operator A is *linear*, if it is additive and homogeneous:

$$A(\alpha_1 x_1 + \alpha_2 x_2) = \alpha_1 A x_1 + \alpha_2 A x_2. \tag{2.1}$$

A linear operator A is *continuous*, if it transforms sequences convergent in norm to sequences convergent in norm. If E_1 and E_2 are Banach spaces, a continuous linear operator also transforms each weakly convergent sequence to a weakly convergent sequence.

We assign to a continuous linear operator A its norm $\|A\|$:

$$\|A\| = \sup_{\|x\|_{E_1} \leqq 1} \|Ax\|_{E_2}. \tag{2.2}$$

[1] General problems in the theory of linear operators in function spaces are discussed in many books on functional analysis (see S. Banach [1], L. A. Ljusternik and V. I. Sobolev [1], L. V. Kantorovic and G. P. Akilov [1], A. Zaanen [3], E. Hille and R. Phillips [1], V. I. Smirnov [1]). See the presentation of these problems and the bibliography in N. Dunford and J. T. Schwartz [1].

Formula (2.2) makes sense in the case of any Banach spaces E_1 and E_2. We will also apply it to the case in which E_1 and E_2 are spaces L_α and L_β, where the numbers α and β can be greater than 1. The norm of a linear operator A, acting from L_α to L_β, will be denoted by $\|A\|_{\alpha\to\beta}$.

In this paragraph we are interested in problems connected with the continuity of linear operators, acting from L_α to L_β. Proving the continuity of such operators is equivalent to giving a proof of the inequality

$$\|Ax\|_\beta \leqq M\|x\|_\alpha \quad (x \in L_\alpha). \tag{2.3}$$

It is clear that the infimum of those M for which (2.3) is fulfilled is equal to $\|A\|_{\alpha\to\beta}$.

It turns out that there does not exist any non-zero continuous linear operator acting from L_α to L_β if $\alpha > 1$ and $\beta < \alpha$. For the proof it suffices to show that any continuous linear operator A acting from L_α to L_β vanishes on all characteristic functions κ_D. Partition the set D into n parts $D_1, \ldots, D_n$ of equal measure. It is clear that

$$\|\kappa_{D_i}\|_\alpha = n^{-\alpha}(\text{mes } D)^\alpha \quad (i = 1, 2, \ldots, n),$$

hence

$$\|A\kappa_{D_i}\|_\beta \leqq \|A\|_{\alpha\to\beta} n^{-\alpha}(\text{mes } D)^\alpha \quad (i = 1, 2, \ldots, n).$$

First let $\beta \leqq 1$. Then

$$\|A\kappa_D\|_\beta = \|\sum_{i=1}^n A\kappa_{D_i}\|_\beta \leqq \sum_{i=1}^n \|A\kappa_{D_i}\|_\beta \leqq \|A\|_{\alpha\to\beta}(\text{mes } D)^\alpha \cdot n^{1-\alpha}.$$

If $\beta > 1$, then

$$\|A\kappa_D\|_\beta = \|\sum_{i=1}^n A\kappa_{D_i}\|_\beta \leqq (\sum_{i=1}^n \|A\kappa_{D_i}\|_\beta^{1/\beta})^\beta \leqq \|A\|_{\alpha\to\beta}(\text{mes } D)^\alpha \cdot n^{\beta-\alpha}.$$

These inequalities and the arbitrariness of n imply that $A\kappa_D = 0$.

This assertion means that the L-characteristic $L(A; \text{cont.})$ of each non-zero linear operator A is situated in the domain hatched in Fig. 2.1. In the sequel, we usually study only those parts of L-characteristics, which

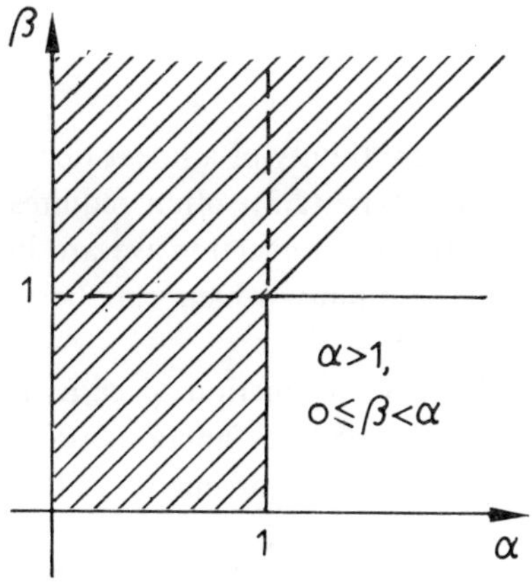

Figure 2.1

lie in the strip $0 \leqq \alpha \leqq 1$, $0 \leqq \beta < \infty$. Furthermore in many important cases only those parts of L-characteristics situated in the unit square are of interest.

In the study of a linear operator A acting from one Banach space E_1 to another Banach space E_2 the *adjoint* operator A^*, which acts from E_2^* to E_1^*, plays an important role. This operator is defined by the relation

$$[A^*f](x) = f(Ax) \qquad (x \in E_1,\ f \in E_2^*). \tag{2.4}$$

In case of a linear operator A acting from L_α to L_β, $0 < \alpha$, $\beta \leqq 1$, the operator A^* acts from $L_{1-\beta}$ to $L_{1-\alpha}$ and the relation (2.4) has the form

$$(Ax, y) = (x, A^*y) \qquad (x \in L_\alpha,\ y \in L_{1-\beta}). \tag{2.5}$$

If $0 < \alpha \leqq 1$, $\beta = 0$, then the adjoint operator A^* will act from $(L_0)^*$ to $L_{1-\alpha}$. Since L_1 is a subspace of $(L_0)^*$, A^* can be treated as an operator from L_1 to $L_{1-\alpha}$; this operator will also be denoted by A^*.

Finally, we recall that

$$\|A\|_{E_1 \to E_2} = \|A^*\|_{E^*{}_2 \to E^*{}_1}. \tag{2.6}$$

This relation holds even in the case when A^* is an operator from L_1 to $L_{1-\alpha}$.

2.2 *Regular operators*

A direct verification of the continuity of a linear operator (i.e. the establishment of inequalities of the type (2.3)) often requires that one overcomes considerable difficulties. In this connection we are interested in obtaining sufficient conditions for continuity which can be expressed in terms of simpler properties.

An operator A acting from a function space E_1 to a space E_2, is called *positive*, if it transforms non-negative functions to non-negative functions. It is easy to establish for positive linear operators the inequality:

$$|Ax| \leqq A|x|. \tag{2.7}$$

THEOREM 2.1: *Let A be a positive linear operator from L_α to L_β. Then A is continuous.*

PROOF: Let us show first that for some constant M the following inequality is fulfilled:

$$\|Ax\|_\beta \leqq M\,\|x\|_\alpha \qquad (x \in L_\alpha,\ x \geqq 0). \tag{2.8}$$

In the contrary case there exists a sequence $x_n \in L_\alpha$, $\|x_n\|_\alpha \leqq 1$, $x_n \geqq 0$, for which:

$$\|Ax_n\|_\beta \geqq n \cdot 2^n. \tag{2.9}$$

Put

$$u_0 = \sum_{n=1}^{\infty} \frac{1}{2^n}\, x_n. \tag{2.10}$$

It is easy to see that $u_0 \in L_\alpha$. For $\alpha \leqq 1$, this follows from the absolute convergence of the series (2.10), while for $\alpha > 1$ it is proved directly:

$$\int_\Omega \left[\sum_{n=1}^{\infty} \frac{1}{2^n}\, x_n(t)\right]^{1/\alpha} \mathrm{d}t \leqq \sum_{n=1}^{\infty} 2^{-n/\alpha} \int_\Omega |x_n(t)|^{1/\alpha} \mathrm{d}t \leqq \sum_{n=1}^{\infty} (2^{-1/\alpha})^n < \infty.$$

On the other hand, $u_0 \geqq x_n/2$, hence it follows that

$$Au_0 \geqq \frac{1}{2^n} Ax_n$$

and consequently

$$\|Au_0\|_\beta \geqq \frac{1}{2^n} \|Ax_n\|_\beta \qquad (n = 1, 2, \ldots).$$

Hence by (2.9)

$$\|Au_0\|_\beta \geqq n \qquad (n = 1, 2, \ldots).$$

and we arrive at a contradiction.

Now let $x(t)$ be an arbitrary function in L_α. By (2.7) and (2.8)

$$\|Ax(t)\|_\beta \leqq \|A(|x|)\|_\beta \leqq M \|x\|_\alpha.$$

The theorem has been proved[1].

Theorem 2.1 means that for positive operators the sets $L(A;\text{def.})$ and $L(A;\text{cont.})$ coincide.

A linear operator A is called *regular*[2], if it can be written in the form

$$A = A_1 - A_2,$$

where A_1 and A_2 are positive operators. Regular operators acting from L_α to L_β are continuous by Theorem 2.1. In the subsequent chapters we mainly deal with regular operators.

---

[1] Theorem 2.1 (for $0 \leqq \alpha, \beta \leqq 1$) becomes a special case of a more general assertion proved in the paper of I. A. Bakhtin, M. A. Krasnoselskii and V. Y. Stecenko [1]. In this paper it is proved that a linear operator A acting from a Banach space E_1 with a reproducing cone K_1 to a Banach space E_2 with a normal cone K_2, is continuous if $AK_1 \subset K_2$. This assertion becomes, on the one hand, a generalization of Banach's theorem [1] on continuity of an integral operator and, on the other hand, of Krein's theorem (see M. G. Krein and M. A. Rutman [1]) on continuity of a positive functional on a conic body.

[2] A detailed analysis of regular linear operators is developed in the monograph of L. V. Kantorovic, B. Z. Vulikh and A. G. Pinsker [1]. In this monograph a number of assertions more general than Theorem 2.2 and 2.3 are proved.

For the regularity of an operator $\boldsymbol{A}$ it is necessary and sufficient that there exists a positive linear operator $\boldsymbol{B}$ satisfying the inequality

$$|Ax| \leqq \boldsymbol{B}(|x|). \tag{2.11}$$

An operator, satisfying the condition (2.11), will be called a *positive majorant* of the operator $\boldsymbol{A}$.

THEOREM 2.2: *A linear operator A, acting from L_α to L_β is regular if and only if for each non-negative function $u(t) \in L_\alpha$ it is possible to take a non-negative function $v(t) \in L_\beta$ such that*

$$Ax(t) \leqq v(t) \qquad (0 \leqq x(t) \leqq u(t)). \tag{2.12}$$

PROOF: Assume that the operator A is regular. Then it is possible to take a positive majorant $\boldsymbol{B}$ of the operator A.

For each $u(t) \in L_\alpha$, $u(t) \geqq 0$, put $v(t) = \boldsymbol{B}u(t)$. Then (2.11) implies that, for $x(t)$ as in (2.12),

$$Ax(t) \leqq |Ax(t)| \leqq \boldsymbol{B}(|x(t)|) \leqq \boldsymbol{B}u(t) = v(t).$$

Thus necessity has been proved.

Let us prove sufficiency. By (2.12) the set of functions $\{Ax(t); 0 \leqq x(t) \leqq u(t)\}$ is bounded above by some element $v(t)$ in L_β. Hence there exists $\sup_{0 \leqq x \leqq u} Ax$. Define the operator A_1 for non-negative $u(t) \in L_\alpha$ by the relation:

$$A_1 u(t) = \sup_{0 \leqq x(t) \leqq u(t)} Ax(t). \tag{2.13}$$

It is clear that A_1 transforms non-negative functions $u(t)$ to non-negative functions and satisfies $A_1 u(t) \geqq Au(t)$ $(u(t) \geqq 0)$.

Let us show that A_1 is an additive operator on non-negative functions in L_α. In fact, if $x = x_1 + x_2$, $0 \leqq y_1 \leqq x_1$, $0 \leqq y_2 \leqq x_2$, then

$$A_1 x = \sup_{0 \leqq y \leqq x} Ay \geqq \sup_{0 \leqq y_i \leqq x_i} A(y_1 + y_2)$$

$$= \sup_{0 \leqq y_i \leqq x_i} (Ay_1 + Ay_2) = \sup_{0 \leqq y_1 \leqq x_1} Ay_1 + \sup_{0 \leqq y_2 \leqq x_2} Ay_2 = A_1 x_1 + A_1 x_2.$$

On the other hand, if $0 \leqq y \leqq x_1 + x_2$, there exist elements y_1 and y_2[1] for which $0 \leqq y_i \leqq x_i$ $(i = 1, 2)$, $y_1 + y_2 = y$. Hence $Ay = Ay_1 + Ay_2$ implies that

$$Ay \leqq \sup_{0 \leqq y_1 \leqq x_1} Ay_1 + \sup_{0 \leqq y_2 \leqq x_2} Ay_2$$

and consequently

$$A_1 x \leqq A_1 x_1 + A_1 x_2.$$

Additivity of the operator A on non-negative functions has been proved. By (2.13) the operator A_1 is also homogeneous:

$$A_1(\lambda u(t)) = \lambda A_1 u(t) \quad (\lambda \geqq 0).$$

Extend the operator A_1 by the relation

$$A_1 x = A_1 x_+ - A_1 x_-$$

where

$$x_+ = \sup\{x, 0\}, \ \ x_- = -\inf\{x, 0\},$$

to a linear[2] operator, defined on the whole of L_α. By Theorem 2.1 A_1 is continuous. (2.13) implies that the operator $A_2 = A_1 - A$ is also positive and continuous. Since $A = A_1 - A_2$, A is regular. The theorem has been proved.

Observe that the operator $B = A_1 + A_2$ becomes a positive majorant of the operator A. Furthermore it can be shown that B becomes the 'least' majorant of the operator A. This means that for any other positive majorant B_1 of the operator A the following relation is fulfilled:

$$Bx(t) \leqq B_1 x(t) \quad (x(t) \geqq 0).$$

---

[1] It suffices, for instance, to put $y_1 = \inf\{x_1, y\}$, $y_2 = y - y_1$.

[2] Linearity follows upon noting that, by the additivity on non-negative functions, $A_1x' - A_1x'' = A_1x_+ - A_1x_-$ whenever $x' - x'' = x$, $x', x'' \geqq 0$. (Ed.)

Thus, it has also been proved that each regular operator has a least majorant.

Let A be a regular operator acting from L_α to L_β $(0 < \alpha \leqq 1, 0 \leqq \beta \leqq 1)$, i.e.

$$A = A_1 - A_2,$$

where A_1 and A_2 are positive linear operators. Then the adjoint operator A^* acting from $L_{1-\beta}$ to $L_{1-\alpha}$ will also be regular because

$$A^* = A_1^* - A_2^*,$$

and the adjoint operator of a positive operator is positive.

THEOREM 2.3: *Each continuous linear operator acting from L_α to L_0 or from L_1 to L_β $(0 \leqq \beta < 1)$, is regular.*

PROOF: Consider first an operator A acting from L_α to L_0. Since A is continuous, for any $u(t) \geqq 0$ the set of functions $\{Ax(t); 0 \leqq x(t) \leqq u(t)\}$ is bounded in norm: $\|Ax(t)\|_0 \leqq \|A\|_{\alpha\to 0} \|u\|_\alpha$. Consequently

$$Ax(t) \leqq v(t) \qquad (0 \leqq x(t) \leqq u(t)),$$

where $v(t) = \|A\|_{\alpha\to 0} \|u\|_\alpha \in L_0$. Regularity of the operator follows now from Theorem 2.2.

Let A be a continuous operator, acting from L_1 to L_β $(0 \leq \beta < 1)$. Then the adjoint operator A^* acting from $L_{1-\beta}$ to L_0, is regular, and consequently the operator A, which is adjoint to the operator A^*, is also regular. The theorem has been proved.

2.3 *The M. Riesz interpolation theorem*

In this section we study operators acting from $L_\alpha(\Omega)$ to $L_\beta(\Omega^*)$, where α and β are arbitrary non-negative numbers.

The remarkable theorem that L-characteristics of linear operators A acting from L_α to L_β are convex sets is due to M. Riesz. More precisely, the L-characteristics $L(A; \text{cont.})$ contain, together with any two points[1]

---

[1] M. Riesz [1] considered only values α_0, β_0, α_1 and β_1, in the interval [0, 1]. The general theorem was obtained by A. Calderon and A. Zygmund [1, 2].

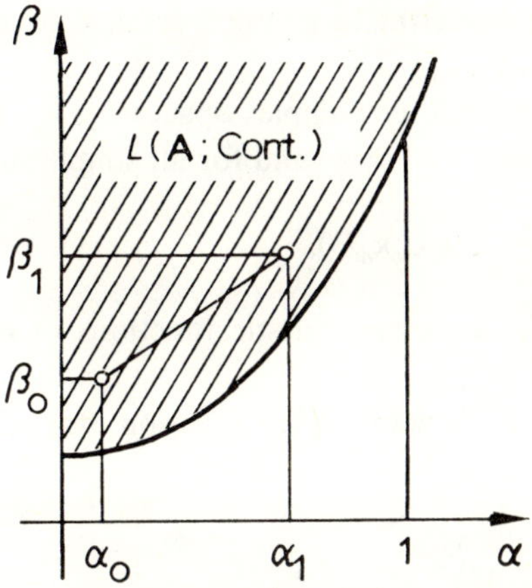

Figure 2.2

$\{\alpha_0, \beta_0\}$ and $\{\alpha_1, \beta_1\}$, the whole segment joining them (see Fig. 2.2). This theorem was established by M. Riesz in the form of the so-called interpolation theorem.

THEOREM 2.4: *Let A be a linear operator, which is continuous simultaneously from L_{α_0} to L_{β_0} and from L_{α_1} to L_{β_1}:*

$$\|Ax\|_{\beta_0} \leqq M_0 \|x\|_{\alpha_0}, \tag{2.14}$$

$$\|Ax\|_{\beta_1} \leqq M_1 \|x\|_{\alpha_1}. \tag{2.15}$$

Then for any $\tau \in (0, 1)$ the operator A acts continuously from $L_{\alpha(\tau)}$ to $L_{\beta(\tau)}$ where

$$\alpha(\tau) = (1 - \tau)\alpha_0 + \tau\alpha_1, \;\; \beta(\tau) = (1 - \tau)\beta_0 + \tau\beta_1, \tag{2.16}$$

moreover in the case of complex spaces L_α,

$$\|Ax\|_{\beta(\tau)} \leqq M_0^{1-\tau} \cdot M_1^{\tau} \|x\|_{\alpha(\tau)}, \tag{2.17}$$

while in case of real spaces L_α,

$$\|Ax\|_{\beta(\tau)} \leqq C(\beta_0, \beta_1, \tau) M_0^{1-\tau} M_1^{\tau} \|x\|_{\alpha(\tau)}, \tag{2.18}$$

where the constant $C(\beta_0, \beta_1, \tau)$ does not depend on the operator A.

PROOF: We confine ourselves to giving a detailed proof for indices α_0, β_0, α_1, β_1 in the interval [0, 1].

Consider first the case of complex spaces L_α, L_β. Let $\alpha_0 > \alpha_1$. Let us show that the inequality (2.17) is valid for all simple functions

$$x(s) = c_1\kappa_1(s) + \ldots + c_m\kappa_m(s),$$

where $\kappa_1(s), \ldots, \kappa_m(s)$ are characteristic functions of disjoint sets $\Omega_i \subset \Omega$ $(i = 1, 2, \ldots, m)$.

The inequality (2.17) with a fixed function $x(s) \in L_{\alpha(\tau)}$ is equivalent to the inequality

$$\sup \frac{|(Ax, y)|}{\|y\|_{1-\beta(\tau)}} \leqq M_0^{1-\tau} M_1^{\tau} \|x\|_{\alpha(\tau)}, \tag{2.19}$$

where the supremum is taken over all $y(t)$ in some set dense in $L_{1-\beta(\tau)}$. Hence to prove (2.17), it suffices to establish the inequality

$$|(Ax, y)| \leqq M_0^{1-\tau} M_1^{\tau} \|x\|_{\alpha(\tau)} \|y\|_{1-\beta(\tau)}$$

for every simple function $y(t)$:

$$y(t) = d_1\chi_1(t) + \ldots + d_n\chi_n(t),$$

where $\chi_1(t), \ldots, \chi_n(t)$ are characteristic functions of disjoint sets $\Omega_j^* \subset \Omega^*$ $(j = 1, \ldots, n)$.

Consider the function $\Phi(z)$ of a complex variable $z = u + iv$ which is defined by the relation

$$\Phi(z) = (A[|x|^{\alpha(z)-1}x], \, |y|^{\beta(z)-1}y)$$

$$= \sum_{\substack{i=1,\ldots,m \\ j=1,\ldots,n}} c_i \bar{d}_j |c_i|^{\alpha(z)-1} |d_j|^{\overline{\beta(z)}-1} [\int_{\Omega^*} A\kappa_i(t)\chi_j(t)\,\mathrm{d}t], \tag{2.20}$$

where

$$\alpha(z) = \frac{(1-z)\alpha_0 + z\alpha_1}{\alpha(\tau)}, \quad \beta(z) = \frac{1-(1-z)\beta_0 - z\beta_1}{1-\beta(\tau)}. \tag{2.21}$$

It is clear that $\Phi(z)$ is analytic in the strip $0 \leqq \operatorname{Re} z \leqq 1$.

By Hadamard's three lines theorem[1] the following inequality is valid for each $\tau \in (0, 1)$:

$$|\Phi(\tau)| \leqq \sup_v |\Phi(iv)|^{1-\tau} \cdot \sup_v |\Phi(1 + iv)|^\tau. \tag{2.22}$$

Let us estimate the expressions $\Phi(iv)$ and $\Phi(1 + iv)$.

For $\operatorname{Re} z = 0$ we have the relation

$$\alpha(z) = \frac{\alpha_0}{\alpha(\tau)} + iv\,\frac{\alpha_1 - \alpha_0}{\alpha(\tau)},\quad \beta(z) = \frac{1 - \beta_0}{1 - \beta(\tau)} + iv\,\frac{\beta_0 - \beta_1}{1 - \beta(\tau)},$$

whence

$$\begin{aligned}
&|\Phi(iv)| \leqq \\
&|(A[|x|^{(\alpha_0/\alpha(\tau)) - 1 + iv(\alpha_1 - \alpha_0)/\alpha(\tau)} x],\quad |y|^{(1-\beta_0)/(1-\beta(\tau)) - 1 + iv(\beta_0 - \beta_1)/(1-\beta(\tau))} \cdot y)| \\
&\leqq \|A[|x|^{(\alpha_0/\alpha(\tau)) - 1 + iv(\alpha_1 - \alpha_0)/\alpha(\tau)} \cdot x]\|_{\beta_0} \\
&\times \|\,|y|^{(1-\beta_0)/(1-\beta(\tau)) - 1 + iv(\beta_0 - \beta_1)/(1-\beta(\tau))} \cdot y\|_{1-\beta_0} \\
&\leqq \|A[|x|^{(\alpha_0/\alpha(\tau)) - 1 + iv(\alpha_1 - \alpha_0)/\alpha(\tau)} \cdot x]\|_{\beta_0} \cdot \|\,|y|^{(1-\beta_0)/(1-\beta(\tau))}\|_{1-\beta_0}.
\end{aligned}$$

By (2.14) it follows from this inequality that

$$\begin{aligned}
|\Phi(iv)| &\leqq M_0 \|\,|x|^{(\alpha_0/\alpha(\tau)) - 1 + iv(\alpha_1 - \alpha_0)/\alpha(\tau)} \cdot x\|_{\alpha_0} \|\,|y|^{(1-\beta_0)/(1-\beta(\tau))}\|_{1-\beta_0} \\
&= M_0 \|\,|x|^{\alpha_0/\alpha(\tau)}\|_{\alpha_0} \cdot \|\,|y|^{(1-\beta_0)/(1-\beta(\tau))}\|_{1-\beta_0},
\end{aligned}$$

or

$$|\Phi(iv)| \leqq M_0 \|x\|_{\alpha(\tau)}^{\alpha_0/\alpha(\tau)} \cdot \|y\|_{1-\beta(\tau)}^{(1-\beta_0)/(1-\beta(\tau))}. \tag{2.23}$$

---

[1] For any function $f(z)$, analytic in the strip $0 < \operatorname{Re}(z) < 1$ and continuous in the strip $0 \leqq \operatorname{Re}(z) \leqq 1$, the following inequality is valid:

$$|f(\tau)| \leqq \sup_y |f(iy)|^{1-\tau} \sup_y |f(1 + iy)|^\tau \quad (0 < \tau < 1)$$

(see, for instance, A. I. Markuschevic [1]).

Similar arguments serve to estimate the second factor in the right side of the inequality (2.22):

$$|\Phi(1 + iv)| \leqq M_1 \|x\|_{\alpha(\tau)}^{\alpha_1/\alpha(\tau)} \|y\|_{1-\beta(\tau)}^{(1-\beta_1)/(1-\beta(\tau))}. \tag{2.24}$$

The inequalities (2.22), (2.23) and (2.24) imply that

$$|(Ax, y)| \leqq M_0^{1-\tau} M_1^{\tau} \|x\|_{\alpha(\tau)} \|y\|_{1-\beta(\tau)}$$

and consequently that the inequality (2.17) is fulfilled for every simple function $x(s) \in L_{\alpha(\tau)}$.

The set of all simple functions is dense in $L_{\alpha(\tau)}$. Let $x(s) \in L_{\alpha(\tau)}$ and let $x_1(s), x_2(s), \ldots$ be a sequence of simple functions converging to $x(s)$ in $L_{\alpha(\tau)}$. The inequality (2.17), valid for all simple functions, then implies that the sequence Ax_n converges in $L_{\beta(\tau)}$-norm to some element $y \in L_{\beta(\tau)}$. On the other hand, the sequence of functions $x_n(s)$ converges to $x(s)$ in L_{α_0}-norm, because $\alpha_0 > \alpha_1$. Hence the sequence $Ax_n(s)$ converges to the element Ax in L_{β_0}-norm and it follows that $y = Ax$.

Proceeding to the limit $n \to \infty$ in the inequality

$$\|Ax_n\|_{\beta(\tau)} \leqq M_0^{1-\tau} M_1^{\tau} \|x_n\|_{\alpha(\tau)},$$

we obtain the inequality (2.17) for the function $x(s) \in L_{\alpha(\tau)}$.

Now let $\alpha_0 = \alpha_1$. Then the inequality (2.17) follows immediately from (1.7):

$$\|Ax\|_{\beta(\tau)} \leqq \|Ax\|_{\beta_0}^{1-\tau} \|Ax\|_{\beta_1}^{\tau} \leqq M_0^{1-\tau} \cdot M_1^{\tau} \|x\|_{\alpha(\tau)}.$$

The theorem for the case of complex spaces L_α has been proved.

Now let the spaces L_α be real and let A be a given operator. Denote by $\tilde{A}$ the extension of the operator A to complex valued functions $z(t) = x(t) + iy(t)$ which is defined by the relation

$$\tilde{A}(x + iy) = Ax + iAy.$$

The inequalities

$$\|\tilde{A}(x + iy)\|_{\beta_0} \leqq \|Ax\|_{\beta_0} + \|Ay\|_{\beta_0}$$

$$\leqq M_0(\|x\|_{\beta_0} + \|y\|_{\beta_0}) \leqq 2M_0 \|x + iy\|_{\alpha_0}$$

imply that

$$\|\tilde{A}z\|_{\beta_0} \leqq 2M_0 \|z\|_{\alpha_0}.$$

Similarly

$$\|\tilde{A}z\|_{\beta_1} \leqq 2M_1 \|z\|_{\alpha_1}.$$

Applying to the operator $\tilde{A}$ the part of the theorem, already proved for the complex cases, we deduce that the operator $\tilde{A}$ acts from $L_{\alpha(\tau)}$ to $L_{\beta(\tau)}$ and that

$$\|\tilde{A}z\|_{\beta(\tau)} \leqq 2M_0^{1-\tau} M_1^{\tau} \|z\|_{\alpha(\tau)},$$

whence follows the inequality

$$\|Ax\|_{\beta(\tau)} \leqq 2M_0^{1-\tau} M_1^{\tau} \|x\|_{\alpha(\tau)}. \tag{2.25}$$

The theorem has been proved for indices α_0, β_0, α_1 and β_1 in the interval [0, 1].

This proof is borrowed from the paper of A. Calderon and A. Zygmund. Their proof, as we have already noted, embraces also the case of indices greater than 1. The basic change made in the proof is that then instead of the analytic function (2.20) the following subharmonic function is considered:

$$\Phi(z) = (|A[|x|^{\alpha(z)-1}x]|^k,\ |y|^{\beta(z)-1}y),$$

where

$$\alpha(z) = \frac{(1-z)\alpha_0 + z\alpha_1}{\alpha(\tau)}$$

$$\beta(z) = \frac{1-(1-z)k\beta_0 - zk\beta_1}{1-k\beta(\tau)}$$

and k satisfies the inequality

$$\beta_0 k < 1,\ \beta_1 k < 1.$$

Here instead of Hadamard's three lines theorem its generalization to subharmonic functions is used.

As was noted earlier, the assertion of Theorem 2.3 means that the L-characteristic $L(A; \text{cont.})$ is a convex set. Hence it follows that the function $\xi(\alpha; A; \text{cont.})$ is convex and, in particular, continuous in α.

In concluding this section we note that the inequality (2.17) is in general invalid in the case of real spaces. But even for real spaces the inequality (2.17) is fulfilled for special pairs of indices α_i, β_i (the inequality (2.17) is trivially fulfilled if $\alpha_0 = \alpha_1$; non-trivial assertions are presented in the sequel in 10.4°). In the next section it will be shown that for special classes of operators the inequality (2.17) is always valid in the case of real spaces.

2.4 *Interpolation theorems for regular operators*

In this section interpolation theorems for positive and regular operators are proved.

THEOREM 2.5:[1] *Let A be a positive linear operator acting simultaneously from a space L_{α_0} to a space L_{β_0} and from a space L_{α_1} to a space L_{β_1}:*

$$\|Ax\|_{\beta_0} \leqq M_0 \|x\|_{\alpha_0}, \quad \|Ax\|_{\beta_1} \leqq M_1 \|x\|_{\alpha_1},$$

where β_0 and β_1 can be any non-negative numbers. Then for any $\tau \in (0, 1)$ the operator A acts from $L_{\alpha(\tau)}$ to $L_{\beta(\tau)}$, where

$$\alpha(\tau) = (1 - \tau)\alpha_0 + \tau\alpha_1, \quad \beta(\tau) = (1 - \tau)\beta_0 + \tau\beta_1,$$

and

$$\|Ax\|_{\beta(\tau)} \leqq M_0^{1-\tau} M_1^{\tau} \|x\|_{\alpha(\tau)}. \tag{2.26}$$

This shows that for positive operators the inequality (2.17) is valid for real spaces as well as for complex spaces.

---

[1] A matrix analogue of Theorem 2.5 is indicated in the book of G. Hardy, D. Littlewood and G. Polya [1]. In general form the theorem was proved in the paper of P. P. Zabreiko and E. I. Pustylnik [2]. A related assertion was previously obtained by E. I. Pustylnik [4].

PROOF: It suffices to prove the theorem for the case $\alpha_0 > \alpha_1$. As in the proof of Theorem 2.4, it suffices to establish the inequality (2.26) for simple functions

$$x(s) = c_1\kappa_1(s) + \ldots + c_m\kappa_m(s) \tag{2.27}$$

where the $\kappa_i(s)$ are characteristic functions of disjoint sets $\Omega_i \subset \Omega$ $(i = 1, 2, \ldots, m)$.

Let us first prove the inequality:

$$|Ax(t)| \leqq [A(|x|^{\alpha_0/\alpha(\tau)})(t)]^{1-\tau}[A(|x|^{\alpha_1/\alpha(\tau)})(t)]^{\tau}. \tag{2.28}$$

By (2.27) this inequality can be rewritten in the form

$$|\sum_{i=1}^{m} c_i A\kappa_i(t)| \leqq [\sum_{i=1}^{m} |c_i|^{\alpha_0/\alpha(\tau)} A\kappa_i(t)]^{1-\tau}[\sum_{i=1}^{m} |c_i|^{\alpha_1/\alpha(\tau)} A\kappa_i(t)]^{\tau}. \tag{2.29}$$

Let $t = t_0$ be such that all the numbers $(A\kappa_i)(t_0)$ are finite. Put $\Delta_i = (A\kappa_i)(t)|_{t=t_0}$ $(i = 1, 2, \ldots, m)$. It is clear that all the Δ_i are non-negative. Hence Hölder's inequality is valid:

$$|\sum_{i=1}^{m} u_i v_i \Delta_i| \leqq (\sum_{i=1}^{m} |u_i|^{1/(1-\tau)} \Delta_i)^{1-\tau} \cdot (\sum_{i=1}^{m} |v_i|^{1/\tau} \Delta_i)^{\tau} \quad (0 < \tau < 1).$$

Setting $u_i = |c_i|^{(1-\tau)(\alpha_0/\alpha(\tau))} \cdot \operatorname{sgn} c_i$, $v_i = |c_i|^{\tau\alpha_1/\alpha(\tau)}$, we obtain the inequality

$$|\sum_{i=1}^{m} c_i \Delta_i| \leqq (\sum_{i=1}^{m} |c_i|^{\alpha_0/\alpha(\tau)} \Delta_i)^{1-\tau} \cdot (\sum_{i=1}^{m} |c_i|^{\alpha_1/\alpha(\tau)} \Delta_i)^{\tau}$$

which, in view of linearity of the operator A, is equivalent to the inequality

$$|Ax(t)| \leqq [A(|x|^{\alpha_0/\alpha(\tau)})(t)]^{1-\tau}[A(|x|^{\alpha_1/\alpha(\tau)})(t)]^{\tau}$$

for $t = t_0$.

Since for almost all $t \in \Omega^*$ the functions $A\kappa_i(t)$ are finite, the inequality (2.28) is valid for almost all $t \in \Omega^*$. Hence

$$\|Ax\|_{\beta(\tau)} \leqq \|[A(|x|^{\alpha_0/\alpha(\tau)})]^{1-\tau} \cdot [A(|x|^{\alpha_1/\alpha(\tau)})]^{\tau}\|_{\beta(\tau)},$$

whence by Hölder's inequality (1.5)

$$\|Ax\|_{\beta(\tau)} \leqq \|A(|x|^{\alpha_0/\alpha(\tau)})\|_{\beta_0}^{1-\tau} \cdot \|A(|x|^{\alpha_1/\alpha(\tau)})\|_{\beta_1}^{\tau}. \tag{2.30}$$

From (2.14), (2.15) and (2.30) it follows immediately that

$$\|Ax\|_{\beta(\tau)} \leqq M_0^{1-\tau} \| \, |x|^{\alpha_0/\alpha(\tau)}\|_{\alpha_0}^{1-\tau} \cdot M_1^{\tau} \| \, |x|^{\alpha_1/\alpha(\tau)}\|_{\alpha_1}^{\tau}$$

$$\leqq M_0^{1-\tau} M_1^{\tau} \|x\|_{\alpha(\tau)}.$$

The inequality (2.26) has been proved for all step functions $x(t)$. Hence it follows (see the proof of Theorem 2.4) that it is valid for all functions $x(t) \in L_{\alpha(\tau)}$. The theorem has been proved.

Theorem 2.5 and the definition of regular operators imply immediately:

THEOREM 2.6: *Let an operator A admit a positive majorant B:*

$$|Ax| \leqq B(|x|). \tag{2.31}$$

Let the positive operator B act simultaneously from L_{α_0} to L_{β_0} and from L_{α_1} to L_{β_1}:

$$\|Bx\|_{\beta_0} \leqq N_0 \|x\|_{\alpha_0}, \quad \|Bx\|_{\beta_1} \leqq N_1 \|x\|_{\alpha_1}.$$

Then for any $\tau \in (0, 1)$ the operator A acts continuously from $L_{\alpha(\tau)}$ to $L_{\beta(\tau)}$, where

$$\alpha(\tau) = (1-\tau)\alpha_0 + \tau\alpha_1, \quad \beta(\tau) = (1-\tau)\beta_0 + \tau\beta_1,$$

and

$$\|Ax\|_{\beta(\tau)} \leqq N_0^{1-\tau} N_1^{\tau} \|x\|_{\alpha(\tau)}.$$

Let A be a regular linear operator, acting from L_{α_0} to L_{β_0} and from L_{α_1} to L_{β_1}, $\alpha_0 \geqq \alpha_1$. Denote by B the 'least' positive majorant of A as an operator from L_{α_0} to L_{β_0} (see 2.2°). By (2.13) the operator B will also be the 'least' positive majorant of A as an operator from L_{α_1} to L_{β_1}. Then by Theorem 2.5 it follows that B is a positive operator acting from $L_{\alpha(\tau)}$ to $L_{\beta(\tau)}$

$(0 < \tau < 1)$. Further $\boldsymbol{B}$ becomes a positive majorant of $\boldsymbol{A}$ as an operator from $L_{\alpha(\tau)}$ to $L_{\beta(\tau)}$.

These arguments show that the *L-characteristic* $L(\boldsymbol{A}; \text{reg.})$ *of each linear operator* $\boldsymbol{A}$ *is convex.*

2.5 *Classes of L-characteristics of linear operators*

It was shown earlier that the L-characteristic $L(\boldsymbol{A}; \text{cont.})$ of a linear operator $\boldsymbol{A}$ becomes a convex set possessing the extrapolation property (see 1.6°). It is natural to raise the question of whether L-characteristics of linear operators possess other general properties. This problem has not been completely investigated.[1] However, as Theorem 2.7, proved below shows, each convex set with the extrapolation property differs from the L-characteristic of some linear operator $\boldsymbol{A}$ only by boundary points.

In this section we consider, for simplicity, spaces of functions defined on [0, 1].

THEOREM 2.7:[2] *Let a convex open set L in the strip* $0 \leqq \alpha \leqq 1$, $\beta \geqq 0$ *possess the property that it contains together with a point* $\{\alpha_0, \beta_0\}$ *all points* $\{\alpha, \beta\}$ *for which* $\alpha \leqq \alpha_0$, $\beta \geqq \beta_0$. *Then there exists a linear operator* $\boldsymbol{A}$ *for which the set of interior points of the L-characteristic* $L(\boldsymbol{A}; \text{cont.})$ *coincides with the set L.*

PROOF: Consider an integral operator

$$\boldsymbol{A}x(t) = \int_0^1 K(t, s)x(s)\,ds, \tag{2.32}$$

where

$$K(t, s) = \frac{1}{t^{a_1}s^{b_1} + t^{a_2}s^{b_2}}, \tag{2.33}$$

and the numbers a_1, a_2, b_1, b_2 satisfy the inequalities

$$0 \leqq a_1 < a_2 < \infty, \quad 0 < b_2 < b_1 \leqq 1.$$

---

[1] See however S. D. Riemenschnieder [1] (Ed.).

[2] B. S. Mitjagin indicated the basic idea of the proof of this theorem.

Equation (2.33) implies that

$$|K(t, s)| \leqq t^{-a_i} s^{-b_i} \quad (i = 1, 2). \tag{2.34}$$

Let i be fixed. Then the operator A acts from L_α to L_β whenever $\alpha < 1 - b_i$, $\beta > a_i$. In fact, when $x(s) \in L_\alpha$, $\alpha < 1 - b_i$

$$|Ax(t)| = |\int_0^1 K(t, s) x(s) \mathrm{d}s| \leqq t^{-a_i} \int_0^1 s^{-b_i} |x(s)| \mathrm{d}s$$

$$\leqq t^{-a_i} (\int_0^1 s^{-b_i/(1-\alpha)} \mathrm{d}s)^{1-\alpha} \|x\|_\alpha.$$

Consequently

$$\|Ax\|_\beta \leqq \frac{1}{[1 - a_i/\beta]^\beta [1 - b_i/(1 - \alpha)]^{1-\alpha}} \|x\|_\alpha. \tag{2.35}$$

Applying Theorem 2.4 to the operator A, we conclude that it acts from L_α to L_β if the numbers α, β satisfy the inequalities

$$\alpha < 1 - b_2, \ \beta > \frac{a_2 - a_1}{b_1 - b_2} \alpha + a_1 - (1 - b_1) \frac{a_2 - a_1}{b_1 - b_2}, \ \beta > a_1. \tag{2.36}$$

Denote by Γ the boundary of the set of points, whose coordinates satisfy these inequalities. The inequality (2.35) and the Riesz interpolation theorem imply that

$$\|Ax\|_\beta \leqq M \|x\|_\alpha, \tag{2.37}$$

where M is some constant, depending, for given α, β, only on the distance of the point $\{\alpha, \beta\}$ from Γ. Thus the set $L(A; \text{cont.})$ contains all points $\{\alpha, \beta\}$, for which three inequalities of (2.36) are fulfilled.

It turns out that the set of interior points of the L-characteristic of the operator A coincides with the set of points $\{\alpha, \beta\}$ for which the inequalities (2.36) are fulfilled (Fig. 2.3). For the proof it suffices to show that any point $\{\alpha, \beta\}$, for which one of the inequalities

$$\alpha > 1 - b_2$$

$$\beta < \frac{a_2 - a_1}{b_1 - b_2}\,\alpha + a_1 - (1 - b_1)\,\frac{a_2 - a_1}{b_1 - b_2},$$

$$\beta < a_1,$$

is fulfilled, does not belong to $L(A; \text{cont.})$.

A function $x_0(s) = s^{-\nu}$ belongs to L_α for $\alpha > \nu$. Consider the function

$$Ax_0(t) = \int_0^1 \frac{s^{-\nu}\,\mathrm{d}s}{t^{a_1}s^{b_1} + t^{a_2}s^{b_2}}. \tag{2.38}$$

To estimate $Ax_0(t)$, perform the change of variables $s = t^{(a_2 - a_1)/(b_1 - b_2)}u$ in (2.38):

$$Ax_0(t) = t^{-[a_1 + (\nu + b_1 - 1)(a_2 - a_1)/(b_1 - b_2)]} \int_0^{t^{-(a_2-a_1)/(b_1-b_2)}} \frac{u^{-b_2-\nu}}{1 + u^{b_1 - b_2}}\,\mathrm{d}u. \tag{2.39}$$

For $\nu \geqq 1 - b_2$ the integral in the right side of (2.39) diverges and the operator A does not act from the space L_α, $\alpha > 1 - b_2$, to any L_β. In other

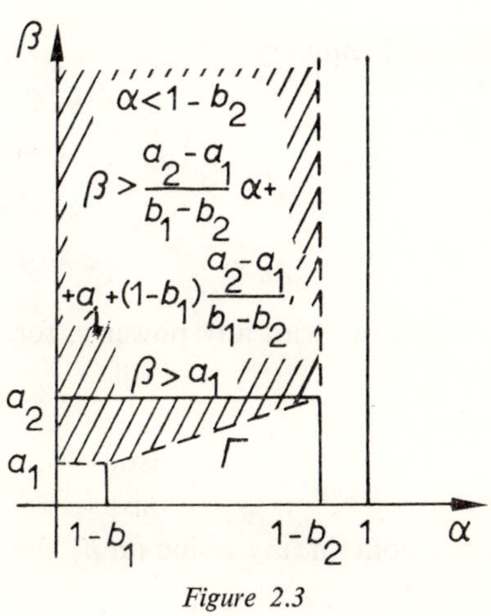

Figure 2.3

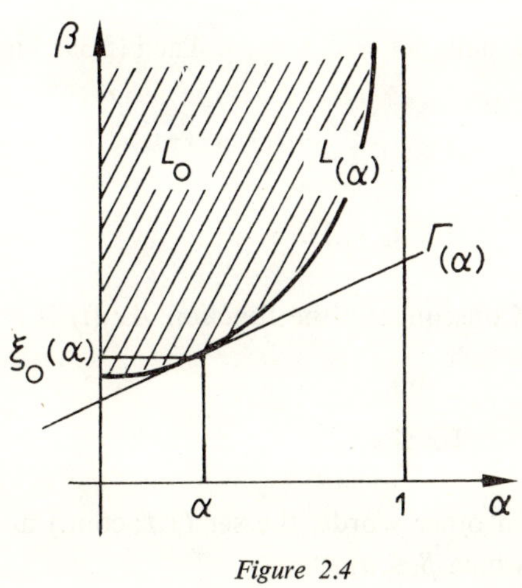

Figure 2.4

words, the set $L(A; \text{cont.})$ does not contain any point $\{\alpha, \beta\}$ with $\alpha > 1 - b_2$.

Let $1 - b_2 > v > 1 - b_1$. (2.39) implies that

$$Ax_0(t) \geqq c_0 t^{-[a_1 + (b_1 + v - 1)(a_2 - a_1)/(b_1 - b_2)]}$$

where

$$c_0 = \int_0^1 \frac{u^{-b_2 - v}}{1 + u^{b_1 - b_2}}\, du.$$

Hence $Ax_0(t)$ is not summable with power $1/\beta$, if

$$\beta \leqq a_1 + \frac{a_2 - a_1}{b_1 - b_2}(b_1 + v - 1). \tag{2.40}$$

This means that the operator A does not act from L_α, $1 - b_1 < \alpha < 1 - b_2$, to any space L_β if β satisfies the inequality (2.40). In other words the set $L(A; \text{cont.})$ does not contain any point $\{\alpha, \beta\}$ for which

$$\beta < \frac{a_2 - a_1}{b_1 - b_2}\alpha + a_1 - (1 - b_1)\frac{a_2 - a_1}{b_1 - b_2}, \quad 1 - b_1 < \alpha < 1 - b_2.$$

Finally let $v < 1 - b_1$. Then (2.39) implies the inequality

$$Ax_0(t) \geqq ct^{-[a_1 + (b_1 + v - 1)(a_2 - a_1)/(b_1 - b_2)]} \int_0^{t^{-(a_2 - a_1)/(b_1 - b_2)}} u^{-b_1 - v} du \geqq c_1 t^{-a_1}$$

$$(t \leqq t_0 < 1).$$

Consequently the function $Ax_0(t)$ is not summable with any power r, for which

$$1/r \leqq a_1.$$

In other words, the set $L(A; \text{cont.})$ does not contain any point $\{\alpha, \beta\}$ for which $\beta \leqq a_1$.

Consider now an arbitrary open set L, satisfying the conditions of the theorem; denote by $\xi(\alpha)$ the convex continuous function

$$\xi(\alpha) = \inf_{\{\alpha, \beta\} \in L} \beta.$$

Denote by $\Gamma_{(\alpha)}$ the support line to the set L at a point $\{\alpha, \xi(\alpha)\}$ and by $L^{(\alpha)}$ the set of points in the strip $0 \leqq \alpha \leqq 1, \beta \geqq 0$, lying above $\Gamma_{(\alpha)}$ (Fig. 2.4).

Take a countable set of numbers $\alpha_1, \alpha_2, \ldots$, dense in the interval $[0, 1]$. It is clear that

$$L = \bigcap_{n=1}^{\infty} L^{(\alpha_n)}.$$

Let A_n be an operator of the form (2.32) with a kernel $K_n(t, s)$ of the form (2.33), the interior part of whose L-characteristic coincides with the set $L^{(\alpha_n)}$.

Consider the positive operator B, defined by the relation

$$B = \sum_{i=1}^{\infty} \frac{1}{2^i} A_i$$

and let us show that it satisfies the conditions of the theorem.

Let $\{\alpha, \beta\} \in L$. Since the distances from this point to the boundaries $\Gamma_{(\alpha_n)}$ of the sets $L^{(\alpha_n)}$ are bounded from below, inequality (2.37) implies that the following inequalities are fulfilled:

$$\|A_i x\|_\beta \leqq M \|x\|_\alpha \quad (i = 1, 2, \ldots).$$

Hence for any $x \in L_\alpha$ the series

$$\sum_{i=1}^{\infty} \frac{1}{2^i} A_i x$$

converges in L_β-norm, so $Bx \in L_\beta$. Thus, $\{\alpha, \beta\} \in L(B; \text{cont.})$.

Now let a point $\{\alpha, \beta\}$ not belong to the closure of L. Then obviously there exists a number i_0 such that $\{\alpha, \beta\} \notin L^{(\alpha_{i_0})}$. This means that A_{i_0} does not act from L_α to L_β. The inequalities

$$Bx \geqq \frac{1}{2^{i_0}} A_{i_0} x,$$

valid for positive functions in L_α, imply that the operator $\boldsymbol{B}$ does not act from L_α to L_β, either, i.e. $\{\alpha, \beta\} \notin L(\boldsymbol{B}; \text{cont.})$. The theorem has been proved.

2.6 *On a property of regular operators*

THEOREM 2.8:[1] *Let $\boldsymbol{A}$ be a regular linear operator acting from L_α to L_β ($0 \leqq \alpha < \infty$, $0 < \beta < \infty$). Then each set of functions $\mathfrak{M}$ with equi-absolutely continuous norms in L_α is transformed by the operator $\boldsymbol{A}$ to a set of functions with equi-absolutely continuous norms in L_β.*

PROOF: The assertion of the theorem is obvious in the case of $\alpha = 0$. Let $\alpha > 0$. It suffices to consider positive operators $\boldsymbol{A}$. It can also be assumed that the functions $x(s) \in \mathfrak{M}$ are non-negative.

From the hypothesis of the theorem it follows that

$$\|x(s)\|_\alpha \leqq a \qquad (x(s) \in \mathfrak{M}).$$

Let $\varepsilon > 0$ be given. Equi-absolute continuity of the norms of the functions $x(s) \in \mathfrak{M}$ in L_α implies the existence of a δ_0 such that

$$\|\boldsymbol{P}_D x(s)\|_\alpha < \frac{\varepsilon}{2^{\beta+1}\,\|\boldsymbol{A}\|_{\alpha\to\beta}} \qquad (x \in \mathfrak{M})$$

whenever $\operatorname{mes}(D) < \delta_0$. Put

$$h_0 = a/\delta_0^\alpha.$$

Write each function $x(s) \in \mathfrak{M}$ in the form

$$x(s) = x_1(s) + x_2(s)$$

where

$$x_1(s) = \begin{cases} x(s), & \text{if } |x(s)| \leqq h_0 \\ 0, & \text{if } |x(s)| > h_0. \end{cases}$$

---

[1] This theorem was proved by P. P. Zabreiko and E. I. Pustylnik (P. P. Zabreiko [2]).

It is clear that $|x_1(s)| \leqq h_0$ and that the measure of the set $D(\varepsilon; x)$ of points s for which $x_2(s) \neq 0$, is less than δ_0. Hence it follows that

$$\|x_2(s)\|_\alpha = \|P_{D(\varepsilon, x)} x(s)\|_\alpha \leqq \frac{\varepsilon}{2^{\beta+1} \|A\|_{\alpha\to\beta}}.$$

From this inequality it follows that for an arbitrary set $D^* \subset \Omega^*$

$$\|P_{D^*}Ax\|_\beta \leqq 2^\beta h_0 \|P_{D^*}Au_0\|_\beta + \varepsilon/2 \quad (x(s) \in \mathfrak{M}),$$

where $u_0(s) \equiv 1$. Hence there exists a $\delta > 0$ such that

$$\|P_{D^*}Ax(t)\|_\beta < \varepsilon \quad (x(s) \in \mathfrak{M}),$$

whenever $\operatorname{mes}(D^*) < \delta$. The theorem has been proved.

2.7 *The Marcinkiewicz interpolation theorem*[1]

Denote by $\lambda(h) = \lambda(z; h)$, where $z(t)$ is a measurable function, the measure of the set of points t for which $|z(t)| \geqq h$:

$$\lambda(z; h) = \operatorname{mes}\{t\colon |z(t)| \geqq h\}. \tag{2.42}$$

The following relation is valid for each measurable function:

$$\int_\Omega |z(t)|\,\mathrm{d}t = \int_0^\infty \lambda(z; h)\,\mathrm{d}h.$$

This implies that for $r > 0$

$$\int_\Omega |z(t)|^r\,\mathrm{d}t = \int_0^\infty \lambda[|z(t)|^r; h]\,\mathrm{d}h = \int_0^\infty \lambda(|z|; h^{1/r})\,\mathrm{d}h$$

$$= \int_0^\infty \lambda(z; h)\,\mathrm{d}h^r = r\int_0^\infty h^{r-1}\lambda(z; h)\,\mathrm{d}h.$$

---

[1] J. Marcinkiewicz [1].

Hence for $\beta > 0$

$$\|z\|_\beta = \left\{\frac{1}{\beta}\int_0^\infty h^{(1-\beta)/\beta}\lambda(z;h)\,\mathrm{d}h\right\}^\beta \qquad (z(t) \in L_\beta). \tag{2.43}$$

The function

$$\mu(h) = h\cdot[\lambda(z;h)]^\beta = [\int_{\{t:|z(t)|\geqq h\}} h^{1/\beta}\,\mathrm{d}t]^\beta \tag{2.44}$$

is obviously bounded if $z(t) \in L_\beta$. The converse does not hold; for instance the function $z(t) = t^{-\beta}$ $(0 \leqq t \leqq 1)$ does not belong to L_β, but the function (2.44) is bounded. Boundedness of the function (2.44) implies only that the function $z(t)$ belongs to all spaces $L_{\beta+\varepsilon}$ for $\varepsilon > 0$.

Denote by M_β $(\beta > 0)$ the totality of all functions $z(t)$, for which the function (2.44) is bounded, and put

$$\|z(t)\|^*_{M_\beta} = \sup_{0<h<\infty} h\cdot[\lambda(z;h)]^\beta.$$

It is convenient to assume that $M_0 = L_0$ and to set

$$\|z(t)\|^*_{M_0} = \|z(t)\|_0.$$

Let a linear operator A be defined on a space L_α, where α is a number in the interval $[0, 1]$. The operator A is said to satisfy the *Marcinkiewicz condition* $LM(\alpha, \beta)$, if the following inequality is fulfilled for all functions $x(s) \in L_\alpha$:

$$\|Ax\|^*_{M_\beta} \leqq C\,\|x\|_\alpha \tag{2.45}$$

where C is some constant.

It is not difficult to show that an operator A satisfies the Marcinkiewicz condition $LM(\alpha, \beta)$, if the inequality (2.45) is fulfilled for all functions in some dense set in L_α.

THEOREM 2.9: *Let a linear operator A satisfy the Marcinkiewicz conditions $LM(\alpha_0, \beta_0)$, $LM(\alpha_1, \beta_1)$:*

$$\|Ax\|^*_{M_{\beta_0}} \leqq C_0 \|x\|_{\alpha_0}, \tag{2.46}$$

$$\|Ax\|^*_{M_{\beta_1}} \leqq C_1 \|x\|_{\alpha_1}. \tag{2.47}$$

Suppose the following inequalities hold:

$$0 \leqq \beta_0 \leqq \alpha_0 \leqq 1, \quad 0 \leqq \beta_1 \leqq \alpha_1 \leqq 1, \tag{2.48}$$

$$\beta_0 \neq \beta_1. \tag{2.49}$$

Then for any $\tau \in (0, 1)$ the operator A acts from the space $L_{\alpha(\tau)}$ to the space $L_{\beta(\tau)}$, where

$$\alpha(\tau) = (1 - \tau)\alpha_0 + \tau\alpha_1, \quad \beta(\tau) = (1 - \tau)\beta_0 + \tau\beta_1, \tag{2.50}$$

and is continuous. Moreover its norm satisfies:

$$\|A\|_{\alpha\to\beta} \leqq \frac{2}{|\beta_0 - \beta_1|\tau(1 - \tau)} C_0^{1-\tau} C_1^{\tau}. \tag{2.51}$$

The inequalities (2.48) and (2.49) mean (Fig. 2.5) that the segment joining the points $\{\alpha_0, \beta_0\}$ and $\{\alpha_1, \beta_1\}$ lies below the bisector $\beta = \alpha$ of the first quadrant and forms a non-zero angle with the horizontal axis. The assertion of Theorem 2.9 means that all interior points of this segment belong to

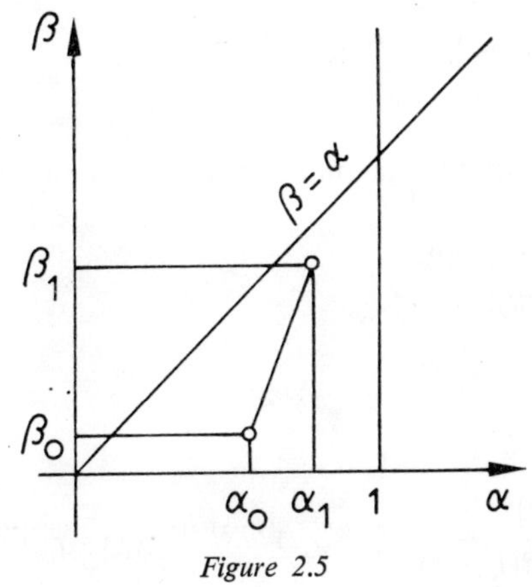

Figure 2.5

$L(A;$ cont.). In this sense the Marcinkiewicz theorem becomes an essential supplement to the Riesz interpolation theorem.

Let us confine ourselves to the proof of Theorem 2.9 for the case in which both numbers β_0 and β_1 are positive. Assume for definiteness that $\beta_0 < \beta_1$.

The proof given below is due to A. Zygmund [1].

Let $y(t) \in M_{\beta_0}$. Consider the integral

$$J = \int_0^\infty h^{(1-\beta(\tau))/\beta(\tau)} \lambda(y; h)\,\mathrm{d}h.$$

For any positive c

$$J = \int_0^c h^{(1-\beta(\tau))/\beta(\tau)} \lambda(y; h)\,\mathrm{d}h + \int_c^\infty h^{(1-\beta(\tau))/\beta(\tau)} \lambda(y; h)\,\mathrm{d}h$$

$$\leqq (\|y\|^*_{M_{\beta_1}})^{1/\beta_1} \int_0^c h^{(1-\beta(\tau))/(\beta(\tau))-(1/\beta_1)}\,\mathrm{d}h$$

$$+ (\|y\|^*_{M_{\beta_0}})^{1/\beta_0} \int_c^\infty h^{(1-\beta(\tau))/(\beta(\tau))-(1/\beta_0)}\,\mathrm{d}h$$

$$= \frac{\beta_1 \cdot \beta(\tau)}{(1-\tau)(\beta_1 - \beta_0)} (\|y\|^*_{M_{\beta_1}})^{1/\beta_1} \times c^{1/\beta(\tau) - 1/\beta_1}$$

$$+ \frac{\beta_0 \cdot \beta(\tau)}{\tau(\beta_1 - \beta_0)} (\|y\|^*_{M_{\beta_0}})^{1/\beta_0} \times c^{1/\beta(\tau) - 1/\beta_0}.$$

The right hand side of this inequality takes its minimum value at

$$c = \frac{(\|y\|^*_{M_{\beta_0}})^{\beta_1/(\beta_1-\beta_0)}}{(\|y\|^*_{M_{\beta_1}})^{\beta_0/(\beta_1-\beta_0)}}.$$

Hence

$$J \leqq \frac{[\beta(\tau)]^2}{(\beta_1 - \beta_0)\tau(1-\tau)} (\|y\|^*_{M_{\beta_0}})^{(1-\tau)/\beta(\tau)} \cdot (\|y\|^*_{M_{\beta_1}})^{\tau/\beta(\tau)}$$

and (2.43) implies

$$\|y\|_{\beta(\tau)} \leqq \frac{1}{|\beta_0 - \beta_1|\tau(1-\tau)} (\|y\|^*_{M_{\beta_0}})^{1-\tau} \cdot (\|y\|^*_{M_{\beta_1}})^{\tau}. \tag{2.52}$$

Obviously this inequality implies the assertion of Theorem 2.9 with $\alpha_0 = \alpha_1$.

Now let $\alpha_0 \neq \alpha_1$. Let $x(s)$ be a simple function.

Put

$$u_c(s) = \begin{cases} x(s), & \text{if } |x(s)| \leqq c \\ c \cdot \operatorname{sgn} x(s), & \text{if } |x(s)| > c \end{cases}$$

and put

$$v_c(s) = x(s) - u_c(s).$$

It is clear that

$$\lambda(u_c; h) = \begin{cases} \lambda(x; h), & \text{if } 0 \leqq h \leqq c \\ 0, & \text{if } c < h < \infty \end{cases} \tag{2.53}$$

and

$$\lambda(v_c; h) = \lambda(x; c+h) \qquad (h > 0). \tag{2.54}$$

Now (2.46) implies that

$$\lambda(Au_c; \tfrac{1}{2}h) \leqq (2C_0)^{1/\beta_0} h^{-1/\beta_0} (\|u_c\|_{\alpha_0})^{1/\beta_0}$$

$$= (2C_0)^{1/\beta_0} h^{-1/\beta_0} \left[\frac{1}{\alpha_0} \int_0^\infty k^{(1-\alpha_0)/\alpha_0} \cdot \lambda(u_c; k)\, \mathrm{d}k \right]^{\alpha_0/\beta_0}$$

and, by (2.53),

$$\lambda(Au_c; \tfrac{1}{2}h) \leqq (2C_0)^{1/\beta_0} \alpha_0^{-\alpha_0/\beta_0} h^{-1/\beta_0} [\int_0^c k^{(1-\alpha_0)/\alpha_0} \cdot \lambda(x; k)\, \mathrm{d}k]^{\alpha_0/\beta_0}. \tag{2.55}$$

Similarly (2.47) implies that

$$\lambda\left(Av_c; \frac{h}{2}\right) \leqq (2C_1)^{1/\beta_1} h^{-1/\beta_1} \cdot (\|v_c\|_{\alpha_1})^{1/\beta_1}$$

$$= (2C_1)^{1/\beta_1} h^{-1/\beta_1} \left[\frac{1}{\alpha_1} \int_0^\infty k^{(1-\alpha_1)/\alpha_1} \lambda(v_c; k)\,\mathrm{d}k\right]^{\alpha_1/\beta_1},$$

and hence from (2.54) and from

$$\int_0^\infty k^{(1-\alpha_1)/\alpha_1} \lambda(x; c+k)\,\mathrm{d}k \leqq \int_c^\infty k^{(1-\alpha_1)/\alpha_1} \lambda(x; k)\,\mathrm{d}k$$

it follows that

$$\lambda(Av_c; \tfrac{1}{2}h) \leqq (2C_1)^{1/\beta_1} \alpha_1^{-\alpha_1/\beta_1} \cdot h^{-1/\beta_1} [\int_c^\infty k^{(1-\alpha_1)/\alpha_1} \lambda(x; k)\,\mathrm{d}k]^{\alpha_1/\beta_1}. \tag{2.56}$$

Let us estimate the function $\lambda(Ax; h)$. The relation

$$Ax = Au_c + Av_c$$

implies that

$$\lambda(Ax; h) \leqq \lambda(Au_c; \tfrac{1}{2}h) + \lambda(Av_c; \tfrac{1}{2}h);$$

hence by (2.55) and (2.56)

$$\lambda(Ax; h) \leqq (2C_0)^{1/\beta_0} \alpha_0^{-\alpha_0/\beta_0} h^{-1/\beta_0} [\int_0^c k^{(1-\alpha_0)/\alpha_0} \lambda(x; k)\,\mathrm{d}k]^{\alpha_0/\beta_0}$$

$$+ (2C_1)^{1/\beta_1} \alpha_1^{-\alpha_1/\beta_1} h^{-1/\beta_1} [\int_c^\infty k^{(1-\alpha_1)/\alpha_1} \lambda(x; k)\,\mathrm{d}k]^{\alpha_1/\beta_1}. \tag{2.57}$$

Equation (2.57) implies the inequality

$$J(x) = \frac{1}{\beta(\tau)} \int_0^\infty h^{(1-\beta(\tau))/\beta(\tau)} \lambda(Ax; h)\,\mathrm{d}h$$

$$\leqq \frac{(2C_0)^{1/\beta_0}}{\alpha_0^{\alpha_0/\beta_0}\beta(\tau)} \int_0^\infty h^{(1-\beta(\tau))/\beta(\tau)-1/\beta_0}\left[\int_0^c k^{(1-\alpha_0)/\alpha_0}\lambda(x;k)\,\mathrm{d}k\right]^{\alpha_0/\beta_0}\mathrm{d}h$$

$$+ \frac{(2C_1)^{1/\beta_1}}{\alpha_1^{\alpha_1/\beta_1}\beta(\tau)} \int_0^\infty h^{(1-\beta(\tau))/\beta(\tau)-1/\beta_1}\left[\int_c^\infty k^{(1-\alpha_1)/\alpha_1}\lambda(x;k)\,\mathrm{d}k\right]^{\alpha_1/\beta_1}\mathrm{d}h. \qquad (2.58)$$

In these inequalities it may be assumed that c depends on h. Let us assume that

$$c = \left(\frac{1}{b}h\right)^{1/\xi}$$

where

$$\xi = \frac{\alpha_1-\alpha_0}{\beta_1-\beta_0}\,\frac{\beta(\tau)}{\alpha(\tau)}$$

and b is an arbitrary positive number.

To estimate each of the integrals on the right side of the inequality (2.58), apply the generalized Minkowski inequality[1]

---

[1] The inequality (2.59) results from the chain of the obvious relations:

$$\int_h \left|\int_k \varphi(h,k)\,\mathrm{d}k\right|^q \mathrm{d}h \leqq \int_h \left[\int_k |\varphi(h,k)|\,\mathrm{d}k\right]^q \mathrm{d}h$$

$$= \int_h \left[\int_k |\varphi(h,k)|\,\mathrm{d}k\right]^{q-1} \int_{k_1} |\varphi(h,k_1)|\,\mathrm{d}k_1\,\mathrm{d}h$$

$$= \int_{k_1}\int_h \left\{\left[\int_k |\varphi(h,k)|\,\mathrm{d}k\right]^{q-1} |\varphi(h,k_1)|\right\}\mathrm{d}h\,\mathrm{d}k_1$$

$$\leqq \int_{k_1}\left\{\int_h\left[\int_k |\varphi(h,k)|\,\mathrm{d}k\right]^q \mathrm{d}h\right\}^{1-1/q}\left\{\int_h |\varphi(h,k_1)|^q\,\mathrm{d}h\right\}^{1/q}\mathrm{d}k_1$$

$$= \left\{\int_h\left[\int_k |\varphi(h,k)|\,\mathrm{d}k\right]^q \mathrm{d}h\right\}^{1-1/q}\cdot\int_{k_1}\left\{\int_h |\varphi(h,k_1)|^q\,\mathrm{d}h\right\}^{1/q}\mathrm{d}k_1.$$

Changes of order of integrations in this chain are guaranteed by the Fubini theorem.

$$\int_h |\int_k \varphi(h, k)\,\mathrm{d}k|^q \mathrm{d}h \leqq \{\int_k [\int_h |\varphi(h, k)|^q \mathrm{d}h]^{1/q} \mathrm{d}k\}^q \quad (q \geqq 1) \tag{2.59}$$

to the first integral with $q = \alpha_0/\beta_0$ and to the second with $q = \alpha_1/\beta_1$. We obtain as result

$$J(x) \leqq$$

$$\frac{(2C_0)^{1/\beta_0}}{\alpha_0^{\alpha_0/\beta_0}\beta(\tau)} \left\{ \int_0^\infty k^{(1-\alpha_0)/\alpha_0} \left[\int_{\psi(k)}^\infty h^{(1-\beta(\tau))/\beta(\tau) - 1/\beta_0} \mathrm{d}h \right]^{\beta_0/\alpha_0} \lambda(x; k)\,\mathrm{d}k \right\}^{\alpha_0/\beta_0}$$

$$+ \frac{(2C_1)^{1/\beta_1}}{\alpha_1^{\alpha_1/\beta_1}\beta(\tau)} \left\{ \int_0^\infty k^{(1-\alpha_1)/\alpha_1} \left[\int_0^{\psi(k)} h^{(1-\beta(\tau))/\beta(\tau) - 1/\beta_1} \mathrm{d}h \right]^{\beta_1/\alpha_1} \lambda(x; k)\,\mathrm{d}k \right\}^{\alpha_1/\beta_1}, \tag{2.60}$$

where $\psi(k) = bk^\xi$.

Calculating the inner integrals in (2.60), we obtain the inequality

$$J(x) \leqq$$

$$\frac{(2C_0)^{1/\beta_0}}{\alpha_0^{\alpha_0/\beta_0}} \frac{\beta_0}{(\beta_1 - \beta_0)\tau} \left\{ \int_0^\infty k^{(1-\alpha(\tau))/\alpha(\tau)} \lambda(x; k)\,\mathrm{d}k \right\}^{\alpha_0/\beta_0} \cdot b^{1/\beta(\tau) - 1/\beta_0}$$

$$+ \frac{(2C_1)^{1/\beta_1}}{\alpha_1^{\alpha_1/\beta_1}} \frac{\beta_1}{(\beta_1 - \beta_0)(1-\tau)} \left\{ \int_0^\infty k^{(1-\alpha(\tau))/\alpha(\tau)} \lambda(x; k)\,\mathrm{d}k \right\}^{\alpha_1/\beta_1} \cdot b^{1/\beta(\tau) - 1/\beta_1}$$

or

$$J(x) \leqq \frac{(2C_0)^{1/\beta_0}[\alpha(\tau)]^{\alpha_0/\beta_0}\beta_0}{\alpha_0^{\alpha_0/\beta_0}(\beta_1 - \beta_0)\tau} (\|x\|_{\alpha(\tau)})^{\alpha_0/(\beta_0\alpha(\tau))} \cdot b^{1/\beta(\tau) - 1/\beta_0}$$

$$+ \frac{(2C_1)^{1/\beta_1}[\alpha(\tau)]^{\alpha_1/\beta_1}\beta_1}{\alpha_1^{\alpha_1/\beta_1}(\beta_1 - \beta_0)(1-\tau)} (\|x\|_{\alpha(\tau)})^{\alpha_1/(\beta_1\alpha(\tau))} \cdot b^{1/\beta(\tau) - 1/\beta_1}. \tag{2.61}$$

The inequality (2.61) is fulfilled for all b. In particular it is fulfilled for

$$b = \left\{ \frac{(2C_0)^{1/\beta_0}\alpha_0^{-\alpha_0/\beta_0}}{(2C_1)^{1/\beta_1}\alpha_1^{-\alpha_1/\beta_1}} \times [\alpha(\tau)]^{\alpha_0/\beta_0 - \alpha_1/\beta_1}(\|x\|_{\alpha(\tau)})^{1/\alpha(\tau)(\alpha_0/\beta_0 - \alpha_1/\beta_1)} \right\}^{\beta_0\beta_1/(\beta_1-\beta_0)}.$$

Substituting this value for (2.61), we obtain the inequality

$$J(x) \leqq \frac{[\alpha(\tau)]^{\alpha(\tau)/\beta(\tau)}}{\alpha_0^{(1-\tau)\alpha_0/\beta(\tau)}\alpha_1^{\tau\alpha_1/\beta(\tau)}} \times \frac{\beta(\tau)}{(\beta_1-\beta_2)\tau(1-\tau)}(2C_0)^{(1-\tau)/\alpha(\tau)}(2C_1)^{\tau/\beta(\tau)}(\|x\|_{\alpha(\tau)})^{1/\beta(\tau)}$$

whence it follows that

$$\|Ax\|_{\beta(\tau)} \leqq \frac{2[\alpha(\tau)]^{\alpha(\tau)}}{\alpha_0^{(1-\tau)\alpha_0}\alpha_1^{\tau\alpha_1}} \times \frac{[\beta(\tau)]^{\beta(\tau)}}{[(\beta_1-\beta_0)\tau(1-\tau)]^{\beta(\tau)}} C_0^{1-\tau}C_1^{\tau}\|x\|_{\alpha(\tau)}. \tag{2.62}$$

The set of step functions is dense in the space $L_{\alpha(\tau)}$. Hence the operator A admits an extension to a continuous operator $\tilde{A}$, acting from $L_{\alpha(\tau)}$ to $L_{\beta(\tau)}$. It is not difficult to see that

$$\tilde{A}x = Ax \quad (x \in L_{\alpha(\tau)}).$$

Hence the operator A acts from $L_{\alpha(\tau)}$ to $L_{\beta(\tau)}$ and is continuous. Its norm $\|A\|_{\alpha(\tau)\to\beta(\tau)}$ satisfies the inequality (2.51), which follows from (2.62) and the obvious inequalities

$$[\alpha(\tau)]^{\alpha(\tau)} \leqq \alpha_0^{(1-\tau)\alpha_0}\alpha_1^{\tau\alpha_1}, \quad [\beta(\tau)]^{\beta(\tau)} \leqq 1$$

$$[(\beta_1-\beta_0)\tau(1-\tau)]^{\beta(\tau)} \geqq (\beta_1-\beta_0)\tau(1-\tau).$$

Theorem 2.9 has been proved.

§ 3 Compact linear operators

3.1 *Compact linear operators*[1]

An operator is called *compact*, if it transforms bounded sets to compact sets. Compactness implies continuity in the case of linear operators. Compact linear operators are often called *completely continuous.*

An operator A acting from L_α to L_β will be called *compact in measure,* if the set of elements $\{Ax; \|x\|_\alpha \leqq 1\}$ is compact in measure. As it turns out, the most important operators acting in the spaces L_α—integral operators—usually possess the property of compactness in measure.

THEOREM 3.1: *Let A be a continuous linear operator acting from L_α to L_β $(0 \leqq \alpha < \infty, 0 < \beta < \infty)$. It is necessary and sufficient for compactness of the operator A that A be compact in measure and satisfy the following relation:*

$$\lim_{\text{mes}\, D^* \to 0} \|P_{D^*}A\|_{\alpha \to \beta} = 0. \tag{3.1}$$

PROOF: (3.1) implies that the range of the operator A on each ball has equi-absolutely continuous norms. Hence sufficiency of the conditions of the theorem results from Lemma 1.1.

If A is compact as an operator from L_α to L_β, it is compact in measure. It remains to show that (3.1) is fulfilled.

Suppose the contrary. Then there exist elements $x_1, x_2, \ldots \|x_n\|_\alpha \leqq 1$ and a sequence of sets D_n with measure converging to zero as $n \to \infty$ such that

$$\|P_{D^*_n} \cdot Ax_n\|_\beta \geqq \varepsilon_0 > 0 \quad (n = 1, 2, \ldots).$$

But this inequality contradicts equi-absolute continuity of the norms of the elements Ax_n. The theorem has been proved.

---

[1] Sections 3.1–3.3 are based on a number of theorems, previously established in the papers of M. A. Krasnoselskii and E. I. Pustylnik [2], M. A. Krasnoselskii and Y. B. Rutickii [1, 3, 5], T. Ando [1], and P. P. Zabreiko [2]. Note especially that the basic ideas of the proofs of Theorems 3.5 and 3.9 are due to T. Ando; their complete formulations were indicated by P. P. Zabreiko [2].

For regular operators acting from L_α to L_β ($\beta > 0$), a stronger assertion holds.

THEOREM 3.2: *Let A be a regular linear operator acting from L_α to L_β ($0 \leqq \alpha < \infty$, $0 < \beta < \infty$). It is necessary and sufficient for compactness of the operator A that A is compact in measure and satisfy the following relation:*

$$\lim_{\text{mes } D^* + \text{mes } D \to 0} \|P_{D^*} \cdot AP_D\|_{\alpha \to \beta} = 0. \tag{3.2}$$

PROOF: By Theorem 3.1 we need only to show the fact that for regular operators (3.2) implies (3.1).

Let $\varepsilon > 0$ be given. Take $\delta_0 > 0$ such that when mes $D < \delta_0$ ($D \subset \Omega$) and mes $D^* < \delta_0$ ($D^* \subset \Omega^*$) then the following inequality is fulfilled:

$$\|P_{D^*}AP_D\|_{\alpha \to \beta} < \frac{\varepsilon}{2^{\beta+1}}.$$

Let $u_0(s) \equiv 1$. Take a number $\delta \leqq \delta_0$ such that $\text{mes}(D^*) < \delta$ implies the inequality

$$\|P_{D^*}Bu_0\|_\beta < \frac{\varepsilon \delta_0^\alpha}{2^{\beta+1}}$$

where B is a positive majorant of the operator A.

Let $x(s)$ be a fixed function in the unit ball of the space L_α. Denote by D_x the set of points s, at which the inequality $|x(s)| \geqq \delta_0^{-\alpha}$ is fulfilled. Let mes $D^* < \delta$. Then

$$\|P_{D^*}Ax\|_\beta \leqq 2^\beta \|P_{D^*}A(x - P_{D_x}x)\|_\beta + 2^\beta \|P_{D^*}AP_{D_x}x\|_\beta$$

$$\leqq 2^\beta \|P_{D^*}B(|x - P_{D_x}x|)\|_\beta + 2^\beta \|P_{D^*}AP_{D_x}\|_{\alpha \to \beta}$$

$$\leqq 2^\beta \delta_0^{-\alpha} \|P_{D^*}Bu_0\|_\beta + 2^\beta \|P_{D^*}AP_{D_x}\|_{\alpha \to \beta}.$$

Consequently, $\|P_{D^*}A\|_{\alpha \to \beta} < \varepsilon$. The theorem has been proved.

Note that the assertions of Theorems 3.1 and 3.2 are not true for operators with values in the space L_0.

3.2 *Compactness and adjoint operators*

In this section we consider operators acting from L_α to L_β, where $0 \leqq \alpha, \beta \leqq 1$.

In applications it is sometimes more convenient to study the adjoint operator A^* than the operator A. Recall that an operator A is compact if and only if the operator A^* is compact. This assertion remains valid even when A acts from L_α to L_0, and A^* is considered as an operator from L_1 to $L_{1-\alpha}$ (see 2.1°). The above assertion also holds for those operators A which act from L_0 to some space L_β $(0 \leqq \beta \leqq 1)$ and possess adjoint operators which transform functions in $L_{1-\beta}$ to L_1. Note that most operators which are important in applications belong to this latter class (for instance, regular integral operators; see § 4).

In our subsequent constructions, products of linear operators A with projection operators of the form $\boldsymbol{P}_D$ and $\boldsymbol{P}_{D^*}$ $(D \subset \Omega, D^* \subset \Omega^*)$ will occur. We will often use the trivial relations

$$(\boldsymbol{P}_{D^*}A)^* = A^*\boldsymbol{P}_{D^*}, \quad (A\boldsymbol{P}_D)^* = \boldsymbol{P}_D A^*.$$

Theorems 3.1 and 3.2 imply the following assertions.

THEOREM 3.3: *Let A be a continuous linear operator acting from L_α to L_β $(0 < \alpha < 1,\ 0 \leqq \beta \leqq 1)$. It is necessary and sufficient for compactness of the operator A that the operator A^* be compact in measure and that A satisfy the following relation:*

$$\lim_{\operatorname{mes} D \to 0} \|A\boldsymbol{P}_D\|_{\alpha \to \beta} = 0. \tag{3.3}$$

THEOREM 3.4: *Let A be a regular linear operator acting from L_α to L_β $(0 < \alpha < 1,\ 0 \leqq \beta \leqq 1)$. It is necessary and sufficient for compactness of the operator A that the operator A^* be compact in measure and that A satisfy the following relation:*

$$\lim_{\operatorname{mes} D + \operatorname{mes} D^* \to 0} \|\boldsymbol{P}_{D^*}A\boldsymbol{P}_D\|_{\alpha \to \beta} = 0. \tag{3.4}$$

In the conditions of this theorem α may be equal to zero if the operator A^* acts from $L_{1-\beta}$ to L_1.

THEOREM 3.5: *Let A be a continuous linear operator acting from L_α to L_β $(0 < \alpha < 1,\ 0 < \beta < 1)$. It is necessary and sufficient for compactness of the operator A that the operators A and A^* be compact in measure and that the following relation be fulfilled:*

$$\lim_{\operatorname{mes} D + \operatorname{mes} D^* \to 0} \|P_{D^*} A P_D\|_{\alpha \to \beta} = 0. \tag{3.4^1}$$

To prove sufficiency, we establish that the operator A transforms weakly convergent sequences to strongly convergent sequences. Then it will follow that A transforms the unit ball $\|x\|_\alpha \leqq 1$ to a compact set (for each sequence Ax_n, $\|x_n\|_\alpha \leqq 1$ it is possible to take a weakly convergent subsequence x_{n_i}, because the spaces L_α for $0 < \alpha < 1$ have weakly compact spheres).

Suppose that A transforms some weakly convergent sequence to a sequence which is not convergent in norm.

Then there exists a sequence $x_n \in L_\alpha$, $\|x_n\|_\alpha \leqq 1$ which is weakly convergent to zero, for which

$$\|Ax_n\|_\beta \geqq c_0 > 0 \quad (n = 1, 2, \ldots). \tag{3.5}$$

The sequence Ax_n also converges weakly to zero (in L_β). Compactness (in measure) of the operator A thus implies that the sequence Ax_n converges to zero in measure. It can even be assumed without loss of generality that Ax_n converges to zero almost everywhere.

By Egorov's theorem it is possible for each $\varepsilon > 0$ to obtain a set $D_\varepsilon^* \subset \Omega^*$ such that $\operatorname{mes}(D_\varepsilon^*) < \varepsilon$ and such that on $\Omega^* - D_\varepsilon^*$ the sequence Ax_n converges to zero uniformly. This implies that

$$\lim_{n \to \infty} \|P_{\Omega^* - D^*_\varepsilon} Ax_n\|_\beta = 0. \tag{3.6}$$

Assume that D_ε^* is a fixed set. Inequality (3.5) implies the existence of functions $y_n \in L_{1-\beta}$, $\|y\|_{1-\beta} \leqq 1$ such that

$$(Ax_n, y_n) \geqq c_0 > 0 \quad (n = 1, 2, \ldots).$$

Here it may be assumed without loss of generality that the sequence y_n converges weakly to some function y_0, $\|y_0\|_{1-\beta} \leqq 1$. It is clear that

$$(Ax_n, y_n - y_0) = (Ax_n, y_n) - (x_n, A^*y_0).$$

The second member in the right side converges to zero as $n \to \infty$. Hence it can be assumed that for all n the following inequality is fulfilled:

$$(Ax_n, y_n - y_0) > \tfrac{1}{2}c_0 \quad (n = 1, 2, \ldots).$$

The sequence $A^* P_{D^*_\varepsilon}(y_n - y_0)$ converges weakly to zero. Compactness (in measure) of the operator A^* implies that the sequence $A^* P_{D^*_\varepsilon}(y_n - y_0)$ also converges to zero in measure (see the end of 1.3°). It can be assumed without loss of generality that the sequence $A^* P_{D^*_\varepsilon}(y_n - y_0)$ converges to zero almost everywhere. Hence for each $\varepsilon_1 > 0$ it is possible to take a set $D_{\varepsilon_1} \subset \Omega$ such that $\operatorname{mes}(D_\varepsilon) < \varepsilon_1$ and on $\Omega - D_{\varepsilon_1}$ the sequence $A^* P_{D^*_\varepsilon}(y_n - y_0)$ converges to zero uniformly. Consequently

$$\lim_{n \to \infty} \| P_{\Omega - D_{\varepsilon_1}} A^* P_{D^*_\varepsilon}(y_n - y_0) \|_{1-\alpha} = 0. \tag{3.7}$$

Consider the identity

$$(Ax_n, y_n - y_0) = (P_{\Omega^* - D^*_\varepsilon} A x_n, y_n - y_0)$$

$$+ (P_{D^*_\varepsilon} A P_{D_{\varepsilon_1}} x_n, y_n - y_0) + (x_n, P_{\Omega - D_{\varepsilon_1}} A^* P_{D^*_\varepsilon}(y_n - y_0)).$$

The first and third summands converge to zero by (3.6) and (3.7). Hence for sufficiently large n the following inequality is fulfilled:

$$(P_{D^*_\varepsilon} A P_{D_{\varepsilon_1}} x_n, y_n - y_0) \geqq c_0/4.$$

Hence

$$\| P_{D^*_\varepsilon} A P_{D_{\varepsilon_1}} x_n \|_\beta \| y_n - y_0 \|_{1-\beta} \geqq c_0/4,$$

and since $\|x_n\|_\alpha \leqq 1$ and $\|y_n - y_0\|_{1-\beta} \leqq 2$,

$$\| P_{D^*_\varepsilon} A P_{D_{\varepsilon_1}} \|_{\alpha \to \beta} \geqq c_0/8.$$

This inequality contradicts (3.4). Thus sufficiency of the conditions of the theorem has been proved. Necessity results, for instance, from Theorem 3.1. The theorem has been proved.

3.3 *Properties of operators compact in measure*

In the next chapter it will be shown that under natural assumptions linear integral operators are compact in measure. Hence an independent study of operators which are compact in measure is of interest.

THEOREM 3.6: *A continuous linear operator A acting from L_α to L_β $(0 \leqq \alpha < 1, 0 < \beta \leqq 1)$ is compact in measure if and only if it transforms each weakly convergent sequence $x_n \in L_\alpha$ to a sequence Ax_n which is convergent in measure.*

PROOF: Sufficiency results from the weak compactness of the unit ball of the space L_α $(0 \leqq \alpha < 1)$.

Now let the operator A be compact in measure and consider a sequence x_n converging weakly to x_0. Then continuity of the operator A implies that the sequence Ax_n converges weakly to Ax_0. On the other hand, the sequence of elements Ax_n is compact in measure. Hence Ax_n converges to Ax_0 in measure. The theorem has been proved.

In many cases compactness of an operator in measure implies its compactness.

For instance let A be a regular operator, compact in measure, which acts from L_0 to L_β, where $\beta > 0$. Then the operator A satisfies the inequality $|Ax| \leqq \boldsymbol{B}(|x|)$ (see (2.10)), where $\boldsymbol{B}$ is a positive linear operator. Functions $x(t)$ in the ball $\|x(t)\|_0 \leqq 1$ satisfy the inequality $|x(t)| \leqq 1$. Hence

$$|Ax(t)| \leqq \boldsymbol{B}u_0(t) \qquad (\|x\|_0 \leqq 1),$$

where $u_0(s) \equiv 1$. This inequality implies that the functions $Ax(t)$ $(\|x\|_0 \leqq 1)$ have equi-absolutely continuous norms. Lemma 1.1 now implies that the set of functions $Ax(t)$ $(\|x\|_0 \leqq 1)$ is compact in L_β. That is, A is compact. These observations show that a linear operator A, acting from L_α to L_1, where $\alpha < 1$, is compact, if its adjoint A^* is compact in measure.

THEOREM 3.7: *Let A be a continuous operator, compact in measure, acting from L_α to L_β. Then A is compact as an operator acting from L_α to L_{β_1}, whenever $\beta_1 > \beta$.*

THEOREM 3.8: *Let A be a regular operator, compact in measure, acting from L_α to L_β ($\beta > 0$). Then A is compact as an operator acting from L_{α_1} to L_β, whenever $\alpha_1 < \alpha$.*

The proof of Theorem 3.7 is obvious: the operator A transforms the ball $\|x\|_\alpha \leqq 1$ to a set of functions, which is compact in measure and has equi-absolutely continuous norms in L_{β_1}. It remains only to use Lemma 1.1.

To prove Theorem 3.8, it is necessary to remark that the unit ball of the space L_{α_1} becomes a set of functions with equi-absolutely continuous norms in L_α. Hence by Theorem 2.8 the functions Ax, $\|x\|_{\alpha_1} \leqq 1$, also have equi-absolutely continuous norms, and consequently A is compact. Theorems 3.7 and 3.8 have been proved.

Assertions similar to Theorems 3.7 and 3.8 are valid for those linear operators A, whose adjoint operators are compact in measure.

In the following our aim is a proof of an interesting theorem of T. Ando [1] which improves the assertions of Theorems 3.7 and 3.8.

LEMMA 3.1: *Let A be a continuous linear operator acting from L_α to L_β, where $0 < \alpha < \beta \leqq 1$. Then*

$$\lim_{\operatorname{mes} D + \operatorname{mes} D^* \to 0} \|P_{D^*} A P_D\|_{\alpha\to\beta} = 0. \tag{3.8}$$

PROOF: Under the contrary assumption it is possible to construct sequences of sets $D_n^* \subset \Omega^*$, $D_n \subset \Omega$ such that $\operatorname{mes}(D_n) + \operatorname{mes}(D_n^*) \to 0$ and

$$\|P_{D^*_n} A P_{D_n}\|_{\alpha\to\beta} > c_0 > 0 \quad (n = 1, 2, \ldots).$$

It can be assumed that

$$\operatorname{mes}(D_n^*) + \operatorname{mes}(D_n) \leqq 1/2^n \quad (n = 1, 2, \ldots).$$

Now consider the sets

$$E_n = \bigcup_{k=n}^{\infty} D_k, \quad E_n^* = \bigcup_{k=n}^{\infty} D_n^*.$$

It is clear that $\operatorname{mes}(E_n) + \operatorname{mes}(E_n^*) \to 0$ and

$$\|P_{E^*_n} A P_{E_n}\|_{\alpha\to\beta} \geqq \|P_{D^*_n} A P_{D_n}\|_{\alpha\to\beta} > c_0 \quad (n = 1, 2, \ldots).$$

The easily derived relation

$$\|P_{E^*_n} A P_{E_n}\|_{\alpha\to\beta} = \lim_{m\to\infty} \|P_{E^*_n - E^*_m} \cdot A P_{E_n - E_m}\|_{\alpha\to\beta}$$

implies the existence of a subsequence of indices n_k such that

$$\|P_{E^*_{n_k} - E^*_{n_{k+1}}} \cdot A P_{E_{n_k} - E_{n_{k+1}}}\|_{\alpha\to\beta} > c_0 \quad (k = 1, 2, \ldots).$$

In other words, there exist functions $x_k \in L_\alpha$, $y_k \in L_{1-\beta}$ such that $\|x_k\|_\alpha \leqq 1$, $\|y_k\|_{1-\beta} \leqq 1$ and

$$(P_{E^*_{n_k} - E^*_{n_{k+1}}} \cdot A P_{E_{n_k} - E_{n_{k+1}}} x_k, y_k) > c_0 \quad (k = 1, 2, \ldots).$$

The functions

$$u_k = P_{E_{n_k} - E_{n_{k+1}}} x_k, \quad v_k = P_{E^*_{n_k} - E^*_{n_{k+1}}} y_k \quad (k = 1, 2, \ldots)$$

have disjoint supports[1], their norms do not exceed 1 in the spaces L_α and $L_{1-\beta}$ respectively, and

$$(Au_k, v_k) > c_0 \quad (k = 1, 2, \ldots). \tag{3.9}$$

The relation

$$\lim_{k\to\infty} [\operatorname{mes}(E^*_{n_k} - E^*_{n_{k+1}}) + \operatorname{mes}(E_{n_k} - E_{n_{k+1}})] = 0$$

implies that

$$\lim_{l\to\infty} (Au_k, v_l) = \lim_{l\to\infty} (Au_k, P_{E^*_{n_l} - E^*_{n_{l+1}}} \cdot v_l)$$

$$= \lim_{l\to\infty} (P_{E^*_{n_l} - E^*_{n_{l+1}}} Au_k, v_l) = 0. \tag{3.10}$$

Similarly

$$\lim_{k\to\infty} (Au_k, v_l) = 0. \tag{3.11}$$

[1] This means that $u_i(s)u_j(s) = 0$, $v_i(t)v_j(t) = 0$ for $i \neq j$.

(3.10) and (3.11) imply the existence of a sequence of indices k_i such that

$$(Au_{k_i}, v_{k_j}) < \frac{1}{2^{i+j}} \quad (i \neq j). \tag{3.12}$$

Define now the functions $\xi(s)$, $\eta(t)$ by the relations

$$\xi(s) = \sum_{i=1}^{\infty} a_i u_{k_i}(s), \quad \eta(t) = \sum_{i=1}^{\infty} b_i v_{k_i}(t),$$

where the numbers a_i, b_i are non-negative and

$$\sum_{i=1}^{\infty} a_i^{1/\alpha} \leqq 1, \quad \sum_{i=1}^{\infty} b_i^{1/(1-\beta)} \leqq 1. \tag{3.13}$$

It is clear that $\xi(s) \in L_\alpha$, $\eta(t) \in L_{1-\beta}$. Hence

$$|(A\xi, \eta)| = | \sum_{i,j=1}^{\infty} a_i b_j (Au_{k_i}, v_{k_j})| < \infty.$$

Hence by (3.9) and (3.12)

$$\sum_{i=1}^{\infty} a_i b_i \leqq \frac{1}{c_0} \sum_{i=1}^{\infty} a_i b_i (Au_{k_i}, v_{k_i})$$

$$\leqq \frac{1}{c_0} (A\xi, \eta) + \sum_{i,j=1}^{\infty} \frac{1}{2^{i+j}} < \infty.$$

On the other hand, since $\alpha < \beta$, it is possible to choose sequences $\{a_i\} \in l_\alpha$ and $\{b_i\} \in l_{1-\beta}$ such that the series $\sum a_i b_i$ diverges. We arrive at a contradiction. The lemma has been proved.

THEOREM 3.9: *Let A be a continuous linear operator acting from L_α to L_β, where $0 < \alpha < \beta < 1$. Let one of the following conditions be fulfilled:*
a) *A is compact in measure and regular,*
b) *A^* is compact in measure and regular,*
c) *A and A^* are compact in measure.*
Then A is a compact operator.

PROOF: Consider case a). Utilizing Theorem 3.2, we see that[1] compactness follows from the relation

$$\lim_{\text{mes } D^* + \text{mes } D \to 0} \|\boldsymbol{P}_{D^*}\boldsymbol{B}\boldsymbol{P}_D\|_{\alpha\to\beta} = 0, \tag{3.14}$$

which was established in Lemma 3.1.

In case b) the operator A^* acting from $L_{1-\beta}$ to $L_{1-\alpha}$ is compact. Consequently the operator A is compact.

In case c) it is necessary to use Theorem 3.5. The theorem has been proved.

3.4 *Interpolation properties of compactness*

The compactness property of a linear operator A can also be interpolated. The relevant assertions[2] are consequences of the inequalities (2.17)–(2.18) of the interpolation theorems for continuity.

THEOREM 3.10: *Let A be a continuous linear operator acting from L_{α_0} to L_{β_0} and from L_{α_1} to L_{β_1} $(0 \leqq \alpha_0, \beta_0, \alpha_1, \beta_1 \leqq 1)$. Let A be compact as an operator from L_{α_0} to L_{β_0}. Then for any $\tau \in (0, 1)$ A is compact as an operator from $L_{\alpha(\tau)}$ to $L_{\beta(\tau)}$, where*

$$\alpha(\tau) = (1-\tau)\alpha_0 + \tau\alpha_1, \quad \beta(\tau) = (1-\tau)\beta_0 + \tau\beta_1. \tag{3.15}$$

PROOF: Denote by $\boldsymbol{P}_1, \boldsymbol{P}_2, \ldots$ a regular sequence of projections (see 1.3°), which converges strongly to the identity operator $\boldsymbol{I}$ on the subspace $E \subset L_{\beta_0}$, the closed linear span of the range of A on L_{α_0}. If $\beta_0 = 0$, the possibility of construction of such a sequence of projections follows from Lemma 1.5. If $\beta_0 > 0$ then by Lemma 1.4 every regular sequence of projections possesses this property.

Since A is a compact operator from L_{α_0} to L_{β_0} and since

$$\|Ax - \boldsymbol{P}_n Ax\|_{\beta_0} \to 0,$$

[1] We are indebted to Prof. H. Flaschka for mentioning this abbreviation of the original argument. (Ed.)

[2] Theorem 3.10 was proved by M. A. Krasnoselskii [8]. The more general Theorem 3.11 was established in the paper of P. P. Zabreiko and E. I. Pustylnik [2].

there can be found for each $\varepsilon > 0$ an n_0 such that for $n \geqq n_0$ the following inequality is fulfilled:

$$\|(A - P_nA)x\|_{\beta_0} < \varepsilon \|x\|_{\alpha_0} \quad (x \in L_{\alpha_0}). \tag{3.16}$$

In fact, in the contrary case there exists a sequence x_k, $\|x_k\|_\alpha \leqq 1$, for which

$$\|(I - P_{n_k})Ax_k\| \geqq \varepsilon_0 \tag{3.17}$$

By compactness of A it can be assumed without loss of generality that Ax_k converges to some y_0. The inequalities

$$\|(I - P_{n_k})Ax_k\|_{\beta_0} \leqq \|(I - P_{n_k})(Ax_k - y_0)\|_{\beta_0}$$

$$+ \|(I - P_{n_k})y_0\|_{\beta_0} \leqq 2\|Ax_{n_k} - y_0\|_{\beta_0} + \|(I - P_{n_k})y_0\|_{\beta_0}$$

imply the relation

$$\lim_{n\to\infty} \|(I - P_{n_k})Ax_k\|_{\beta_0} = 0,$$

which contradicts (3.17).

On the other hand, the following inequality is fulfilled:

$$\|(A - P_nA)x\|_{\beta_1} \leqq 2\|A\|_{\alpha_1\to\beta_1}\|x\|_{\alpha_1}. \tag{3.18}$$

Now for each n the operator P_nA is obviously compact and acts from $L_{\alpha(\tau)}$ to $L_{\beta(\tau)}$. By Theorem 2.3 (inequalities (2.16) and (2.17)) and the estimates (3.16) and (3.18), there follows the inequality

$$\|(I - P_n)Ax\|_{\beta(\tau)} \leqq 2\varepsilon^{1-\tau}\|A\|^{\tau}_{\alpha_1\to\beta_1}\|x\|_{\alpha(\tau)}.$$

Thus, for any $\tau \in (0, 1)$ the operator A acting from $L_{\alpha(\tau)}$ to $L_{\beta(\tau)}$ is the limit (in operator norm) of the compact operators P_nA. Hence the operator A is compact. The theorem has been proved.

THEOREM 3.11: *Let A be a linear operator acting from L_{α_0} to L_{β_0} and from L_{α_1} to L_{β_1} $(0 \leqq \alpha_0, \alpha_1 \leqq 1, 0 \leqq \beta_0, \beta_1 < \infty)$. Let A be a compact operator* *from* L_{α_0} *to* L_{β_0}. *Then for any* $\tau \in (0, 1)$ *the operator* A *is a compact operator from* $L_{\alpha(\tau)}$ *to* $L_{\beta(\tau)}$, *where*

$$\alpha(\tau) = (1 - \tau)\alpha_0 + \tau\alpha_1, \quad \beta(\tau) = (1 - \tau)\beta_0 + \tau\beta_1.$$

PROOF: Let first $\beta_0 > 0$. Compactness of the operator A as an operator from L_{α_0} to L_{β_0} implies by Theorem 3.1 that

$$\lim_{\text{mes}\, D^* \to 0} \|\boldsymbol{P}_{D^*}A\|_{\alpha_0 \to \beta_0} = 0.$$

It is clear that

$$\|\boldsymbol{P}_{D^*}A\|_{\alpha_1 \to \beta_1} \leqq \|A\|_{\alpha_1 \to \beta_1}.$$

These two relations and interpolation Theorem 2.4 imply the relation

$$\lim_{\text{mes}\, D^* \to 0} \|\boldsymbol{P}_{D^*}A\|_{\alpha(\tau) \to \beta(\tau)} = 0.$$

By Theorem 3.1 in order to prove the compactness of A as an operator from $L_{\alpha(\tau)}$ to $L_{\beta(\tau)}$ it suffices to show that A is compact in measure.

Compactness in measure of the operator A acting from $L_{\alpha(\tau)}$ to $L_{\beta(\tau)}$, is obvious if $\alpha_0 \geqq \alpha_1$. It remains to consider the case $\alpha_0 < \alpha_1$.

Consider the operator

$$\boldsymbol{T}_h x(s) = \begin{cases} x(s), & \text{if } |x(s)| \leqq h \\ h \cdot \operatorname{sgn}[x(s)], & \text{if } |x(s)| > h. \end{cases}$$

This operator transforms each set of functions to a uniformly bounded set of functions, which lies in some ball of the space L_{α_0}. Compactness of A as an operator L_{α_0} to L_{β_0} implies that the range of the operator $A\boldsymbol{T}_h$ (for each fixed h) is compact in measure. Hence compactness in measure of A as an operator from $L_{\alpha(\tau)}$ to $L_{\beta(\tau)}$ will be established if we show that for some β^*, $\|Ax - A\boldsymbol{T}_h x\|_{\beta^*}$ converges to zero as $h \to \infty$, uniformly with respect to functions $x(s)$ in the ball $\|x\|_{\alpha(\tau)} \leqq 1$.

Put $\beta^* = \beta_1$. Then

$$\|Ax - AT_hx\|_{\beta_1} \leqq \|A\|_{\alpha_1\to\beta_1}\|x - T_hx\|_{\alpha_1}$$

$$= \|A\|_{\alpha_1\to\beta_1}\|(x - T_hx)\cdot\kappa(h; x)\|_{\alpha_1}$$

$$\leqq \|A\|_{\alpha_1\to\beta_1}\|x - T_hx\|_{\alpha(\tau)}\|\kappa(h; x)\|_{\alpha_1-\alpha(\tau)}, \tag{3.19}$$

where $\kappa(h; x)$ is the characteristic function of the set of points, for which $|x(s)| \geqq h$. It is clear that

$$\operatorname{mes}\{x\colon |x(s)| \geqq h\} \leqq h^{-1/\alpha(\tau)},$$

whence

$$\|\kappa(h; x)\|_{\alpha_1-\alpha(\tau)} \leqq h^{(\alpha(\tau)-\alpha_1)/\alpha(\tau)}.$$

Then it follows from (3.19) that

$$\|Ax - AT_hx\|_{\beta_1} \leqq \|A\|_{\alpha_1\to\beta_1}\cdot h^{(\alpha(\tau)-\alpha_1)/\alpha(\tau)}.$$

The assertion has been proved for the case $\beta_0 > 0$.

Let $\beta_0 = 0$. If $\beta_1 \leqq 1$, we are led to the conditions of Theorem 3.10, and hence the compactness of A as an operator from $L_{\alpha(\tau)}$ to $L_{\beta(\tau)}$ follows. It remains to consider the case $\beta_1 > 1$.

Put $\alpha^* = \alpha_0(1 - 1/\beta_1) + \alpha_1/\beta_1$. Theorem 2.4 implies that A is continuous as an operator from L_{α^*} to L_1. Then it follows from Theorem 3.10 that A is compact as an operator from $L_{\alpha(\tau)}$ to $L_{\beta(\tau)}$ for all τ such that $\beta(\tau) < 1$.

Now let τ_0 be a fixed number in (0, 1). Since $\beta_0 = 0$, it is possible to find a τ_1 such that $0 \leqq \beta(\tau_1) \leqq 1$, $\beta(\tau_1) < \beta(\tau_0)$. Now consider A as a continuous operator acting from L_{α_1} to L_{β_1} and as a compact operator acting from $L_{\alpha(\tau_1)}$ to $L_{\beta(\tau_1)}$. It is not difficult to see that

$$\alpha(\tau_0) = (1 - s)\alpha(\tau_1) + s\alpha_1, \quad \beta(\tau_0) = (1 - s)\beta(\tau_1) + s\beta_1,$$

where $s \in (0, 1)$. Since $\beta(\tau_1) > 0$, it follows by the part of the theorem already proved that A is compact as an operator from $L_{\alpha(\tau_0)}$ to $L_{\beta(\tau_0)}$. The theorem has been completely proved.

Compactness of an operator A was seen to involve two properties: compactness in measure and the relation

$$\lim_{\text{mes } D^* \to 0} \|P_{D^*}A\|_{\alpha \to \beta} = 0.$$

In the proof of Theorem 3.11 we actually showed that each of these properties admits interpolation. Hence compactness of A as an operator from $L_{\alpha(\tau)}$ to $L_{\beta(\tau)}$ $(0 < \tau < 1)$ can be obtained as a consequence of the compactness of A in measure as an operator from L_{α_0} to L_{β_0} and of the property

$$\lim_{\text{mes } D^* \to 0} \|P_{D^*}A\|_{\alpha_1 \to \beta_1} = 0$$

(if $\beta_1 > 0$).

Theorem 3.11 implies, in particular, that the part of the set $L(A; \text{comp.})$, lying in the strip $0 \leqq \alpha \leqq 1$, $0 \leqq \beta < \infty$, is convex; hence the function $\xi(\alpha; A; \text{comp.})$ is convex and continuous. From this there follows a stronger assertion concerning the set $L(A; \text{comp.})$.

Suppose that A is compact as an operator, acting from L_{α_0} to L_{β_0}. Then by Theorem 3.11 the L-characteristic $L(A; \text{comp.})$ contains all those points on rays from $\{\alpha_0, \beta_0\}$, which are interior points of the L-characteristic $L(A; \text{cont.})$. Hence it follows that $L(A; \text{comp.})$ contains all interior points of $L(A; \text{cont.})$ if $L(A; \text{comp.})$ contains at least one point. The last assumption is essential—for the identity operator I the set $L(I; \text{cont.})$ contains all points $\{\alpha, \beta\}$, for which $\beta \geqq \alpha$, but the set $L(I; \text{comp.})$ is empty.

In the following we shall study operators A for which it is easy to establish compactness as an operator from L_0 to L_1. For such operators all interior points of $L(A; \text{cont.})$ become points in $L(A; \text{comp.})$.

The proof of Theorem 3.11 (with non-essential modification) remains valid for the case when one or both of the numbers α_0, α_1 is greater than 1. We did not give the corresponding more general theorem, because we do not know of any compact operators which act from some space L_α with $\alpha > 1$, to a space L_β.

3.5 *Strongly continuous linear operators*

An operator A, acting from a Banach space E_1 to a Banach space E_2, is called *strongly continuous* if it transforms each sequence x_n weakly convergent to an x_0 in E_1 to a sequence Ax_n which is strongly convergent to Ax_0 in E_2.

For operators A acting from L_α to L_β, where $0 < \alpha < 1$, $0 \leqq \beta \leqq 1$, strong continuity is equivalent to compactness. In fact, if A is compact, it transforms weakly convergent sequences to weakly convergent sequences, which are compact and hence strongly convergent. Conversely, if A is strongly continuous then compactness of its range in L_β on each ball of the space L_α results from the weak compactness of this ball.

For operators acting from L_1 to L_β, where $\beta \in [0, 1]$, compactness implies strong continuity. The converse is not true.

Now consider linear operators A acting from L_α, where $0 < \alpha < 1$, to L_β with $\beta > 1$. In this case strong continuity implies compactness.

Denote by $\boldsymbol{P}_n$ $(n = 1, 2, \ldots)$ projections on finite dimensional subspaces E_n of a Banach space E. Assume that $E_1 \subset E_2 \subset \ldots \subset E_n \subset \ldots$ and further that $\boldsymbol{P}_i\boldsymbol{P}_j = \boldsymbol{P}_j\boldsymbol{P}_i = \boldsymbol{P}_j$ for $i \geqq j$. A sequence of such projections will be called *complete*, if it converges strongly to the identity operator, i.e. if for each fixed $x \in E$ the following relation holds: $\lim_{n\to\infty} \|x - \boldsymbol{P}_n x\| = 0$. It is easy to construct a regular sequence (see 1.5°) of projections $\boldsymbol{P}_n$ such that they form a complete sequence in all spaces L_α $(0 < \alpha \leqq 1)$.

LEMMA 3.2: *A continuous linear operator A acting from L_α to L_β, where $0 < \alpha < 1, 0 \leqq \beta < \infty$, is strongly continuous, if for some complete sequence of projections $\boldsymbol{P}_n$ in L_α, the following relation is fulfilled:*

$$\lim_{n\to\infty} \|A(\boldsymbol{I} - \boldsymbol{P}_n)\|_{\alpha\to\beta} = 0. \tag{3.20}$$

Conversely, if A is strongly continuous, the relation (3.20) *is valid for each complete sequence of projections in L_α.*

PROOF: Let us begin by establishing the first assertion of the lemma. Let x_n, with $\|x_n\|_\alpha \leqq M$, converge weakly to x_0 and let $\varepsilon > 0$. Take a number n_0 for which the following inequality is fulfilled: $\|A(\boldsymbol{I} - \boldsymbol{P}_{n_0})\|_{\alpha\to\beta} \leqq \varepsilon/4 \times 2^\beta M$. The sequence $\boldsymbol{P}_{n_0}x_n$ converges weakly to $\boldsymbol{P}_{n_0}x_0$. But since the sequence lies in a finite dimensional space, it converges strongly to $\boldsymbol{P}_{n_0}x_0$. Hence there exists a number $n_1 \geqq n_0$, such that for $n \geqq n_1$ the following inequality holds: $\|\boldsymbol{P}_{n_0}x_n - \boldsymbol{P}_{n_0}x_0\| \leqq \varepsilon/2^{\beta+1}\|A\|_{\alpha\to\beta}$. From the inequality

$$\|Ax_n - Ax_0\|_\beta \leqq 2^\beta \|A(\boldsymbol{I} - \boldsymbol{P}_{n_0})(x_n - x_0)\|_\beta + 2^\beta \|A\boldsymbol{P}_{n_0}(x_n - x_0)\|_\beta$$

it follows that, for $n \geqq n_1$,

$$\|Ax_n - Ax_0\|_\beta < \varepsilon.$$

Now let the operator A be strongly continuous, but suppose that $\|A(I - P_n)\|_{\alpha\to\beta}$ does not converge to 0 as $n \to \infty$. Then there exists a sequence x_n, $\|x_n\|_\alpha \leqq 1$ such that

$$\|A(x_n - P_n x_n)\|_\beta \geqq \varepsilon_0 > 0 \quad (n = 1, 2, \ldots). \tag{3.21}$$

It can be assumed without loss of generality that the sequences x_n and $P_n x_n$ converge weakly to elements z_1 and z_2. The relations

$$P_{n_0} z_1 = P_{n_0}(\lim_{n\to\infty} x_n) = \lim_{n\to\infty} P_{n_0} x_n$$

$$= \lim_{n\to\infty} P_{n_0} P_n x_n = P_{n_0} \lim_{n\to\infty} P_n x_n = P_{n_0} z_2$$

(here the limit is taken in the sense of weak convergence) imply $z_1 = z_2$. Thus the sequence $x_n - P_n x_n$ converges weakly to zero and consequently the sequence $A(x_n - P_n x_n)$ must converge to zero strongly, which contradicts (3.21). The lemma has been proved.

The assertions of this lemma are valid for any reflexive Banach space (the first assertion for arbitrary Banach spaces).

Note that the property of strong continuity of a linear operator also admits interpolation. This fact is interesting, of course, only for the case of operators with values in L_β where $\beta > 1$. The proof of this assertion can be obtained by the same method as is used in the proof of Theorem 3.10. Here it is only necessary to replace the operators $(I - P_n)A$ by the operators $A(I - P_n)$ and to use Lemma 3.2.

2

Continuity and compactness of linear integral operators

§ 4 General theorems on continuity of integral operators[1]

4.1 *Linear integral operators*

In the sequel Ω and Ω^* denote sets of finite measure in finite dimensional spaces.

In this chapter we examine operators of the form

$$Kx(t) = \int_{\Omega} K(t, s)x(s)\,\mathrm{d}s \tag{4.1}$$

and give conditions under which they act from one given space $L_\alpha = L_\alpha(\Omega)$ to another given space $L_\beta = L_\beta(\Omega^*)$ and possess sufficiently good properties (continuity, compactness etc.).

The integral used in the definition of the operator in (4.1) is taken in the sense of Lebesgue. In this connection it is always assumed that the kernel

---

[1] Problems on the continuity and compactness of integral operators in various function spaces have been considered by a number of authors (see, for instance, S. Banach [1], S. L. Sobolev [1, 2], L. V. Kantorovic [1], L. V. Kantorovic and G. P. Akilov [1], A. C. Zaanen [1–3], T. Ando [1]. M. A. Krasnoselskii [7–9], V. P. Ilin [1–3], M. A. Krasnoselskii - Y. B. Pustylnik [1, 2], E. N. Pustylnik [3, 6], P. P. Zabreiko [2, 3], A. Calderon and A. Zygmund [1, 2] and Y. B. Rutickii [2, 4, 5]). A considerable portion of the results presented in § 4 (sometimes with other terminologies or in other forms) has been applied by other authors, as well.

$K(t, s)$ is measurable in s on Ω for almost all $t \in \Omega^*$. Usually we will assume that a more severe restriction is satisfied—the kernel $K(t, s)$ is to be jointly measurable in the variables $t \in \Omega^*$, $s \in \Omega$. In this case the function $\boldsymbol{K}x(t)$ is measurable for any measurable function $x(s)$ such that the integral (4.1) is finite for almost all $t \in \Omega^*$.

Incidentally, we note that for integral operators (4.1) the L-characteristics $L(\boldsymbol{K}$; def.) and $L(\boldsymbol{K}$; cont.) coincide. Let us sketch a proof of this interesting theorem of S. Banach [1]. Let $\boldsymbol{K}$ act from L_α to L_β. Denote by $\mathfrak{M}_n$ the set of all functions in L_α such that $\|\boldsymbol{K}x\|_\beta \leqq n$. If $x_k \in \mathfrak{M}_n$ and $\|x_k - x^*\|_\alpha \to 0$, then the sequence $x_k(s)$ converges to $x^*(s)$ in measure and thus, for a subsequence, $x_{k'}(s) \to x^*(s)$ a.e. while $\sum \|x_{k'} - x''\|_\alpha < \infty$. Hence it can be assumed that the inequality $|x_k(s) - x^*(s)| \leqq u(s)$ is fulfilled for almost all $s \in \Omega$, where $u(s) \in L_\alpha$. For almost all $t \in \Omega^*$ the function

$$v(t) = \int_\Omega |K(t, s)| u(s) \, ds$$

takes finite values. For such values of t

$$\lim_{k \to \infty} \int_\Omega K(t, s) x_k(s) \, ds = \int_\Omega K(t, s) x^*(s) \, ds$$

and Fatou's lemma (see I. P. Natanson [1]) implies that

$$\|\boldsymbol{K}x^*\|_\beta \leqq \underline{\lim}_{k \to \infty} \|\boldsymbol{K}x_k\|_\beta \leqq n.$$

Hence each of the sets $\mathfrak{M}_n$ is closed in L_α. The space L_α is of second category, because it is complete. Consequently, one of the sets $\mathfrak{M}_n$ contains some ball; from this there follows the boundedness of the range of the operator $\boldsymbol{K}$ on each ball.

We are interested in conditions under which $\boldsymbol{K}$ acts from L_α to L_β and is continuous (as seen above, the continuity is automatic). In other words, we are interested in conditions under which, for all functions $x(s)$ in L_α, the following inequality is valid:

$$\|\boldsymbol{K}x\|_\beta \leqq M \|x\|_\alpha. \tag{4.2}$$

Sometimes this inequality is proved easily for functions $x(s)$ in some set $\mathfrak{M}$ dense in L_α (for instance, for bounded functions). The problem of identifying those kernels $K(t, s)$ for which the inequality then follows for all $x(s) \in L_\alpha$ has not been adequately investigated. We thus present only partial results below. These at least cover cases which are basic for applications.

THEOREM 4.1: *Let the kernel $K(t, s)$ be jointly measurable in its variables and non-negative. Let the inequality* (4.2) *be fulfilled for all bounded functions $x(s)$, where α and β are fixed numbers. Then the operator* (4.1) *is defined on all functions $x(s) \in L_\alpha$ and the inequality* (4.2) *is fulfilled for all these functions.*

PROOF: It suffices to prove the inequality (4.2) for non-negative functions $x(s) \in L_\alpha$. For each such function it is possible to construct a monotonically increasing sequence $x_n(s)$ of step functions which converges to $x(s)$. By (4.2)

$$\|\boldsymbol{K}x_n - \boldsymbol{K}x_m\|_\beta \leqq M\,\|x_n - x_m\|_\alpha.$$

Hence the sequence $\boldsymbol{K}x_n$ converges to some function $y(t)$ in L_β. Since the sequence $\boldsymbol{K}x_n(t)$ is nondecreasing, it converges to $y(t)$ for almost all $t \in \Omega^*$. By B. Levi's theorem (see I. P. Natanson [1]) on the interchange of limits under an integral sign, this implies that for almost all t

$$\int_\Omega K(t, s)x(s)\,\mathrm{d}s = \lim_{n\to\infty} \int_\Omega K(t, s)x_n(s)\,\mathrm{d}s = y(t).$$

Hence

$$\int_{\Omega^*} \Big| \int_\Omega K(t, s)x(s)\,\mathrm{d}s\Big|^{1/\beta}\mathrm{d}t = \|y\|_\beta^{1/\beta} = \lim_{n\to\infty} \|\boldsymbol{K}x_n\|_\beta^{1/\beta}$$

$$\leqq M^{1/\beta}\,\overline{\lim_{n\to\infty}}\,\|x_n\|_\alpha^{1/\beta} = M^{1/\beta}\,\|x\|_\alpha^{1/\beta}.$$

The theorem has been proved.

4.2 *Regular operators*

$$K_+(t, s) = \begin{cases} K(t, s), & \text{if } K(t, s) \geqq 0 \\ 0, & \text{if } K(t, s) < 0, \end{cases}$$

$$K_-(t, s) = \begin{cases} -K(t, s), & \text{if } K(t, s) \leqq 0 \\ 0, & \text{if } K(t, s) > 0. \end{cases}$$

It is clear that

$$K(t, s) = K_+(t, s) - K_-(t, s)$$

and

$$|K(t, s)| = K_+(t, s) + K_-(t, s).$$

Denote by $\boldsymbol{K}_+$, $\boldsymbol{K}_-$, $|\boldsymbol{K}|$ the linear operators:

$$\boldsymbol{K}_+ x(t) = \int_\Omega K_+(t, s) x(s) \mathrm{d}s, \tag{4.3}$$

$$\boldsymbol{K}_- x(t) = \int_\Omega K_-(t, s) x(s) \mathrm{d}s, \tag{4.4}$$

$$|\boldsymbol{K}| x(t) = \int_\Omega |K(t, s)| x(s) \mathrm{d}s. \tag{4.5}$$

Suppose that the operator $\boldsymbol{K}$ is defined on some fixed function $x(s)$. This means that for almost all $t \in \Omega^*$ the function $K(t, s) x(s)$ is summable in s. Consequently, the functions

$$|K(t, s)|\, |x(s)|, \quad K_+(t, s)\, |x(s)|, \quad K_-(t, s)\, |x(s)|$$

are summable in s. In other words, the operators (4.3)–(4.5) are also defined on the function $|x(s)|$. This implies that these operators are defined also on the function $x(s)$. Hence the following relations hold:

$$\boldsymbol{K} x = \boldsymbol{K}_+ x - \boldsymbol{K}_- x \tag{4.6}$$

$$|\boldsymbol{K}| x = \boldsymbol{K}_+ x + \boldsymbol{K}_- x. \tag{4.7}$$

We emphasize that the relations (4.6) and (4.7) can not always be considered as relations for operators acting in fixed spaces. For instance it may happen that $\boldsymbol{K}$ acts from L_α to L_β, while the values of the operators $\boldsymbol{K}_+$ and $\boldsymbol{K}_-$ on L_α do not belong to L_β. An example of this type will be furnished in the next section.

Recall that a linear operator $\boldsymbol{A}$, acting from L_α to L_β, is called *regular*, if it can be written in the form

$$\boldsymbol{A} = \boldsymbol{A}_1 - \boldsymbol{A}_2 \tag{4.8}$$

where $\boldsymbol{A}_1$ and $\boldsymbol{A}_2$ are positive linear operators, acting from L_α to L_β (see 2.2°).

THEOREM 4.2: *A linear integral operator $\boldsymbol{K}$ is a regular operator acting from L_α to L_β if and only if the integral operator $|\boldsymbol{K}|$ acts from L_α to L_β.*

PROOF OF SUFFICIENCY: Since the following inequalities hold:

$$0 \leqq K_+(t, s), \;\; K_-(t, s) \leqq |K(t, s)|,$$

the operators $\boldsymbol{K}_+$ and $\boldsymbol{K}_-$ also act from L_α to L_β. Hence the operator

$$\boldsymbol{K}_1 = \boldsymbol{K}_+ - \boldsymbol{K}_-$$

is a regular operator acting from L_α to L_β. It remains only to observe that $\boldsymbol{K}_1 = \boldsymbol{K}$ (by (4.6)).

PROOF OF NECESSITY: Let an integral operator $\boldsymbol{K}$ with kernel $K(t, s)$ be regular. Then (see 2.2°) there exists a positive operator $\boldsymbol{B}$ acting from L_α to L_β such that:

$$|\boldsymbol{K}x| \leqq \boldsymbol{B}|x|. \tag{4.9}$$

In particular, (4.9) implies that for any non-negative function $x \in L_\alpha$ we have:

$$\sup_{|u(s)| \leqq x(s)} |\boldsymbol{K}u| \leqq \boldsymbol{B}x$$

or in other words

$$\sup_{|u(s)| \leqq x(s)} \Big|\int_\Omega K(t, s)u(s)\,\mathrm{d}s\Big| \leqq \boldsymbol{B}x(t). \tag{4.10}$$

The function $\boldsymbol{B}x(t)$ belongs to L_β and hence is finite almost everywhere. Let $\boldsymbol{B}x(t_0) < \infty$. Consider the linear functional F_{t_0} on the space L_0, defined by the relation

$$F_{t_0}(h) = \int_\Omega K(t_0, s)x(s)h(s)\,\mathrm{d}s.$$

The number

$$\sup_{|h(s)|\leqq 1} \int_\Omega K(t_0, s)x(s)h(s)\,\mathrm{d}s$$

$$= \sup_{|u(s)|\leqq x(s)} \left|\int_\Omega K(t_0, s)u(s)\,\mathrm{d}s\right|$$

is the norm of this functional. However (see 1.1°)

$$\sup_{|u(s)|\leqq x(s)} \left|\int_\Omega K(t_0, s)u(s)\,\mathrm{d}s\right| = \int_\Omega |K(t_0, s)|\,x(s)\,\mathrm{d}s,$$

whence by (4.10)

$$|\boldsymbol{K}|\,x(t_0) = \int_\Omega |K(t, s)|\,x(s)\,\mathrm{d}s \leqq \boldsymbol{B}x(t_0). \tag{4.11}$$

Equation (4.11) implies that $|\boldsymbol{K}|\,x \in L_\beta$ for $x \geqq 0$. But each function $x(s) \in L_\alpha$ can be written in the form of a difference of non-negative functions

$$x(s) = x_+(s) - x_-(s)$$

where

$$x_+(s) = \sup\{x(s), 0\}, \quad x_-(s) = -\inf\{x(s), 0\}.$$

Since $|\boldsymbol{K}|$ is an additive operator,

$$|\boldsymbol{K}|\,x = |\boldsymbol{K}|\,x_+ - |\boldsymbol{K}|\,x_- \in L_\beta.$$

Thus it has been proved that the operator $\boldsymbol{K}$ acts from L_α to L_β. The theorem has been proved.

Theorem 4.2 means that for regular operators the relation (4.6) can be considered as a representation of the form (4.8). Moreover (4.6) and (4.7) imply that

$$\|K\|_{\alpha\to\beta} \leqq \| |K| \|_{\alpha\to\beta}. \tag{4.12}$$

The assertion of Theorem 4.2 can be formulated in terms of L-characteristics. It means that $L(|K|; \text{def.}) = L(K; \text{reg.})$ and since the L-characteristics $L(A; \text{def.})$ and $L(A; \text{cont.})$ of integral operators coincide

$$L(K; \text{reg.}) = L(|K|; \text{cont.}). \tag{4.13}$$

In studying the operator (4.1), estimates of the following form are often used:

$$|K(t, s)| \leqq K_0(t, s) \quad (t \in \Omega^*,\ s \in \Omega). \tag{4.14}$$

These estimates imply immediately that for any non-negative function $x(s)$ the following inequality holds:

$$|Kx(t)| \leqq |K|x(t) \leqq K_0x(t), \tag{4.15}$$

where

$$K_0x(t) = \int_\Omega K_0(t, s)x(s)\,ds. \tag{4.16}$$

Thus the following result is valid.

THEOREM 4.3: *Let the inequality* (4.14) *be fulfilled. Then*

$$L(K; \text{reg.}) \supset L(K_0; \text{cont.}). \tag{4.17}$$

Let us show that the L-characteristic $L(K; \text{reg.})$ of each non-zero integral operator K is contained in the half-strip $0 \leqq \alpha \leqq 1$, $\beta \geqq 0$. Otherwise there is an integral operator K_0 with a non-negative kernel $K_0(t, s)$, acting from some L_{α_0}, where $\alpha_0 > 1$, to some L_{β_0}. As was shown in 2.1°, the L-characteristics of all linear operators are distributed in the domain hatched in Fig. 4.1, and hence $\beta_0 \geqq \alpha_0$. Consider an auxiliary kernel

$$K_1(t, s) = \min\{1, K_0(t, s)\}.$$

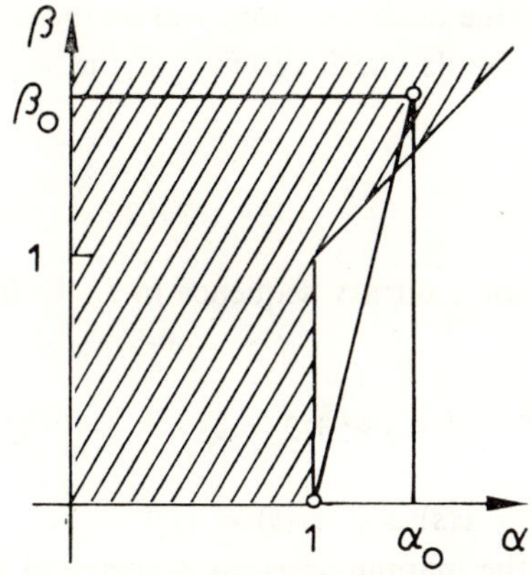

Figure 4.1

The integral operator $\boldsymbol{K}_1$ with the kernel $K_1(t, s)$ also acts from L_{α_0} to L_{β_0}. Furthermore, the boundedness of the kernel $K_1(t, s)$ implies that $\boldsymbol{K}_1$ acts from L_1 to L_0 and is continuous. Theorem 2.4 then implies that $L(\boldsymbol{K}_1; \text{cont.})$ contains the entire segment joining $\{1, 0\}$ and $\{\alpha_0, \beta_0\}$. A part of this segment lies outside the domain hatched in Fig. 4.1—we have arrived at a contradiction:

4.3 *Example of a non-regular operator*

We confines ourselves for simplicity to an example (due to B. S. Mityagin) of an operator in the space $L_{\frac{1}{2}}$ of functions defined on the set $\Omega = [0, 1]$.

Put

$$\boldsymbol{P}x(s) = 2^i \int_{2^{-i}}^{2^{-i+1}} x(\sigma)\mathrm{d}\sigma, \text{ if } 1/2^i \leqq s < 1/2^{i-1}. \tag{4.18}$$

It is not difficult to show that this operator acts in $L_{\frac{1}{2}}$ and is continuous.

The range E of the operator $\boldsymbol{P}$ consists of countable valued functions $y(s)$, taking constant values on the intervals $1/2^i \leqq s < 1/2^{i-1}$ $(i = 1, 2, \dots)$. Assign to each function $y(s) \in E$ a numerical sequence

$$\tau y(s) = \left\{\frac{1}{\sqrt{2}}\, c_1, \frac{1}{2}\, c_2, \dots, \frac{1}{(\sqrt{2})^i}\, c_i, \dots\right\}, \tag{4.19}$$

where c_i denotes the value of the function $y(s)$ on the corresponding interval $1/2^i \leqq s < 1/2^{i-1}$. A simple check shows that the sequence (4.19) belongs to $l_{\frac{1}{2}}$[1] and moreover that

$$\|\tau y(s)\|_{\frac{1}{2}} = \|y(s)\|_{\frac{1}{2}}. \tag{4.20}$$

Now let $\xi = \{\xi_i\}$ be an arbitrary sequence in $l_{\frac{1}{2}}$. Define the function $y(s)$ by the relation

$$y(s) = (\sqrt{2})^i \xi_i, \text{ if } 1/2^i \leqq s < 1/2^{i-1}.$$

It is clear that $\boldsymbol{P}y(s) = y(s)$ and $\tau y(s) = \{\xi_i\}$.

This shows that the mapping τ is an isometry of E onto $l_{\frac{1}{2}}$.

Now consider the operator

$$\boldsymbol{T}_0\{\xi_i\} = \left\{\sum_{j=1}^{\infty} \frac{\xi_j}{i - j + \frac{1}{2}}\right\}. \tag{4.21}$$

As D. Hilbert showed (see, for instance, G. Hardy, D. Littlewood, G. Polya [1]), $\boldsymbol{T}_0$ is a continuous operator acting in $l_{\frac{1}{2}}$ but the operator

$$\boldsymbol{T}_1\{\xi_i\} = \left\{\sum_{j=1}^{\infty} \frac{\xi_j}{|i - j + \frac{1}{2}|}\right\} \tag{4.22}$$

does not act in $l_{\frac{1}{2}}$.

Now consider the operator

$$\boldsymbol{K}x = \tau^{-1}\boldsymbol{T}_0\tau\boldsymbol{P}x. \tag{4.23}$$

This operator obviously acts in $L_{\frac{1}{2}}$ and is continuous. Here

$$\boldsymbol{K}x = \tau^{-1}\boldsymbol{T}_0\left\{(\sqrt{2})^i \int_{2^{-i}}^{2^{-i+1}} x(\sigma)\,\mathrm{d}\sigma\right\} = \tau^{-1}\left\{\sum_{j=1}^{\infty} \frac{(\sqrt{2})^j \int_{2^{-j}}^{2^{-j+1}} x(\sigma)\,\mathrm{d}\sigma}{i - j + \frac{1}{2}}\right\}.$$

[1] In analogy to our notation for spaces of functions we denote by l_α the space consisting of numerical sequences $\{\xi_i\}$ for which the following norm is finite:

$$\|\{\xi_i\}\|_\alpha = \{\sum_{i=1}^{\infty} |\xi_i|^{1/\alpha}\}^\alpha.$$

Consequently

$$Kx(t) = \sum_{j=1}^{\infty} \frac{(\sqrt{2})^{i+j} \int\limits_{2^{-j}}^{2^{-j+1}} x(\sigma)\,d\sigma}{i-j+\frac{1}{2}} \qquad (2^{-i} \leqq t < 2^{-i+1}).$$

This means that $\boldsymbol{K}$ is an integral operator

$$Kx(t) = \int\limits_0^1 K(t,s)x(s)\,ds \tag{4.24}$$

with kernel

$$K(t,s) = \frac{(\sqrt{2})^{i+j}}{i-j+\frac{1}{2}}, \text{ if } 2^{-i} \leqq t < 2^{-i+1},\ 2^{-j} \leqq s < 2^{-j+1}.$$

Let us show that this operator $\boldsymbol{K}$ does not possess the property of regularity. Otherwise, the integral operator $|\boldsymbol{K}|$ with the kernel $|K(t,s)|$ would act in $L_{\frac{1}{2}}$ and be continuous. The range of the operator $|\boldsymbol{K}|$ is then contained in the subspace E. Hence the operator $\tau|\boldsymbol{K}|\tau^{-1}$ would act in $l_{\frac{1}{2}}$ and be continuous.

On the other hand, the function $x(s) = \tau^{-1}\{\xi_i\}$ $(\{\xi_i\} \in l_{\frac{1}{2}})$ is defined by the relation

$$x(s) = (\sqrt{2})^j \xi_j, \text{ if } 2^{-j} \leqq s < 2^{-j+1}.$$

Hence for $2^{-j} \leqq t < 2^{-j+1}$ the following identity is valid:

$$|\boldsymbol{K}|x(t) = \int\limits_0^1 |K(t,s)|x(s)\,ds$$

$$= \sum_{j=1}^{\infty} \frac{(\sqrt{2})^{i+j} \int\limits_{2^{-j}}^{2^{-j+1}} x(s)\,ds}{|i-j+\frac{1}{2}|} = \sum_{j=1}^{\infty} \frac{(\sqrt{2})^i \xi_j}{|i-j+\frac{1}{2}|},$$

whence it follows that

$$\tau|\boldsymbol{K}|\tau^{-1}\xi = \boldsymbol{T}_1\xi \qquad (\xi \in l_{\frac{1}{2}}).$$

As was already mentioned, the operator $\boldsymbol{T}_1$ does not act in $l_{\frac{1}{2}}$. We have arrived at a contradiction.

4.4 *The adjoint operator*[1]

The symbol $\boldsymbol{K}^{\#}$ will denote the integral operator

$$\boldsymbol{K}^{\#}y(s) = \int_{\Omega^*} K(t, s)y(t)\mathrm{d}t. \tag{4.25}$$

In contrast to the operator $\boldsymbol{K}$ with kernel $K(t, s)$, the operator $\boldsymbol{K}^{\#}$ acts from spaces of functions defined on Ω^* to spaces of functions defined on Ω. The operator $\boldsymbol{K}^{\#}$ will be called *transposed* (with respect to the operator $\boldsymbol{K}$). Under natural assumption $\boldsymbol{K}^{\#}$ coincides with the operator $\boldsymbol{K}^*$ which is adjoint to the operator $\boldsymbol{K}$.

THEOREM 4.4: *Suppose that the operator*

$$\boldsymbol{K}x(t) = \int_{\Omega} K(t, s)x(s)\mathrm{d}s \tag{4.26}$$

acts continuously from L_α to L_β $(0 \leqq \alpha, \beta \leqq 1)$ and that the operator

$$\boldsymbol{K}^{\#}y(s) = \int_{\Omega^*} K(t, s)y(t)\mathrm{d}t$$

acts continuously from $L_{1-\beta}$ to $L_{1-\alpha}$. Let the kernel $K(t, s)$ be summable:

$$\int_{\Omega^*}\int_{\Omega} |K(t, s)|\,\mathrm{d}s\mathrm{d}t < \infty. \tag{4.27}$$

Then $\boldsymbol{K}^{\#}$ is adjoint to the operator $\boldsymbol{K}$.

PROOF: For bounded functions $x(s)$ and $y(t)$ the following relation is valid by Fubini's Theorem:

$$\int_{\Omega^*}\Big[\int_{\Omega} K(t, s)x(s)\mathrm{d}s\Big]y(t)\mathrm{d}t = \int_{\Omega} x(s)\Big[\int_{\Omega^*} K(t, s)y(t)\mathrm{d}t\Big]\mathrm{d}s,$$

[1] A problem on the relation between transposed and adjoint operators was considered even by S. Banach. In this connection, see also the paper of Y. B. Rutickii [3].

i.e.

$$(\boldsymbol{K}x, y) = (x, \boldsymbol{K}^{\#}y).$$

On the other hand,

$$(\boldsymbol{K}x, y) = (x, \boldsymbol{K}^{*}y).$$

Hence for any bounded functions $x(s)$ and $y(t)$

$$(x, (\boldsymbol{K}^{*} - \boldsymbol{K}^{\#})y) = 0.$$

Since the set of bounded functions is dense in L_{α}, the above relation implies that $\boldsymbol{K}^{*}y = \boldsymbol{K}^{\#}y$ for bounded functions $y(t)$. From continuity of the operators $\boldsymbol{K}^{*}$ and $\boldsymbol{K}^{\#}$ and from the fact that the set of bounded functions is dense in $L_{1-\beta}$ it follows that $\boldsymbol{K}^{\#} = \boldsymbol{K}^{*}$. The theorem has been proved.

Let us now consider regular operators.

THEOREM 4.5: *Let the operator* (4.26) *be regular as an operator acting from* L_{α} *to* L_{β} $(0 \leqq \alpha, \beta \leqq 1)$. *Then the transposed operator* $\boldsymbol{K}^{\#}$ *acts from* $L_{1-\beta}$ *to* $L_{1-\alpha}$, *coincides with the adjoint operator* $\boldsymbol{K}^{*}$, *and is regular.*

PROOF: Without loss of generality it may be assumed that the kernel $K(t, s)$ is non-negative. It is only necessary to prove that the transposed operator coincides with the adjoint operator. On account of Theorem 4.4 it suffices to show that the transposed operator $\boldsymbol{K}^{\#}$ acts from $L_{1-\beta}$ to $L_{1-\alpha}$, since (4.27) necessarily holds.

To prove this last assertion we use Fubini's theorem according to which

$$\int_{\Omega} x(s)\boldsymbol{K}^{\#}y(s)\,\mathrm{d}s = \int_{\Omega^*} \boldsymbol{K}x(t)y(t)\,\mathrm{d}t \qquad (x \in L_{\alpha},\ y \in L_{1-\beta}).$$

Hence

$$\int_{\Omega} x(s)\boldsymbol{K}^{\#}y(s)\,\mathrm{d}s < \infty \qquad (x \in L_{\alpha},\ y \in L_{1-\beta}),$$

which implies that $\boldsymbol{K}^{\#}y(s) \in L_{1-\alpha}$ if $y(t) \in L_{1-\beta}$. The theorem has been proved.

We emphasize that Theorem 4.5 covers the case when the operator $\boldsymbol{K}$ acts from L_0 to L_β ($0 \leqq \beta \leqq 1$). Thus for a regular integral operator $\boldsymbol{K}$ acting from L_0 to L_β, the adjoint operator is also integral and acts from $L_{1-\beta}$ to L_1 (cf. 2.1°).

Theorem 4.5 means that a point $\{\alpha, \beta\}$, where $0 \leqq \alpha, \beta \leqq 1$, belongs to the L-characteristic $L(\boldsymbol{K}^\#; \text{reg.})$ if and only if $\{1-\beta, 1-\alpha\} \in L(\boldsymbol{K}; \text{reg.})$. In other words, the parts of the L-characteristics $L(\boldsymbol{K}; \text{reg.})$ and $L(\boldsymbol{K}^\#; \text{reg.})$ located in the unit square are symmetric with respect to the line $\alpha + \beta = 1$ (see Fig. 4.2).

4.5 *Operators with symmetric kernels*

Suppose that $\Omega^* = \Omega$. A kernel $K(t, s)$ is called *symmetric* if

$$K(t, s) \equiv K(s, t) \qquad (s, t \in \Omega). \tag{4.28}$$

In the case of symmetric kernels the operators $\boldsymbol{K}$ and $\boldsymbol{K}^\#$ coincide. Then Theorem 4.5 implies that the part of the L-characteristic $L(\boldsymbol{K}; \text{reg.})$ located in the unit square is symmetric with respect to the line $\alpha + \beta = 1$ (see Fig. 4.3).

It is possible to find other classes of kernels $K(t, s)$ such that the L-characteristic $L(\boldsymbol{K}; \text{reg.})$ is symmetric with respect to the line $\alpha + \beta = 1$. There are, for instance, the *skew-symmetric* kernels $K(t, s)$, i.e. kernels for

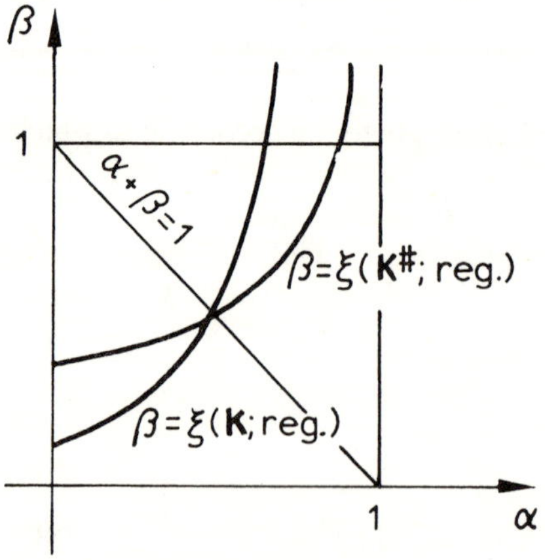

Figure 4.2

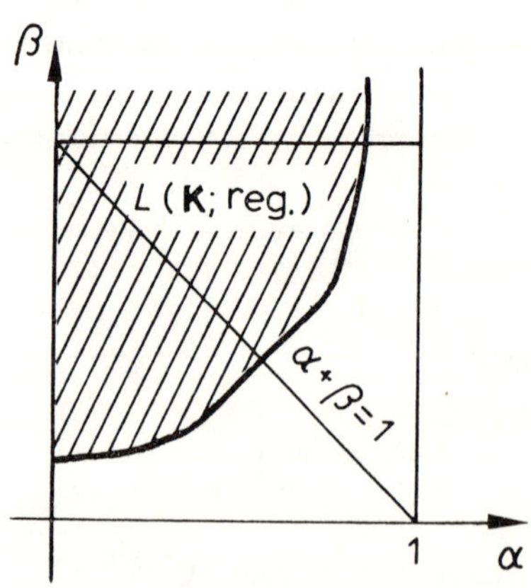

Figure 4.3

which

$$K(t, s) \equiv -K(s, t) \qquad (t, s \in \Omega). \tag{4.29}$$

Note also that the L-characteristic $L(\boldsymbol{K}; \text{reg.})$ is symmetric with respect to the line $\alpha + \beta = 1$ if the kernel $K(t, s)$ depends on a difference of its arguments

$$K(t, s) = k(t - s),$$

and if Ω is symmetric with respect to the origin.

Suppose that the L-characteristic $L(\boldsymbol{K}; \text{reg.})$ is symmetric with respect to the line $\alpha + \beta = 1$ and let $\{\alpha, \beta\} \in L(\boldsymbol{K}; \text{reg.})$. Then also $\{1 - \beta, 1 - \alpha\} \in L(\boldsymbol{K}; \text{reg.})$. Convexity of the L-characteristic $L(\boldsymbol{K}; \text{reg.})$ then implies that all points $\{\alpha(\tau), \beta(\tau)\}$ of the segment

$$\alpha(\tau) = (1 - \tau)\alpha + \tau(1 - \beta), \; \beta(\tau) = (1 - \tau)\beta + \tau(1 - \alpha) \quad (0 < \tau < 1)$$

also belong to $L(\boldsymbol{K}; \text{reg.})$.

4.6 *Products of integral operators*

Let Ω, Ω^*, Ω^{**} be three sets. Consider the two integral operators

$$\boldsymbol{K}_1 x(t) = \int_{\Omega} K_1(t, s) x(s) \mathrm{d}s \qquad (t \in \Omega^*), \tag{4.30}$$

$$\boldsymbol{K}_2 y(u) = \int_{\Omega^*} K_2(u, t) y(t) \mathrm{d}t \qquad (u \in \Omega^{**}). \tag{4.31}$$

Assume that $\boldsymbol{K}_1$ acts from $L_\alpha(\Omega)$ to $L_\beta(\Omega^*)$ and is continuous, and $\boldsymbol{K}_2$ acts from $L_\beta(\Omega^*)$ to $L_\gamma(\Omega^{**})$ and is continuous. Then the operator

$$\boldsymbol{K} = \boldsymbol{K}_2 \boldsymbol{K}_1 \tag{4.32}$$

acts from $L_\alpha(\Omega)$ to $L_\gamma(\Omega^{**})$ and is continuous. It is natural to raise a question is the operator $\boldsymbol{K}$ integral? This problem is easily solved if the order of integration on the right side of the relation

$$\boldsymbol{K} x(u) = \int_{\Omega^*} K_2(u, t) \Big[\int_{\Omega} K_1(t, s) x(s) \mathrm{d}s \Big] \mathrm{d}t$$

can be reversed. If this is possible,

$$\boldsymbol{K}x(u) = \int_{\Omega} \left[\int_{\Omega^*} K_2(u, t) K_1(t, s) \mathrm{d}t\right] x(s) \mathrm{d}s.$$

In other words,

$$\boldsymbol{K}x(u) = \int_{\Omega} K(u, s) x(s) \mathrm{d}s, \tag{4.33}$$

where

$$K(u, s) = \int_{\Omega^*} K_2(u, t) K_1(t, s) \mathrm{d}t. \tag{4.34}$$

As usual, a change of order of integration is easily justified when Fubini's theorem can be applied.

THEOREM 4.6: *Let the operators* $\boldsymbol{K}_1, \boldsymbol{K}_2$ *defined in* (4.30) *and* (4.31) *be regular. Then the operator* $\boldsymbol{K} = \boldsymbol{K}_2 \boldsymbol{K}_1$ *admits the representation* (4.33) *with the kernel given in* (4.34) *and is regular.*

For positive kernels this theorem follows immediately from Fubini's theorem. In the general case it is only necessary to write the kernels $K_1(t, s)$ and $K_2(t, s)$ in the form

$$K_1(t, s) = K_1^+(t, s) - K_1^-(t, s)$$

$$K_2(u, t) = K_2^+(u, t) - K_2^-(u, t),$$

where

$$K_1^+(t, s) = \max\{K_1(t, s), 0\},\ K_1^-(t, s) = -\min\{K_1(t, s), 0\}$$

$$K_2^+(t, s) = \max\{K_2(u, t), 0\},\ K_2^-(u, t) = -\min\{K_2(u, t), 0\},$$

and to remark that

$$\int_{\Omega^*} K_2(u, t) K_1(t, s) \mathrm{d}t$$

$$= \int_{\Omega^*} K_2^+(u, t) K_1^+(t, s) \mathrm{d}t - \int_{\Omega^*} K_2^+(u, t) K_1^-(t, s) \mathrm{d}t$$

$$- \int_{\Omega^*} K_2^-(u, t) K_1^+(t, s) \mathrm{d}t + \int_{\Omega^*} K_2^-(u, t) K_1^-(t, s) \mathrm{d}t.$$

Suppose now that $\Omega^* = \Omega$. Together with an integral operator

$$\boldsymbol{K}x(t) = \int_{\Omega} K(t, s) x(s) \mathrm{d}s \tag{4.35}$$

it is often necessary to consider its powers $\boldsymbol{K}^n$. Assume that $\boldsymbol{K}$ acts from L_{α_0} to L_{α_1}, from L_{α_1} to L_{α_2}, ..., from $L_{\alpha_{n-1}}$ to L_{α_n}. Then $\boldsymbol{K}^n$ acts from L_{α_0} to L_{α_n}. Theorem 4.6 implies that the operator $\boldsymbol{K}^n$ is regular if the operator (4.35) is regular. Here $\boldsymbol{K}^n$ is the integral operator

$$\boldsymbol{K}^n x(t) = \int_{\Omega} K_{[n]}(t, s) x(s) \mathrm{d}s, \tag{4.36}$$

where

$$K_{[n]}(t, s) = \int_{\Omega} \dots \int_{\Omega} K(t, s_1) K(s_1, s_2) \dots K(s_{n-1}, s) \mathrm{d}s_1 \dots \mathrm{d}s_{n-1}. \tag{4.37}$$

The kernels (4.37) are called *iterates* of the kernel $K(t, s)$ *or iterated* kernels.

It was shown in 1.6° how to determine a part of the $\{\alpha, \beta\}$ plane belonging to the L-characteristic $L(\boldsymbol{K}_2\boldsymbol{K}_1; \text{cont.})$ by using the L-characteristics of the operators $\boldsymbol{K}_1$ and $\boldsymbol{K}_2$. The set obtained in this way is, generally speaking, essentially smaller than the full L-characteristic. However for some important operators (for instance, for operators of potential type, considered in § 8) this set coincides with the full L-characteristic.

4.7 *Truncations of kernels of integral operators*

Let $K(t, s)$ be a kernel. Denote by $K_h(t, s)$, where $h > 0$, the bounded kernel

$$K_h(t, s) = \min\{|K(t, s)|, h\} \cdot \operatorname{sign} K(t, s), \tag{4.38}$$

and by $\boldsymbol{K}_h$ the linear integral operator

$$\boldsymbol{K}_h x(t) = \int_{\Omega} K_h(t, s) x(s) \mathrm{d}s. \tag{4.39}$$

THEOREM 4.7: *Let an integral operator* $\boldsymbol{K}$ *with kernel* $K(t, s)$ *act from* L_α *to* L_β $(0 \leqq \alpha \leqq 1, 0 < \beta < \infty)$ *and be regular. Then the operators* $\boldsymbol{K}_h$ *converge strongly to the operator* $\boldsymbol{K}$ *as* $h \to \infty$, *i.e. for each function* $x(s) \in L_\alpha$

$$\lim_{h \to \infty} \|\boldsymbol{K}_h x - \boldsymbol{K} x\|_\beta = 0. \tag{4.40}$$

PROOF: Let $x(s)$ be a fixed function in L_α. Since the kernels $K_h(t, s)$ are bounded, the functions $\boldsymbol{K}_h x$ $(0 < h < \infty)$ are also bounded and consequently belong to L_β.

It is clear that for almost all $t \in \Omega^*$ the following inequality holds:

$$|K_h(t, s) x(s)| \leqq |K(t, s)|\, |x(s)|$$

and the functions $K_h(t, s) x(s)$ converge to $K(t, s) x(s)$ almost everywhere (in s) as $h \to \infty$. Lebesgue's theorem on limits under the integral sign implies that for almost all $t \in \Omega^*$

$$\lim_{h \to \infty} \int_\Omega K_h(t, s) x(s) \, ds = \int_\Omega K(t, s) x(s) \, ds,$$

i.e. the functions $\boldsymbol{K}_h x(t)$ converge to $\boldsymbol{K} x(t)$ almost everywhere.

The inequality

$$\left| \int_\Omega K_h(t, s) x(s) \, ds \right| \leqq \int_\Omega |K(t, s)|\, |x(s)| \, ds \qquad (t \in \Omega^*)$$

implies that the functions $\boldsymbol{K}_h x(t)$ have equi-uniformly continuous norms in L_β. Hence $\boldsymbol{K}_h x(t)$ converges to $\boldsymbol{K} x(t)$ in the norm of L_β. The theorem has been proved.

The assertion of this theorem does not hold for operators acting from L_0 to L_0, as is shown by the operator $\boldsymbol{K}$ with the kernel:

$$K(t, s) = \begin{cases} 2^n, & \text{if } 2^{-n} \leqq t, s < 2^{-n+1}; \ n = 1, 2, \ldots \\ 0 & \text{for other } t, s. \end{cases}$$

In fact if $x_0(s) = 1$, the functions $\boldsymbol{K}_h x_0(t)$ converge to $\boldsymbol{K} x_0(t)$ almost everywhere, but

$$\|\boldsymbol{K} x_0(t) - \boldsymbol{K}_h x_0(t)\| = 1 \qquad (0 < h < \infty).$$

§ 5 General theorems on compactness of integral operators[1]

5.1 Problem setting

Let us continue our investigation of the integral operator

$$\boldsymbol{K}x(t) = \int_{\Omega} K(t, s)x(s)\,\mathrm{d}s. \tag{5.1}$$

In this and the following paragraph of the present chapter we will be concerned with the problem: Under what conditions is $\boldsymbol{K}$ compact as an operator acting from L_α to L_β? An answer to this problem can be given in various ways: in the form of estimates of the kernel, in terms of various properties of the iterated kernels, in terms of various properties of the L-characteristic $L(\boldsymbol{K}; \text{cont.})$, etc.

In this paragraph it is assumed that one point $\{\alpha_0, \beta_0\}$ in the L-characteristic $L(\boldsymbol{K}; \text{cont.})$ is known in advance. It will be made clear under what general supplementary assumptions this point also belongs to the set $L(\boldsymbol{K}; \text{comp.})$. These general assumptions can be formulated, for instance, in the form of relations between the numbers α_0 and β_0.

In some cases it is deduced from $\{\alpha_0, \beta_0\} \in L(\boldsymbol{K}; \text{cont.})$ that some points $\{\alpha, \beta\}$ different from $\{\alpha_0, \beta_0\}$ belong to the set $L(\boldsymbol{K}; \text{comp.})$.

In the sequel, operators of the form (5.1), acting from L_α to L_β are studied, and it is usually assumed that $\alpha \neq 1$ and $\beta \neq 0$. The cases excluded from consideration in this paragraph will require special analysis, which is provided in § 6.

Let $\{\alpha_0, \beta_0\} \in L(\boldsymbol{K}; \text{reg.})$ $(0 \leqq \alpha_0, \beta_0 \leqq 1)$. In the preceding paragraph it was shown (see Theorem 4.5) that in this case the point $\{1-\beta_0, 1-\alpha_0\}$ belongs to the L-characteristic $L(\boldsymbol{K}^{\#}; \text{reg.})$ of the transposed operator

$$\boldsymbol{K}^{\#}y(s) = \int_{\Omega^*} K(t, s)y(t)\,\mathrm{d}t. \tag{5.2}$$

---

[1] The basic results of this paragraph are taken from the papers of M. A. Krasnoselskii and E. I. Pustylnik [1], P. P. Zabreiko [2], P. P. Zabreiko and M. A. Krasnoselskii [1]. Theorems 5.5 and 5.6 are due to T. Ando [1], who dealt with Orlicz spaces.

THEOREM 5.1: *Let* $\{\alpha_0, \beta_0\} \in L(\boldsymbol{K}; \text{reg.})$. *Then* $\{\alpha_0, \beta_0\} \in L(\boldsymbol{K}; \text{comp.})$ *if and only if* $\{1 - \beta_0, 1 - \alpha_0\} \in L(\boldsymbol{K}^{\#}; \text{comp.})$.

The assertion of this theorem follows from the fact that the transposed operator $\boldsymbol{K}^{\#}$ coincides with the adjoint operator, while the adjoint operator, as is known, is compact if and only if the operator itself is compact.

Theorem 5.1 implies that the parts of the L-characteristics $L(\boldsymbol{K}; \text{reg. and comp.})$ and $L(\boldsymbol{K}^{\#}; \text{reg. and comp.})$ lying in the unit square, are symmetric to each other with respect to the line $\alpha + \beta = 1$. In particular, for operators with symmetric kernels the L-characteristic $L(\boldsymbol{K}; \text{reg. and comp.})$ is symmetric with respect to the line $\alpha + \beta = 1$ (compare with analogous arguments at the end of 4.4°).

5.2 *Regular operators acting from* L_0 *to* L_{β_0} *and from* L_{α_0} *to* L_1

THEOREM 5.2: *Each regular linear integral operator* $\boldsymbol{K}$ *acting from* L_0 *to* L_{β_0}, *where* $\beta_0 > 0$, *is compact.*

PROOF: Introduce the notation

$$\varphi(t) = \int_{\Omega} |K(t, s)| \,\mathrm{d}s.$$

Since $\varphi(t)$ is the value of the operator $|\boldsymbol{K}|$ on the function $u_0 \equiv 1$, $\varphi(t) \in L_{\beta_0}$. The following inequality is obvious:

$$|\boldsymbol{K}x(t)| \leqq \int_{\Omega} |K(t, s)|\, |x(s)| \,\mathrm{d}s \leqq \varphi(t) \,\|x\|_0 \qquad (x \in L_0). \tag{5.3}$$

It follows that for any set $D^* \subset \Omega^*$

$$\|\boldsymbol{P}_{D^*}\boldsymbol{K}\|_{0\to\beta_0} \leqq \|\boldsymbol{P}_{D^*}\varphi(t)\|_{\beta_0}$$

and consequently

$$\lim_{\text{mes}\, D^* \to 0} \|\boldsymbol{P}_{D^*}\boldsymbol{K}\|_{0\to\beta_0} = 0. \tag{5.4}$$

The expression

$$F_t(x) = \int_{\Omega} K(t, s) x(s) \,\mathrm{d}s$$

is a continuous linear functional on L_0 for those t for which $\varphi(t)$ is finite. Hence (see 1.3°) it is possible to choose from each bounded sequence in L_0 a subsequence $x_n(s)$ such that the values of the functionals F_t taken on it converge. This means that for almost all $t \in \Omega^*$ the sequence of numbers $\boldsymbol{K}x_n(t)$ converges. It follows that $\boldsymbol{K}$ transforms each ball of the space L_0 to a set of functions which is compact in measure.

By Theorem 3.1 an operator which is compact in measure and satisfies the condition (5.4) is compact. The theorem has been proved.

THEOREM 5.3: *Each regular linear integral operator acting from L_{α_0} to L_1, where $0 \leqq \alpha_0 < 1$, is compact.*

Theorems 5.1 and 5.3 mean that each point $\{\alpha_0, \beta_0\}$ on the segments, indicated by bold lines in Fig. 5.1 belongs to the L-characteristic $L(\boldsymbol{K}$; comp.) if it belongs to $L(\boldsymbol{K}$; reg.).

One more important fact can be derived from Theorem 5.2. Suppose that the L-characteristic $L(\boldsymbol{K}$; reg.) is non-void and that $\{\alpha_0, \beta_0\} \in L(\boldsymbol{K}$; reg.). Then all points $\{0, \beta\}$ for $\beta \geqq \beta_0$ also belong to the set $L(\boldsymbol{K}$; reg.) and consequently the set $L(\boldsymbol{K}$; comp.) is non-void. Hence the following result holds (see Theorem 3.11).

THEOREM 5.4: *Let $\boldsymbol{K}$ be a regular integral operator acting from L_{α_0} to L_{β_0}, where $0 \leqq \alpha_0 \leqq 1$, $0 \leqq \beta_0 < \infty$. Then $\boldsymbol{K}$ is compact as an operator acting from L_α to L_β if either $\alpha \leqq \alpha_0 < 1$, $\beta > \beta_0$ or $0 \leqq \alpha < \alpha_0$, $\beta \geqq \beta_0 > 0$.*

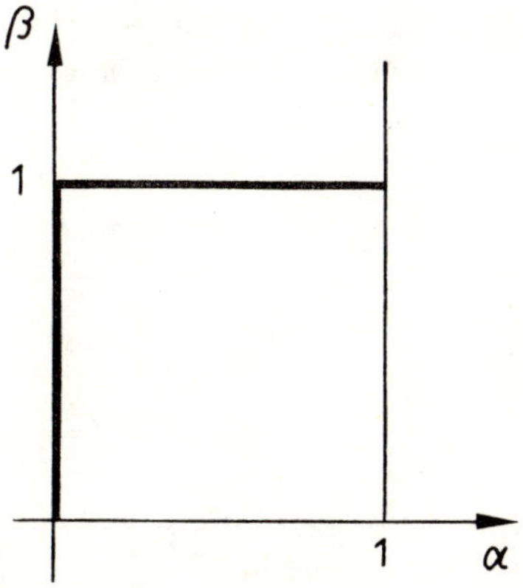

Figure 5.1

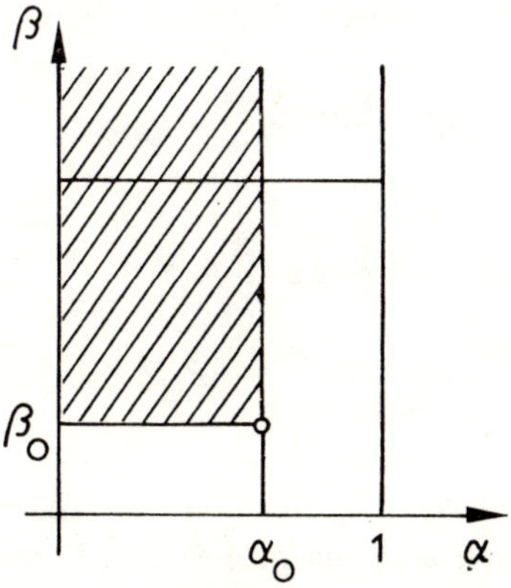

Figure 5.2

If $\{\alpha_0, \beta_0\} \in L(\boldsymbol{K}; \text{reg.})$, Theorem 5.4 implies that $L(\boldsymbol{K}; \text{comp.})$ contains the set hatched in Fig. 5.2. In particular, the set of interior points of the L-characteristic $L(\boldsymbol{K}; \text{reg.})$ is contained in the L-characteristic $L(\boldsymbol{K}; \text{comp.})$.

5.3 *Regular operators acting from L_{α_0} to L_{β_0}, where $0 < \alpha_0 < 1$, $0 < \beta_0 \leqq 1$.*

The cases studied in this section can be considered the most basic ones.

LEMMA 5.1: *Each regular linear integral operator $\boldsymbol{K}$ acting from L_{α_0} to L_{β_0}, where $0 \leqq \alpha_0 < 1$, $0 \leqq \beta_0 \leqq 1$, is compact in measure.*

PROOF: Under the conditions of the lemma the operator $\boldsymbol{K}$ acts from L_{α_0} to L_1. By Theorem 5.3 it is compact and hence compact in measure. The lemma has been proved.

From this lemma and Theorem 3.9 there follows immediately:

THEOREM 5.5: *Each regular linear integral operator $\boldsymbol{K}$ acting from L_{α_0} to L_{β_0}, where $0 < \alpha_0 < \beta_0 \leqq 1$, is compact.*

In terms of L-characteristics this theorem means that the part of the set $L(\boldsymbol{K}; \text{reg.})$ in the unit square lying above the line $\beta = \alpha$ is contained in the L-characteristic $L(\boldsymbol{K}; \text{comp.})$.

Unfortunately the assertion of Theorem 5.5 is not true if $\alpha_0 \geqq \beta_0$. This is shown by examples of so-called operators of potential type, to which § 8 is devoted.

In what follows, the study of compactness of integral operators acting from L_{α_0} to L_{β_0}, where $\alpha_0 \geqq \beta_0$, will be explored with the aid of the following simple theorem.

THEOREM 5.6: *A regular linear integral operator $\boldsymbol{K}$ acting from L_{α_0} to L_{β_0}, where $0 < \beta_0 \leqq \alpha_0 < 1$, is compact if and only if*

$$\lim_{\text{mes } D^* + \text{mes } D \to 0} \|\boldsymbol{P}_{D^*}\boldsymbol{K}\boldsymbol{P}_D\|_{\alpha_0 \to \beta_0} = 0. \tag{5.5}$$

PROOF: Lemma 5.1 and Theorem 4.5 imply that the operators $\boldsymbol{K}$ and $\boldsymbol{K}^*$ are compact in measure. It remains only to apply Theorem 3.5.

In study of concrete operators $\boldsymbol{K}$, it is often possible to establish immediately one of the following relations:

$$\lim_{\text{mes } D^* \to 0} \|P_{D^*}K\|_{\alpha_0 \to \beta_0} = 0 \tag{5.6}$$

or

$$\lim_{\text{mes } D \to 0} \|KP_D\|_{\alpha_0 \to \beta_0} = 0. \tag{5.7}$$

It is clear that (5.5) is a consequence of either of these.

5.4 *Regular operators acting from L_{α_0} to L_{β_0}, where $0 < \alpha_0 < 1$, $\beta_0 \geqq 1$*

In this section we will prove a general assertion which contains Theorem 5.3 and supplements Theorems 5.5 and 5.6.

THEOREM 5.7: *Let the regular linear operator A act from L_{α_0} to L_{β_0}, where $0 < \alpha_0 < 1$ and $\beta_0 \geqq 1$. Then*

$$\lim_{\text{mes } D \to 0} \|AP_D\|_{\alpha_0 \to \beta_0} = 0. \tag{5.8}$$

PROOF: Without loss of generality it may be assumed that A is positive.

We will prove the theorem by contradiction. Suppose that there exists a sequence of sets $D_n \subset \Omega$ such that mes $D_n \leqq 1/2^n$ and

$$\|AP_{D_n}\|_{\alpha_0 \to \beta_0} > \varepsilon_0 > 0 \quad (n = 1, 2, \ldots). \tag{5.9}$$

Consider the sets

$$D'_n = \bigcup_{m \geqq n} D_m.$$

It is clear that $\text{mes}(D'_n) \to 0$ and

$$\|AP_{D'_n}\|_{\alpha_0 \to \beta_0} > \varepsilon_0 \quad (n = 1, 2, \ldots).$$

The inequalities

$$\|AP_{D'_n}\|_{\alpha_0 \to \beta_0} = \sup_{\|x\|_{\alpha_0} \leqq 1, m \geqq n} \|AP_{D'_n - D'_m}x\|_{\beta_0} \quad (n = 1, 2, \ldots)$$

imply that for any n there can be chosen a number m and an element x such that $\|x\|_{\alpha_0} \leqq 1$ and

$$\|AP_{D'_n - D'_m} x\|_{\beta_0} > \varepsilon_0.$$

Hence by induction it is possible to construct a sequence of numbers n_k and a sequence of elements x_k, for which the following inequalities hold:

$$\|AP_{D'_{n_k} - D'_{n_{k+1}}} x_k\|_{\beta_0} > \varepsilon_0 \quad (k = 1, 2, \ldots). \tag{5.10}$$

Put

$$\Omega_k = D'_{n_k} - D'_{n_{k+1}}.$$

The sets Ω_k do not intersect and from (5.10) it follows that

$$\|AP_{\Omega_k} x_k\|_{\beta_0} > \varepsilon_0 \quad (k = 1, 2, \ldots). \tag{5.11}$$

The functions x_k can, without loss of generality, be assumed to be non-negative.

Put

$$u(s) = \sum_{k=1}^{\infty} \frac{1}{k} P_{\Omega_k} x_k(s). \tag{5.12}$$

It is obvious that

$$\int_{\Omega} |u(s)|^{1/\alpha_0} \mathrm{d}s \leqq \sum_{k=1}^{\infty} \left(\frac{1}{k}\right)^{1/\alpha_0} < \infty,$$

and hence $u(s) \in L_{\alpha}$. Consequently $Au(t) \in L_{\beta_0}$.

On the other hand, in the representation

$$Au(t) = \sum_{k=1}^{\infty} \frac{1}{k} AP_{\Omega_k} x_k \tag{5.13}$$

all summands are non-negative. Hence the inverse Minkowski inequality[1] implies that

$$\|Au(t)\|_{\beta_0} \geqq \sum_{k=1}^{\infty} \frac{1}{k} \|AP_{\Omega_k} x_k\|_{\beta_0}$$

and by (5.11) $Au(t) \notin L_{\beta_0}$. We have arrived at a contradiction. The theorem has been proved.

THEOREM 5.8: *Each regular linear integral operator* K *acting from* L_{α_0} *to* L_{β_0}, *where* $0 < \alpha_0 < 1$, $\beta_0 \geqq 1$, *is compact.*

PROOF: Denote by $\mathfrak{M}_h$ the set of functions satisfying the inequality $|x(s)| \leqq h$, where h is an arbitrary positive number.

Consider K as an operator acting from L_0 to L_{β_0}. Theorem 5.2 implies that it is compact. Hence the set $K\mathfrak{M}_h$ is compact in L_{β_0}.

Write each function $x(s)$ in the unit ball of the space L_{α_0} in the form

$$x(s) = x_1(s) + x_2(s)$$

where

$$x_1(s) = \begin{cases} x(s), & \text{if } |x(s)| \leqq h \\ 0, & \text{if } |x(s)| > h. \end{cases}$$

The set of functions $Kx_1(t)$ is compact in L_{β_0}. Hence to prove the theorem, it remains to show that for large h the norms of the functions $Kx_2(t)$ are arbitrarily small, uniformly with respect to functions $x(s)$ in the unit ball of the space L_{α_0}.

Denote by $D(x)$ the set of s, for which $|x(s)| > h$. Then it is clear that

$$\text{mes } D(x) \leqq h^{-1/\alpha_0} \qquad (\|x\|_{\alpha_0} \leqq 1)$$

---

[1] For any non-negative functions $z_i(t) \in L_\beta$ $(\beta \geqq 1)$

$$\|\sum_{i=1}^{\infty} z_i(t)\|_\beta \geqq \sum_{i=1}^{\infty} \|z_i(t)\|_\beta.$$

and

$$x_2(s) = \boldsymbol{P}_{D(x)} x(s).$$

Hence for $\|x(s)\|_{\alpha_0} \leqq 1$

$$\|\boldsymbol{K} x_2(t)\|_{\beta_0} = \|\boldsymbol{K}\boldsymbol{P}_{D(x)} x(t)\|_{\beta_0}$$

$$\leqq \|\boldsymbol{K}\boldsymbol{P}_{D(x)}\|_{\alpha_0 \to \beta_0} \leqq \sup_{\operatorname{mes} D \leqq h^{-1/\alpha_0}} \|\boldsymbol{K}\boldsymbol{P}_D\|_{\alpha_0 \to \beta_0},$$

and by Theorem 5.7, uniformly with respect to functions $x(s)$ in the ball $\|x\|_\alpha \leqq 1$, we have,

$$\lim_{h \to \infty} \|\boldsymbol{K} x_2(t)\|_{\beta_0} = 0.$$

The theorem has been proved.

Theorem 5.8 implies that all points of the L-characteristic $L(\boldsymbol{K}$; reg.) lying in the half-strip $0 \leqq \alpha < 1$, $\beta \geqq 1$, belong to the L-characteristic $L(\boldsymbol{K}$: comp.).

Note that Theorem 5.8 contains an improvement of Lemma 5.1: each regular integral operator acting from L_α to L_β, where $0 \leqq \alpha < 1$, $\beta \geqq 0$, is compact in measure.

5.5 *Regular integral operators acting from L_1 to L_{β_0}*

LEMMA 5.2: *Let a regular integral operator* $\boldsymbol{K}$ *act from* L_1 *to* L_{β_0}, *where* $\beta_0 > 0$. *Then the operator* $\boldsymbol{K}$ *transforms each weakly compact set* $\mathfrak{M}$ *of functions in* L_1 *to a set* $\boldsymbol{K}\mathfrak{M}$ *which is compact in* L_{β_0}.

To prove this lemma, it is necessary to repeat the arguments used to prove Theorem 5.8. The only change is that instead of (5.8) the following relation is used:

$$\lim_{\operatorname{mes} D \to 0} \sup_{x \in \mathfrak{M}} \|\boldsymbol{K}\boldsymbol{P}_D x\|_{\beta_0} = 0.$$

This relation results from the inequality

$$\|\boldsymbol{K}\boldsymbol{P}_D x\|_{\beta_0} \leqq \|\boldsymbol{K}\|_{1 \to \beta_0} \|\boldsymbol{P}_D x\|_1$$

and from the relation

$$\lim_{\operatorname{mes} D \to 0} \sup_{x \in \mathfrak{M}} \|P_D x\|_1 = 0$$

(which is, in view of Lemma 1.3, equivalent to weak compactness of the set $\mathfrak{M} \subset L_1$).

THEOREM 5.9: *Each regular integral operator* $\boldsymbol{K}$ *acting from* L_1 *to* L_{β_0}, *where* $0 < \beta_0 \leqq 1$, *is strongly continuous.*

PROOF: The operator $\boldsymbol{K}$ transforms each weakly convergent sequence x_n in L_1 to a sequence $\boldsymbol{K}x_n$ which converges weakly in L_{β_0} and which is compact in L_{β_0} by Lemma 5.2. Hence the sequence $\boldsymbol{K}x_n$ converges in the norm of L_{β_0}. The theorem has been proved.

In investigating integral operators in the preceding sections we often used compactness in measure and relations of the type (5.8). Neither of these methods is applicable to the investigation of operators acting from L_1 to L_{β_0}. Let us elucidate this.

Suppose that the following relation holds:

$$\lim_{\operatorname{mes} D \to 0} \|\boldsymbol{K}\boldsymbol{P}_D\|_{1 \to \beta_0} = 0$$

where $0 < \beta_0 \leqq 1$. Then for the adjoint operator we have the relation:

$$\lim_{\operatorname{mes} D \to 0} \|\boldsymbol{P}_D \boldsymbol{K}^*\|_{1 - \beta_0 \to 0} = 0.$$

However the norms of non-zero functions in L_0 do not possess the property of absolute continuity. Hence the last relation implies that $\boldsymbol{K}^*$, consequently $\boldsymbol{K}$, is the zero operator.

Let $\varphi_k(t)$ $(k = 1, 2, \ldots)$ be a uniformly bounded sequence of non-negative functions defined on $\Omega = [0, 1]$ which do not possess the property of compactness in measure (for instance, put $\varphi_k(t) = 1 + \sin kt$). Let $\Delta_1, \Delta_2, \ldots$, be a sequence of disjoint intervals in [0, 1]; denote by $\kappa_k(s)$ the characteristic function of the interval Δ_k. Put

$$K(t, s) = \sum_{k=1}^{\infty} \varphi_k(t) \kappa_k(s). \tag{5.14}$$

The kernel $K(t, s)$ is bounded and hence the integral operator $\boldsymbol{K}$ with this kernel acts from L_1 to any L_{β_0} and is continuous. It is clear that

$$\boldsymbol{K}x_k(t) = \varphi_k(t) \qquad (k = 1, 2, \ldots)$$

if

$$x_k(s) = \frac{1}{\text{mes } \Delta_k} \kappa_k(s).$$

The norms of the functions $x_k(s)$ in L_1 are equal to one. At the same time the range of the operator $\boldsymbol{K}$ on these functions forms a set which is not compact in measure.

Thus, even operators with bounded kernels may not possess the property of compactness in measure, when considered as operators on L_1.

5.6 *Operators with compact majorants*

Let $\boldsymbol{K}$ be a linear integral operator

$$\boldsymbol{K}x(t) = \int_{\Omega} K(t, s)x(s)\,\mathrm{d}s. \tag{5.15}$$

In many cases estimates of the kernel $K(t, s)$ of the following type are known

$$|K(t, s)| \leqq K_0(t, s), \tag{5.16}$$

where $K_0(t, s)$ is a simpler kernel. If the operator

$$\boldsymbol{K}_0 x(t) = \int_{\Omega} K_0(t, s)x(s)\,\mathrm{d}s \tag{5.17}$$

acts from L_{α_0} to L_{β_0}, (5.16) implies that the operator $\boldsymbol{K}$ also acts from L_{α_0} to L_{β_0} and is regular, i.e. $L(\boldsymbol{K}_0; \text{reg.}) \subset L(\boldsymbol{K}; \text{reg.})$. It is natural to raise the question: what other properties of the operator $\boldsymbol{K}$ follows from the corresponding properties of the operator $\boldsymbol{K}_0$?

THEOREM 5.10: *Let the condition* (5.16) *be fulfilled and suppose the operator* $\boldsymbol{K}_0$ *acts from* L_{α_0} *to* L_{β_0} *and is compact, where* $0 \leqq \alpha_0 < 1$, $0 < \beta_0 \leqq 1$. *Then* $\boldsymbol{K}$ *is also compact as an operator from* L_{α_0} *to* L_{β_0}.

PROOF: If $\alpha_0 = 0$, the assertion of Theorem 5.10 is contained in the assertion of Theorem 5.2.

Let $0 < \alpha_0 < 1$. Equation (5.16) and Theorem 3.1 imply that

$$\lim_{\text{mes } D^* \to 0} \|P_{D^*} \cdot K\|_{\alpha_0 \to \beta_0} = \lim_{\text{mes } D^* \to 0} \sup_{\|x\|_{\alpha_0} \leqq 1} \|P_{D^*} K x\|_{\beta_0}$$

$$\leqq \lim_{\text{mes } D^* \to 0} \sup_{\|x\|_{\alpha_0} \leqq 1} \|P_{D^*} K_0(|x|)\|_{\beta_0}$$

$$\leqq \lim_{\text{mes } D^* \to 0} \|P_{D^*} K_0\|_{\alpha_0 \to \beta_0} = 0.$$

Moreover, the operator is compact in measure by Lemma 5.1. Theorem 3.1 implies that K is compact. The theorem has been proved.

Note that the assertion of Theorem 5.10 does not hold for operators acting from L_{α_0} to L_0 or from L_1 to L_{β_0}, $0 < \beta_0 < 1$. The operator with the kernel (5.14) can serve as example. This kernel is bounded, i.e. it admits the estimate (5.16), where $K_0(t, s) \equiv M$. The operator K_0 here is compact as an operator, acting from any L_α to any L_β. At the same time the operator K with the kernel (5.14) does not possess property of compactness if it is considered as an operator in L_1. Its transposed operator $K^{\#}$ also becomes an operator with bounded kernel. It does not possess the property of compactness if it is considered as an operator from L_{α_0} to L_0.

Theorem 5.10 implies[1], in particular, that the operator (5.15) is compact as an operator from L_{α_0} to L_{β_0} ($0 \leqq \alpha_0 < 1$, $0 < \beta_0 \leqq 1$) if the operator $|K|$ possesses this property. Moreover, this assertion remains valid for operators, acting from L_1 to L_{β_0} ($0 < \beta_0 \leqq 1$) or from L_{α_0} to L_0 ($0 \leqq \alpha_0 < 1$).

Theorem 5.10 can be extended to integral operators A acting from L_1 to L_β, where $0 < \beta < \infty$, by taking A to be compact in measure.

Integral operators with bounded kernels should be considered the simplest. Theorem 5.10 implies that such operators are compact, if they are considered as operators acting from L_{α_0} to L_{β_0}, where $0 \leqq \alpha_0 < 1$, $0 < \beta_0 < \infty$.

---

[1] It is not clear whether the converse assertion is true—that compactness of the operator $|K|$ follows from compactness of a regular operator K.

5.7 *The case of kernels with reinforced singularities*

Consider the integral operator

$$\boldsymbol{K}x(t) = \int_{\Omega} K(t, s)x(s)\,\mathrm{d}s \tag{5.18}$$

acting from L_{α_0} to L_{β_0}, where $0 \leqq \alpha_0 \leqq 1$, $0 \leqq \beta_0 < \infty$.

Consider, together with (5.18), the integral operators

$$\boldsymbol{K}_\varphi x(t) = \int_{\Omega} K_\varphi(t, s)x(s)\,\mathrm{d}s \tag{5.19}$$

whose kernels have the form

$$K_\varphi(t, s) = K(t, s)\varphi(t, s), \tag{5.20}$$

where the $\varphi(t, s)$ are jointly measurable functions in the variables $t \in \Omega^*$, $s \in \Omega$.

If the operator in (5.18) is regular and the function $\varphi(t, s)$ is bounded then, as follows from Theorem 4.3, the operator in (5.19) also acts from L_{α_0} to L_{β_0} and is regular. A similar assertion does not hold if the operator in (5.18) does not possess the property of regularity.

The operator in (5.19) acts from L_{α_0} to L_{β_0} even for some unbounded functions $\varphi(t, s)$. It turns out that the operator of (5.18) is compact if such unbounded functions $\varphi(t, s)$ possess special properties.

The kernel $K(t, s)$ of the operator (5.18) acting from L_{α_0} to L_{β_0} is said to *admit reinforcement of its singularities*, if it is possible to take a function $\varphi(t, s)$ satisfying the following three conditions:

1°) $\varphi(t, s) \geqq 1$ $(t \in \Omega^*,\ s \in \Omega)$.
2°) *The operator* (5.19) *acts from* L_{α_0} *to* L_{β_0} *and is regular.*[1]
3°) *For any* h *the integral operator*

$$\boldsymbol{K}_h x(t) = \int_{\Omega} K(t, s)\kappa_h(t, s)x(s)\,\mathrm{d}s, \tag{5.21}$$

[1] 1° and 2° imply that the operator (5.18) acts from L_{α_0} to L_{β_0} and is regular.

where

$$\kappa_h(t, s) = \begin{cases} 1, & \text{if } \varphi(t, s) \leqq h \\ 0, & \text{if } \varphi(t, s) > h \end{cases}$$

acts from L_{α_0} to L_{β_0} and is compact.

THEOREM 5.11: *If the kernel $K(t, s)$ of the operator in* (5.18) *acting from L_{α_0} to L_{β_0} admits reinforcement of its singularities, then the operator* (5.18) *is compact.*

PROOF: For each function $x \in L_{\alpha_0}$

$$\|(K - K_h)x\|_{\beta_0} = \|\int_\Omega K(t, s)[1 - \kappa_h(t, s)]x(s)\mathrm{d}s\|_{\beta_0}$$

$$\leqq \frac{1}{h}\left\|\int_\Omega |K(t, s)|[1 - \kappa_h(t, s)]\varphi(t, s)|x(s)|\mathrm{d}s\right\|_{\beta_0}$$

$$\leqq \frac{1}{h}\left\|\int_\Omega |K(t, s)|\varphi(t, s)|x(s)|\mathrm{d}s\right\|_{\beta_0}.$$

It follows that

$$\|K - K_h\|_{\alpha_0 \to \beta_0} \leqq \frac{1}{h}\| |K_\varphi| \|_{\alpha_0 \to \beta_0}.$$

Consequently, the operator K can be approximated with arbitrary accuracy in operator norm by compact operators. The theorem has been proved.

Note the following consequence of Theorem 5.11. Let an even, positive function $M(u)$ satisfy the condition

$$\lim_{u \to \infty} \frac{M(u)}{u} = \infty. \tag{5.22}$$

Suppose, further, that the integral operator

$$K_M x(t) = \int_\Omega M[K(t, s)]x(s)\mathrm{d}s \tag{5.23}$$

acts from L_{α_0} to L_{β_0}. If $\alpha_0 < 1$, $\beta_0 > 0$, the kernel $K(t, s)$ admits reinforcement of its singularities—it suffices to put

$$\varphi(t, s) = \begin{cases} 1, & \text{if } |K(t, s)| \leqq 1 \\ \dfrac{M[K(t, s)]}{K(t, s)}, & \text{if } |K(t, s)| > 1. \end{cases}$$

Thus, if the operator (5.23) acts from L_{α_0} to L_{β_0} the operator (5.18) is compact as an operator from L_{α_0} to L_{β_0}.

Note finally that *a kernel of a regular compact operator always admits reinforcement of its singularities.*

5.8 *Truncations of kernels of integral operators*

Here we use the notation introduced in 4.7°.

THEOREM 5.12: *Let $|\boldsymbol{K}|$ be compact as an operator acting from L_{α_0} to L_{β_0}, where $0 \leqq \alpha_0 < 1$, $0 < \beta_0 \leqq 1$. Then the operators*

$$\boldsymbol{K}_h x(t) = \int_\Omega K_h(t, s) x(s) \mathrm{d}s,$$

where

$$K_h(t, s) = \begin{cases} K(t, s), & \text{if } |K(t, s)| \leqq h \\ h \cdot \operatorname{sign}[K(t, s)], & \text{if } |K(t, s)| > h, \end{cases}$$

converge to $\boldsymbol{K}$ in operator norm.

PROOF: Consider first the case, when $\alpha_0 = 0$. Let $x(s) \in L_0$, $\|x\|_0 \leqq 1$. Then

$$\left|\int_\Omega [K(t, s) - K_h(t, s)] x(s) \mathrm{d}s\right| \leqq \int_\Omega [|K(t, s)| - |K_h(t, s)|] \mathrm{d}s$$

whence

$$\|(\boldsymbol{K} - \boldsymbol{K}_h)x\|_{\beta_0} \leqq \|(|\boldsymbol{K}| - |\boldsymbol{K}_h|)u_0\|_{\beta_0}, \tag{5.24}$$

where $u_0(s) \equiv 1$, and

$$|\boldsymbol{K}|_h x(t) = |\boldsymbol{K}_h|\, x(t) = \int_\Omega |K_h(t, s)|\, x(s)\,\mathrm{d}s.$$

Theorem 4.7 implies that the right side of the inequality (5.24) converges to zero as $h \to \infty$. Consequently

$$\lim_{h\to\infty} \|\boldsymbol{K} - \boldsymbol{K}_h\|_{0\to\beta_0} = 0. \tag{5.25}$$

Note that this argument is valid for all $\beta_0 > 0$.

Now let $\alpha_0 < 1$, $\beta_0 = 1$. The trivial relation

$$(\boldsymbol{K}_h)^{\#} = (\boldsymbol{K}^{\#})_h$$

and Theorem 4.5 imply that

$$\|\boldsymbol{K} - \boldsymbol{K}_h\|_{\alpha_0\to 1} = \|\boldsymbol{K}^{\#} - (\boldsymbol{K}^{\#})_h\|_{0\to 1-\alpha_0}$$

and by (5.25)

$$\lim_{h\to\infty} \|\boldsymbol{K} - \boldsymbol{K}_h\|_{\alpha_0\to 1} = 0. \tag{5.26}$$

Let us examine the principal case: $0 < \alpha_0 < 1$, $0 < \beta_0 < 1$. If the assertion of the theorem is not true, there can be found a sequence $x_n \in L_{\alpha_0}$, $\|x_n\|_{\alpha_0} = 1$ and a sequence of numbers $h_n \to \infty$, such that

$$\|(\boldsymbol{K} - \boldsymbol{K}_{h_n})x_n\|_{\beta_0} \geqq \delta_0 > 0 \qquad (n = 1, 2, \ldots) \tag{5.27}$$

By the results already proved, the sequence $(\boldsymbol{K} - \boldsymbol{K}_{h_n})x_n$ converges to zero in norm of L_1. Hence $(\boldsymbol{K} - \boldsymbol{K}_{h_n})x_n$ converges to zero in measure. Moreover,

$$|(\boldsymbol{K} - \boldsymbol{K}_{h_n})x_n| \leqq |\boldsymbol{K}|\cdot(|x_n|),$$

whence it follows that functions $(\boldsymbol{K} - \boldsymbol{K}_{h_n})x_n$ have equi-absolutely continuous norms in L_{β_0}. This means that

$$\lim_{n\to\infty} \|(\boldsymbol{K} - \boldsymbol{K}_{h_n})x_n\|_{\beta_0} = 0,$$

which contradicts (5.27). The theorem has been proved.

5.9 *Products of integral operators*

We shall use the notations of 4.6°.

In order that the product $\boldsymbol{K}_2\boldsymbol{K}_1$ be a compact operator, it is sufficient that one of the operators $\boldsymbol{K}_1$, $\boldsymbol{K}_2$ be compact and the other be continuous. However the product can be a compact operator even when neither of the operators $\boldsymbol{K}_1$, $\boldsymbol{K}_2$ possesses the property of compactness (for instance, a product of two non-compact operator can be equal to zero).

5.10 *Compactness of non-regular operators*

For non-regular integral operators the results which can be obtained are weaker than those established for regular operators in the preceding sections.

LEMMA 5.3: *Each integral operator $\boldsymbol{K}$ acting from L_0 to L_{β_0} is compact in measure.*

The proof of this fact coincides with the corresponding part of the proof of Theorem 5.2.

THEOREM 5.13: *Let an integral operator $\boldsymbol{K}$ act from L_0 to L_{β_0} and be continuous. Then $\boldsymbol{K}$ is compact as an operator, acting from L_0 to L_β for $\beta > \beta_0$.*

For the proof, it suffices to use Lemma 5.3 and Theorem 3.1.

THEOREM 5.14: *Let an integral operator $\boldsymbol{K}$ act from L_{α_0} to L_{β_0}, where $0 < \alpha_0 \leqq 1$, $\beta_0 \geqq 0$, and be continuous. Then $\boldsymbol{K}$ is compact as an operator acting from L_α to L_β for $\alpha < \alpha_0$ and $\beta > \beta_0$.*

PROOF: Let $\alpha < \alpha_0$ and $\beta > \beta_0$. It is obvious that $\|\boldsymbol{P}_{D^*}\boldsymbol{K}\|_{\alpha\to\beta} \leqq (\text{mes } D^*)^{\beta-\beta_0}\|\boldsymbol{K}\|_{\alpha\to\beta_0}$, consequently

$$\lim_{\text{mes } D^*\to 0} \|\boldsymbol{P}_{D^*}\boldsymbol{K}\|_{\alpha\to\beta} = 0.$$

Theorem 3.1 implies that it suffices to show compactness in measure of the operator $\boldsymbol{K}$ on L_α.

Denote by T_h the operator, defined by the relation

$$T_h x(s) = \begin{cases} x(s), & \text{if } |x(s)| \leqq h \\ 0, & \text{if } |x(s)| > h. \end{cases}$$

Lemma 5.3 implies that the set of functions $KT_h x(\|x\|_\alpha \leqq 1)$ for any h is is compact in measure. Hence it suffices to show that the following relation is fulfilled uniformly with respect to functions $x(s)$ in the unit ball of the space L_α:

$$\lim_{h\to\infty} \|Kx - KT_h x\|_{\beta_0} = 0. \tag{5.28}$$

The function $T_h x(s)$ fails to coincide with the function $x(s)$ $(\|x(s)\|_\alpha \leqq 1)$ only on the set $D(x; h)$ whose measure does not exceed $h^{-1/\alpha}$. Hence

$$\|x - T_h x\|_{\alpha_0} \leqq \|x - T_h x\|_\alpha \cdot h^{(\alpha-\alpha_0)/\alpha}.$$

Therefore

$$\|Kx - KT_h x\|_\beta \leqq \|K\|_{\alpha_0\to\beta} \|x - T_h x\|_\alpha \cdot h^{(\alpha-\alpha_0)/\alpha},$$

whence follows (5.28). The theorem has been proved.

As a consequence of this theorem, interior points (if they exist) of the L-characteristic $L(K;\text{cont.})$ of a linear integral operator belong to $L(K;\text{comp.})$. A similar fact was previously established for the set $L(K;\text{reg.})$ (Theorem 5.4).

§ 6 Linear u_0-bounded operators[1]

6.1 *Simplest criteria for continuity of integral operators*

Ordinarily, one is given the kernel of an integral operator

$$Kx(t) = \int_\Omega K(t, s)x(s)\,ds, \tag{6.1}$$

---

[1] The classes of integral operators, discussed in this paragraph, play a basic role in many applications of the theory of cones (M. A. Krasnoselskii [9]). Many general theorems on properties of solutions of equations with u_0-bounded operators were established by

and is required either to determine whether it acts from L_α to L_β for given α and β or else to draw its L-characteristics more completely.

Assume that for almost all $t \in \Omega^*$ the kernel $K(t, s)$, as a function of s, belongs to a space L_r $(0 \leqq r \leqq 1)$ and put

$$\varphi(t) = \|K(t, s)\|_r. \tag{6.2}$$

In many cases it is easily established that $\varphi(t) \in L_q$ where $q \geqq 0$.

Let $r > 0$. Then $\varphi(t) \in L_q$ for $q > 0$ if

$$\int_{\Omega^*} [\int_{\Omega} |K(t, s)|^{1/r} \mathrm{d}s]^{r/q} \mathrm{d}t < \infty, \tag{6.3}$$

and $\varphi(t) \in L_0$, if

$$\operatorname*{ess\,sup}_{t \in \Omega^*} \int_{\Omega} |K(t, s)|^{1/r} \mathrm{d}s < \infty. \tag{6.4}$$

In particular, inequality (6.3) holds if

$$\int_{\Omega^*} \int_{\Omega} |K(t, s)|^m \mathrm{d}s \, \mathrm{d}t < \infty \tag{6.5}$$

where

$$m \geqq \max\left\{\frac{1}{r}, \frac{1}{q}\right\}.$$

Let $r = 0$. Then $\varphi(t) \in L_q$ for $q > 0$, if

$$\int_{\Omega^*} [\operatorname*{ess\,sup}_{s \in \Omega} |K(t, s)|^{1/q}] \mathrm{d}t < \infty, \tag{6.7}$$

L. A. Ladyzenskii and M. A. Krasnoselskii [2]. Classes of v_0-cobounded operators were investigated in the paper of P. P. Zabreiko [3].

The important theorem 6.2 for the case $\beta = 0$ was proved by I. M. Gelfand [1]; the general case was considered (in other terms) by L. V. Kantorovic, B. Z. Vulikh and A. G. Pinsker [1]. Theorems 6.4–6.6, in similar format, were established by I. M. Gelfand [1] (see also N. Dunford and J. T. Schwartz [1]). In N. Dunford and J. T. Schwartz [1] proved many assertions containing Theorem 6.10 are. Operators with values in the space C were investigated by F. Riesz [2].

and $\varphi(t) \in L_0$, if

$$\operatorname*{ess\,sup}_{t \in \Omega^*} [\operatorname*{ess\,sup}_{s \in \Omega} |K(t, s)|] < \infty. \tag{6.8}$$

THEOREM 6.1: *Let $\varphi(t) = \|K(t, s)\|_r \in L_q$. Then the integral operator $\boldsymbol{K}$ with kernel $K(t, s)$ acts from L_α to L_β, where $\alpha = 1 - r$, $\beta = q$, and*

$$\|\boldsymbol{K}\|_{\alpha \to \beta} \leqq \|\varphi(t)\|_q. \tag{6.9}$$

The proof is almost trivial. Under the conditions of the theorem, the value of the function $\boldsymbol{K}x(t)$ admits for almost all $t \in \Omega^*$ the estimate:

$$|\boldsymbol{K}x(t)| \leqq \varphi(t) \|x\|_\alpha, \tag{6.10}$$

whence follows the inequality (6.9).

As will be seen in the following, the hypothesis in Theorem 6.1 covers only special classes of integral operators acting from L_α to L_β. In the sequel there will be established significantly weaker restrictions under which a kernel $K(t, s)$ defines a continuous operator acting from L_α to L_β.

If the operator (6.1) is regular and acts from L_0 to some L_β, then the function

$$\varphi(t) = \int_\Omega |K(t, s)| \, \mathrm{d}s$$

belongs to L_β. This means that condition (6.3) holds with $r = 1$, $q = \beta$. Thus, the condition (6.3) with $r = 1$, $q = \beta$ is not only sufficient but necessary in order that the integral operator (6.1) acts from L_0 to L_β and is regular. A similar assertion is valid for operators acting L_α to L_0; this fact will be established in the subsequent sections.

Let us make the following observations concerning operators whose kernels satisfy the conditions of Theorem 6.1.

First, these operators are regular.

Second, these operators transform each sequence of functions $x_n(t)$ converging in norm of L_α to a sequence of functions $\boldsymbol{K}x_n(t)$ which converges not only in norm of L_β, but at the same time almost everywhere. For the proof it suffices to remark that (6.10) implies the inequality

$$|\boldsymbol{K}x_n(t) - \boldsymbol{K}x_m(t)| \leqq \varphi(t) \cdot \|x_n - x_m\|_\alpha.$$

6.2 Spaces E_{u_0}

Let E be a linear space. We do not assume that E is topologized.

Denote by $\mathscr{K}$ a convex, linearly closed[1] set of elements which contains with each non-zero point all points λx, where $0 \leqq \lambda < \infty$, but which does not contain $-x$. The set $\mathscr{K}$ is called a *cone*. A cone $\mathscr{K}$ permits the introduction in E of an order-relation (see 1.4°): $x \prec y$ means $y - x \in \mathscr{K}$. The symbol $\prec$ possesses the usual properties of the sign $\leqq$.

Let u_0 be some fixed non-zero element in $\mathscr{K}$. Denote by E_{u_0} the totality of $x \in E$, which satisfy for some μ (depending, of course, on x) the inequality

$$-\mu u_0 \prec x \prec \mu u_0. \tag{6.11}$$

Put

$$\|x\|_{u_0} = \inf \mu, \tag{6.12}$$

where the infimum runs over all μ for which the inequality (6.11) is fulfilled. A simple check shows that the functional (6.12) satisfies all the axioms for a norm. The norm (6.12) is called the *u_0-norm.*

For our applications the important case is that in which E is some space L_α, $\mathscr{K}$ is the set of all non-negative functions on Ω and u_0 is some fixed non-zero function in $\mathscr{K}$. In this case $E_{u_0} = L_{\alpha, u_0}$ is the totality of functions $x(t)$ satisfying almost everywhere the inequality

$$|x(t)| \leqq \mu u_0(t), \tag{6.13}$$

for some $\mu = \mu(x)$. The space of functions satisfying the inequality (6.13), is complete in u_0-norm. The proof is left to the reader.

Note that the norm $\|x\|_\alpha$ on L_{α, u_0} is subordinate to the u_0-norm $\|x\|_{\alpha, u_0}$ in the sense that

$$\|x\|_\alpha \leqq \|x\|_{\alpha, u_0} \cdot \|u_0\|_\alpha. \tag{6.14}$$

The space L_{α, u_0} is a proper subset of the space L_α, if $\alpha > 0$.

---

[1] A convex set is called *linearly closed* if its intersection with any line is closed.

Let $u_0(t)$ be a fixed non-negative function in L_β. A linear operator A acting from L_α to L_β is called *u_0-bounded* if there exists a number k such that

$$-k\,\|x\|_\alpha u_0 \leqslant Ax \leqslant k\,\|x\|_\alpha u_0. \tag{6.15}$$

This condition implies that A acts from L_α to L_{β,u_0} and is a continuous operator, and

$$\|Ax\|_{\beta,u_0} \leqq k\,\|x\|_\alpha \qquad (x \in L_\alpha).$$

It is easy to see that the integral operator (6.1) is u_0-bounded if its kernel satisfies the condition of Theorem 6.1.

6.3 *General form of u_0-bounded operators*

Denote by $\Omega^*(t;\varepsilon)$ the intersection of the set Ω^* with the sphere of radius ε and center at t.

It is possible to define for each function $z(t) \in L_0$ a function

$$\lim_{\varepsilon\to 0} \frac{1}{\operatorname{mes}[\Omega^*(t,\varepsilon)]} \int\limits_{\Omega^*(t,\varepsilon)} z(\tau)\,\mathrm{d}\tau = f_t(z). \tag{6.16}$$

It is defined for almost all t and coincides with $z(t)$ for almost all t. For smooth functions this fact is obvious; in the general case it is necessary to use the notion of approximate continuity points (see I. P. Natanson [1]). The set of t for which the function (6.16) is defined depends, of course, on the function $z(t)$.

LEMMA 6.1: *Let E_0 be a separable subspace of the space L_0. Then the function* (6.16) *becomes a linear functional on E_0 for almost all $t \in \Omega^*$.*

PROOF: Let a sequence of functions $z_1(t), z_2(t), \ldots$ form a dense set in E_0. Denote by $\Omega_0^* \subset \Omega^*$ the set of full measure ($\operatorname{mes}\Omega_0^* = \operatorname{mes}\Omega^*$) at whose points the following relation holds:

$$z_i(t) = \lim_{\varepsilon\to 0} \frac{1}{\operatorname{mes}[\Omega^*(t;\varepsilon)]} \int\limits_{\Omega^*(t;\varepsilon)} z_i(\tau)\,\mathrm{d}\tau \qquad (i = 1, 2, \ldots). \tag{6.17}$$

Equation (6.17) implies that for $t \in \Omega_0^*$ the value of the function $f_t(x)$ becomes a bounded linear functional on the linear span of the sequence of functions $z_1(t), z_2(t), \ldots$. This functional can be continuously extended to all of E_0; the extended functional is denoted by $\tilde{f}_t(y)$. It remains to show that the value of the extended functional coincides with the value of the function $f_t(y)$. This fact follows immediately from the inequality

$$\left| \lim_{\varepsilon \to 0} \frac{1}{\operatorname{mes}[\Omega^*(t;\varepsilon)]} \int_{\Omega^*(t,\varepsilon)} y(\tau)\,\mathrm{d}\tau - \tilde{f}_t(y) \right|$$

$$\leqq \left| \lim_{\varepsilon \to 0} \frac{1}{\operatorname{mes}[\Omega^*(t;\varepsilon)]} \int_{\Omega^*(t;\tau)} [y(\tau) - y_n(\tau)]\,\mathrm{d}\tau \right| + |\tilde{f}_t(y_n) - \tilde{f}_t(y)|,$$

where the sequence of functions $y_n(t)$ converges to $y(t)$ in L_0 and possesses the property that $f_t(y_n) = y_n(t)$. The lemma has been proved.

THEOREM 6.2: *Each linear u_0-bounded operator A acting from L_α to L_β $(0 < \alpha \leqq 1,\ 0 \leqq \beta < \infty)$ is a linear integral operator*

$$Ax(t) = \int_\Omega K(t, s)x(s)\,\mathrm{d}s,$$

whose kernel $K(t, s)$ satisfies the condition

$$\varphi(t) = \|K(t, s)\|_{1-\alpha} \leqq k \cdot u_0(t).$$

PROOF: Let us prove the theorem first for operators acting from L_α to L_0 in the case $u_0(t) \equiv 1$.

The space L_α $(0 < \alpha \leqq 1)$ is separable. Hence the elements Ax lie in some separable subspace E_0 of the space L_0. By Lemma 6.1 it is possible to take a subset $\Omega_0^* \subset \Omega^*$ of full measure, such that for $t \in \Omega_0^*$ the expression

$$f_t(y) = \lim_{\varepsilon \to 0} \frac{1}{\operatorname{mes}\Omega^*(t;\varepsilon)} \int_{\Omega^*(t;\varepsilon)} y(\tau)\,\mathrm{d}\tau$$

is a continuous linear functional on E_0.

Let $t_0 \in \Omega_0^*$. By continuity of the operator A and continuity of the functional $f_{t_0}(x)$ it follows that the functional $g_{t_0}(x) = f_{t_0}(Ax)$ is a continuous linear functional on L_α; hence

$$g_{t_0}(x) = \int_\Omega K(t_0, s)x(s)\,\mathrm{d}s, \tag{6.18}$$

where $K(t_0, s)$ is a measurable function in $L_{1-\alpha}$. Since

$$Ax(t) = g_t(x),$$

it follows that

$$Ax(t) = \int_\Omega K(t, s)x(s)\,\mathrm{d}s. \tag{6.19}$$

Therefore the integral operator (6.19) acts from L_α to L_0.

Denote by $x_1(s), x_2(s), \ldots$, a dense sequence in the ball $\|x\|_\alpha \leqq 1$ of the space L_α. Then

$$\sup_{n=1,2,\ldots} \left|\int_\Omega K(t, s)x_n(s)\,\mathrm{d}s\right| = \|K(t, s)\|_{1-\alpha} = \varphi(t). \tag{6.20}$$

On the other hand, from the u_0-boundedness of the operator A it follows that

$$|Ax| \leqq \|A\|_{\alpha\to 0},$$

that is,

$$\sup_n |Ax_n| \leqq \|A\|_{\alpha\to 0}. \tag{6.21}$$

Equations (6.20) and (6.21) imply that

$$\varphi(t) = \|K(t, s)\|_{1-\alpha} \leqq \|A\|_{\alpha\to 0}\cdot u_0(t).$$

Now let A be an arbitrary u_0-bounded operator acting from L_α to a space L_β, $\beta \geqq 0$.

Consider the operator

$$\boldsymbol{B}x(t) = u_1(t)\boldsymbol{A}x(t),$$

where

$$u_1(t) = \begin{cases} \dfrac{1}{u_0(t)}, & \text{if } u_0(t) > 0 \\ 0, & \text{if } u_0(t) = 0. \end{cases}$$

The operator $\boldsymbol{B}$ is obviously a continuous linear operator acting from L_α to L_0. By the part of the theorem proved above, it is an integral operator

$$\boldsymbol{B}x(t) = \int_\Omega K_0(t, s)x(s)\,\mathrm{d}s$$

with a kernel $K_0(t, s)$ satisfying the condition

$$\varphi_0(t) = \|K_0(t, s)\|_{1-\alpha} \leqq k.$$

Hence the operator

$$\boldsymbol{A}x = u_0(t)\cdot \boldsymbol{B}x(t)$$

admits a representation in which the kernel $K(t, s) = u_0(t)K_0(t, s)$, satisfies the condition

$$\varphi(t) = \|K(t, s)\|_{1-\alpha} \leqq k\cdot u_0(t).$$

The theorem has been proved.

Note that Theorem 6.2 does not hold for u_0-bounded operators acting from L_0 to some L_β. It suffices to consider as an example the operator $\boldsymbol{A}x = x$.

It is useful to recall that each linear operator acting from L_α to L_0 is u_0-bounded, because it is possible to take $u_0(t) \equiv 1$. Theorem 6.2 thus implies that *each continuous linear operator acting from* L_α $(0 < \alpha \leqq 1)$ *to* L_0 *is an integral operator.*

It is sometimes convenient to study linear operators A acting from L_α to L_β and satisfying a condition similar to (6.15)

$$-k\|x\|_\alpha \cdot u_0 \leqq Ax \leqq k \cdot \|x\|_\alpha u_0, \tag{6.22}$$

where $u_0 \notin L_\beta$. It is easy to see that such operators also admit an integral representation. This fact will be used in the proof of Theorem 9.4.

6.4 *Compactness of u_0-bounded operators*

THEOREM 6.3: *Let a linear integral operator $\boldsymbol{K}$ act from L_α to L_β ($0 \leqq \alpha < 1$, $0 < \beta < \infty$) and be u_0-bounded. Then $\boldsymbol{K}$ is compact.*

PROOF: Let

$$\boldsymbol{K}x(t) = \int_\Omega K(t, s)x(s)\,\mathrm{d}s,$$

then by u_0-boundedness,

$$\varphi(t) = \{\int_\Omega |K(t, s)|^{1/(1-\alpha)}\mathrm{d}s\}^{1-\alpha} \in L_\beta. \tag{6.23}$$

Let a sequence $x_n(t)$ be weakly convergent to $x_0(t)$ in the space L_α[1]. Then for almost all $t \in \Omega^*$ the numbers $\boldsymbol{K}x_n(t)$ converge to $\boldsymbol{K}x_0(t)$. In other words, the sequence $\boldsymbol{K}x_n(t)$ converges almost everywhere to the function $\boldsymbol{K}x_0(t)$. Weak convergence of the sequence $x_n(t)$ implies that it is bounded in norm: $\|x_n(t)\|_\alpha \leqq M$ ($n = 1, 2, \ldots$). Hence the inequality (6.10) implies that

$$|\boldsymbol{K}x_n(t)| \leqq M\varphi(t) \quad (n = 1, 2, \ldots),$$

and consequently the functions $\boldsymbol{K}x_n(t)$ have equi-absolutely continuous norms in L_β. This means that the sequence $\boldsymbol{K}x_n(t)$ converges not only almost everywhere but in norm.

We have shown that the operator $\boldsymbol{K}$ is strongly continuous in the sense that it transforms weakly convergent sequences in L_α to norm convergent sequences in L_β.

---

[1] If $\alpha = 0$, L_1-weak convergence is meant (see 1.3°).

Sets which are bounded in the norm of L_α for $0 \leqq \alpha < 1$ are weakly compact[1]. Hence the strong continuity of the operator $\boldsymbol{K}$ implies its compactness. The theorem has been proved.

It will be a useful exercise for readers to seek other proofs of Theorem 6.3. It is possible, for instance, to use results of § 3 and 5.

6.5 *Compactness of u_0-bounded operators acting from L_α to L_0*

Consider a linear integral operator, acting from L_α $(0 \leqq \alpha < 1)$ to L_0. As was noted at the end of the preceding section, these operators are u_0-bounded and comprise the entire class of operators acting from L_α to L_0 (if $\alpha > 0$).

Note further that such operators need not be compact. As an example[2], it suffices to consider the operator $\boldsymbol{K}_1$ with the kernel

$$K_1(t, s) = \begin{cases} 2^{n(1-\alpha)}, & \text{if } 1/2^n \leqq t,\, s < 1/2^{n-1} \\ 0 & \text{for other } t, s \in [0, 1] \end{cases}$$

(here $\Omega^* = \Omega = [0, 1]$). The trivial inequality

$$\int_0^1 |K_1(t, s)|^{1/(1-\alpha)} ds \leqq 1$$

implies that $\boldsymbol{K}_1$ acts from L_α to L_0 and is u_0-bounded (with $u_0(t) \equiv 1$). At the same time the operator $\boldsymbol{K}_1$ transforms the norm bounded sequence of functions

$$x_n(s) = \begin{cases} 2^{n\alpha}, & \text{if } 1/2^n \leqq s < 1/2^{n-1} \\ 0, & \text{if } s < 1/2^n \text{ or } s \geqq 1/2^{n-1}, \end{cases}$$

to the sequence of functions

$$\boldsymbol{K}_1 x_n(t) = \begin{cases} 1, & \text{if } 1/2^n \leqq t < 1/2^{n-1} \\ 0, & \text{if } t < 1/2^n \text{ or } t \geqq 1/2^{n-1} \end{cases}$$

---

1 If $\alpha = 0$, L_1-weak convergence is meant (see 1.3°).

2 The idea of this example is due to A. I. Perov.

which is not compact in L_0, because

$$\|Kx_n - Kx_m\|_0 = 1 \qquad (n \neq m).$$

We remark that the u_0-bounded operator (6.1) acting from L_α to L_0, transforms each weakly convergent sequence in L_α to a sequence, which is convergent almost everywhere.

A vector-function $w(t)$ $(t \in \Omega^*)$ with values in some $L_r = L_r(\Omega)$ is called *essentially compact* if there is a subset $\Omega_0^* \subset \Omega^*$, of full measure, on which values of the function $w(t)$ are contained in a compact set in L_r.

If K is a u_0-bounded integral operator acting from L_α to L_0, the kernel $K(t, s)$ can be considered as a vector-function $w(t) = K(t, s)$, defined for $t \in \Omega^*$, and with values in the space $L_{1-\alpha} = L_{1-\alpha}(\Omega)$.

THEOREM 6.4: *Let a linear integral operator K act from L_α $(0 \leqq \alpha \leqq 1)$ to L_0. Then it is compact if and only if the vector-function*

$$w(t) = K(t, s) \qquad (t \in \Omega^*) \tag{6.24}$$

with values in the space $L_{1-\alpha}(\Omega)$ is essentially compact.

PROOF: Let the function (6.24) be essentially compact. Denote by Ω_0^* a set of full measure on which it takes values belonging to a compact set $\mathfrak{M} \subset L_{1-\alpha}$.

Let $\varepsilon > 0$ be given. Partition the set Ω_0^* into a finite number of parts $\Omega_{(1)}^*, \ldots, \Omega_{(n)}^*$ such that the range of the function (6.24) on each of these parts forms a set in $L_{1-\alpha}$ whose diameter is less than ε. If t_1 and t_2 belong to a fixed set $\Omega_{(i)}^*$,

$$|Kx(t_1) - Kx(t_2)| = |\int_\Omega [K(t_1, s) - K(t_2, s)]x(s)\,ds|$$

$$\leqq \|w(t_1) - w(t_2)\|_{1-\alpha} \|x\|_\alpha \leqq \varepsilon \|x\|_\alpha.$$

Lemma 1.2 implies that the range of the operator K on each ball of the space L_α is compact. This means that K is compact.

Let us enter the second part of the theorem. Let the operator K be compact. Then its range is contained in a separable subspace of L_0. Lemma 6.1 implies the existence of a subset $\Omega_0^* \subset \Omega^*$ with full measure, such that the value of $Kx(t)$ for each fixed $t \in \Omega_0^*$ is a continuous linear functional on L_α. Hence for any $t_1, t_2 \in \Omega_0^*$ the following relation holds

$$\|w(t_1) - w(t_2)\|_{1-\alpha} = \sup_{\|x\|_\alpha \leqq 1} |\int_\Omega [K(t_1, s) - K(t_2, s)]x(s)\,ds|. \tag{6.25}$$

Let $\varepsilon > 0$ be given. Lemma 1.2 implies that the set Ω_0^* can be split into a finite number of parts $\Omega_{(1)}^*, \Omega_{(2)}^*, \ldots, \Omega_{(n)}^*$, such that for t_1, t_2 in a fixed $\Omega_{(i)}^*$ one has:

$$|Kx(t_1) - Kx(t_2)| < \varepsilon$$

for all functions $x(t)$ in the unit ball of the space L_α. Consequently for each i

$$\|w(t_1) - w(t_2)\|_{1-\alpha} \leqq \varepsilon \quad (t_1, t_2 \in \Omega_{(i)}^*).$$

The theorem has been proved.

In many cases compactness of the range of a function $w(t)$ is checked without difficulty. For instance, if the set Ω^* is bounded and closed, it is sufficient that the vector-function $w(t)$ is continuous. Hence it follows that an integral operator with a kernel $K(t, s)$ which acts from L_α ($\alpha > 0$) to L_0 is compact if

$$\int_\Omega |K(t, s)|^{1/(1-\alpha)}\,ds \leqq c_1 < \infty \tag{6.27}$$

and if for each $t \in \Omega^*$

$$\lim_{h \to 0} \int_\Omega |K(t + h, s) - K(t, s)|^{1/(1-\alpha)}\,ds = 0.$$

If Ω^* is bounded but not closed the compactness of range of a function $w(t)$ results, for instance, from its uniform continuity. Consequently, an integral operator with a kernel $K(t, s)$ acts from L_α to L_0 and is compact if the condition (6.27) holds and if

$$\lim_{h \to 0} \sup_{\substack{t_1, t_2 \in \Omega^* \\ |t_1 - t_2| < h}} \int_\Omega |K(t_1, s) - K(t_2, s)|^{1/(1-\alpha)}\,ds = 0. \tag{6.29}$$

6.6 *Compactness of u_0-bounded operators acting from L_1 to L_β ($\beta > 0$)*

Strong continuity is a general property of u_0-bounded operators acting from L_1 to L_β ($\beta > 0$) (this is proved by using the same argument as was used in the proof of Theorem 6.3). But strong continuity of these operators does not imply their compactness, because a ball of L_1 does not possess the property of weak compactness. The operator $\boldsymbol{K}$ with the kernel (5.14) can serve as an example.

Let $w(t)$ $(t \in \Omega^*)$ be a vector-function with values in the space L_0. Suppose that the values of $w(t)$ for almost all t belong to a subspace $F \subset L_0$ such that the unit ball of the space L_1 becomes an F-weakly compact set (see 1.3°). This condition is, for instance, fulfilled if the values of $w(t)$ belong to a separable subspace $F \subset L_0$ (in particular, if $w(t)$ is continuous). Then the function $w(t)$ is called *essentially co-compact.*

THEOREM 6.5: *Let a u_0-bounded linear integral operator* $\boldsymbol{K}$ *with a kernel $K(t, s)$ act from L_1 to L_β ($\beta > 0$). Then it is compact if and only if the vector-function*

$$w(t) = K(t, s) \quad (t \in \Omega^*)$$

with values in the space $L_0(\Omega)$ is essentially co-compact.

The proof is almost obvious.

If $\beta \in [0, 1]$, a more convenient condition for compactness can be given. In this case the operator $\boldsymbol{K}^*$ adjoint to $\boldsymbol{K}$ will coincide with the transposed operator and will be a continuous operator acting from $L_{1-\beta}$ to L_0. Problems on the compactness of the operators $\boldsymbol{K}$ and $\boldsymbol{K}^*$ are equivalent. The operator $\boldsymbol{K}^*$ is, obviously, u_0-bounded and hence a condition for its compactness can be obtained from Theorem 6.4. Thus the following holds:

THEOREM 6.6: *Let a u_0-bounded linear integral operator* $\boldsymbol{K}$ *with kernel $K(t, s)$ act from L_1 to L_β ($0 < \beta \leqq 1$). Then it is compact if and only if the vector-function*

$$w^*(s) = K(t, s) \quad (s \in \Omega)$$

with values in the space $L_\beta(\Omega^*)$ is essentially compact.

6.7 *Integral operators acting from L_α to C*

In this section we will assume that the set Ω^* is bounded and closed.

If a continuous linear operator A acts from L_α $(0 < \alpha \leqq 1)$ to C, then the formula

$$f_t(x) = Ax(t) \qquad (x \in L_\alpha) \tag{6.30}$$

defines a linear continuous functional for each $t \in \Omega^*$. The theorem on the general form of linear functionals implies that A admits an integral representation

$$Ax(t) = \int_\Omega K(t, s)x(s)\,ds. \tag{6.31}$$

Here the kernel $K(t, s)$ for each fixed $t \in \Omega^*$ becomes a function in $L_{1-\alpha}$. Theorem 6.2 implies that the function $\varphi(t) = \|K(t, s)\|_{1-\alpha}$ belongs to L_0. If the kernel $K(t, s)$ is considered as a vector-function

$$w(t) = K(t, s) \qquad (t \in \Omega^*) \tag{6.32}$$

with values in $L_{1-\alpha}$, it is weakly continuous.

It is easy to see that these properties of a kernel $K(t, s)$ are also sufficient for that the integral operator with this kernel act from L_α $(0 < \alpha \leqq 1)$ to C and be continuous.

Weak continuity of the vector-function (6.32) in $L_{1-\alpha}$ $(0 < \alpha < 1)$ is equivalent to the requirement that the kernel $K(t, s)$ satisfy the conditions:

1°) $\int_\Omega |K(t, s)|^{1/(1-\alpha)}\,ds \leqq M$,

2°) For each measurable subset $D \subset \Omega$ one has:

$$\lim_{t \to t_0} \int_D K(t, s)\,ds = \int_D K(t_0, s)\,ds.$$

It remains to consider the case of linear operators acting from L_0 to C. Such operators are not necessarily integral. But for an integral operator with a kernel $K(t, s)$ the conditions 1° and 2° above are necessary and sufficient for that this operator acts from L_0 to C.

If in the above conditions the requirement of weak continuity for the vector-function (6.32) is replaced by the assumption of strong continuity, we obtain a criterion for compactness of the integral operator. The vector-function (6.32) will be strongly continuous in the space $L_{1-\alpha}$ if it satisfies the condition 1°, and if instead of the condition 2°, it satisfies the condition

3°) $\lim_{t\to t_0} \int_\Omega |K(t, s) - K(t_0, s)|^{1/(1-\alpha)} \mathrm{d}s = 0.$

6.8 v_0-*Cobounded linear operators*

In some cases it is easier to prove compactness of the operator A^* which is adjoint to a given linear operator A. Compactness of the operator A^* implies, as is known, compactness of the operator A.

If A acts from L_α to L_β and $0 < \alpha \leqq 1$, $0 \leqq \beta \leqq 1$, then the adjoint operator A^* acts from $L_{1-\beta}$ to $L_{1-\alpha}$. If A acts from L_0 to L_β, then A^* acts from $L_{1-\beta}$ to the dual space of L_0; it is assumed throughout this section that the values of A^* actually belong to L_1 (recall that regular integral operators possess this property).

Let $v_0(s)$ be a fixed non-negative function in a space $L_{1-\alpha}$. A continuous linear operator A acting from a space L_α $(0 \leqq \alpha \leqq 1)$ to a space L_β is called v_0-*cobounded* if

$$\|Ax\|_\beta \leqq k f_0(|x|) \qquad (x \in L_\alpha), \tag{6.33}$$

where

$$f_0(x) = \int_\Omega x(s) v_0(s) \mathrm{d}s. \tag{6.34}$$

Note that each continuous linear operator acting from L_1 to L_β is v_0-cobounded with $v_0(s) \equiv 1$. For operators acting from L_α to L_β with $\alpha < 1$, the assumption of v_0-coboundedness is an additional restriction.

THEOREM 6.7: *A continuous linear operator A acting from L_α to L_β $(0 \leqq \alpha, \beta \leqq 1)$ is v_0-cobounded if and only if the adjoint operator A^* acting from $L_{1-\beta}$ to $L_{1-\alpha}$, is v_0-bounded.*

PROOF: Formula (6.33) implies that for any functional $g \in L_{1-\beta}$ one has:

$$-k\|g\| f_0(|x|) \leqq g(Ax) \leqq k\|g\| f_0(|x|),$$

which can be written in the form

$$-k\|g\|f_0(|x|) \leqq [A^*g](x) \leqq k\|g\|f_0(|x|)$$

or

$$-k\|g\| \int_\Omega |x(s)|\, v_0(s)\,\mathrm{d}s \leqq [A^*g](x) \leqq k\|g\| \int_\Omega |x(s)|\, v_0(s)\,\mathrm{d}s.$$

Consequently,

$$-k\|g\|\, v_0(s) \leqq A^*g(s) \leqq k\|g\| \cdot v_0(s).$$

We have proved that v_0-coboundedness of the operator $\boldsymbol{A}$ implies v_0-boundedness of the adjoint operator. The converse assertion is established similarly. The theorem has been proved.

THEOREM 6.8: *Let*

$$\psi(s) = \|K(t, s)\|_r \in L_q, \tag{6.35}$$

where $0 \leqq q,\ r \leqq 1$. *Then the integral operator* $\boldsymbol{K}$ *with the kernel* $K(t, s)$ *acts from* L_α *to* L_β, *where* $\alpha = 1 - q$, $\beta = r$ *and is* v_0*-cobounded.*

PROOF: Condition (6.35) and Theorem 6.1 imply that the transposed operator

$$\boldsymbol{K}^{\#} y(s) = \int_{\Omega^*} K(t, s) y(t)\,\mathrm{d}t \tag{6.36}$$

acts from L_{1-r} to L_q is regular and is v_0-bounded, where $v_0(s) = \psi(s)$. Now use the preceding theorem. The theorem has been proved.

Condition (6.35) can be written in the form of inequalities as follows:

$$1^\circ)\quad \int_\Omega \Big[\int_{\Omega^*} |K(t, s)|^{1/r}\,\mathrm{d}t\Big]^{r/q}\mathrm{d}s < \infty, \quad \text{if } r, q > 0. \tag{6.37}$$

$$2^\circ)\quad \int_\Omega \operatorname*{ess\,sup}_{t\in\Omega^*} [|K(t, s)|^{1/q}]\,\mathrm{d}s < \infty, \quad \text{if } r = 0,\ q > 0. \tag{6.38}$$

3°) $\operatorname{ess\,sup}_{s \in \Omega} \int_{\Omega^*} |K(t, s)|^{1/r} dt < \infty,$

if $r > 0,\ q = 0.$ (6.39)

4°) $\operatorname{ess\,sup}_{s \in \Omega},\ \operatorname{ess\,sup}_{t \in \Omega^*} \{|K(t, s)|\} < \infty,$

if $r = q = 0.$ (6.40)

If a v_0-cobounded operator A acts from L_α to L_β, where either $\alpha = 0$ or $\beta = 1$, it may not be integral—a simple example is the operator $Ax = x$. But if $0 < \alpha \leqq 1$, $0 \leqq \beta < 1$, a v_0-cobounded operator A is integral. For the proof use Theorem 6.7. This implies that the adjoint operator A^* acts from $L_{1-\beta}$ to $L_{1-\alpha}$ and is v_0-bounded. By Theorem 6.2 the operator A^* admits an integral representation

$$A^*y(s) = \int_{\Omega^*} K_0(t, s) y(t) dt,$$

and the kernel $K_0(t, s)$ satisfies the condition

$$\psi(s) = \|K_0(t, s)\|_\beta \in L_{1-\alpha}.$$

The operator A^* is regular and by Theorem 4.4 its adjoint operator coincides with its transposed operator. Consequently

$$Ax(t) = \int_{\Omega} K_0(t, s) x(s) ds.$$

6.9 *Compactness of v_0-cobounded operators*

THEOREM 6.9: *Let a continuous linear operator A act from L_α to L_β $(0 < \alpha < 1,\ 0 < \beta < 1)$ and be v_0-cobounded. Then it is compact.*

PROOF: Theorem 6.7 implies that the operator A^* is v_0-bounded. By Theorem 6.3 it is compact. Hence the operator A is compact. The theorem has been proved.

Theorem 6.9 is an analogue of Theorem 6.3. We leave it to the reader to formulate analogues of Theorems 6.5 and 6.6.

6.10 *Interpolation properties of u_0-boundedness*

Denote by $L(\boldsymbol{K}; *\text{-bound.})$ the totality of points $\{\alpha, \beta\}$, such that the operator $\boldsymbol{K}$ acts from L_α to L_β and is u_0-bounded. Here the element u_0 may depend on the point $\{\alpha, \beta\}$.

LEMMA 6.2: *Let*

$$\varphi_0(t) = \|K(t, s)\|_{r_0} \in L_{q_0}, \quad \varphi_1(t) = \|K(t, s)\|_{r_1} \in L_{q_1} \tag{6.41}$$

where r_0, q_0, r_1, q_1 are non-negative numbers. Then for any $\tau \in (0, 1)$

$$\varphi_\tau(t) = \|K(t, s)\|_{r(\tau)} \in L_{q(\tau)}, \tag{6.42}$$

where

$$r(\tau) = (1 - \tau)r_0 + \tau r_1, \quad q(\tau) = (1 - r)q_0 + \tau q_1. \tag{6.43}$$

PROOF: Logarithmic convexity of norms in L_α (see formula (1.6)) implies the inequality

$$\varphi_\tau(t) \leqq [\varphi_0(t)]^{1-\tau}[\varphi_1(t)]^\tau.$$

Hence from Hölder's inequality it follows that $\varphi_\tau(t) \in L_{q(\tau)}$ and

$$\|\varphi_\tau(t)\|_{q(\tau)} \leqq \|\varphi_0(t)\|_{q_0}^{1-\tau}\|\varphi_1(t)\|_{q_1}^{\tau}.$$

The lemma has been proved.

Lemma 6.2 implies convexity of the L-characteristic $L(\boldsymbol{K}; *\text{-bound.})$.

Let condition (6.41) be fulfilled and let

$$\alpha_0 = 1 - r_0, \ \alpha_1 = 1 - r_1, \ \beta_0 = q_0, \ \beta_1 = q_1.$$

Consider the points $M_0 = \{\alpha_0, \beta_0\}$ and $M_1 = \{\alpha_1, \beta_1\}$ in the $\{\alpha, \beta\}$ plane. If both of these lie in the half-strip $0 \leqq \alpha \leqq 1$, $\beta \geqq 0$ (Fig. 6.1, a), then the segment joining the points M_0 and M_1 belongs completely to $L(\boldsymbol{K}; *\text{-bound.})$ and interior points of the segment belong to $L(\boldsymbol{K}; \text{comp.})$; if the points

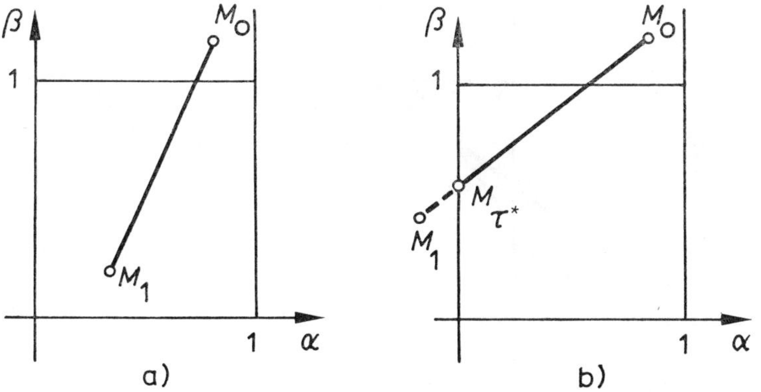

Figure 6.1

M_0 and M_1 lie in interior of the half-strip, they too belong to $L(\boldsymbol{K}$; comp.). If the point M_0 lies in the half-strip $0 \leqq \alpha \leqq 1$, $\beta \geqq 0$ but the point M_1 lies outside it (Fig. 6.1, b), then the L-characteristic $L(\boldsymbol{K}$; $*$-bound.) contains all points in the part $[M_0, M_{\tau^*}]$ of the segment, joining M_0 and M_1, which lies in the half-strip; points on the segment $[M_0, M_{\tau^*}]$ (except, possibly, M_0) belong to $L(\boldsymbol{K}$; comp.).

Let us consider v_0-coboundedness of linear integral operators. The L-characteristic $L(A$; $*$-cobd.) is defined in the obvious way. In 6.8° there was given a condition for v_0-coboundedness in terms of adjoint operators. Hence Lemma 6.2 can be used to deduce the convexity of the part of the L-characteristic $L(A$; $*$-cobd.) which lies in the unit square $0 \leqq \alpha, \beta \leqq 1$.

Let the following condition the fulfilled:

$$\psi_0(s) = \|K(t, s)\|_{r_0} \in L_{q_0}$$

$$\psi_1(s) = \|K(t, s)\|_{r_1} \in L_{q_1}, \tag{6.44}$$

where r_0, r_1, q_0, q_1 are non-negative numbers. Then by using the prescription given in the beginning of this section it is possible to draw a part of the L-characteristic $L(\boldsymbol{K}^{\#}$; $*$-bound.) of the transposed operator $\boldsymbol{K}^{\#}$. Consider the intersection $M_0(\boldsymbol{K}^{\#}$; $*$-bound.) of this part with the unit square. The set N_0, symmetric to $M_0(\boldsymbol{K}^{\#}$; $*$-bound.) with respect to the line $\alpha + \beta = 1$, will belong to the L-characteristic $L(\boldsymbol{K}$; $*$-cobd.) (Fig. 6.2). To form the set N_0 take first the points $P_0 = \{1 - q_0, r_0\}$, $Q_0 =$

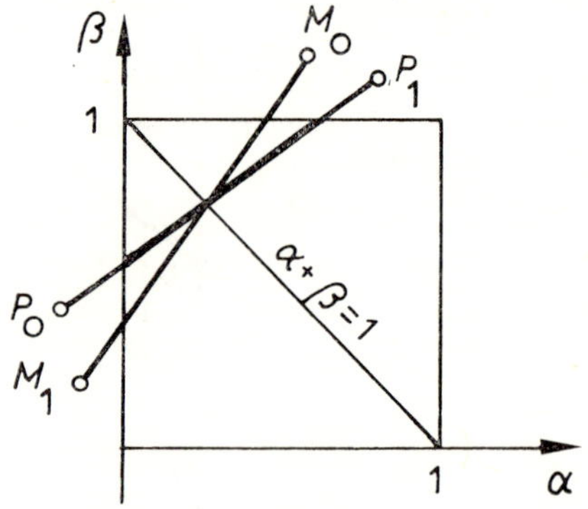

Figure 6.2

$\{1 - q_1, r_1\}$ and join them by a segment; the part of this segment, lying in the unit square $0 \leqq \alpha, \beta \leqq 1$ will belong to $L(\boldsymbol{K}; *\text{-cobd.})$.

6.11 *On weakly compact operators in L_1*

Recall that a linear operator A acting from Banach space E_1 to a Banach space E_2 is called weakly compact if the set of elements Ax, $\|x\|_{E_1} \leqq 1$, is weakly compact in E_2. Weakly compact operators are continuous.

Linear operators, acting from L_α to L_β, are always weakly compact if one of the numbers α, β lies in the interval (0, 1). Integral operators acting from L_0 to L_1 are also weakly compact (they are even compact).

Let us consider weakly compact operators acting in the space L_1. Each such operator (see N. Dunford and J. T. Schwartz [1]) is regular and is an integral operator. We do not prove this theorem because we are here interested only in integral operators.

Recall that an integral operator with a kernel $K(t, s)$ acts and is regular in the space L_1, if and only if

$$\psi(s) = \int_{\Omega^*} |K(t, s)| \,\mathrm{d}t \in L_0. \tag{6.45}$$

A vector-function $w^*(s) = K(t, s)$ $(s \in \Omega)$ is called *essentially weakly compact*, if it is possible to take a set $\Omega_0 \subset \Omega$ with full measure such that the values of $w^*(s)$ on Ω_0 lie in a weakly compact set. By Lemma 1.3 essential weak compactness means that

$$\lim_{\text{mes } D^* \to 0} \sup_{s \in \Omega_0} \int_{D^*} |K(t, s)| \, dt = 0. \tag{6.46}$$

THEOREM 6.10: *Let **K** be a linear integral operator with kernel $K(t, s)$ acting from L_1 to L_1. Then it is weakly compact if and only if the vector-function*

$$w^*(s) = K(t, s) \qquad (s \in \Omega)$$

with values in the space $L_1(\Omega^*)$ is essentially weakly compact.

PROOF: Weak compactness of the operator $\boldsymbol{K}$ means that

$$\lim_{\text{mes } D^* \to 0} \|\boldsymbol{P}_{D^*}\boldsymbol{K}\|_{1 \to 1} = 0.$$

Hence it suffices to show that

$$\|\boldsymbol{P}_{D^*}\boldsymbol{K}\|_{1 \to 1} = \operatorname*{ess\,sup}_{s \in \Omega} \int_{D^*} |K(t, s)| \, dt. \tag{6.47}$$

Since for each function $x(s) \in L_1$ we have:

$$\|\boldsymbol{P}_{D^*}\boldsymbol{K}x\|_1 = \int_{D^*} \left| \int_{\Omega} K(t, s) x(s) \, ds \right| dt$$

$$\leqq \int_{D^*} \int_{\Omega} |K(t, s)| \, |x(s)| \, ds \, dt \leqq \int_{\Omega} \left[\int_{D^*} |K(t, s)| \, dt \right] |x(s)| \, ds$$

$$\leqq \left\| \int_{D^*} |K(t, s)| \, dt \right\|_0 \|x\|_1,$$

$$\|\boldsymbol{P}_{D^*}\boldsymbol{K}\|_{1 \to 1} \leqq \left\| \int_{D^*} |K(t, s)| \, dt \right\|_0. \tag{6.48}$$

On the other hand, for each $s \in \Omega$

$$\int_{D^*} |K(t, s)| \, dt = \sup_{\|y\|_0 \leqq 1} \int_{D^*} K(t, s) y(t) \, dt,$$

and this means

$$\|\int_{D^*} |K(t, s)| \, dt\|_0 \leqq \sup_{\|x\|_1 \leqq 1, \|y\|_0 \leqq 1} \int_{\Omega} \int_{D^*} K(t, s) y(t) x(s) \, dt \, ds$$

$$= \sup \|\boldsymbol{P}_{D^*} \boldsymbol{K} x\|_1 = \|\boldsymbol{P}_{D^*} \boldsymbol{K}\|_{1 \to 1}.$$

Hence it follows that

$$\|\int_{D^*} |K(t, s)| \, dt\|_0 \leqq \|\boldsymbol{P}_{D^*} \boldsymbol{K}\|_{1 \to 1}. \tag{6.49}$$

From (6.48) and (6.49) there follows (6.47). The theorem has been proved.

An interesting property of a weakly compact operator $\boldsymbol{K}$ acting in L_1 is that its square $\boldsymbol{K}^2$ is compact. This follows from Lemma 5.2, according to which the range of the operator $\boldsymbol{K}$ on each weakly compact set is compact.

§ 7 Integral operators with kernels satisfying conditions of Kantorovic type[1]

7.1 *Simplest criteria*

In the preceding paragraph two classes of integral operators acting from L_α to L_β were studied: u_0-bounded and v_0-cobounded operators. The operators considered in this paragraph form classes which are intermediate to the above in some sense. Throughout this paragraph only regular operators are investigated.

---

[1] The theorems presented here are obtained on developing ideas due to L. V. Kantorovic [1] (see also F. L. Smolizkii [1], L. V. Kantorovic and G. P. Akilov [1]). In one of his reports S. G. Krein focused attention on the 'interpolation character' of Kantorovic' theorems on the continuity of integral operators. The connection of Kantorovic' theorem on the continuity of integral operators with theorems on interpolation properties of compactness was studied by M. A. Krasnoselskii [8].

The change from Kantorovic' conditions to conditions of the type (7.1) and (7.2) was given in the book of M. A. Krasnoselskii and Y. B. Rutickii [3] (see also the papers of Y. B. Rutickii [4, 5]).

New results, which are given in this paragraph, were obtained in essence by P. P. Zabreiko and E. I. Pustylnik.

§ 7 *Integral operators with kernels, satisfying conditions of Kantorovic type*

The theorems to be proved in the sequel contain, as a rule, two different kinds of restrictions on a kernel $K(t, s)$. The simplest (and most important) of these theorems follows immediately from the interpolation Theorems 2.4 and 3.10. Let us present one such result.

THEOREM 7.1: *Let a kernel $K(t, s)$ satisfy the conditions*

$$\varphi(t) = \|K(t, s)\|_r \in L_q, \tag{7.1}$$

$$\psi(s) = \|K(t, s)\|_{r^*} \in L_{q^*}, \tag{7.2}$$

where $0 \leqq r^*, q^*, r \leqq 1$, $0 \leqq q < \infty$. Then for any $\tau \in (0, 1)$ the integral operator $\boldsymbol{K}$ with kernel $K(t, s)$ acts from $L_{\alpha(\tau)}$ to $L_{\beta(\tau)}$, where

$$\alpha(\tau) = 1 - (1 - \tau)r - \tau q^*, \quad \beta(\tau) = (1 - \tau)q + \tau r^*, \tag{7.3}$$

and

$$\|\boldsymbol{K}\|_{\alpha(\tau)\to\beta(\tau)} \leqq \|\varphi\|_q^{1-\tau} \|\psi\|_{q^*}^{\tau}. \tag{7.4}$$

If one of the following supplementary conditions is fulfilled:

a) $q^* > 0, \ q + r^* > 0$

b) $q > 0, \ q^* + r > 0,$

then $\boldsymbol{K}$, as an operator from $L_{\alpha(\tau)}$ to $L_{\beta(\tau)}$, is compact.

PROOF: Condition (7.1) and Theorem 6.1 imply that the operator $\boldsymbol{K}$ acts from L_{1-r} to L_q, while condition (7.2) and Theorem 6.8 imply that $\boldsymbol{K}$ acts from L_{1-q^*} to L_{r^*}. The assertion on continuity and the estimate (7.4) now follow from interpolation theorems. Note that the interpolation theorems actually imply a more general estimate

$$\|\boldsymbol{P}_{D^*}\boldsymbol{K}\boldsymbol{P}_D\|_{\alpha(\tau)\to\beta(\tau)} \leqq \|\boldsymbol{P}_{D^*}\varphi\|_q^{1-\tau} \cdot \|\boldsymbol{P}_D\psi\|_{q^*}^{\tau}. \tag{7.5}$$

Let us now give the proof of the compactness of the operator $\boldsymbol{K}$.

If one of the conditions a) or b) is fulfilled then $\alpha(\tau) < 1$ and $\beta(\tau) > 0$. According to Theorem 5.8 it is necessary only to consider the case $\beta(\tau) \leqq 1$.

If condition a) is fulfilled, (7.5) implies

$$\lim_{\text{mes}\, D \to 0} \|\boldsymbol{K}\boldsymbol{P}_D\|_{\alpha(\tau) \to \beta(\tau)} = 0,$$

while if condition b) is fulfilled then (7.5) implies that

$$\lim_{\text{mes}\, D^* \to 0} \|\boldsymbol{P}_{D^*}\boldsymbol{K}\|_{\alpha(\tau) \to \beta(\tau)} = 0.$$

Hence compactness of the operator $\boldsymbol{K}$ follows from Theorem 5.6. The theorem has been proved.

If neither of the conditions a), b) of Theorem 7.1 is fulfilled $\boldsymbol{K}$ as an operator from $L_{\alpha(\tau)}$ to $L_{\beta(\tau)}$ may not possess the property of compactness. Consider, for example, the integral operator $\boldsymbol{K}_0$ with the kernel

$$K_0(t, s) = \begin{cases} 2^{nrr^*}, & \text{if } t \in \Omega_n^*,\ s \in \Omega_n \\ 0 & \text{for other } t \in \Omega^*,\ s \in \Omega, \end{cases} \tag{7.6}$$

where

$$\Omega^* = \left[0, \frac{1}{1 - 2^{-r}}\right], \quad \Omega = \left[0, \frac{1}{1 - 2^{-r^*}}\right],$$

and

$$\Omega_n^* = \left[\frac{1 - 2^{-r(n-1)}}{1 - 2^{-r}}, \frac{1 - 2^{-rn}}{1 - 2^{-r}}\right)$$

$$\Omega_n = \left[\frac{1 - 2^{-r^*(n-1)}}{1 - 2^{-r}}, \frac{1 - 2^{-r^*n}}{1 - 2^{-r}}\right) \quad (n = 1, 2, \ldots).$$

This kernel satisfies all the conditions of Theorem 7.1 (with $q = q^* = 0$), except for the conditions a), b). Hence the integral operator $\boldsymbol{K}_0$ will act from $L_{\alpha(\tau)}$ to $L_{\beta(\tau)}$ for all $\tau \in (0, 1)$, where

$$\alpha(\tau) = 1 - (1 - \tau)r, \quad \beta(\tau) = \tau r^*.$$

At the same time this operator is not compact—the sequence of functions $x_n(t)$, bounded in $L_{\alpha(\tau)}$ given by

$$x_n(s) = \begin{cases} 2^{nr^*\alpha(\tau)}, & \text{if } s \in \Omega_n \\ 0, & \text{if } s \notin \Omega_n, \end{cases}$$

is transformed into a sequence of functions which is not compact in $L_{\beta(\tau)}$:

$$K_0 x_n(t) = \begin{cases} 2^{nr\beta(\tau)}, & \text{if } t \in \Omega_n^* \\ 0, & \text{if } t \notin \Omega_n^*. \end{cases}$$

Let us return to Theorem 7.1. Let a kernel $K(t, s)$ satisfy the conditions (7.1) and (7.2), but suppose the conditions a) and b) are not fulfilled. Assume that $q = q^* = 0$; $r, r^* > 0$. In this case K will be compact[1] as an operator from $L_{\alpha(\tau)}$ to $L_{\beta(\tau)}$ $(0 < \tau < 1)$ if either one of the following conditions is holds:

$$\lim_{\text{mes}\, D \to 0} \| \int_D |K(t, s)|^{1/r} \text{d}s \|_0 = 0,$$

or

$$\lim_{\text{mes}\, D^* \to 0} \| \int_{D^*} |K(t, s)|^{1/r^*} \text{d}t \|_0 = 0.$$

For the proof it suffices to derive the inequality

$$\|P_{D^*} K P_D\|_{\alpha(\tau) \to \beta(\tau)} \leqq \|\{\int_D |K(t, s)|^{1/r} \text{d}s\}^r\|_0^{1-\tau} \cdot \|\{\int_{D^*} |K(t, s)|^{1/r^*} \text{d}t\}^{r^*}\|_0^\tau$$

and to use Theorem 5.6.

Now assume that $\Omega^* = \Omega$ and consider the class of integral operators with kernels possessing a property of 'symmetry':

$$|K(t, s)| = |K(s, t)| \quad (t, s \in \Omega). \tag{7.7}$$

---

[1] This assertion was proved by Y. B. Rutickii.

Suppose that

$$\varphi(t) = \|K(t, s)\|_r \in L_q, \tag{7.8}$$

for some $0 \leqq r, q \leqq 1$. Then from (7.7) it follows that

$$\psi(t) = \|K(t, s)\|_r \in L_q.$$

Thus, conditions (7.1) and (7.2) are fulfilled with $r = r^*$ and $q = q^*$. Hence Theorem 7.1 implies

THEOREM 7.2: *Let a kernel* $K(t, s)$ *satisfy the conditions* (7.7) *and* (7.8). *Then for any* $\tau \in (0, 1)$ *the integral operator* $\boldsymbol{K}$ *with the kernel* $K(t, s)$ *acts from* $L_{\alpha(\tau)}$ *to* $L_{\beta(\tau)}$, *where*

$$\alpha(\tau) = 1 - (1 - \tau)r - \tau q, \quad \beta(\tau) = (1 - \tau)q + \tau r$$

and

$$\|\boldsymbol{K}\|_{\alpha(\tau)\to\beta(\tau)} \leqq \|\varphi\|_q.$$

If $q > 0$, $\boldsymbol{K}$ *is compact as an operator from* $L_{\alpha(\tau)}$ *to* $L_{\beta(\tau)}$.

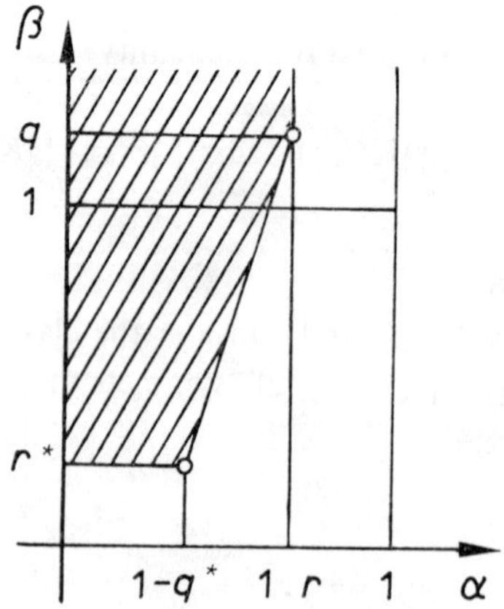

Figure 7.1

Theorems 7.1 and 7.2 permit one to draw some parts of the L-characteristics of the operator $\boldsymbol{K}$. Let, for instance, the conditions of Theorem 7.1 be fulfilled with $q^* > r$, $q > r^*$; then $L(\boldsymbol{K}; \text{reg.})$ contains the polygon, hatched in Fig. 7.1.

It is important to note that Theorems 7.1 and 7.2 are 'exact'—it is possible to construct kernels $K(t, s)$ such that the L-characteristics of the corresponding operators are completely determined by these theorems.

7.2 *Theorems with intermediate conditions*

In this section it is assumed that the kernel $K(t, s)$ satisfies the condition

$$\psi(s) = \| |K(t, s)|^{\nu} \cdot \varphi(t)^{1-\nu} \|_{\theta} \in L_{q^*}, \tag{7.9}$$

where $0 \leqq \nu \leqq 1$ and

$$\varphi(t) = \|K(t, s)\|_r. \tag{7.10}$$

The condition (7.9) is intermediate between conditions of u_0-boundedness and v_0-coboundedness,—it reduces to the former for $\nu = 0$ and to the latter for $\nu = 1$. If the numbers r, θ, q^* are positive, the condition (7.9) can be written in the form of inequality:

$$\int_{\Omega} \{ \int_{\Omega^*} |K(t, s)|^{\nu/\theta} [\int_{\Omega} |K(t, \sigma)|^{1/r} \mathrm{d}\sigma]^{r(1-\nu)/\theta} \mathrm{d}t \}^{\theta/q^*} \mathrm{d}s < \infty.$$

THEOREM 7.3: *Let a kernel* $K(t, s)$ *satisfy the condition* (7.9) *and let*

$$\theta \geqq q^*, \ (1 - \nu)r + \theta \leqq 1. \tag{7.11}$$

Then the integral operator $\boldsymbol{K}$ *with kernel* $K(t, s)$ *is a continuous operator acting from* L_{α} *to* L_{β}*, where*

$$\alpha = 1 - (1 - \nu)r - q^*, \ \beta = \theta, \tag{7.12}$$

and

$$\|\boldsymbol{K}\|_{\alpha \to \beta} \leqq \|\psi(s)\|_{q^*}. \tag{7.13}$$

If further

$$q^* > 0, \tag{7.14}$$

then $\boldsymbol{K}$ *is compact as an operator from* L_α *to* L_β.

PROOF: Consider the number

$$b = \frac{1 - (1 - v)r - \theta}{1 - (1 - v)r - q^*}.$$

By (7.11) we have:

$$0 \leqq b \leqq 1$$

and

$$1 - (1 - v)r - b\alpha = \theta.$$

It is clear that

$$|\boldsymbol{K}x(t)| \leqq \int_\Omega |K(t, s)|^{1-v} |K(t, s)|^v |x(s)|^{1-b} |x(s)|^b \mathrm{d}s.$$

Apply the following form of Hölder's inequality to the right side:

$$|\int_\Omega u(s) v(s) w(s) \mathrm{d}s|$$

$$\leqq \{\int_\Omega |u(s)|^{1/v_1} \mathrm{d}s\}^{v_1} \{\int_\Omega |v(s)|^{1/v_2} \mathrm{d}s\}^{v_2} \{\int_\Omega |w(s)|^{1/v_3} \mathrm{d}s\}^{v_3} \tag{7.15}$$

$$(v_i \geqq 0,\ v_1 + v_2 + v_3 = 1),$$

in which

$$u(s) = |K(t, s)|^{1-v},\ v(s) = |K(t, s)|^v |x(s)|^{1-b},\ w(s) = |x(s)|^b$$

and

$$v_1 = r(1 - v),\ v_2 = \theta,\ v_3 = b\alpha.$$

We obtain as a result the inequality

$$|\boldsymbol{K}x(t)|$$

$$\leqq \{\int_{\Omega} |K(t,s)|^{1/r} \mathrm{d}s\}^{r(1-\nu)} \cdot \{\int_{\Omega} |K(t,s)|^{\nu/\theta} |x(s)|^{(1-b)/\theta} \mathrm{d}s\}^{\theta} \times$$

$$\times \{\int_{\Omega} |x(s)|^{1/\alpha} \mathrm{d}s\}^{b\alpha}$$

$$= \varphi(t)^{1-\nu} [\int_{\Omega} |K(t,s)|^{\nu/\theta} |x(s)|^{(1-b)/\theta} \mathrm{d}s]^{\theta} \cdot \|x\|_{\alpha}^{b}.$$

This in turn implies an estimate for $\|\boldsymbol{K}x\|_{\beta}$:

$$\|\boldsymbol{K}x\|_{\beta} \leqq \{\int_{\Omega^*} \varphi(t)^{(1-\nu)/\theta} [\int_{\Omega} |K(t,s)|^{\nu/\theta} |x(s)|^{(1-b)/\theta} \mathrm{d}s] \mathrm{d}t\}^{\theta} \|x\|_{\alpha}^{b}. \tag{7.16}$$

The order of integration on the right side can be changed by appeal to Fubini's theorem. Hence from (7.16) it follows that

$$\|\boldsymbol{K}x\|_{\beta} \leqq \{\int_{\Omega} |x(s)|^{(1-b)/\theta} [\int_{\Omega^*} |K(t,s)|^{\nu/\theta} \varphi(t)^{(1-\nu)/\theta} \mathrm{d}t] \mathrm{d}s\}^{\theta} \|x\|_{\alpha}^{b},$$

or, equivalently

$$\|\boldsymbol{K}x\|_{\beta} \leqq \{\int_{\Omega} |x(s)|^{(1-b)/\theta} \psi(s)^{1/\theta} \mathrm{d}s\}^{\theta} \|x\|_{\alpha}^{b}. \tag{7.17}$$

To complete the estimates of $\|\boldsymbol{K}x\|_{\beta}$, use the form of Hölder's inequality:

$$|\int_{\Omega} u(s) v(s) \mathrm{d}s| \leqq \{\int_{\Omega} |u(s)|^{1/\nu_1} \mathrm{d}s\}^{\nu_1} \{\int_{\Omega} |v(s)|^{1/\nu_2} \mathrm{d}s\}^{\nu_2} \tag{7.18}$$

$$(\nu_i \geqq 0;\ \nu_1 + \nu_2 = 1),$$

in which

$$u(s) = |x(s)|^{(1-b)/\theta}, \quad v(s) = \psi(s)^{1/\theta}$$

and

$$\nu_1 = \frac{\alpha(1-b)}{\theta}, \quad \nu_2 = \frac{q^*}{\theta}.$$

We arrive at the final estimate

$$\|Kx\|_\beta \leqq \|\psi(s)\|_{q^*}\|x\|_\alpha.$$

The assertions of the theorem on continuity and the estimate (7.13) have been proved.

Compactness of the operator K follows from Theorem 5.8 and the estimate

$$\|KP_D\|_{\alpha\to\beta} \leqq \|P_D\psi\|_{q^*},$$

which is proved exactly as the estimate (7.13). The theorem has been proved.

Assume now that condition (7.14) is not fulfilled, i.e. $q^* = 0$. In this case, as the following example shows, the integral operator may not possess the property of compactness as an operator from L_α to L_β. Put

$$K(t, s) = \begin{cases} 2^{n(\theta+r-\nu r)}, & \text{if } 1/2^n \leqq t,\ s < 1/2^{n-1} \\ 0 & \text{for other } t,\ s. \end{cases} \tag{7.19}$$

Here $\Omega^* = \Omega = [0, 1]$, and numbers ν, r, θ are fixed and satisfy the relations

$$0 \leqq \nu \leqq 1,\ \ \theta \geqq 0,\ \ (1-\nu)r + \theta \leqq 1.$$

The function $\varphi(t)$ in this case is given by the relation

$$\varphi(t) = 2^{n(\theta-\nu r)},\ \text{ if }\ 1/2^n \leqq t < 1/2^{n-1}.$$

Hence

$$\|\,|K(t, s)|^\nu|\varphi(t)|^{1-\nu}\|_\theta \equiv 1 \qquad (s \in [0, 1])$$

that is,

$$\psi(s) = \|\,|K(t, s)|^\nu|\varphi(t)|^{1-\nu}\|_\theta \in L_0.$$

The conditions of Theorem 7.3 are fulfilled and the operator K with kernel (7.19) acts from L_α, $\alpha = 1 - r(1-\nu)$, to L_β, $\beta = \theta$. However this operator is not compact because it transforms the bounded sequence of functions in L_α given by

$$x_n(s) = \begin{cases} 2^{n\alpha}, & \text{if } 1/2^n \leqq s < 1/2^{n-1} \\ 0 & \text{for other } s \in [0, 1] \end{cases}$$

to the sequence of functions

$$\boldsymbol{K}x_n(t) = \begin{cases} 2^{n\beta}, & \text{if } 1/2^n \leqq t < 1/2^{n-1} \\ 0 & \text{for other } t \in [0, 1], \end{cases}$$

which is obviously not compact in L_β.

Theorem 7.2 and theorems on adjoint operators imply one more proposition on conditions of continuity and compactness for integral operators.

THEOREM 7.4: *Let a kernel $K(t, s)$ satisfy the condition*

$$\varphi(t) = \| |K(t, s)|^\nu |\psi(s)|^{1-\nu} \|_{\theta^*} \in L_q, \tag{7.20}$$

where

$$\psi(s) = \|K(t, s)\|_{r^*}, \tag{7.21}$$

and suppose the following holds:

$$\theta^* \geqq q, \ (1 - \nu)r^* + \theta^* \leqq 1. \tag{7.22}$$

Then the integral operator $\boldsymbol{K}$ with kernel $K(t, s)$ is a continuous operator acting from L_α to L_β, where

$$\alpha = 1 - \theta^*, \ \beta = (1 - \nu)r^* + q, \tag{7.23}$$

and

$$\|\boldsymbol{K}\|_{\alpha\to\beta} \leqq \|\varphi(t)\|_q. \tag{7.24}$$

If further

$$q > 0,$$

then $\boldsymbol{K}$ *is a compact operator acting from* L_α *to* L_β.

7.3 *Lemmas*

In this section a class of integral operators $\boldsymbol{K}$ is studied for which the kernels $K(t, s)$ satisfy the following two conditions simultaneously:

$$\varphi(t) = \|K(t, s)\|_r \in L_q, \tag{7.25}$$

$$\psi(s) = \| |K(t, s)|^\nu |\varphi(t)|^{1-\nu}\|_\theta \in L_p \quad (0 < \nu \leqq 1). \tag{7.26}$$

If at some point $s_0 \in \Omega$[1] the following inequality holds for almost all $t \in \Omega^*$

$$|K(t, s)| \geqq c_0 > 0,$$

then (7.26) implies that $\varphi(t) \in L_{\theta/(1-\nu)}$. Hence condition (7.25) is a consequence of the condition (7.26) if $\theta \leqq (1 - \nu)q$.

In those cases where the conditions (7.25) and (7.26) yield independent criteria for regularity (or compactness) of $\boldsymbol{K}$ as operators from L_{α_0} to L_{β_0} and from L_{α_1} to L_{β_1} respectively, interpolation theorems can be applied to $\boldsymbol{K}$ to obtain additional points belonging to the L-characteristic $L(\boldsymbol{K}; \text{reg.})$ (or $L(\boldsymbol{K}; \text{comp.})$). In other cases it is convenient to consider the conditions (7.25) and (7.26) simultaneously.

Lemma 7.1: *Let a function* $K(t, s)$ *satisfy the conditions* (7.25) *and* (7.26), *with* $\theta \geqq p$. *Let* τ *be a number satisfying:*

$$1 - \nu \leqq \tau \leqq 1, \tag{7.27}$$

$$\tau r + \frac{(1 - \tau)}{\nu} \theta \leqq 1. \tag{7.28}$$

[1] Translator's note. It should be changed as: 'on some set of positive measure'.

Then the integral operator $\boldsymbol{K}$ *with the kernel* $K(t, s)$ *acts from* $L_{\alpha(\tau)}$ *to* $L_{\beta(\tau)}$, *where*

$$\alpha(\tau) = 1 - \tau r - \frac{1-\tau}{v} p, \quad \beta(\tau) = \frac{1-\tau}{v} \theta + \frac{\tau + v - 1}{v} q, \tag{7.29}$$

and the following inequality holds:

$$\|\boldsymbol{K}\|_{\alpha(\tau) \to \beta(\tau)} \leqq \|\varphi\|_q^{(\tau+v-1)/v} \|\psi\|_p^{(1-\tau)/v}. \tag{7.30}$$

The proof of this lemma is similar to the proof of Theorem 7.3. Consider the number

$$b = \frac{1 - \tau r - \dfrac{1-\tau}{v}\theta}{1 - \tau r - \dfrac{1-\tau}{v} p} = \frac{1}{\alpha(\tau)}\left(1 - \tau r - \frac{1-\tau}{v}\theta\right).$$

Inequality (7.28) implies that

$$1 - \tau r - \frac{1-\tau}{v}\theta \geqq 0;$$

hence

$$1 - \tau r - \frac{1-\tau}{v} p \geqq 1 - \tau r - \frac{1-\tau}{v}\theta \geqq 0.$$

Hence it follows that

$$0 \leqq b \leqq 1.$$

It is clear that

$$|\boldsymbol{K}x(t)| \leqq \int_\Omega |K(t, s)|^\tau |K(t, s)|^{1-\tau} |x(s)|^{1-b} |x(s)|^b ds. \tag{7.31}$$

Apply to the right side Hölder's inequality in the form (7.15), in which

$$u(s) = |K(t, s)|^{\tau}, \; v(s) = |K(t, s)|^{1-\tau}|x(s)|^{1-b}$$

$$w(s) = |x(s)|^{b},$$

$$\nu_1 = \tau r, \; \nu_2 = \frac{(1-\tau)\theta}{\nu}, \; \nu_3 = b\alpha(\tau).$$

This yields the estimate

$$|\boldsymbol{K}x(t)| \leqq \{\int_{\Omega} |K(t, s)|^{1/r} \mathrm{d}s\}^{\tau r}$$

$$\times \{\int_{\Omega} |K(t, s)|^{\nu/\theta} |x(s)|^{(\nu(1-b))/\theta(1-\tau)} \mathrm{d}s\}^{((1-\tau)\theta)/\nu} \{\int_{\Omega} |x(s)|^{1/\alpha(\tau)} \mathrm{d}s\}^{b\alpha(\tau)}$$

$$= \varphi(t)^{\tau} [\int_{\Omega} |K(t, s)|^{\nu/\theta} |x(s)|^{(\nu(1-b))/\theta(1-\tau)} \mathrm{d}s]^{((1-\tau)\theta)/\nu} \|x\|^{b}_{\alpha(\tau)}.$$

Further we obtain the following estimate for $\|\boldsymbol{K}x\|_{\beta(\tau)}$:

$$\|\boldsymbol{K}x\|_{\beta(\tau)}$$

$$\leqq \{\int_{\Omega^*} \varphi(t)^{\tau/\beta(\tau)} [\int_{\Omega} |K(t, s)|^{\nu/\theta} |x(s)|^{(\nu(1-b))/\nu\beta(\tau)} \mathrm{d}s]^{((1-\tau)\theta)/\nu\beta(\tau)} \mathrm{d}t\}^{\beta(\tau)} \|x\|^{b}_{\alpha(\tau)}. \tag{7.32}$$

Put

$$c = \frac{1-\nu}{\nu} \cdot \frac{1-\tau}{\tau}.$$

Obviously (7.27) implies that $0 \leqq c \leqq 1$ and

$$\frac{\beta(\tau)}{q\tau(1-c)} = 1 + \frac{\theta(1-\tau)}{q(\tau+\nu-1)} \geqq 1.$$

Rewrite inequality (7.32) in the form

$$\|Kx\|_{\beta(\tau)} \leqq \{\int_{\Omega^*} \varphi(t)^{(\tau(1-c))/\beta(\tau)} \cdot \varphi(t)^{\tau c/\beta(\tau)}$$

$$\times [\int_{\Omega} |K(t,s)|^{\nu/\theta} |x(s)|^{(\nu(1-b))/\theta(1-\tau)} \mathrm{d}s]^{(\theta(1-\tau))/\nu\beta(\tau)} \mathrm{d}t\}^{\beta(\tau)} \|x\|^b_{\alpha(\tau)}. \tag{7.33}$$

Apply to the inner integral in (7.33) Hölder's inequality in the form

$$|\int_{\Omega^*} u(t)v(t)\mathrm{d}t| \leqq \{\int_{\Omega^*} |u(t)|^{1/\nu_1}\mathrm{d}t\}^{\nu_1}\{\int_{\Omega^*} |v(t)|^{1/\nu_2}\mathrm{d}t\}^{\nu_2}$$

$$(\nu_i \geqq 0,\ \nu_1 + \nu_2 = 1),$$

in which

$$u(t) = \varphi(t)^{(\tau(1-c))/\beta(\tau)},$$

$$v(t) = \varphi(t)^{\tau c/\beta(\tau)}[\int_{\Omega} |K(t,s)|^{\nu/\theta} |x(s)|^{(\nu(1-b))/\theta(1-\tau)} \mathrm{d}s]^{(\theta(1-\tau))/\nu\beta(\tau)}$$

and

$$\nu_1 = \frac{q\tau(1-c)}{\beta(\tau)},\quad \nu_2 = \frac{\theta(1-\tau)}{\nu\beta(\tau)}.$$

This yields

$$\|Kx\|_{\beta(\tau)} \leqq \{\int_{\Omega^*} \varphi(t)^{1/q}\mathrm{d}t\}^{q\tau(1-c)}$$

$$\times \{\int_{\Omega^*} \varphi(t)^{\tau c\nu/\theta(1-\tau)}[\int_{\Omega} |K(t,s)|^{\nu/\theta} |x(s)|^{(\nu(1-b))/\theta(1-\tau)} \mathrm{d}s]\mathrm{d}t\}^{\theta(1-\tau)/\nu} \|x\|^b_{\alpha(\tau)}$$

$$= \|\varphi\|_q^{\tau(1-c)}\{\int_{\Omega^*} \varphi(t)^{(1-\nu)/\theta}$$

$$\times [\int_{\Omega} |K(t,s)|^{\nu/\theta} |x(s)|^{(\nu(1-b))/\theta(1-\tau)} \mathrm{d}s]\mathrm{d}t\}^{\theta(1-\tau)/\nu} \|x\|^b_{\alpha(\tau)}.$$

Changing the order of integration on the right side, we obtain

$$\|Kx\|_{\beta(\tau)} \leqq \|\varphi\|_q^{\tau(1-c)}\{\int_{\Omega} |x(s)|^{(\nu/\theta)\cdot(1-b)/(1-\tau)}$$

$$\times [\int_{\Omega^*} |K(t,s)|^{\nu/\theta} \cdot \varphi(t)^{(1-\nu)/\theta}\mathrm{d}t]\mathrm{d}s\}^{\theta(1-\tau)/\nu} \|x\|^b_{\alpha(\tau)}$$

or equivalently

$$\|\boldsymbol{K}x\|_{\beta(\tau)} \leqq \|\varphi\|_q^{\tau(1-c)}\{\int_\Omega |x(s)|^{(v/\theta)\cdot(1-b)/(1-\tau)}\psi(s)^{1/\theta}\mathrm{d}s\}^{\theta(1-\tau)/v}\|x\|_{\alpha(\tau)}^b. \tag{7.34}$$

The final estimate can be obtained from (7.34). To this end, apply to the integral on the right side of (7.34) Hölder's inequality in the form (7.18), putting

$$u(s) = |x(s)|^{(v/\theta)\cdot(1-b)/(1-\tau)}, \quad v(s) = \psi(s)^{1/\theta}$$

and

$$v_1 = \frac{\alpha(\tau)v(1-b)}{\theta(1-\tau)}, \quad v_2 = \frac{p}{\theta}.$$

This yields

$$\|\boldsymbol{K}x\|_{\beta(\tau)} \leqq \|\varphi\|_q^{\tau(1-c)}\|\psi(s)\|_p^{(1-\tau)/v}\cdot\|x\|_{\alpha(\tau)}. \tag{7.35}$$

Now (7.35) implies the inequality (7.30). The lemma has been proved.

Note that it is possible to prove the following inequality by means of the same arguments:

$$\|\boldsymbol{P}_{D^*}\boldsymbol{K}\boldsymbol{P}_D\|_{\alpha(\tau)\to\beta(\tau)} \leqq \|\boldsymbol{P}_{D^*}\varphi\|_q^{\tau(1-c)}\cdot\|\boldsymbol{P}_D\psi\|_p^{(1-\tau)/v}. \tag{7.36}$$

LEMMA 7.2: *Let the conditions of Lemma* 7.1 *be fulfilled and suppose one of the following conditions holds:*

a) $\tau < 1,\ p > 0,$

b) $1 - v < \tau,\ q > 0,\ \tau r + p\dfrac{1-\tau}{v} > 0.$

Then $\boldsymbol{K}$ *is compact as an operator from* $L_{\alpha(\tau)}$ *to* $L_{\beta(\tau)}$.

PROOF: Under the conditions of the lemma the following inequalities are clear:

$$\alpha(\tau) < 1, \quad \beta(\tau) > 0.$$

If $\beta(\tau) \geqq 1$, the compactness of the operator $\boldsymbol{K}$ results from Theorem 5.8.

Let $\beta(\tau) < 1$. If condition a) is fulfilled, (7.36) implies the relation

$$\lim_{\text{mes } D \to 0} \|\boldsymbol{K}\boldsymbol{P}_D\|_{\alpha(\tau) \to \beta(\tau)} = 0.$$

If the condition b) is fulfilled, (7.36) implies the relation

$$\lim_{\text{mes } D^* \to 0} \|\boldsymbol{P}_{D^*}\boldsymbol{K}\|_{\alpha(\tau) \to \beta(\tau)} = 0.$$

In both cases compactness of the operator $\boldsymbol{K}$ results from Theorem 5.6. The lemma has been proved.

7.4 *Applications of theorems on adjoint operators*

In determining the L-characteristics $L(\boldsymbol{K};$ reg.) and $L(\boldsymbol{K};$ reg. and comp.) of a linear integral operator

$$\boldsymbol{K}x(t) = \int_{\Omega} K(t, s)x(s)\,\mathrm{d}s$$

it is often convenient to examine first the transposed operator

$$\boldsymbol{K}^{\#}y(s) = \int_{\Omega^*} K(t, s)y(t)\,\mathrm{d}t.$$

Let a point $\{\alpha_0, \beta_0\}$ belong to the L-characteristic $L(\boldsymbol{K}^{\#};$ reg.) (or the L-characteristic $L(\boldsymbol{K}^{\#};$ reg. and comp.)). Then the point $\{1-\beta_0, 1-\alpha_0\}$ belongs to the L-characteristic $L(\boldsymbol{K};$ reg.) (or the L-characteristic $L(\boldsymbol{K};$ reg. and comp.)). These considerations have already been used more than once (see, for instance, 5.1°, 5.2° and 6.2°). It follows that to each criterion for regularity (or compactness) of an integral operator $\boldsymbol{K}$ there corresponds a dual criterion obtained by information on the transposed operator.

In the following we formulate lemmas dual to Lemma 7.1 and 7.2 in the above sense. In these lemmas we consider a class of integral operators with kernels $K(t, s)$, satisfying the conditions

$$\varphi^*(s) = \|K(t, s)\|_{r^*} \in L_{q^*}, \tag{7.37}$$

$$\psi^*(t) = \| |K(t, s)|^\nu \varphi^*(s)^{1-\nu}\|_{\theta^*} \in L_{p^*} \quad (0 < \nu \leqq 1). \tag{7.38}$$

LEMMA 7.3: *Let a function $K(t, s)$ satisfy the conditions (7.37) and (7.38), with $\theta^* \geqq p^*$. Let τ be a number satisfying the inequalities*

$$1 - \nu \leqq \tau \leqq 1 \tag{7.39}$$

$$\tau r^* + \frac{1 - \tau}{\nu}\theta^* \leqq 1, \tag{7.40}$$

$$\frac{1 - \tau}{\nu}\theta^* + \frac{\tau + \nu - 1}{\nu}q^* \leqq 1. \tag{7.41}$$

Then the integral operator $\boldsymbol{K}$ with kernel $K(t, s)$ acts from $L_{\alpha(\tau)}$ to $L_{\beta(\tau)}$, where

$$\alpha(\tau) = 1 - \frac{1 - \tau}{\nu}\theta^* - \frac{\tau + \nu - 1}{\nu}q^*, \ \beta(\tau) = \tau r^* + \frac{1 - \tau}{\nu}p^*, \tag{7.42}$$

and the following inequality holds:

$$\|\boldsymbol{K}\|_{\alpha(\tau)\to\beta(\tau)} \leqq \|\varphi^*\|_{q^*}^{(\tau+\nu-1)/\nu} \|\psi^*\|_{p^*}^{(1-\tau)/\nu}. \tag{7.43}$$

LEMMA 7.4: *Suppose the conditions of Lemma 7.3 hold. Let one of the following conditions be fulfilled:*

a) $\tau < 1, \ p^* > 0;$

b) $1 - \nu < \tau, \ q^* > 0, \ \tau r^* + \dfrac{1 - \tau}{\nu}p^* > 0.$

Then $\boldsymbol{K}$ is compact as an operator acting from $L_{\alpha(\tau)}$ to $L_{\beta(\tau)}$.

7.5 *Fundamental theorems*

In this section several theorems will be formulated which follow from Lemmas 7.1 to 7.4. They are obtained in the following way.

In the conditions of the above lemmas there is a free parameter τ. This parameter can vary on some interval. Given the quantities v, r, q, θ, p ($v, r^*, q^*, \theta^*, p^*$) the ends of this interval can be determined. This permits us to formulate several simple criteria for regularity and compactness of linear integral operators.

Suppose that a kernel $K(t, s)$ satisfies the conditions (7.25) and (7.26) with $vr \geqq \theta$. In this case inequality (7.28) can be written in the form

$$\tau \leqq \frac{v - \theta}{vr - \theta}.$$

Consequently, a number τ satisfies the conditions of Lemma 7.1, if the following inequality holds:

$$1 - v \leqq \tau \leqq \min\left\{1, \frac{v - \theta}{vr - \theta}\right\}.$$

These inequalities are consistent if

$$1 - v \leqq \frac{v - \theta}{vr - \theta}$$

or equivalently

$$\theta + r(1 - v) \leqq 1.$$

It is also clear that

$$\min\left\{1, \frac{v - \theta}{vr - \theta}\right\} = \begin{cases} 1, & \text{if } r \leqq 1 \\ \dfrac{v - \theta}{rv - \theta}, & \text{if } r \geqq 1. \end{cases}$$

Hence Lemmas 7.1 and 7.2 imply the following assertion.

THEOREM 7.5: *Let the function* $K(t, s)$ *satisfy the conditions*

$$\varphi(t) = \|K(t, s)\|_r \in L_q$$

$$\psi(s) = \| |K(t, s)|^v \varphi(t)^{1-v} \|_\theta \in L_p,$$

and suppose the following inequalities hold:

$$0 < v \leqq 1, \; vr \geqq \theta \geqq p, \; \theta + r(1 - v) \leqq 1.$$

Then for any τ *in the interval* $[1 - v, 1]$ *if* $r \leqq 1$, *or in the interval* $\left[1 - v, \dfrac{v - \theta}{vr - \theta}\right]$, *if* $r > 1$, *the operator* $\boldsymbol{K}$ *acts from* $L_{\alpha(\tau)}$ *to* $L_{\beta(\tau)}$, *where*

$$\alpha(\tau) = 1 - \tau r - \frac{1 - \tau}{v} p, \; \beta(\tau) = \frac{1 - \tau}{v} \theta + \frac{\tau + v - 1}{v} q,$$

and

$$\|\boldsymbol{K}\|_{\alpha(\tau) \to \beta(\tau)} \leqq \|\varphi(t)\|_q^{(\tau + v - 1)/v} \|\psi(s)\|_p^{(1 - \tau)/v}.$$

If one of the following supplementary conditions holds:

a) $p > 0, \; \tau < 1,$

b) $1 - v < \tau, \; q > 0, \; \tau r + \dfrac{1 - \tau}{v} p > 0,$

then $\boldsymbol{K}$ *is compact as an operator from* $L_{\alpha(\tau)}$ *to* $L_{\beta(\tau)}$.

The case $vr < \theta$ is treated similarly. In this case inequality (7.28) can be written in the form

$$\tau \geqq \frac{\theta - v}{\theta - rv}.$$

Hence the conditions of Lemma 7.1 are fulfilled if the number τ satisfies the inequalities

$$\max\left\{1 - v, \frac{\theta - v}{\theta - rv}\right\} \leqq \tau \leqq 1.$$

It is necessary for consistency of these inequalities that one has:

$$\frac{\theta - v}{\theta - rv} \leqq 1,$$

or equivalently

$$r \leqq 1.$$

Then it is clear that

$$\max\left\{1 - v, \frac{\theta - v}{\theta - rv}\right\} = \begin{cases} 1 - v, & \text{if } \theta + r(1 - v) \leqq 1 \\ \dfrac{\theta - v}{\theta - rv}, & \text{if } \theta + r(1 - v) \geqq 1. \end{cases}$$

Lemmas 7.1 and 7.2 imply:

THEOREM 7.6: *Let the function $K(t, s)$ satisfy the conditions*

$$\varphi(t) = \|K(t, s)\|_r \in L_q,$$

$$\psi(s) = \|\,|K(t, s)|^v \varphi(t)^{1-v}\|_\theta \in L_p,$$

and suppose the following inequalities hold:

$$0 < v \leqq 1, \ \theta \geqq p, \ \theta > vr, \ r < 1.$$

Then for any τ in the interval $[1 - v, 1]$, if $\theta + r(1 - v) \leqq 1$, or in the interval $\left[\dfrac{\theta - v}{\theta - rv}, 1\right]$, if $\theta + r(1 - v) \geqq 1$, the operator $\boldsymbol{K}$ *acts from* $L_{\alpha(\tau)}$

to $L_{\beta(\tau)}$, where

$$\alpha(\tau) = 1 - \tau r - \frac{1-\tau}{\nu} p, \ \beta(\tau) = \frac{1-\tau}{\nu}\theta + \frac{r+\nu-1}{\nu} q;$$

moreover one has:

$$\|\boldsymbol{K}\|_{\alpha(\tau)\to\beta(\tau)} \leqq \|\varphi(t)\|_q^{(\tau+\nu-1)/\nu} \|\psi(s)\|_p^{(1-\tau)/\nu}.$$

If one of the following supplementary conditions holds:

a) $p > 0, \ \tau < 1;$

b) $1 - \nu < \tau, \ q > 0, \ \tau r + \dfrac{1-\tau}{\nu} p > 0,$

*then **K** is compact as an operator from $L_{\alpha(\tau)}$ to $L_{\beta(\tau)}$.*

A similar analysis can be made in the case of a kernel $K(t, s)$ satisfying the conditions (7.37) and (7.38). We leave it to the reader to formulate the corresponding theorems.

7.6 *Conditions of 'Kantorovic' type*

In this section we consider the class of integral operators with kernels satisfying the following two conditions:

$$\varphi(t) = \|K(t, s)\|_r \in L_q, \tag{7.44}$$

and

$$\psi(s) = \|K(t, s)\|_{r^*} \in L_{q^*}. \tag{7.45}$$

Introduce the notations

$$\alpha(\tau) = 1 - (1-\tau)r - \tau q^*, \ \beta(\tau) = \tau r^* + (1-\tau)q.$$

In 7.1° it was shown that the operator $\boldsymbol{K}$ with the kernel $K(t, s)$ acts from $L_{\alpha(\tau)}$ to $L_{\beta(\tau)}$ and is continuous (and compact under some supplementary assumption on the numbers r, r^*, q, q^*) for all τ in the interval (0, 1). Here the assumption that numbers r and r^* are not greater than 1 played a decisive role.

In this section the case, in which one of the numbers r or r^* is greater than 1, will be considered. It turns out that in this case similar assertions on continuity and compactness of the operator $\boldsymbol{K}$ are valid, but only for values τ lying in certain parts of the interval (0, 1). These parts are determined by supplementary relations among the numbers r, r^*, q, q^*.

Put

$$d = \begin{cases} \dfrac{1-r}{r^*-r}, & \text{if } r^* \geqq q^*,\ r^* > 1 \geqq r \\[2ex] \dfrac{1-r}{q^*-r}, & \text{if } q^* > r^* > 1 \geqq r \geqq q \\[2ex] \dfrac{r-1}{r-r^*}, & \text{if } r > 1 \geqq r^* \geqq q^* \\[2ex] \dfrac{r-1}{r-q^*}, & \text{if } r^* < q^* \leqq 1 < r,\ r \geqq q \end{cases} \tag{7.46}$$

then clearly $d \in [0, 1]$.

THEOREM 7.7: *Let conditions* (7.44) *and* (7.45) *hold. Then the integral operator* $\boldsymbol{K}$ *with kernel* $K(t, s)$ *acts from* $L_{\alpha(\tau)}$ *to* $L_{\beta(\tau)}$ *and is regular for values of* τ *in the interval* $[0, d]$, *if*

$$r^* > 1 \geqq r, \tag{7.47}$$

and for values of τ *in the interval* $[d, 1]$, *if*

$$r > 1 \geqq r^*. \tag{7.48}$$

Moreover:

$$\|\boldsymbol{K}\|_{\alpha(\tau)\to\beta(\tau)} \leqq \|\varphi\|_q^{1-\tau} \|\psi\|_{q^*}^{\tau}.$$

THEOREM 7.8: *Let the conditions of Theorem 7.7 be fulfilled and suppose one of the following supplementary conditions holds:*

$$q^* > 0, \; q + r^* > 0$$

or

$$q > 0, \; q^* + r > 0.$$

Then $\boldsymbol{K}$ is compact as an operator from $L_{\alpha(\tau)}$ to $L_{\beta(\tau)}$ for $\tau \in [0, d]$, if the inequality (7.47) *is fulfilled, and for $\tau \in [d, 1]$, if the inequality* (7.48) *is fulfilled.*

Both of these theorems follow from Lemmas 7.1–7.4. More precisely, one applies Lemmas 7.1 and 7.2 when $r^* \geqq q^*$, but Lemmas 7.3 and 7.4 when $r \geqq q$.

Note that Theorems 7.7 and 7.8 do not cover all possible relations among the numbers r, r^*, q, q^*. Here we have not considered the cases $r < q$, $r^* < q^*$ and $r \geqq q$, $q^* > 1 > r^*$.

7.7 *Summability of kernels of integral operators*[1,2]

Let an integral operator

$$\boldsymbol{K}x(t) = \int_{\Omega} K(t, s)\, x(s)\, \mathrm{d}s \tag{7.49}$$

with non-negative kernel $K(t, s)$ act from L_α to L_β, i.e. for some C

$$\|\boldsymbol{K}x(t)\|_\beta \leqq C\, \|x\|_\alpha \quad (x \in L_\alpha). \tag{7.50}$$

We are interested in whether this implies that some power of the kernel $K(t, s)$ is summable in the joint variables $t \in \Omega^*$, $s \in \Omega$.

In the preceding sections, what is essentially a converse problem was

[1] The problem of summability of kernels of integral operators was considered in the papers of D. V. Salefov [1] and E. P. Pustylnik [6].

[2] Translator's note. D. V. Salefov [1] is not in the bibliography.

considered—to determine, from summability properties of the kernel $K(t, s)$, from which space L_α to which space L_β the operator $\boldsymbol{K}$ acts.

We confine ourselves to operators acting from L_α to L_β for $0 \leqq \alpha, \beta \leqq 1$. In this case inequality (7.50) can be written in the form

$$|\int_{\Omega^*}\int_{\Omega} K(t, s)x(s)y(t)\mathrm{d}s\,\mathrm{d}t| \leqq C\|x\|_\alpha\|y\|_{1-\beta}. \tag{7.51}$$

This implies at once that the kernel $K(t, s)$ is summable. It turns out however that the kernel $K(t, s)$ may not be summable to any higher power.

Consider, for instance, the integral operator $\boldsymbol{K}_0$ with kernel

$$K_0(t, s) = \frac{1}{|t-s|\ln^2\left|\dfrac{t-s}{2}\right|}$$

(here $\Omega = \Omega^* = [0, 1]$). Since the kernel $K_0(t, s)$ is symmetric and satisfies the condition

$$\varphi(t) = \|K_0(t, s)\|_1 \in L_0,$$

it follows by Theorem 7.1 that the operator $\boldsymbol{K}_0$ acts from any L_α to any L_β whenever $\beta \geqq \alpha$. But this function $K_0(t, s)$ is not summable to any power $r > 1$.

Let us consider in more detail the case of an operator $\boldsymbol{K}$ which acts from L_α to L_β, where $\alpha > \beta$. In this case under some restriction on the Lebesgue sets

$$\hat{\Omega}(h) = \{\{t, s\} : |K(t, s)| \geqq h\}$$

it will be proved that the kernel is summable to some power $r > 1$. In the sequel it is assumed that numbers α and β ($\alpha > \beta$) are fixed.

Let us introduce some new notions. A set $\hat{D} \subset \Omega^* \times \Omega$ is called a *rectangle with sides a, b*, if

$$\hat{D} = D^* \times D, \ D^* \subset \Omega^*, \ D \subset \Omega$$

and $\operatorname{mes}(D) = a$, $\operatorname{mes}(D^*) = b$. It is clear that the measure of a rectangle $\hat{D}$

is equal to ab. The quantity

$$\chi_{\alpha,\beta}(\hat{D}) = a^{\alpha} b^{1-\beta}$$

is called the *quasi-measure* of the rectangle $\hat{D}$.

Let $\hat{Q}$ be an arbitrary measurable subset of $\Omega^* \times \Omega$. Consider an arbitrary countable covering of the set $\hat{Q}$ consisting of rectangles $\hat{D}_n$ with sides a_n, b_n. Define the quasi-measure of the set $\hat{Q}$ by the formula

$$\chi_{\alpha,\beta}(\hat{Q}) = \inf \sum_{n=1}^{\infty} a_n^{\alpha} b_n^{1-\beta},$$

where the infimum is taken over all such coverings.

An exact calculation of the quasi-measure of even relatively simple sets is often difficult, but in the following it suffices to know only upper estimates of $\chi_{\alpha,\beta}(\hat{Q})$.

Consider, for instance, the set hatched in Fig. 7.2 (here $\Omega^* = \Omega = [0, 1]$). For the quasi-measure of this set the following estimate is valid:

$$\chi_{\alpha,\beta}(\hat{Q}) \leqq C(\text{mes}\, \hat{Q})^{\alpha-\beta}. \tag{7.52}$$

To see this, it suffices to cover the set Q by squares with sides $a = 1 - \sqrt{1 - \text{mes}\, \hat{Q}}$, as is shown in Fig. 7.2.

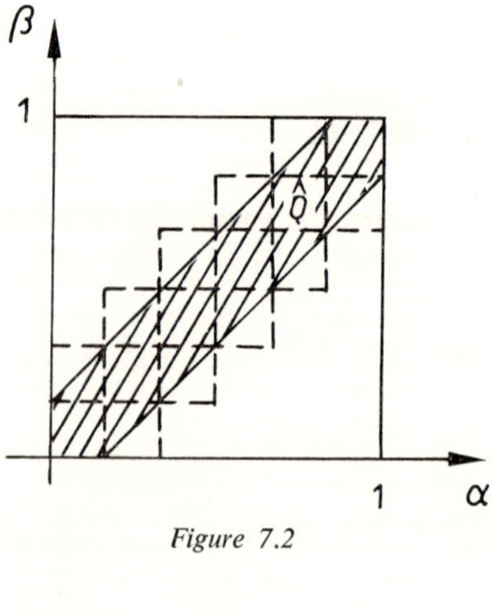

Figure 7.2

THEOREM 7.9: *Let an integral operator* $\boldsymbol{K}$ *with a non-negative kernel* $K(t, s)$ *act from* L_α *to* L_β *where* $\alpha > \beta$. *Suppose the quasi-measures* $\chi_{\alpha,\beta}[\hat{\Omega}(h)]$ *of Lebesgue sets of the kernel* $K(t, s)$ *satisfy the inequality*

$$\chi_{\alpha,\beta}[\hat{\Omega}(h)] \leqq C_0[\text{mes } \hat{\Omega}(h)]^k, \tag{7.53}$$

where $k \in (0, 1)$. *Then the kernel* $K(t, s)$ *is summable in its joint variables to any power* $r < r_0$, *where*

$$r_0 = \frac{1}{1-k}.$$

PROOF: Let us show that

$$\text{mes}[\hat{\Omega}(h)] \leqq Ch^{-r_0}.$$

Then the assertion of the theorem will follow from (2.43).

Let $\hat{D}_1, \hat{D}_2, \ldots, \hat{D}_n, \ldots$ be an arbitrary covering of the set $\hat{\Omega}(h)$ by rectangles. Then (7.51) implies that

$$\int_{\hat{D}_i}\int K(t, s)\,ds\,dt \leqq \|K\|_{\alpha\to\beta} a_i^\alpha b_i^{1-\beta} \quad (i = 1, 2, \ldots)$$

where a_i, b_i are the sides of rectangles $\hat{D}_i$. Hence

$$\int_{\hat{\Omega}(h)}\int K(t, s)\,ds\,dt \leqq \|K\|_{\alpha\to\beta}\chi_{\alpha,\beta}[\hat{\Omega}(h)]$$

and by (7.53)

$$\int_{\hat{\Omega}(h)}\int K(t, s)\,ds\,dt \leqq C_1[\text{mes } \hat{\Omega}(h)]^k.$$

But on the other hand,

$$\int_{\hat{\Omega}(h)}\int K(t, s)\,ds\,dt \geqq h\cdot\text{mes } \hat{\Omega}(h),$$

whence it follows that

$$\text{mes } \hat{\Omega}(h) \leqq Ch^{1/(k-1)} = Ch^{-r_0}.$$

The theorem has been proved.

Consider an example. Let $\Omega = \Omega^* = [0, 1]$. Assume that a function $K(t, s)$ has the form

$$K(t, s) = k(|t - s|),$$

where $k(u)$ is a monotone decreasing function defined for $0 \leqq u \leqq 1$. Obviously the Lebesgue sets of this function have the form shown in Fig. 7.2. Hence inequality (7.52) is valid. Theorem 7.9 then implies that the kernel $K(t, s)$ is summable to any power r which satisfies the inequality

$$r < \frac{1}{1 - \alpha + \beta},$$

if the operator $\boldsymbol{K}$ with kernel $K(t, s)$ acts from L_α to L_β.

§ 8 Operators of potential type[1]

8.1 *Definitions*

In the sequel, denote by $|t - s|$, the distance between points t and s in the n-dimensional Euclidean space R_n.

In this paragraph we investigate integral operators

$$A_\lambda x(t) = \int_\Omega K_\lambda(t, s) x(s) \, ds \tag{8.1}$$

---

[1] Homogeneous operators of potential type were studied in detail in connection with theory of Fourier series, fractional differentiation etc. (see the satisfactorily complete bibliography in the monograph of G. Hardy, D. Littlewood and K. G. Polya [1]). Basic work on the theory of multidimensional operators of potential type is due to S. L. Sobolev [1, 2]. He clarified the role of such operators in the theory of spaces of differentiable functions and in boundary value problems of mathematical physics. The first theorems on compactness of operators of potential type were indicated by V. I. Kondrashov. Important results in the theory of operators of potential type are due to V. P. Ilin [1–3] and other authors.

The technique used in this paragraph in applying the theorems of Stein-Weiss [1] to study some classes of operators of potential type is pointed out in the paper of S. G. Krein and E. M. Semenov [1]. The proof of Ilin's theorem, presented in 8.8, seems to be new.

with the special kernels

$$K_\lambda(t, s) = |t - s|^{-\lambda}. \tag{8.2}$$

Here Ω is a bounded set of the space R_n, with non-zero Lebesgue measure. Operators of the form (8.1) are called *potential*[1]; sometimes the number λ is called the *exponent* of the potential (or exponent of the operator A_λ).

Our main interest is in determining the L-characteristics $L(A_\lambda;\text{ cont.})$ and $L(A_\lambda;\text{ comp.})$ of the potentials A_λ.

The operator

$$A_{(\ln)}x(t) = \int_\Omega K_{(\ln)}(t, s)x(s)\,ds \tag{8.3}$$

with kernel

$$K_{(\ln)}(t, s) = |\ln|t - s|| \tag{8.4}$$

is called the *logarithmic potential.*

Immediate generalizations of the operators (8.1) are the operators of *potential type*

$$Ax(t) = \int_\Omega \frac{Q(t, s)}{|t - s|^\lambda}\,x(s)\,ds. \tag{8.5}$$

Here $Q(t, s)$ is, as a rule, a bounded function. Note that boundary problems for equations of elliptic type are reduced to integral equations with these operators.

8.2 *Simplest theorems on continuity and compactness of potentials*

Consider the function

$$\varphi_r(t) = \|K_\lambda(t, s)\|_r = \{\int_\Omega |t - s|^{-\lambda/r}\,ds\}^r. \tag{8.6}$$

This function is considered to be defined for all points $t \in R_n$.

---

[1] If $n = 3$, $\lambda = 1$, then the right side of (8.1) represents the potential, at a point t, of charge distributed on the domain Ω with density $x(s)$.

LEMMA 8.1: *If*

$$\lambda/n < r < \infty, \tag{8.7}$$

then the function (8.6) *is bounded on* R_n.

PROOF: The following inequality is obvious:

$$\int_{\Omega} |t - s|^{-\lambda/r} \mathrm{d}s \leqq \int_{T(t,R)} |t - s|^{-\lambda/r} \mathrm{d}s,$$

where $T(t, R)$ is the sphere with center at t and radius R, equal to the diameter of the set Ω. The integral J on the right side of the above inequality does not depend on t and is easily calculated by changing to polar coordinates:

$$J = \int_{T(t,R)} |t - s|^{-\lambda/r} \mathrm{d}s = v_n \int_0^R \rho^{n-\lambda/r-1} \mathrm{d}\rho = \frac{rv_n}{nr - \lambda} R^{n-\lambda/r},$$

where v_n is the area of the unit surface in R_n. Thus

$$\int_{\Omega} |K_\lambda(t, s)|^{1/r} \mathrm{d}s \leqq \frac{rv_n}{nr - \lambda} R^{n-\lambda/r}. \tag{8.8}$$

The lemma has been proved.

The symmetry of the kernel $K_\lambda(t, s)$ together with Lemma 8.1 and Theorem 7.2 imply the basic part of the assertion of the following theorem.

THEOREM 8.1: *Let* $0 < \lambda < n$. *Then the potential is compact as an operator acting from* $L_\alpha = L_\alpha(\Omega)$ *to* $L_0 = L_0(\Omega)$ *whenever*

$$0 \leqq \alpha < 1 - \lambda/n, \tag{8.9}$$

and as an operator from $L_\alpha = L_\alpha(\Omega)$ *to* $L_\beta = L_\beta(\Omega)$ *whenever*

$$1 - \lambda/n < \alpha \leqq 1, \ \beta > \alpha - 1 + \lambda/n.$$

The proof requires only that we verify that A_λ is compact as an operator from L_α to L_0 when α satisfies the inequality (8.9) and as an operator from L_1 to L_β when $\beta > \lambda/n$.

Compactness of A_λ as an operator from L_α $(0 \leqq \alpha < 1 - \lambda/n)$ to L_0 follows immediately from Theorem 6.4 because the range of the vector-function

$$w(t) = |t - s|^{-\lambda} \qquad (t \in \Omega)$$

in the space $L_{1-\alpha}$ forms a compact set (that the range of $w(t)$ for $t \in \Omega$ is compact in measure is obvious; equi-absolute continuity of $w(t)$ in $L_{1-\alpha}$ follows from the fact that $w(t) \in L_\delta$, where δ is some number in the interval $(\lambda/n, 1 - \alpha)$).

Compactness of A_λ as an operator from L_1 to L_β, $\beta > \lambda/n$, follows from symmetry of the kernel $K_\lambda(t, s)$ and Theorem 5.1.

The assertion of Theorem 8.1 means that the L-characteristics $L(A_\lambda;$ cont.) and $L(A_\lambda;$ comp.) of the potential A_λ contain all points $\{\alpha, \beta\}$, for which the following inequalities hold:

$$0 \leqq \alpha \leqq 1, \ \beta \geqq 0, \ \beta > \alpha - 1 + \lambda/n \tag{8.11}$$

(Fig. 8.1). It turns out that the L-characteristic $L(A_\lambda;$ comp.) of the potential A_λ does not contain any other point. The L-characteristic $L(A_\lambda;$ cont.) does contain other points—in 8.4° it will be shown that all interior points of the segment joining the points $\{1 - \lambda/n, 0\}$ and $\{1, \lambda/n\}$ also belong to this L-characteristic.

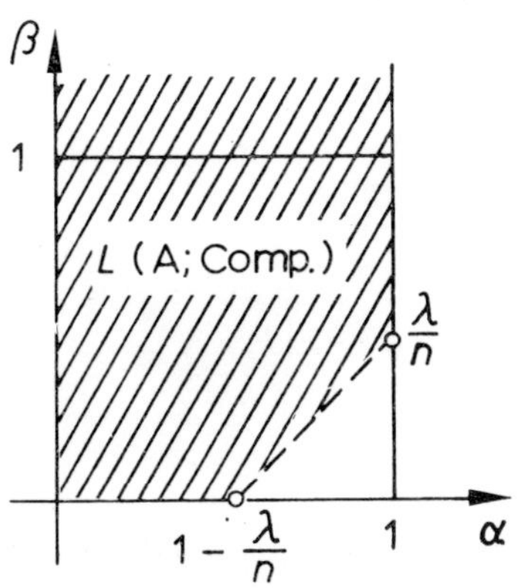

Figure 8.1

8.3 *Interpolation theorem of Stein-Weiss*

In section 2.7° the interpolation theorem of Marcinkiewicz was proved. To prove further theorems on potentials we need an essential improvement of this theorem, due to Stein and Weiss.

Recall (see 2.7°) that $\lambda(x; h)$ denotes the measure of the set of points $t \in \Omega$, at which $|x(t)| \geqq h$, and M_α $(0 < \alpha \leqq 1)$ denotes the linear space of functions $x(t)$ with finite quasi-norm:

$$\|x\|^*_{M_\alpha} = \sup_{0<h<\infty} h \cdot \lambda(x; h)^\alpha, \tag{8.12}$$

while M_0 denotes the space L_0, and

$$\|x\|^*_{M_0} = \|x\|_{L_0}. \tag{8.13}$$

Let $0 < \alpha < 1$. Put

$$\|x\|_{M_\alpha} = \sup_{D \subset \Omega} \left\{ \frac{1}{(\text{mes } D)^{1-\alpha}} \int_D |x(t)| \, dt \right\} \tag{8.14}$$

and let us show that the class of functions for which $\|x\|_{M_\alpha} < \infty$ coincides with M_α.

Let $\|x\|_{M_\alpha} < \infty$, i.e. for each D

$$\int_D |x(t)| \, dt \leqq \|x\|_{M_\alpha} \cdot (\text{mes } D)^{1-\alpha}.$$

Then, in particular,

$$\int_{\{t: |x(t)| \geqq h\}} |x(t)| \, dt \leqq \|x\|_{M_\alpha} \cdot \lambda(x; h)^{1-\alpha}.$$

But, on the other hand,

$$\int_{\{t: |x(t)| \geqq h\}} |x(t)| \, dt \geqq h \cdot \lambda(x; h).$$

Hence

$$h \cdot \lambda(x; h)^{\alpha} \leqq \|x\|_{M_\alpha} \quad (0 < h < \infty) \tag{8.15}$$

and consequently, $x \in M_\alpha$.

Now let $x \in M_\alpha$, i.e. $\|x\|^*_{M_\alpha} < \infty$. Then for any set $D \subset \Omega$

$$\int_D |x(t)|\,\mathrm{d}t \leqq \int_0^{\mathrm{mes}\,D} \tilde{x}(\tau)\,\mathrm{d}\tau$$

where $\tilde{x}(\tau)$ $(0 \leqq \tau \leqq \mathrm{mes}\,D)$ is the monotone decreasing function such that

$$\lambda(x; h) = \lambda(\tilde{x}; h).$$

From (8.12) it follows that

$$|\tilde{x}(\tau)| \leqq \frac{\|x\|^*_{M_\alpha}}{\tau^\alpha}$$

and hence

$$\int_D |x(t)|\,\mathrm{d}t \leqq \|x\|^*_{M_\alpha} \int_0^{\mathrm{mes}\,D} \tau^{-\alpha}\,\mathrm{d}\tau = \frac{\|x\|^*_{M_\alpha}}{1-\alpha}(\mathrm{mes}\,D)^{1-\alpha}, \tag{8.16}$$

whence it follows that $\|x\|_{M_\alpha} < \infty$.

From (8.15) and (8.16) follow the important inequalities:

$$(1-\alpha)\|x\|_{M_\alpha} \leqq \|x\|^*_{M_\alpha} \leqq \|x\|_{M_\alpha}. \tag{8.17}$$

It is easy to see that M_α $(0 < \alpha < 1)$ becomes a Banach space if the norm is defined by the relation (8.14). The space M_α will be called a *Marcinkiewicz space*.

A linear operator A is said to satisfy the condition $\Lambda M(\alpha, \beta)$, if it is defined on all characteristic functions κ_D of measurable sets $D \subset \Omega$ and if

$$\|A\kappa_D\|^*_{M_\beta} \leqq C(\mathrm{mes}\,D)^\alpha, \tag{8.18}$$

where C is some constant not depending on D. The condition $\Lambda M(\alpha, \beta)$ is, of course, fulfilled if A acts from L_α to L_β and is continuous, or if A satisfies the Marcinkiewicz condition $LM(\alpha, \beta)$ (see 2.7°).

THEOREM 8.2: *Let a linear operator A satisfy the conditions $\Lambda M(\alpha_0, \beta_0)$ and $\Lambda M(\alpha_1, \beta_1)$:*

$$\begin{aligned} \|A\kappa_D\|^*_{M_{\beta_0}} &\leqq C_0(\text{mes } D)^{\alpha_0} \quad (D \subset \Omega), \\ \|A\kappa_D\|^*_{M_{\beta_1}} &\leqq C_1(\text{mes } D)^{\alpha_1} \quad (D \subset \Omega) \end{aligned} \tag{8.19}$$

where

$$0 \leqq \beta_0 \leqq \alpha_0 \leqq 1, \; 0 \leqq \beta_1 \leqq \alpha_1 \leqq 1 \tag{8.20}$$

and

$$\beta_0 \neq \beta_1, \; \alpha_0 \neq \alpha_1. \tag{8.21}$$

Then for each $\tau \in (0, 1)$ the operator A acts from the space $L_{\alpha(\tau)}$ to the space $L_{\beta(\tau)}$ where

$$\alpha(\tau) = (1-\tau)\alpha_0 + \tau\alpha_1, \; \beta(\tau) = (1-\tau)\beta_0 + \tau\beta_1, \tag{8.22}$$

and is continuous. Here

$$\|A\|_{L_{\alpha(\tau)} \to L_{\beta(\tau)}} \leqq k(\alpha_0, \beta_0, \alpha_1, \beta_1) C_0^{1-\tau} C_1^{\tau}. \tag{8.23}$$

PROOF: The inequality (2.52) implies that for each $\tau \in (0, 1)$ we have:

$$\|A\kappa_D\|_{L_{\beta(\tau)}} \leqq \frac{1}{|\beta_0 - \beta_1| \cdot \tau(1-\tau)} \|A\kappa_D\|^{*1-\tau}_{M_{\beta_0}} \|A\kappa_D\|^{*\tau}_{M_{\beta_1}},$$

whence by (8.19)

$$\|A\kappa_D\|_{L_{\beta(\tau)}} \leqq \frac{1}{|\beta_0 - \beta_1| \cdot \tau(1-\tau)} C_0^{1-\tau} C_1^{\tau} (\text{mes } D)^{\alpha(\tau)}. \tag{8.24}$$

Let $y(t)$ be a step-function. On account of inequality (8.24) and the fact that $\alpha(\tau) \neq 0$ the additive set function

$$\psi(D) = \int_{\Omega} A\kappa_D(t)y(t)\,dt \tag{8.25}$$

is absolutely continuous; hence by the Radon-Nikodym theorem (see, for instance, N. Dunford and J. T. Schwartz [1]) it can be written in the form

$$\psi(D) = \int_{D} y^*(t)\,dt \quad (D \subset \Omega). \tag{8.26}$$

Define the operator $\boldsymbol{B}$ on step-functions by the relation

$$\boldsymbol{B}y(s) = y^*(s). \tag{8.27}$$

It is obvious that the operator $\boldsymbol{B}$ is linear on the set of step functions.

Let us show that for each step function $y(t)$ the following inequality holds:

$$\|\boldsymbol{B}y\|_{M_{1-\alpha(\tau)}} \leqq \frac{2C_0^{1-\tau}C_1^{\tau}}{|\beta_0 - \beta_1|\tau(1-\tau)} \|y\|_{L_{1-\beta(\tau)}}. \tag{8.28}$$

Let $D \subset \Omega$, then

$$\int_D |\boldsymbol{B}y(s)|\,ds = \int_{D^+} \boldsymbol{B}y(s)\,ds - \int_{D^-} \boldsymbol{B}y(s)\,ds,$$

where D^+ is the set of points $s \in D$ at which $\boldsymbol{B}y(s)$ is non-negative, and $D^- = D - D^+$. By (8.25)–(8.27) the above relation can be written in the form

$$\int_D |\boldsymbol{B}y(s)|\,ds = \int_{\Omega} A\kappa_{D^+}(t)y(t)\,dt - \int_{\Omega} A\kappa_{D^-}(t)y(t)\,dt.$$

Applying Hölder's inequality to each integral on the right side and using the estimate (8.24), we obtain

$$\int_D |\boldsymbol{B}y(s)|\,ds \leqq \frac{2}{|\beta_0 - \beta_1|\tau(1-\tau)} C_0^{1-\tau}C_1^{\tau} \|y\|_{L_{1-\beta(\tau)}} (\text{mes } D)^{\alpha(\tau)},$$

whence follows (8.28).

From (8.28) it follows, in particular, that $\boldsymbol{B}$ can be extended to an operator $\boldsymbol{B}_1$ which acts from each $L_{1-\beta(\tau)}$ $(0 < \tau < 1)$ to the corresponding $M_{1-\alpha(\tau)}$ and is continuous. This means that the operator $\boldsymbol{B}_1$ satisfies the Marcinkiewicz condition $LM[1 - \beta(\tau), 1 - \alpha(\tau)]$ for all $\tau \in (0, 1)$. From (8.28) and (8.17) it follows that

$$\|\boldsymbol{B}_1 y(s)\|_{M_{1-\alpha(\tau)}} \leqq \frac{2}{|\beta_0 - \beta_1|\beta(\tau)\tau(1-\tau)} C_0^{1-\tau} C_1^{\tau} \|y(t)\|_{L_{1-\beta(\tau)}}. \tag{8.29}$$

Now apply Marcinkiewicz's interpolation Theorem 2.9. It implies that $\boldsymbol{B}_1$ acts from $L_{1-\beta(\tau)}$ to $L_{1-\alpha(\tau)}$ (for all $\tau \in (0, 1)$) and is continuous. The inequalities (2.51) and (8.29) imply that whenever

$$0 < \tau_0 < \tau < \tau_1 < 1$$

the following inequality holds:

$$\|\boldsymbol{B}_1\|_{L_{1-\beta(\tau)} \to L_{1-\alpha(\tau)}} \leqq \frac{4(\tau_1 - \tau_0)}{|\beta(\tau_0) - \beta(\tau_1)|(\tau_1 - \tau)(\tau - \tau_0)}$$

$$\times \left[\frac{C_0^{1-\tau_0} C_1^{\tau_0}}{|\beta_0 - \beta_1|\beta(\tau_0)\tau_0(1 - \tau_0)}\right]^{(\tau_1 - \tau)/(\tau_1 - \tau_0)}$$

$$\times \left[\frac{C_0^{1-\tau_1} C_1^{\tau_1}}{|\beta_0 - \beta_1|\beta(\tau_1)\tau_1(1 - \tau_1)}\right]^{(\tau - \tau_0)/(\tau_1 - \tau_0)}$$

$$= k(\alpha_0, \beta_0, \alpha_1, \beta_1, \tau, \tau_0, \tau_1) C_0^{1-\tau} C_1^{\tau}.$$

Hence there follows the estimate

$$\|\boldsymbol{B}_1\|_{L_{1-\beta(\tau)} \to L_{1-\alpha(\tau)}} \leqq k(\alpha_0, \beta_0, \alpha_1, \beta_1, \tau) C_0^{1-\tau} C_1^{\tau} \tag{8.30}$$

if we put, for instance, $\tau_0 = \frac{1}{2}\tau$, $\tau_1 = \frac{1}{2}(1 + \tau)$.

From (8.25)–(8.27) it follows that for any step functions $x(s)$ and $y(t)$ we have the relation:

$$(\boldsymbol{A}x, y) = (x, \boldsymbol{B}y).$$

This implies that $A = B_1^*$. Hence the operator A acts from $L_{\alpha(\tau)}$ to $L_{\beta(\tau)}$ (for $0 < \tau < 1$) and is continuous. The estimate (8.23) follows from the estimate (8.30). The theorem has been proved.

8.4 *Limit theorems on continuity of potentials*

THEOREM 8.3 *Let $0 < \lambda < n$. Then for $1 - \lambda/n < \alpha < 1$ the potential A_λ acts from L_α to $L_{\alpha-1+\lambda/n}$ and is continuous.*

PROOF: Let us show that the operator A_λ satisfies the conditions $\Lambda M(1 - \lambda/n, 0)$ and $\Lambda M(1, \lambda/n)$.

Let $\kappa_D(s)$ be the characteristic function of a set $D \subset \Omega$. Then

$$A_\lambda \kappa_D(t) = \int_D |t - s|^{-\lambda} ds.$$

Let us estimate $A_\lambda \kappa_D(t)$.

Denote by $T(t, r)$ the sphere with center at the point t and radius r, such that

$$\text{mes } T(t, r) = \text{mes } D. \tag{8.31}$$

It is clear that

$$v_n r^n = \text{mes } D, \tag{8.32}$$

where v_n is the volume of the unit sphere in R_n.

Let

$$T_1 = T(t, r) \cap D.$$

By (8.31)

$$\text{mes}(D - T_1) = \text{mes}(T(t, r) - T_1)$$

and hence

$$\int_{T(t;r)} \frac{ds}{|t-s|^\lambda} - \int_D \frac{ds}{|t-s|^\lambda} = \int_{T(t,r)-T_1} \frac{ds}{|t-s|^\lambda} - \int_{D-T_1} \frac{ds}{|t-s|^\lambda}$$

$$\geqq \frac{1}{r^\lambda}\{\text{mes}[T(t,r) - T_1] - \text{mes}(D - T_1)\} = 0.$$

Thus

$$\int_D \frac{ds}{|t-s|^\lambda} \leqq \int_{T(t,r)} \frac{ds}{|t-s|^\lambda} = \int_{T(0,r)} \frac{ds}{|s|^\lambda}. \tag{8.33}$$

We use polar coordinate to evaluate the last integral in (8.33). Then

$$\int_{T(0,r)} \frac{ds}{|s|^\lambda} = nv_n \int_0^r \rho^{n-\lambda-1}\,d\rho = \frac{nv_n}{n-\lambda} r^{n-\lambda}$$

and by (8.32)

$$\int_{T(0,r)} \frac{ds}{|s|^\lambda} = \frac{n v_n^{\lambda/n}}{n-\lambda} (\text{mes } D)^{1-\lambda/n}. \tag{8.34}$$

Comparing (8.33) with (8.34) we deduce that for any $t \in \Omega$

$$|A_\lambda \kappa_D(t)| \leqq C(\text{mes } D)^{1-\lambda/n},$$

where C is a constant. This means that

$$\|A_\lambda \kappa_D(t)\|_0 \leqq C(\text{mes } D)^{1-\lambda/n}. \tag{8.35}$$

Thus we have proved that the operator A_λ satisfies the condition $\Lambda M(1 - \lambda/n, 0)$.

Now let $x(s)$ be an arbitrary summable function. Then for each measurable set $D \subset \Omega$

$$\int_D |A_\lambda x(t)|\,dt = \int_D \left|\int_\Omega \frac{x(s)}{|t-s|^\lambda}\,ds\right| dt$$

$$\leqq \int_D \int_\Omega \frac{|x(s)|}{|t-s|^\lambda}\,ds\,dt = \int_\Omega \left[\int_D \frac{dt}{|t-s|^\lambda}\right] |x(s)|\,ds.$$

From (8.35) it follows that

$$\int_D \frac{dt}{|t-s|^\lambda} \leqq C(\text{mes } D)^{1-\lambda/n}.$$

Hence

$$\int_D |A_\lambda x(t)|\,dt \leqq C\,\|x(s)\|_1 \cdot (\text{mes } D)^{1-\lambda/n}. \tag{8.36}$$

Consequently the operator A_λ satisfies the condition $LM(1, \lambda/n)$ and, a fortiori, the condition $\Lambda M(1, \lambda/n)$.

Now it remains only to apply the interpolation theorem of Stein-Weiss. The theorem has been proved.

Let us show that A_λ, as an operator from L_α, $1 - \lambda/n < \alpha < 1$, to $L_{\alpha-1+\lambda/n}$ is not compact[1].

Consider a sequence of sets $\Omega_k \subset \Omega$ $(k = 1, 2, \ldots)$, the diameters of which are respectively less than $1/k$ and the measures of which are equal to $c_k k^{-n}$, where $0 < a \leqq c_k \leqq b < \infty$[2]. Put

$$x_k(s) = \begin{cases} k^{n\alpha}, & \text{if } s \in \Omega_k \\ 0, & \text{if } s \notin \Omega_k \end{cases} \quad (k = 1, 2, \ldots).$$

---

[1] This fact was first proved, perhaps, by V. M. Babic [1].

[2] The existence of such sets follows, for instance, from the fact that almost all points of the set Ω are density points (see Vallée-Poussin [1]). Let s_0 be a density point of the set Ω and let Ω_k be the intersection of Ω and the sphere with radius $1/2k$ and center at the point s_0. It is clear that the diameter of Ω_k is equal to $1/k$, and the measure is equal to $c_k \cdot k^{-n}$, where c_k is bounded from above and below.

It is clear that

$$\|x_k(s)\|_{L_\alpha} = k^{n\alpha}(\text{mes } \Omega_k)^\alpha \leqq b^\alpha.$$

Hence the functions

$$A_\lambda x_k(t) = k^{n\alpha} \int\limits_{\Omega_k} \frac{ds}{|t-s|^\lambda}$$

belong to the space $L_{\alpha-1+\lambda/n}$. Let us show that these functions do not possess equi-absolutely continuous norm—whence by Theorem 3.1 it follows that A_λ, as an operator from L_α to $L_{\alpha-1+\lambda/n}$, is not compact.

Since the diameter of Ω_k is less than $1/k$,

$$A_\lambda x_k(t) \geqq k^{n\alpha}k^\lambda \text{ mes } \Omega_k \geqq a \cdot k^{n(\alpha-1+\lambda/n)} \qquad (t \in \Omega_k)$$

and consequently

$$\|P_{\Omega_k} A_\lambda x_k(t)\|_{\alpha-1+\lambda/n} \geqq a^{\alpha+\lambda/n}.$$

This inequality means that the norms of the functions $A_\lambda x_k$ do not possess the property of equi-absolute continuity, because mes $\Omega_k \to 0$ as $k \to \infty$.

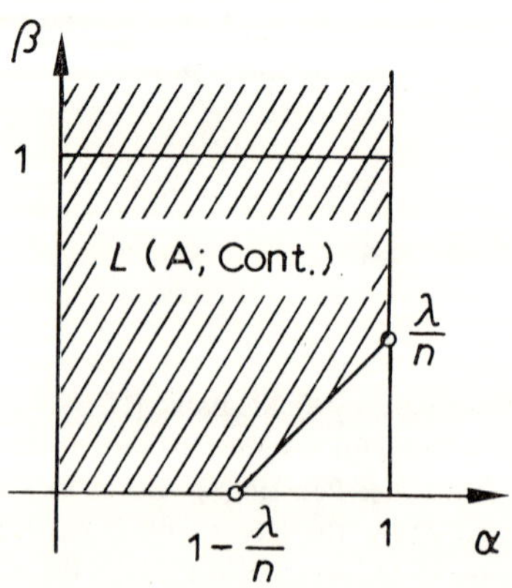

Figure 8.2

We have proved that points $\{\alpha, \alpha - 1 + \lambda/n\}$ for $1 - \lambda/n < \alpha < 1$ do not belong to the L-characteristic $L(A_\lambda;\text{ comp.})$. Hence it follows (see Theorem 5.4) that no point $\{\alpha, \beta\}$, for which

$$\beta < \alpha - 1 + \lambda/n, \; 1 - \lambda/n < \alpha < 1$$

belongs to the L-characteristic $L(A_\lambda;\text{ cont.})$. This assertion also follows from Theorem 7.9.

Note further that from Theorems 7.2 and 7.7 it follows that the points $\{1 - \lambda/n, 0\}$ and $\{1, \lambda/n\}$ do not belong to the L-characteristic $L(A_\lambda;\text{ cont.})$.

The assertions of Theorems 8.1 and 8.3 give a complete description of the L-characteristics $L(A_\lambda;\text{ comp.})$ and $L(A_\lambda;\text{ cont.})$ (see Fig. 8.1 and Fig. 8.2).

8.5 *Operators of potential type*

Let us begin an investigation of operators of the form

$$Ax(t) = \int_\Omega \frac{Q(t, s)}{|t - s|^\lambda} x(s)\,ds. \tag{8.37}$$

Some simple but very important theorems on the L-characteristics of this operator can be obtained if the function $Q(t, s)$ is bounded. In this case Theorems 4.2, 5.10 and 6.4 imply:

THEOREM 8.4: *Let $0 < \lambda < n$ and let the function $Q(t, s)$ be bounded. Then* $L(A;\text{ cont.})$ *contains* $L(A_\lambda;\text{ comp.})$.

THEOREM 8.5: *Let $0 < \lambda < n$ and let the function $Q(t, s)$ be bounded. Then* $L(A;\text{ comp.})$ *contains all points of the L-characteristic* $L(A_\lambda;\text{ comp.})$ *with the possible exception of some boundary points.*

THEOREM 8.6: *Let $0 < \lambda < n$ and let the function $Q(t, s)$ be bounded and jointly continuous for $t \neq s$. Then* $L(A;\text{ comp.})$ *contains* $L(A_\lambda;\text{ comp.})$.

With appropriate modifications, these theorems can be applied to cases in which the function $Q(t, s)$ defining the operator has weak singularities. Let, for instance, a function $Q(t, s)$ be bounded in each domain $|t - s| \geqq c_0$ and suppose that for each $\varepsilon > 0$ one has:

$$\lim_{|t-s|\to 0} |t - s|^{\varepsilon} Q(t, s) = 0.$$

Then for each $\varepsilon > 0$ the following inequality is fulfilled:

$$\frac{Q(t, s)}{|t - s|^{\lambda}} \leqq c(\varepsilon)|t - s|^{-(\lambda+\varepsilon)}.$$

This inequality implies that the L-characteristic $L(A;\text{cont.})$ of the operator (8.37) contains all points of the L-characteristic $L(A_\lambda;\text{cont.})$, with the possible exception of points on the line $\beta = \alpha - 1 + \lambda/n$. Moreover the L-characteristic $L(A;\text{comp.})$ contains all interior points of the L-characteristic $L(A_\lambda;\text{comp.})$ and under supplementary assumption on continuity of $Q(t, s)$ for $t \neq s$, all points of $L(A_\lambda;\text{comp.})$.

Consider, in more detail, integral operators of the form

$$A_{\lambda,\nu}x(t) = \int_{\Omega} K_{\lambda,\nu}(t, s)x(s)\,\mathrm{d}s \tag{8.38}$$

with kernels

$$K_{\lambda,\nu}(t, s) = |t - s|^{-\lambda}|\ln|t - s||^{\nu}. \tag{8.39}$$

If $\nu > 0$, the operators $A_{\lambda,\nu}$ are operators of the form (8.37) in which the functions $Q(t, s)$ have a weak singularity. In this case $L(A_{\lambda,\nu};\text{cont.}) = L(A_{\lambda,\nu};\text{comp.}) = L(A_\lambda;\text{comp.})$. If $\nu < 0$, then $L(A_{\lambda,\nu};\text{cont.}) = L(A_{\lambda,\nu};\text{comp.})$, and when $0 < \nu \leqq \lambda n$, the L-characteristic $L(A_{\lambda,\nu};\text{comp.})$ coincides with the L-characteristic $L(A_\lambda;\text{cont.})$, while when $\lambda n < \nu$ the L-characteristic $L(A_{\lambda,\nu};\text{comp.})$ consists of the L-characteristic $L(A_\lambda;\text{cont.})$ plus the two points $\{1 - \lambda/n, 0\}$ and $\{1, \lambda/n\}$. The proofs are left to the reader. We note only that to study operators $A_{\lambda,\nu}$, it is convenient to use Theorem 5.11.

Note in conclusion that these methods permit us to investigate integral operators with kernels $K(t, s)$ whose singularities are distributed not only on the 'diagonal' $t = s$ but on some other 'smooth surface'.

8.6 *The logarithmic potential*

Consider the operator

$$A_{(\ln)}x(t) = \int_{\Omega} K_{(\ln)}(t, s)x(s)\,\mathrm{d}s, \tag{8.40}$$

where

$$K_{(\ln)}(t, s) = |\ln|t - s||. \tag{8.41}$$

THEOREM 8.7: *The L-characteristics* $L(A_{(\ln)};$ cont.) *and* $L(A_{(\ln)};$ comp.) *both coincide with the half-strip* $0 \leqq \alpha \leqq 1$, $\beta \geqq 0$, *excluding the point* $\{1, 0\}$.

The proof is trivial.

It is easy to formulate similar assertions for operators whose kernels $K(t, s)$ satisfy inequalities of the form

$$|K(t, s)| \leqq M|\ln|t - s|| + N. \tag{8.42}$$

8.7 *Iterates of operators of potential type*

Let two operators of potential type be given:

$$A_1x(t) = \int_{\Omega} \frac{Q_1(t, s)}{|t - s|^{\lambda_1}}\, x(s)\,\mathrm{d}s,$$

$$A_2x(t) = \int_{\Omega} \frac{Q_2(t, s)}{|t - s|^{\lambda_2}}\, x(s)\,\mathrm{d}s,$$

where $Q_1(t, s)$, $Q_2(t, s)$ are bounded functions. We are interested in the integral operator

$$\boldsymbol{K}x(t) = \int_{\Omega} K(t, s)x(s)\,\mathrm{d}s$$

with the kernel

$$K(t, s) = \int_{\Omega} \frac{Q_1(t, u) Q_2(u, s)}{|t - u|^{\lambda_2} |u - s|^{\lambda_1}} \, du. \tag{8.43}$$

If the operator A_1 acts from L_α to L_β and the operator A_2 acts from L_β to L_γ, then by Theorem 4.3 the operator $\boldsymbol{K}$ acts from L_α to L_γ and

$$\boldsymbol{K} = A_2 A_1. \tag{8.44}$$

THEOREM 8.8: *If* $\lambda_1 + \lambda_2 > n$, $0 < \lambda_1, \lambda_2 < n$, *then the kernel* (8.43) *satisfies the inequality*

$$|K(t, s)| \leqq \frac{M}{|t - s|^{\lambda_1 + \lambda_2 - n}} \qquad (t, s \in \Omega). \tag{8.45}$$

If $\lambda_1 + \lambda_2 = n$, *then the kernel* (8.43) *satisfies the inequality* (8.42):

$$|K(t, s)| \leqq M |\ln |t - s|| + N \qquad (t, s \in \Omega). \tag{8.46}$$

If $\lambda_1 + \lambda_2 < n$, *then the kernel* (8.43) *is bounded:*

$$|K(t, s)| \leqq M \qquad (t, s \in \Omega). \tag{8.47}$$

PROOF: Boundedness of the functions $Q_1(t, s)$ and $Q_2(t, s)$ implies that

$$|K(t, s)| \leqq c \int_{\Omega} \frac{du}{|t - u|^{\lambda_1} |u - s|^{\lambda_2}}$$

where c is a constant. Hence the proof of the theorem is reduced to estimation of the integral

$$J(\lambda_1, \lambda_2) = \int_{\Omega} \frac{du}{|t - u|^{\lambda_1} |u - s|^{\lambda_2}}. \tag{8.48}$$

First let $\lambda_1 + \lambda_2 > n$. Use the trivial inequality

$$J(\lambda_1, \lambda_2) \leqq \int\limits_{R_n} \frac{du}{|t-u|^{\lambda_1}|u-s|^{\lambda_2}}, \tag{8.49}$$

where R_n is the Euclidean space in which Ω lies. Changing variables (for $t \neq s$):

$$u = t - v|t-s|,$$

we obtain

$$\int\limits_{R_n} \frac{du}{|t-u|^{\lambda_1}|u-s|^{\lambda_2}} = |t-s|^{n-\lambda_1-\lambda_2} \int\limits_{R_n} \frac{dv}{|v|^{\lambda_1}|v_0-v|^{\lambda_2}}$$

where $v_0 = (t-s)/|t-s|$. Since $|v_0| = 1$, the integral on the right side does not depend on v_0. Hence

$$\int\limits_{R_n} \frac{du}{|t-u|^{\lambda_1}|u-s|^{\lambda_2}} = c_0|t-s|^{n-\lambda_1-\lambda_2}$$

and the expected estimate results from (8.49).

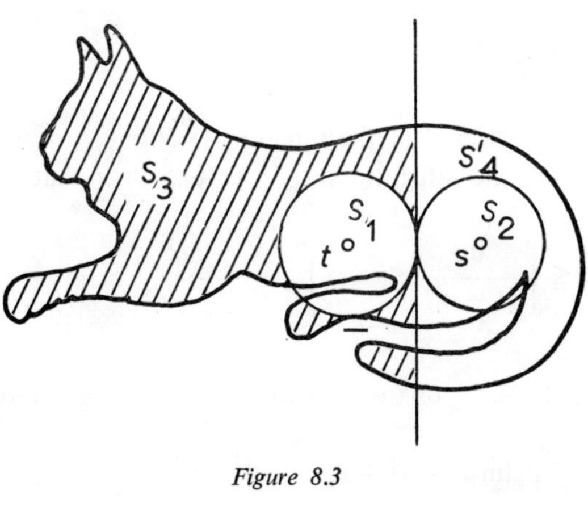

Figure 8.3

Let now $\lambda_1 + \lambda_2 = n$. Assume that the points $t, s \in \Omega$ are fixed and $t \neq s$. Denote by S_1 and S_2 the spheres in the space R_n with centers at t and s respectively and with radius, equal to $\frac{1}{2}|t - s|$. (Fig. 8.3). Put $S_1' = S_1 \cap \Omega$, $S_2' = S_2 \cap \Omega$. Denote by S_3' the set of points u in Ω, for which $|u - t| \leqq |u - s|$ and u does not belong to the sphere S_1. (This set is hatched in Fig. 8.3). Denote by S_4' the set of points $u \in \Omega$, for which $|u - t| > |u - s|$ and u does not belong to the sphere S_2.

The integral (8.48) becomes the sum of four integrals

$$J_i(\lambda_1, \lambda_2) = \int\limits_{S'_i} \frac{du}{|t-u|^{\lambda_1}|u-s|^{\lambda_2}} \qquad (i = 1, 2, 3, 4). \tag{8.50}$$

For the first of these we have the estimate:

$$J_1(\lambda_1, \lambda_2) \leqq \int\limits_{S'_1} \frac{du}{|t-u|^{\lambda_1}|u-s|^{\lambda_2}} \leqq \frac{2^{\lambda_2}}{|t-s|^{\lambda_2}} \int\limits_{S'_1} \frac{du}{|t-u|^{\lambda_1}}.$$

The integral in the right side of the above inequality is easily calculated (for this, it suffices to use polar coordinates). We obtain as a result

$$J_1(\lambda_1, \lambda_2) \leqq c_1$$

where c_1 does not depend on t and s but on $|t - s|$ alone. Similarly an estimate of the form

$$J_2(\lambda_1, \lambda_2) \leqq c_2$$

is obtained for the second of the integrals (8.50).

Consider next the integral $J_3(\lambda_1, \lambda_2)$. It is clear that

$$J_3(\lambda_1, \lambda_2) \leqq \int\limits_{S'_3} \frac{du}{|t-u|^n} \leqq \int\limits_{\frac{1}{2}|t-s| \leqq |t-u| \leqq \Delta} \frac{du}{|t-u|^n}$$

where Δ is the diameter of the domain Ω. Evaluating, we obtain

$$J_3(\lambda_1, \lambda_2) \leqq c_3 |\ln|t - s|| + c_4.$$

Similarly it is proved that

$$J_4(\lambda_1, \lambda_2) \leqq c_3 |\ln|t - s|| + c_4.$$

From these estimates for the integrals (8.50) follows the inequality (8.46).

Let, finally, $\lambda_1 + \lambda_2 < n$. Let $\delta \in (\lambda_1/n, 1 - \lambda_2/n)$. Then from Hölder's inequality it follows that

$$J(\lambda_1, \lambda_2) \leqq \left\{ \int_\Omega \frac{du}{|t-u|^{\lambda_1/\delta}} \right\}^\delta \left\{ \int_\Omega \frac{du}{|u-s|^{\lambda_2/1-\delta}} \right\}^{1-\delta}.$$

Hence Lemma 8.1 implies that the integral $J(\lambda_1, \lambda_2)$ is bounded. The theorem has been completely proved.

In special cases the estimates contained in Theorem 8.8 may turn out to be very rough. Consider, for instance, the product $A_2 A_1$ of the operators

$$A_1 x(t) = \int_0^t (t-s)^{-\lambda_1} x(s) \, ds$$

and

$$A_2 x(t) = \int_0^t (t-s)^{-\lambda_2} x(s) \, ds$$

where $\lambda_1 + \lambda_2 = 1$. The kernel (8.43) has the form

$$K(t, s) = \begin{cases} \int_s^t (t-u)^{-\lambda_2} (u-s)^{-\lambda_1} \, du, & \text{if } 0 < s \leqq t \\ 0, & \text{if } s > t. \end{cases}$$

Changing variables by $u - s = \xi(t - s)$, we obtain

$$\int_s^t (t-u)^{-\lambda_2} (u-s)^{-\lambda_1} \, du = \int_0^1 (1-\xi)^{-\lambda_2} \xi^{-\lambda_1} \, d\xi = c < \infty.$$

Thus, in this example, the function $K(t, s)$ is bounded while from Theorem 8.8 there follows merely the estimate (8.46).

Let A be an operator of potential type with exponent λ. From Theorem 8.8 it follows that:
1°) When

$$k < \frac{n}{n - \lambda}$$

then A^k is an integral operator of potential type with exponent $k\lambda - (k - 1)n$.
2°) When

$$k = \frac{n}{n - \lambda},$$

then A^k is an integral operator of logarithmic potential type.
3°) When

$$k > \frac{n}{n - \lambda},$$

then A^k is an integral operator with a bounded kernel.

8.8 *Generalizations to the case of distinct dimensions*

In the preceding sections we considered potentials and operators of potential type, as operators whose domains and ranges consist of functions defined on the same set Ω. In many applications other situations occur.

In this section we examine integral operators with kernels

$$K_\lambda(t, s) = |t - s|^{-\lambda},$$

where the variables t and s lie in distinct sets. We confine ourselves here to the simplest cases.

Let R_m $(m < n)$ be a subspace of a space R_n, Ω be a bounded set of non-zero measure in R_n, and Ω^* be some subset of the set Ω lying in the subspace R_m and having non-zero measure in R_m. Consider the integral operator

$$A_\lambda x(t) = \int_\Omega K_\lambda(t, s)x(s)\mathrm{d}s, \tag{8.51}$$

acting from a space of functions defined on Ω to a space of functions defined on Ω^*.

THEOREM 8.9: *Let $0 < \lambda < m < n$. Then the integral operator A_λ in* (8.51) *satisfies:*

a) *for $0 \leqq \alpha < 1 - \lambda/n$, A_λ acts from L_α to L_0 and is compact;*

b) *A_λ acts from $L_{1-\lambda/n}$ to any space L_β where $\beta > 0$, and is compact;*

c) *for $1 - \lambda/n < \alpha < 1$, A_λ acts from L_α to L_β, where $\beta = (n/m)(\alpha - 1) + \lambda/m$, and is continuous;*

d) *A_λ acts from L_1 to L_β, where $\beta > \lambda/m$ and is compact.*

The proof of this theorem is closely related to the proofs of Theorems 8.1 and 8.2.

In the case $0 < \lambda < m < n$, Theorem 8.9 describes the L-characteristics $L(A_\lambda; \text{cont.})$ and $L(A_\lambda; \text{comp.})$ of the operator. A_λ completely (Fig. 8.4).

THEOREM 8.10: *Let $m \leqq \lambda < n$. Then the integral operator A_λ in* (8.51) *satisfies:*

a) *for $0 \leqq \alpha < 1 - \lambda/n$, A_λ acts from L_α to L_0 and is compact;*

b) *A_λ acts from $L_{1-\lambda/n}$ to any space L_β, where $\beta > 0$, and is compact;*

c) *for $\dfrac{n-\lambda}{n} < \alpha < \dfrac{n-\lambda}{n-m}$, A_λ acts from L_α to L_β, where $\beta = (n/m)(\alpha - 1) + \lambda/m$, and is continuous.*

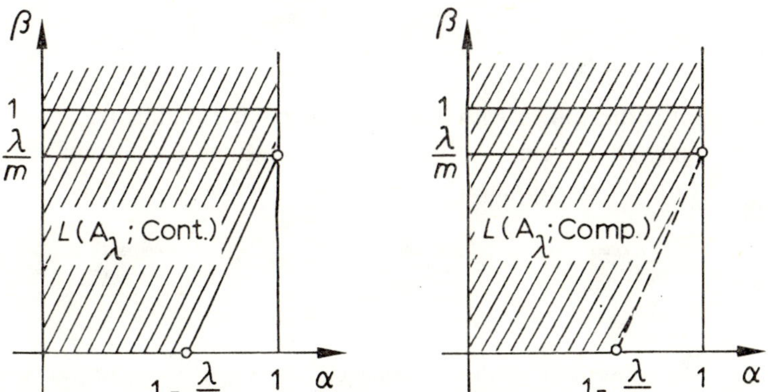

Figure 8.4

Assertion a) of this theorem follows from Theorems 6.1 and 6.4. From Theorem 6.1 it follows that the operator (8.51) acts from L_α with $0 \leqq \alpha < 1 - \lambda/n$ to L_0 and is continuous. Compactness of the operator (8.51) results from Theorem 6.4. (Here it is necessary to use arguments similar to those used in the proof of Theorem 8.1.)

Proof of assertion c) involves an essential difficulty. Suppose that c) has been proved. Then $\{\alpha, \beta\}$ is an interior point of the L-characteristic $L(A_\lambda;\text{cont.})$, whenever $\alpha < (n - \lambda)/(n - m)$ and $\beta > (\alpha - 1)n/m + \lambda/m$, $\beta > 0$. But the interior points of $L(A_\lambda;\text{cont.})$ belong to $L(A_\lambda;\text{comp.})$ by Theorem 5.4. Hence, in particular, there follows assertion b).

In the following we prove that the operator (8.51) satisfies the condition[1] $\Lambda M(\alpha, (\alpha - 1)n/m + \lambda/m)$ for all α in the interval $((n - \lambda)/n, (n - \lambda)/(n - m))$. Then assertion c) will follow from Theorem 8.2.

Let $\kappa_D(s)$ be the characteristic function of a subset $D \subset \Omega$. Then

$$A_\lambda \kappa_D(t) = \int_D |t - s|^{-\lambda} ds = \int_D |t - s|^{-(\lambda - \beta m)} |t - s|^{-\beta m} ds. \tag{8.52}$$

To estimate the last integral in the right side of (8.52), apply Hölder's inequality with exponents $1/(1 - \gamma)$ and $1/\gamma$, where γ is some number in the interval (β, α). We obtain as a result that

$$|A_\lambda \kappa_D(t)| \leqq \{\int_D |t - s|^{-(\lambda - \beta m)/(1-\gamma)} ds\}^{1-\gamma} \{\int_D |t - s|^{-\beta m/\gamma} ds\}^{\gamma}. \tag{8.53}$$

As in the proof of Theorem 8.3 it is possible to show that

$$\int_D |t - s|^{-(\lambda - \beta m)/(1-\gamma)} ds \leqq \int_{T(t,r)} |t - s|^{-(\lambda - \beta m)/(1-\gamma)} ds, \tag{8.54}$$

where $T(t, r)$ is the sphere with center at the point t, radius r and volume $v = v_n r^n = \operatorname{mes} D$ (v_n being the volume of the n-dimensional unit sphere). Changing to polar coordinates, we obtain the inequality

$$\int_D |t - s|^{-(\lambda - \beta m)/(1-\gamma)} ds \leqq v_n^{(\lambda - \beta m)/n(1-\gamma)} (\operatorname{mes} D)^{1-(\lambda - \beta m)/n(1-\gamma)}.$$

[1] We do not know whether it satisfies the condition $\Lambda M((n - \lambda)/(n - m), (n - \lambda)/(n - m))$. It is not difficult to show that it satisfies the condition $\Lambda M(1 - \lambda/n, 0)$ (for this it suffices to repeat, for instance, the first part of the proof of Theorem 8.3).

Thus we arrive at the estimate

$$|A_\lambda \kappa_D(t)| \leqq c(\text{mes } D)^{\alpha-\gamma}(\int_D |t-s|^{-m\beta/\gamma}\mathrm{d}s)^{\gamma}.$$

From this estimate it follows that

$$\|A_\lambda \kappa_D\|_{M_\beta} \leqq c(\text{mes } D)^{\alpha-\gamma}\|(\int_D |t-s|^{-m\beta/\gamma}\mathrm{d}s)^{\gamma}\|_{M_\beta}$$

or (see 8.3°) that

$$\|A_\lambda \kappa_D\|_{M_\beta} \leqq c_1(\text{mes } D)^{\alpha-\gamma}(\|\int_D |t-s|^{-m\beta/\gamma}\mathrm{d}s\|_{M_{\beta/\gamma}})^{\gamma}.$$

The expression $|t-s|^{-m\beta/\gamma}$ can be considered a bounded vector-function defined on D

$$w(s) = |t-s|^{-m\beta/\gamma} \qquad (s \in D)$$

with values in the space $M_{\beta/\gamma}$ of functions of the variable $t \in \Omega^*$. In fact, if s is fixed the inequality $|t-s|^{-m\beta/\gamma} \geqq h$ is equivalent to the inequality $|t-s| \leqq h^{-\gamma/m\beta}$. Hence

$$\text{mes}\{t\colon |t-s|^{-m\beta/\gamma} \geqq h\} \leqq v_m(h^{-\gamma/m\beta})^m = v_m h^{-\gamma/\beta}.$$

But it follows from this that for each $s \in D$ (see 8.3°)

$$\|\,|t-s|^{-m\beta/\gamma}\|^*_{M_{\beta/\gamma}} = \sup_h h\cdot[\text{mes}\{t\colon |t-s|^{-m\beta/\gamma} \geqq h\}]^{\beta/\gamma} \leqq v_m^{\beta/\gamma}.$$

From general properties of integrals[1] of vector-functions it follows that

$$\|\int_D |t-s|^{-m\beta/\gamma}\mathrm{d}s\|_{M_{\beta/\gamma}} \leqq \int_D \|\,|t-s|^{-m\beta/\gamma}\|_{M_{\beta/\gamma}}\mathrm{d}s \leqq c_1\cdot \text{mes } D.$$

---

[1] If E is a Banach space and $w(t)$ is a vector-function, defined on $D \subset R_n$ with values in E, then

$$\|\int_D w(t)\,\mathrm{d}t\| \leqq \int_D \|w(t)\|\,\mathrm{d}t$$

(see also 13.1°).

Thus we arrive at the inequality

$$\|A_\lambda \kappa_D\|_{M_\beta} \leqq c_2(\text{mes } D)^\alpha,$$

which means that the operator (8.51) satisfies the condition

$$AM\left(\alpha, \frac{n}{m}(\alpha - 1) + \frac{\lambda}{n}\right) \text{ for } \frac{n-\lambda}{n} < \alpha < \frac{n-\lambda}{n-m}.$$

Theorem 8.10 means that the set of points $\{\alpha, \beta\}$ for which the following conditions hold:

$$0 \leqq \alpha < \frac{n-\lambda}{n-m}, \ \beta \geqq 0, \ \beta \geqq \frac{n}{m}(\alpha - 1) + \frac{\lambda}{m}$$

(excluding the point $\{1 - \lambda/n, 0\}$), is contained in the L-characteristic $L(A_\lambda;\text{cont.})$ while the set of points $\{\alpha, \beta\}$ for which one has the stronger conditions:

$$0 \leqq \alpha < \frac{n-\lambda}{n-m}, \ \beta \geqq 0, \ \beta > \frac{n}{m}(\alpha - 1) + \frac{\lambda}{m},$$

is contained in the L-characteristic $L(A_\lambda;\text{comp.})$ (Fig. 8.5). We do not know

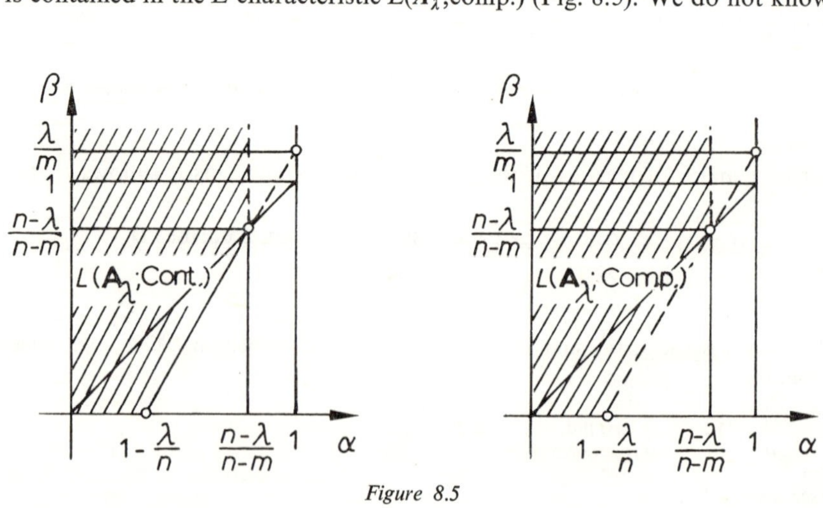

Figure 8.5

whether the points $\{(n-\lambda)/(n-m), \beta\}$, where $\beta > (n-\lambda)/(n-m)$, belong to the L-characteristic $L(A_\lambda; \text{cont.})$.

Now assume that R_n ($n < m$) is a subspace of a space R_m, that Ω^* is a bounded set of non-zero Lebesgue measure in R_m, and that Ω is a subset of Ω^*, lying in R_n and having non-zero measure in R_n. We are interested in the L-characteristics $L(A_\lambda; \text{cont.})$ and $L(A_\lambda; \text{comp.})$ of the operator

$$A_\lambda x(t) = \int_\Omega K_\lambda(t, s) x(s) \mathrm{d}s. \tag{8.55}$$

To construct these L-characteristics, it is possible first by means of Theorems 8.9 and 8.10 to construct the corresponding L-characteristics of the transposed operator

$$A_\lambda^{\#} y(s) = \int_{\Omega^*} K_\lambda(t, s) y(t) \mathrm{d}t. \tag{8.56}$$

After this it remains only to use Theorems 4.5 and 5.1. The assertions obtained in this way are formulated below in a form similar to Theorems 8.9 and 8.10.

THEOREM 8.11: *Let $0 < \lambda < n < m$. Then the integral operator A_λ in* (8.55) *satisfies:*
a) *for $0 \leqq \alpha < 1 - \lambda/n$, A_λ acts from L_α to L_0 and is compact;*
b) *A_λ acts from $L_{1-\lambda/n}$ to any space L_β, where $\beta > 0$ and is compact;*
c) *for $1 - \lambda/n < \alpha < 1$, A_λ acts from L_α to L_β, where $\beta = (\alpha - 1)n/m + \lambda/m$ and is continuous;*
d) *A_λ acts from L_1 to any L_β, where $\beta > \lambda/m$ and is compact.*

THEOREM 8.12: *Let $n \leqq \lambda < m$. Then the integral operator A_λ in* (8.55) *satisfies:*
a) *for $0 \leqq \alpha \leqq (\lambda - n)/(m - n)$, A_λ acts from L_α to any space L_β, where $\beta > (\lambda - n)/(m - n)$, and is compact;*
b) *for $(\lambda - n)/(m - n) < \alpha < 1$, A_λ acts from L_α to L_β, where $\beta = (\alpha - 1)n/m + \lambda/m$ and is continuous;*
c) *A_λ acts from L_1 to any space L_β, where $\beta > \lambda/m$ and is compact.*

The L-characteristics of the operator in (8.55) are shown in Fig. 8.6 (for the case $n < \lambda < m$).

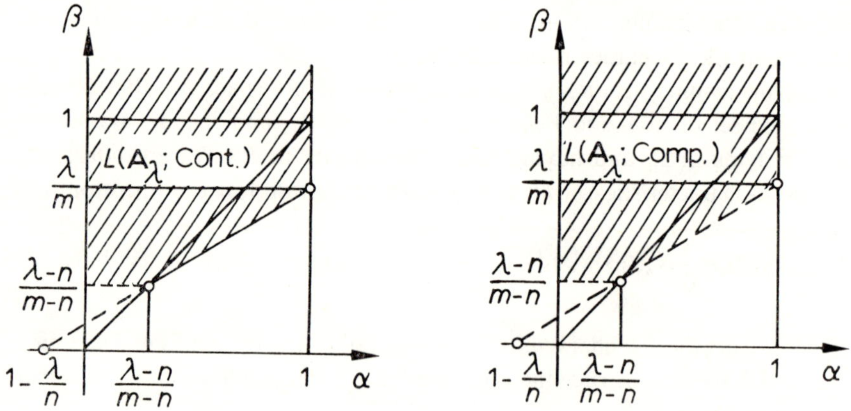

Figure 8.6

We leave it to the reader to study in the present context integral operators with kernels of the form

$$K(t, s) = Q(t, s) \cdot |t - s|^{-\lambda}$$

where $Q(t, s)$ is bounded.

8.9 *Potentials with respect to non-Lebesgue measure*[1]

Let G be a bounded set in an n-dimensional Euclidean space R_n. In the sequel $G(s; \rho)$ will denote the intersection of the set G and the sphere $|u - s| \leqq \rho$.

Let a positive non-Lebesgue measure $\mu = \mu(D)$ $(D \subseteq \bar{G})$ be given on $\bar{G}$, such that $\mu(\bar{G}) < \infty$. Assume that there exist positive numbers k such that for all $s \in R_n$ and for all $\rho > 0$ the following inequality holds:

$$\mu[G(s; \rho)] \leqq c(k)\rho^k. \tag{8.57}$$

If this inequality holds with $k = k_0$, then as is easily seen, it also holds with all $k < k_0$. The supremum of the numbers k for which the inequality

[1] This section is written according to the paper of E. I. Pustylnik [6].

(8.57) holds, is called the *dimension of the set G with respect to the measure* μ. This dimension will be denoted by $\dim_\mu G$.

Dimension with respect to a measure defined in this way is always positive. It can be any number not exceeding n.

It is easy to see that the dimension of each domain of the space R_n with respect to Lebesgue measure is equal to n. The dimension of a hypersurface with respect to the natural measure (i.e. 'area') is equal to $n - 1$. As a further example, consider the measure μ, defined by the relation

$$\mu(D) = \int_D f(s)\,\mathrm{d}s \qquad (D \subset \bar{\Omega}), \tag{8.58}$$

where $f(s)$ is a non-negative and summable (with respect to the usual Lebesgue measure) function. The dimension $\dim_\mu G$ of a set G with respect to this measure is determined by the 'singularity' of the function $f(s)$; if α_0 $(0 \leqq \alpha_0 < 1)$ is the infimum of the numbers α such that $f(s) \in L_\alpha(G)$, then

$$\dim_\mu G = n(1 - \alpha_0).$$

If the function $f(s)$ does not belong to any $L_\alpha(G)$, where $0 \leqq \alpha < 1$, then the dimension of the set G with respect to the measure (8.58) is not determined.

Consider an operator of potential type

$$Ax(t) = \int_\Omega \frac{Q(t, s)}{|t - s|^\lambda}\, x(s)\,\mathrm{d}\mu(s) \qquad (t \in \Omega^*) \tag{8.59}$$

where $Q(t, s)$ is a bounded function. Assume that measures μ and μ^* are given on Ω and Ω^* respectively and that the dimensions $\dim_\mu(\Omega)$ and $\dim_{\mu^*}(\Omega^*)$ are determined.

A considerable part of the theorems on L-characteristics of operators of potential type which have been established in the preceding subsections can be extended to the operators (8.59).

For any set $T \subset R_n$, T_r will denote the closure of the union of all spheres of radius r with centers in the set T. On the sets

$$G_\rho = \Omega \cap \Omega^*_\rho, \;\; G^*_\rho = \Omega^* \cap \Omega_\rho$$

there are given the measures μ and μ^* respectively. The definition of the dimensions $\dim_\mu G_\rho$ and $\dim_{\mu^*} G_\rho^*$ implies that they increase as ρ decreases. Put

$$n_0 = \sup_\rho \dim_\mu G_\rho, \quad m_0 = \sup_\rho \dim_{\mu^*} G_\rho^*$$

(clearly, $m_0 \leqq n$, and $n_0 \leqq n$).

THEOREM 8.13: *If* $0 < \lambda < n_0, m_0$, *then the L-characteristic* $L(A_\lambda;\text{comp.})$ *of the operator* A_λ *contains all points* $\{\alpha, \beta\}$ *for which*

$$0 \leqq \alpha \leqq 1, \quad \beta \geqq 0, \quad \beta > \frac{n_0}{m_0}(\alpha - 1) + \frac{\lambda}{m_0}.$$

If $0 < m_0 \leqq \lambda < n_0$, *then the L-characteristic* $L(A_\lambda;\text{comp.})$ *of the operator* A_λ *contains all points* $\{\alpha, \beta\}$ *for which*

$$0 \leqq \alpha \leqq 1, \quad \alpha \geqq \beta \geqq 0, \quad \beta > \frac{n_0}{m_0}(\alpha - 1) + \frac{\lambda}{m_0}.$$

If $0 < n_0 \leqq \lambda < m_0$, *then the L-characteristic* $L(A_\lambda;\text{comp.})$ *of the operator* A_λ *contains all points* $\{\alpha, \beta\}$ *for which*

$$0 \leqq \alpha \leqq 1, \quad \alpha \geqq \beta \geqq 0, \quad \beta > \frac{n_0}{m_0}(\alpha - 1) + \frac{\lambda}{m_0}.$$

The proof is left to the reader.

3

Fractional powers of selfadjoint operators

§ 9 Splitting of linear operators

9.1 *Square root of selfadjoint operators*

Selfadjoint operators form an important class of linear operators acting in a Hilbert space H. Recall that a continuous linear operator A is called *selfadjoint* if

$$(Ax, y) = (x, Ay) \quad (x, y \in H). \tag{9.1}$$

This definition applies to the case of a real Hilbert space as well as a complex Hilbert space.

An important example of a selfadjoint operator acting in the space $L_{\frac{1}{2}}$ is an integral operator of the form

$$Ax(t) = \int_{\Omega} K(t, s)x(s)\,\mathrm{d}s, \tag{9.2}$$

whose kernel satisfies the condition

$$K(t, s) = \overline{K(s, t)} \quad (t, s \in \Omega). \tag{9.3}$$

In the case of a real kernel condition (9.3) means that K is symmetric:

$$K(t, s) = K(s, t) \quad (t, s \in \Omega). \tag{9.4}$$

Naturally, the kernel $K(s, t)$ has to possess some additional properties, in order that the operator (9.2) be continuous in $L_{\frac{1}{2}}$; such conditions were investigated in the preceding chapter.

A selfadjoint operator $\boldsymbol{A}$ is called *positive definite*[1] if

$$(\boldsymbol{A}x, x) \geqq 0 \qquad (x \in H). \tag{9.5}$$

It is possible[2] to extract from each selfadjoint positive definite operator $\boldsymbol{A}$ a positive square root in the sense that there exists a unique selfadjoint positive definite operator $\boldsymbol{A}^{\frac{1}{2}}$ such that $\boldsymbol{A} = (\boldsymbol{A}^{\frac{1}{2}})^2$.

Consider all operators $\boldsymbol{B}$ of the form

$$\boldsymbol{B} = \boldsymbol{A}^{\frac{1}{2}}\boldsymbol{U}, \tag{9.6}$$

where $\boldsymbol{U}$ is a partial isometry whose range contains the range of the operator $\boldsymbol{A}$. The operator $\boldsymbol{B}^*$ adjoint to $\boldsymbol{B}$ is defined by the relation

$$\boldsymbol{B}^* = (\boldsymbol{A}^{\frac{1}{2}}\boldsymbol{U})^* = \boldsymbol{U}^*(\boldsymbol{A}^{\frac{1}{2}})^* = \boldsymbol{U}^*\boldsymbol{A}^{\frac{1}{2}}.$$

Thus for the operator $\boldsymbol{A}$ the following representation is valid:

$$\boldsymbol{A} = \boldsymbol{B}\boldsymbol{B}^* \tag{9.7}$$

where $\boldsymbol{B}$ is any operator of the form (9.6). It can be shown that the converse assertion is valid too: equation (9.7) implies that the operator $\boldsymbol{B}$ is of the form (9.6).

The norms of a positive definite operator $\boldsymbol{A}$ and its square root $\boldsymbol{A}^{\frac{1}{2}}$ are connected through the relation

$$\|\boldsymbol{A}^{\frac{1}{2}}\| = \|\boldsymbol{A}\|^{\frac{1}{2}},$$

whence it follows immediately that

$$\|\boldsymbol{B}\| = \|\boldsymbol{A}\|^{\frac{1}{2}}, \tag{9.8}$$

where $\boldsymbol{B}$ is any operator of the form (9.6).

---

[1] In English this notion is more often called ‘positive semidefinite’ or sometimes simply ‘positive’.

[2] See, for instance, L. A. Ljusternik and V. I. Sobolev [1], N. I. Akhiezer and I. M. Glazman [1].

9.2 *Splitting of an operator*[1]

Suppose that a Hilbert space H forms a dense set in a Banach space E and that *the imbedding operator of H into E is continuous*:

$$\|x\|_E \leqq C\|x\|_H \quad (x \in H). \tag{9.9}$$

It then follows that the adjoint space E^* is imbedded into H in the sense that a one-to-one correspondence is established between the elements of E^* and the elements of some linear submanifold of H. This means that each functional $y \in E^*$ can be considered as an element of the Hilbert space H, such that the following relation is fulfilled:

$$y(x) = (x, y) \quad (x \in H). \tag{9.10}$$

If this condition is fulfilled, the spaces

$$E^* \subset H \subset E \tag{9.11}$$

are said to constitute a *regular triple of spaces*. For example, the spaces $E^* = L_\alpha$ $(0 \leqq \alpha \leqq \frac{1}{2})$, $H = L_{\frac{1}{2}}$, $E = L_{1-\alpha}$ can be considered a regular triple.

In the sequel it is convenient to extend the use of the scalar product symbol to cases in which one of the components belongs to a space E and the other to the space E^*. If $x \in E$, $y \in E^*$, then (x, y) is the value of the functional y on the element x; if $x \in E^*$, $y \in E$, then $\overline{(x, y)}$ is the value of the functional x on the element y.

Suppose that a linear operator A acts from a space E to the space E^*. Such operators will be called *constrictions*. In many problems the splitting of a constriction into a product of the form

$$A = A_1 A_2 \tag{9.12}$$

---

[1] The problem of splitting a linear operator in various situations was studied in the papers: A. Hammerstein [1], N. Golomb [1, 2], M. A. Krasnoselskii [4–7], M. A. Krasnoselskii and V. I. Sobolev [1], M. M. Vainberg [4, 6], V. I. Sobolev [1], M. A. Krasnoselskii and Y. B. Rutikii [5, 6] and others. Here we present theorems obtained in the paper by M. A. Krasnoselskii and S. G. Krein [1].

is useful, where A_2 acts from E to H and A_1 from H to E^*. The question of whether the representation (9.12) is always available will not be investigated for the general case.

A continuous operator A acting from E to E^* is called *selfadjoint* if

$$(Ax, y) = (x, Ay) \qquad (x, y \in E). \tag{9.13}$$

It can be proved that (9.13) is valid provided that

$$(Ax, y) = (x, Ay) \qquad (x, y \in H). \tag{9.14}$$

Indeed, for any elements $x_0, y_0 \in E$ there are sequences $x_n, y_n \in H$ $(n = 1, 2, \ldots)$, converging to x_0, y_0 in E. Then a passage to the limit in the relations

$$(Ax_n, y_n) = (x_n, Ay_n) \qquad (n = 1, 2, \ldots)$$

is permissible. The limit relation has the form

$$(Ax_0, y_0) = (x_0, Ay_0).$$

Hence (9.13) follows from (9.14).

A selfadjoint operator A acting from E to E^* is called *positive definite*[1] if

$$(Ax, x) \geqq 0 \qquad (x \in E). \tag{9.15}$$

It is easily seen that (9.15) is valid provided that

$$(Ax, x) \geqq 0 \qquad (x \in H). \tag{9.16}$$

The positive square root of A, considered as an operator in H, is denoted by $A^{\frac{1}{2}}$.

THEOREM 9.1: *Let A act from E to E^* and be a continuous, selfadjoint and positive definite operator. Let the range of A be contained in a subspace E_0^* of the space E^*. Then $A^{\frac{1}{2}}$ is a continuous operator acting from H to E_0^*.*

---

[1] See the footnote to equation (9.5) (Ed.).

PROOF: Clearly

$$\|A^{\frac{1}{2}}x\|_H = \sqrt{(A^{\frac{1}{2}}x, A^{\frac{1}{2}}x)} =$$

$$= \sqrt{(Ax, x)} \leqq \sqrt{\|Ax\|_{E^*}\|x\|_E} \quad (x \in H), \tag{9.17}$$

whence it follows that

$$\|A^{\frac{1}{2}}x\|_H \leqq a\|x\|_E \quad (x \in H), \tag{9.18}$$

where

$$a^2 = \|A\|_{E \to E^*}.$$

Inequality (9.18) shows that $A^{\frac{1}{2}}$ admits a continuous extension to an operator C, defined on the entire space E and acting from E to H; moreover

$$\|Cx\|_H \leqq a\|x\|_E \quad (x \in E). \tag{9.19}$$

The adjoint operator C^* acts from H to E^*, and the following inequality holds:

$$\|C^*y\|_{E^*} \leqq a\|y\|_H \quad (y \in H). \tag{9.20}$$

But for any $x, y \in H$ the following relation is valid:

$$(x, A^{\frac{1}{2}}y) = (A^{\frac{1}{2}}x, y) = (Cx, y) = (x, C^*y),$$

whence it follows that $C^*y = A^{\frac{1}{2}}y$ for $y \in H$. Thus inequality (9.20) means that

$$\|A^{\frac{1}{2}}y\|_{E^*} \leqq a\|y\|_H \quad (y \in H). \tag{9.21}$$

We have proved that $A^{\frac{1}{2}}$ acts from H to E^* and is continuous.

Denote by H_0 the closure (in H) of the range of the operator $A^{\frac{1}{2}}$ on H. The selfadjointness of $A^{\frac{1}{2}}$ implies that it vanishes on the orthogonal complement H_1 of H_0 in H. Hence the range of the operator $A^{\frac{1}{2}}$ on H coincides with the range on H_0.

Take $x_0 \in H_0$. A sequence $y_n \in H$ can be chosen, such that elements $A^{\frac{1}{2}}y_n$ converge to x_0 in H. Then the elements Ay_n converge to $A^{\frac{1}{2}}x_0$ in the norm of E^*. But $Ay_n \in E_0^*$, hence $A^{\frac{1}{2}}x_0 \in E_0^*$. The theorem has been proved.

In the following the notation $A^{\frac{1}{2}}$ (instead of $\boldsymbol{C}$) will be retained for the continuous extension of the square root to the entire space E.

Theorem 9.1 gives a positive answer to the question of whether it is possible to split a selfadjoint positive definite operator into a product of the form given in (9.12). Furthermore, this theorem implies that: *the operator* $\boldsymbol{A}$ *can be written in the form*

$$\boldsymbol{A} = \boldsymbol{B}\boldsymbol{B}^*, \tag{9.22}$$

where $\boldsymbol{B}$ *is an operator of the form* (9.6) *which acts from* H *to* E^* *and is continuous.*

THEOREM 9.2: *Let* A *act from* E *to* E^* *and be a positive definite compact selfadjoint operator. Then* $A^{\frac{1}{2}}$ *is a compact operator acting from* H *to* E^*.

PROOF: It will first be shown that $A^{\frac{1}{2}}$ is compact when considered as an operator from E to H. To this end, it suffices to show that it transforms the intersection T of the unit ball $\|x\|_E \leqq 1$ and the space H to a set compact in H.

Given a sequence $x_n \in T$ $(n = 1, 2, \ldots)$, the compactness of A implies that there can be chosen a subsequence $x_{n_k} \in T$ such that the elements Ax_{n_k} converge in E^*. But then

$$\lim_{k,l\to\infty} \|A^{\frac{1}{2}}x_{n_k} - A^{\frac{1}{2}}x_{n_l}\|_H$$

$$= \lim_{k,l\to\infty} \sqrt{(A(x_{n_k} - x_{n_l}), x_{n_k} - x_{n_l})} = 0.$$

Consequently $A^{\frac{1}{2}}$ is compact as an operator from E to H. Its adjoint operator acts from H to E^* and is compact too. But the adjoint operator coincides on H with the operator $A^{\frac{1}{2}}$. The theorem has been proved.

In the following Theorems 9.1 and 9.2 will be applied mainly to spaces $E = L_{1-\alpha}$, $H = L_{\frac{1}{2}}$, $E^* = L_\alpha$ $(0 \leqq \alpha < \frac{1}{2})$. In the case $E^* = L_0$, the space C of continuous functions will usually be selected as the subspace E_0^*.

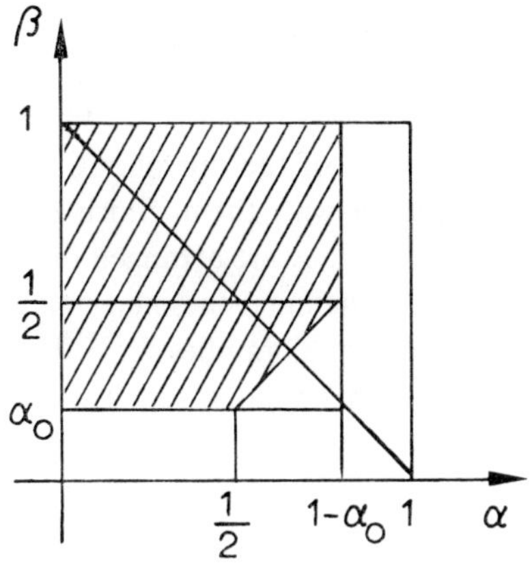

Figure 9.1

9.3 *L-Characteristic of a square root*

Consider a regular triple of spaces

$$L_{\alpha_0} \subset L_{\frac{1}{2}} \subset L_{1-\alpha_0}, \tag{9.23}$$

where $\alpha_0 \in [0, \frac{1}{2})$. Suppose that an operator A satisfies the conditions of Theorem 9.1 with respect to the triple (9.23). This theorem implies that the L-characteristic $L(A^{\frac{1}{2}}; \text{cont.})$ contains two points: $\{\frac{1}{2}, \alpha_0\}$ and $\{1 - \alpha_0, \frac{1}{2}\}$. It then follows from the interpolation Theorem 2.4 that the L-characteristic $L(A^{\frac{1}{2}}; \text{cont.})$ contains the entire polygon, shown in Fig. 9.1.

If the operator A satisfies the conditions of Theorem 9.2, a similar remark can be given concerning the L-characteristic $L(A^{\frac{1}{2}}; \text{comp.})$. It suffices to use Theorem 3.10 to obtain this result.

9.4 *Representation of compact operators*[1]

Each compact selfadjoint operator C acting in a Hilbert space H can be written in the form

---

[1] Such representations have been studied by many authors. A result similar to Theorem 9.3 was established by E. I. Pustylnik [1, 5]. Here we give an essentially simpler proof.

$$Cx = \sum_{k=1}^{\infty} \lambda_k (x, e_k) e_k, \tag{9.24}$$

where the λ_k are the eigenvalues of the operator C and the e_k are the normalized eigenvectors associated with them. Since the eigenvalues λ_k of C are real, it may be assumed that they are arranged in the order of decreasing absolute value; if C is positive definite, these eigenvalues are non-negative. The compactness of the operator C implies $\lambda_k \to 0$. The series (9.24) converges in H, uniformly on each bounded set. This means that the operators C_n

$$C_n x = \sum_{k=1}^{n} \lambda_k (x, e_k) e_k \tag{9.25}$$

converge to C in the uniform operator norm.

The representation (9.24) plays a decisive role in the theory of selfadjoint operators.

LEMMA 9.1: *Let $E^* \subset H \subset E$ be a regular triple of spaces. Let the operator* (9.24) *act from H to E^* and be compact. Then the operators* (9.25) *converge to the operator* (9.24) *in norm, as operators acting from H to E^*.*

PROOF: Introduce the notation

$$Q_n x = \sum_{k=1}^{n} (x, e_k) e_k. \tag{9.26}$$

The operator C^* adjoint to C can be considered as a compact operator acting from E to H. The range of the operator C^* on the unit ball $\|x\|_E \leqq 1$ forms a compact set in H. Hence

$$\lim_{n\to\infty} \sup_{\|x\|_E \leqq 1} \|Q_n C^* x - C^* x\|_H = \lim_{n\to\infty} \|Q_n C^* - C^*\|_{E\to H} = 0. \tag{9.27}$$

The relation

$$\|CQ_n - C\|_{H\to E^*} = \|Q_n C^* - C^*\|_{E\to H}$$

then implies that

$$\lim_{n\to\infty} \|\boldsymbol{C}\boldsymbol{Q}_n - \boldsymbol{C}\|_{H\to E^*} = 0.$$

It remains only to note that $\boldsymbol{C}_n = \boldsymbol{C}\boldsymbol{Q}_n$. The lemma has been proved.

Now let $\boldsymbol{B}$ be an arbitrary compact operator acting from H to E^*. This operator can be written in the form

$$\boldsymbol{B} = \boldsymbol{C}\boldsymbol{U},$$

where $\boldsymbol{U}$ is a partial isometry (see (9.6)), and $\boldsymbol{C}$ is a selfadjoint positive definite operator. The compactness of the operator $\boldsymbol{B}$ implies the compactness of $\boldsymbol{C}$, both as an operator acting in H, and as an operator acting from H to E^*. Let the operator $\boldsymbol{C}$ be written in the form (9.24). Then one obtains for the operator $\boldsymbol{B}$ the representation

$$\boldsymbol{B}x = \sum_{k=1}^{\infty} \lambda_k(x, g_k)e_k, \tag{9.28}$$

where $g_k = \boldsymbol{U}^* e_k$. The representation (9.28) will be called the *expansion of an operator along the associated fundamental functions of Schmidt*. Lemma 9.1 implies that *the series* (9.28) *converges*[1] *in operator norm, as operators acting from H to E**.

A similar assertion can be made concerning the operator $\boldsymbol{B}^*$ adjoint to $\boldsymbol{B}$. Represent this operator in the form

$$\boldsymbol{B}^* = \boldsymbol{U}^*\boldsymbol{C}^*.$$

If $\boldsymbol{B}$ acts compactly from H to E^*, then $\boldsymbol{B}^*$ and $\boldsymbol{C}^*$ act compactly from E to H. It follows from Lemma 9.1[2] that

$$\|\boldsymbol{B}^*\boldsymbol{Q}_n - \boldsymbol{B}^*\|_{E\to H} = \|\boldsymbol{C}^*\boldsymbol{Q}_n - \boldsymbol{C}^*\|_{E\to H} \to 0$$

---

[1] Here and in the following, when the convergence of a series of the type (9.28) in operator norm is mentioned, the convergence of the sequence of operators defined by the partial sums of the series is meant.

[2] Translator's note. A minor modification is necessary.

as $n \to \infty$. But this means that $\boldsymbol{B}^*$ possesses the representation

$$\boldsymbol{B}^* x = \sum_{k=1}^{\infty} \lambda_k(x, e_k) g_k \tag{9.29}$$

where the series converges in operator norm.

Consider next a positive definite selfadjoint operator $\boldsymbol{A}$ acting compactly from E to E^*. On the basis of Theorem 9.2 it is possible to split it, i.e. to represent it in the form

$$\boldsymbol{A} = \boldsymbol{B}\boldsymbol{B}^*,$$

where the operators $\boldsymbol{B}$ and $\boldsymbol{B}^*$ act respectively from H to E^* and from E to H, and are compact. Hence $\boldsymbol{B}$ and $\boldsymbol{B}^*$ can be represented by the series of the form (9.28) and (9.29), respectively. Thus the following representation is obtained for $\boldsymbol{A}$:

$$\boldsymbol{A} x = \sum_{k=1}^{\infty} \lambda_k^2(x, e_k) e_k, \tag{9.30}$$

where the series converges in operator norm, as operators acting from E to E^*. It is not difficult to see that the λ_k^2 are the eigenvalues of the operator $\boldsymbol{A}$, and that the e_k are the corresponding eigenvectors. The series (9.30) obviously converges in operator norm, as operators acting in the Hilbert space H.

The above result is formulated as follows:

THEOREM 9.3: *Let μ_k and e_k be corresponding eigenvalues and eigenvectors of a positive definite compact selfadjoint operator $\boldsymbol{A}$ in a Hilbert space H. Let $\boldsymbol{A}$ also be a compact operator acting from E to E*, where $E^* \subset H \subset E$ is some regular triple of spaces. Then the operators*

$$\boldsymbol{A}_n x = \sum_{k=1}^{n} \mu_k(x, e_k) e_k \tag{9.31}$$

converge to the operator $\boldsymbol{A}$ in operator norm, as operators acting from E to E.*

9.5 *Square root of integral operator*

The following is an interesting problem: under what conditions does a compact selfadjoint operator

$$Ax = \sum_{k=1}^{\infty} \lambda_k (x, e_k) e_k, \tag{9.32}$$

acting in $L_{\frac{1}{2}}$ admit an integral representation

$$Ax(t) = \int_{\Omega} K(t, s) x(s) \mathrm{d}s. \tag{9.33}$$

This problem has not been satisfactorily investigated.[1]

The series (9.32) converges in operator norm, as operators acting in $L_{\frac{1}{2}}$. This means that A is the limit, in norm, of the sequence A_n of integral operators

$$A_n x(t) = \int_{\Omega} K_n(t, s) x(s) \mathrm{d}s, \tag{9.34}$$

with degenerate kernels

$$K_n(t, s) = \sum_{k=1}^{n} \lambda_k e_k(t) e_k(s). \tag{9.35}$$

---

[1] As was shown recently by V. B. Korotkov [1], a selfadjoint operator A satisfies the condition (6.22) when and only when

$$\sum_{k=1}^{\infty} \lambda_k^2 \, |e_k(t)|^2 \leqq u_0^2(t).$$

Hence a remark at the end of 6.3° implies that a selfadjoint operator A is an integral operator if for almost all t

$$\sum_{k=1}^{\infty} \lambda_k^2 \, |e_k(t)|^2 < \infty.$$

It is natural to expect that the operator in (9.32) admits the representation (9.33) in cases where the degenerate kernels (9.35) converge in some sense to a function $K(t, s)$, which is then the kernel of the integral operator A.

It is easy to see that *the degenerate kernels* (9.35) *converge in mean to square summable function* $K(t, s)$ *when and only when*

$$\sum_{k=1}^{\infty} \lambda_k^2 < \infty. \tag{9.36}$$

Then for each function $x(s) \in L_{\frac{1}{2}}$

$$\|\int_{\Omega} K(t, s)x(s)\mathrm{d}s - A_n x(t)\|_{\frac{1}{2}}$$

$$\leqq \sqrt{\int_{\Omega}\int_{\Omega} |K(t, s) - K_n(t, s)|^2 \mathrm{d}s\,\mathrm{d}t} \cdot \|x(s)\|_{\frac{1}{2}}.$$

Hence, in such cases the operator A does admit the representation (9.33).

Now let A be a compact, selfadjoint, positive definite operator. Consider the square root $A^{\frac{1}{2}}$:

$$A^{\frac{1}{2}}x = \sum_{k=1}^{\infty} \sqrt{\lambda_k}(x, e_k)e_k. \tag{9.37}$$

From the above consideration it follows that the operator $A^{\frac{1}{2}}$ is an integral operator if the following condition is fulfilled

$$\sum_{k=1}^{\infty} \lambda_k < \infty. \tag{9.38}$$

Operators whose eigenvalues satisfy the condition (9.38) are called *nuclear*. The square root $A^{\frac{1}{2}}$ is likewise an integral operator when, for almost all $t \in \Omega$, one has the inequality[1]

$$\sum_{k=1}^{\infty} \lambda_k |e_k(t)|^2 < \infty.$$

---

[1] See the footnote on p. 183.

Conditions under which the square root of an integral operator is itself an integral operator can also be formulated in other terms.

THEOREM 9.4: *Let a symmetric positive definite kernel $K(t, s)$ define a continuous operator in $L_{\frac{1}{2}}$. Further let the following inequality be satisfied:*

$$|K(t, s)| \leqq u_0(t)u_0(s) \qquad (t, s \in \Omega) \tag{9.39}$$

where $u_0(t)$ is finite almost everywhere.

Then the square root $A^{\frac{1}{2}}$ of the integral operator A with kernel $K(t, s)$ is also an integral operator.

PROOF: Without loss of generality it may be assumed that $u_0(t)$ is positive. Denote by E_{u_0} (see 6.2°) the totality of functions $y(t)$ $(t \in L_{\frac{1}{2}})$ which for some $\mu > 0$ fulfill the inequality

$$-\mu u_0(t) \leqq y(t) \leqq \mu u_0(t).$$

This is a Banach space when the norm of $y(t)$ is defined to be the infimum of those μ which fulfill the above inequality. Denote by E'_{u_0} the totality of functions $x(t)$, for which

$$\|x\|^*_{u_0} = \int_\Omega |x(t)|\, u_0(t)\,\mathrm{d}t < \infty.$$

Condition (9.39) implies that each function $x(t) \in E'_{u_0}$ satisfies the inequality

$$-u_0(t)\,\|x\|^*_{u_0} \leqq Ax(t) \leqq u_0(t)\,\|x\|^*_{u_0}.$$

This means that A acts from E'_{u_0} to E_{u_0} and is continuous, with norm $\leqq 1$.

Let us now consider the operator $A^{\frac{1}{2}}$. It is clear that each $x(t) \in E'_{u_0} \cap L_{\frac{1}{2}}$ satisfies the inequality

$$\|A^{\frac{1}{2}}x\|_{\frac{1}{2}} = \sqrt{(Ax, x)} \leqq \sqrt{\|Ax\|_{u_0} \cdot \|x\|^*_{u_0}} \leqq \|x\|^*_{u_0}.$$

Thus the operator $A^{\frac{1}{2}}$ can be continuously extended to an operator $\boldsymbol{B}$, acting from E'_{u_0} to $L_{\frac{1}{2}}$. Its adjoint operator $\boldsymbol{B}^*$ acts from $L_{\frac{1}{2}}$ to E_{u_0} and is bounded (its norm is not greater than 1).

The operator $\boldsymbol{B}^*$ admits an integral representation (see a remark at the end of 6.3°). Thus to complete the proof it suffices to show that $\boldsymbol{B}^*$ coincides with $A^{\frac{1}{2}}$. Moreover, it suffices to show that the operators $\boldsymbol{B}^*$ and $A^{\frac{1}{2}}$ take the same values on some set dense in $L_{\frac{1}{2}}$. Clearly one can take $L_{\frac{1}{2}} \cap E'_{u_0}$ as such a set. The theorem has been proved.

In particular, Theorem 9.4 implies that the square root of an integral operator with bounded kernel is also an integral operator. This result also admits a simple direct proof. Boundedness implies that A acts from L_1 to L_0 and is continuous. Then by Theorem 9.1 the square root $A^{\frac{1}{2}}$ is a bounded operator acting from L_1 to $L_{\frac{1}{2}}$, but all such operators are integral operators.

It is well known that the square root of an integral operator is also an integral operator, if the kernel in question is the Green function of some elliptic differential operator of second order.

Suppose that the square root of an integral operator

$$Ax(t) = \int_{\Omega} K(t, s)x(s)\mathrm{d}s \tag{9.40}$$

admits the integral representation

$$A^{\frac{1}{2}}x(t) = \int_{\Omega} K_{\frac{1}{2}}(t, s)x(s)\mathrm{d}s. \tag{9.41}$$

If the operator (9.41) is regular, then from the relation

$$\|A^{\frac{1}{2}}x\|_{\frac{1}{2}}^2 = \int_{\Omega} [\int_{\Omega} K_{\frac{1}{2}}(t, s)x(s)\mathrm{d}s \cdot \int_{\Omega} K_{\frac{1}{2}}(t, \sigma)x(\sigma)\mathrm{d}\sigma]\mathrm{d}t$$

$$= \int_{\Omega} \int_{\Omega} [\int_{\Omega} K_{\frac{1}{2}}(t, s)K_{\frac{1}{2}}(t, \sigma)\mathrm{d}t]x(s)x(\sigma)\mathrm{d}s\,\mathrm{d}\sigma$$

$$= \int_{\Omega} \int_{\Omega} K(s, \sigma)x(\sigma)x(s)\mathrm{d}s\,\mathrm{d}\sigma = (Ax, x) \tag{9.41}$$

it follows that $K(t, s)$ can be considered as the iterated kernel

$$K(t, s) = \int_{\Omega} K_{\frac{1}{2}}(t, \sigma)K_{\frac{1}{2}}(s, \sigma)\mathrm{d}\sigma. \tag{9.43}$$

It is natural to raise the question as to whether a representation of the kernel $K(t, s)$ in the form

$$K(t, s) = \int_{\Omega} R(t, \sigma) R(s, \sigma) d\sigma \tag{9.44}$$

implies that $A^{\frac{1}{2}}$ is an integral operator with kernel $R(t, s)$. A positive answer is easily obtained for this question, when the kernel $R(t, s)$ is non-negative.[1] The derivation is left to the reader.

9.6 *Example*

In this section[2] we show how to construct an integral operator A whose kernel is summable to any power, but whose square root $A^{\frac{1}{2}}$ is not an integral operator.

Denote by T_α $(0 \leqq \alpha \leqq 2\pi)$ the operator defined on functions $x(s)$ $s \in [0, 2\pi]$ by the relation

$$T_\alpha x(s) = \begin{cases} x(s + \alpha) & \text{if } 0 \leqq s \leqq 2\pi - \alpha \\ x(s + \alpha - 2\pi) & \text{if } 2\pi - \alpha \leqq s \leqq 2\pi. \end{cases}$$

Consider on $L_{\frac{1}{2}}[0, 2\pi]$ the operator

$$Ax(t) = \int_0^{2\pi} K(t - s) x(s) ds \tag{9.45}$$

where $K(u)$ is an even, 2π-periodic function defined on the whole axis. It is easy to prove that the operator (9.45) commutes with all operators T_α. Evenness of the function $K(u)$ implies that the operator (9.45) is selfadjoint. Assume further that the operator (9.45) is positive definite. Since the operators T_α commute with the operator A, a general theorem on functions of operators implies that the square root $A^{\frac{1}{2}}$ commutes with all the operators T_α.

Suppose that the operator $A^{\frac{1}{2}}$ is an integral operator with a kernel $Q(t, s)$. Then the relation

$$T_\alpha \int_0^{2\pi} Q(t, s) x(s) ds = \int_0^{2\pi} Q(t, s) T_\alpha x(s) ds \quad (0 \leqq \alpha \leqq 2\pi)$$

---

[1] Translator's note. This assertion needs some modification. See p. 189.

[2] Here we use a proposition of B. S. Mitijagin (who showed us a slightly different example).

implies that $Q(t, s)$ is a symmetric function depending only on the difference $t - s$:

$$Q(t, s) = f(t - s).$$

Now let the Fourier series of the function $K(u)$ have the form

$$K(u) \sim \sum_{k=1}^{\infty} a_k^2 \cos(2^k u), \tag{9.46}$$

where

$$\sum_{k=1}^{\infty} a_k^4 < \infty, \quad \sum_{k=1}^{\infty} a_k^2 = \infty. \tag{9.47}$$

A general theorem on lacunary series (N. K. Bari [1]) implies that the function $K(u)$ is summable to any power.

It can be shown that the non-zero eigenvalues of the operator (9.45) coincide with the a_k^2 ($k = 1, 2, \ldots$). To each of those eigenvalues belong two eigenfunctions $\cos(2^k s)$ and $\sin(2^k s)$. The operator $A^{\frac{1}{2}}$ has the same eigenfunctions, but its eigenvalues are equal to a_k ($k = 1, 2, \ldots$). If the operator $A^{\frac{1}{2}}$ were an integral operator with a kernel $Q(t, s) = f(t - s)$, then the Fourier series of the function $f(u)$ would have the form

$$f(u) \sim \sum_{k=1}^{\infty} a_k \cos(2^k u). \tag{9.48}$$

This series is also lacunary and the sum of the squares of its Fourier coefficients is infinite. Such a series is never the Fourier series of any function (N. K. Bari [1]). Consequently the square root $A^{\frac{1}{2}}$ of the operator (9.45) does not admit an integral representation.

9.7 *Investigation of integral operators by means of properties of iterated kernels*

Consider an integral operator

$$\boldsymbol{K}x(t) = \int_{\Omega} K(t, s)x(s)\,ds \tag{9.49}$$

with a non-negative kernel. Then the iterated kernel

$$K_2(t, s) = \int_{\Omega} K(t, \sigma)K(s, \sigma)\mathrm{d}\sigma \tag{9.50}$$

is symmetric and positive definite, whether the kernel $K(t, s)$ possesses these properties or not.

THEOREM 9.5: *Suppose the operator*

$$\boldsymbol{K}_2 x(t) = \int_{\Omega} K_2(t, s)x(s)\mathrm{d}s \tag{9.51}$$

acts from L_α ($\frac{1}{2} \leqq \alpha \leqq 1$) *to* $L_{1-\alpha}$ *and is continuous. Then the operator* (9.49) *acts from* $L_{\frac{1}{2}}$ *to* $L_{1-\alpha}$ *and is continuous.*

THEOREM 9.6: *Suppose the operator* (9.51) *acts from* L_α ($\frac{1}{2} \leqq \alpha \leqq 1$) *to* $L_{1-\alpha}$ *and is compact. Then the operator* (9.49) *acts from* $L_{\frac{1}{2}}$ *to* $L_{1-\alpha}$ *and is compact.*

Both these assertions result from Theorems 9.1 and 9.2. To apply those theorems, it suffices to show that the operator $\boldsymbol{K}_2$ can be written in the form $\boldsymbol{K}_2 = \boldsymbol{K}\boldsymbol{K}^{\#}$, where

$$\boldsymbol{K}^{\#} x(t) = \int_{\Omega} K(s, t)x(s)\mathrm{d}s,$$

and to establish that $\boldsymbol{K}^{\#}$ is continuous in the space $L_{\frac{1}{2}}$. The first of these assertions results from Fubini's theorem while the second results from the chain of relations

$$\|\boldsymbol{K}^{\#} x\|_{\frac{1}{2}}^2 = \int_{\Omega}\{\int_{\Omega} K(t, \sigma)[\int_{\Omega} K(s, \sigma)x(s)\mathrm{d}s]x(t)\mathrm{d}t\}\mathrm{d}\sigma$$

$$= \int_{\Omega}\{\int_{\Omega}[\int_{\Omega} K(t, \sigma)K(s, \sigma)\mathrm{d}\sigma]x(s)\mathrm{d}s\}x(t)\mathrm{d}t = (\boldsymbol{K}_2 x, x).$$

Theorems 9.5 and 9.6 have been proved.

Theorems 9.5 and 9.6 can be applied to investigate the L-characteristic of the operator $\boldsymbol{K}$. If the kernel $K(t, s)$ is in addition, symmetric, these theorems imply that the L-characteristic contains the whole segment joining $\{\frac{1}{2}, \alpha\}$ to $\{1 - \alpha, \frac{1}{2}\}$ (see 9.3°).

9.8 Remark on Mercer's theorem

THEOREM 9.7: *Let a bounded kernel $K(t, s)$ be symmetric and positive definite, and let*

$$\boldsymbol{K}x(t) = \int_{\Omega} K(t, s) x(s) \mathrm{d}s \tag{9.52}$$

be compact as an operator from L_1 to L_0. Then the bilinear expansion of the kernel

$$K(t, s) \sim \sum_{k=1}^{\infty} \lambda_k e_k(t) e_k(s) \tag{9.53}$$

(where λ_k, e_k are eigenvalues and eigenfunctions of the kernel $K(t, s)$) is a uniformly convergent expansion (up to a set of measure zero).

PROOF: Theorem 9.3 shows that the norms of the integral operators

$$\boldsymbol{F}_n x(t) = \int_{\Omega} \{K(t, s) - \sum_{k=1}^{n} \lambda_k e_k(t) e_k(s)\} x(s) \mathrm{d}s$$

(as operators from L_1 to L_0) converge to zero as $n \to \infty$. But

$$\|\boldsymbol{F}_n\|_{1 \to 0} = \operatorname*{ess\,sup}_{t \in \Omega} \{\operatorname*{ess\,sup}_{s \in \Omega} |K(t, s) - \sum_{k=1}^{n} \lambda_k e_k(t) e_k(s)|\},$$

hence

$$\lim_{n \to \infty} \operatorname*{ess\,sup}_{t \in \Omega} \{\operatorname*{ess\,sup}_{s \in \Omega} |K(s, t) - \sum_{k=1}^{n} \lambda_k e_k(t) e_k(s)|\} = 0.$$

The theorem has been proved.[1]

---

[1] This theorem contains the well-known theorem of Mercer (see, for instance, I. G. Petrovskii [1]). Here we have presented a new proof (P. P. Zabreiko, M. A. Krasnoselskii and E. I. Pustylnik [3]). Another functional analytic scheme of proof of a similar assertion was given by M. G. Krein [1], who developed, in this connection, a theory of Banach spaces with two norms.

The assumption that the operator (9.52) is compact as an operator from L_1 to L_0, is fulfilled, for instance, when the set Ω is bounded and closed and the kernel $K(t, s)$ is continuous. Theorem 6.4 gives more general conditions.

§ 10 Fractional powers of bounded operators[1]

10.1 *The spectral function*

We recall several notions connected with selfadjoint operators, acting in a Hilbert space H. All definitions and assertions of this paragraph apply to the case of a real Hilbert space as well as to a complex Hilbert space.

Let $\boldsymbol{A}$ and $\boldsymbol{B}$ be two selfadjoint operators. In the sequel $\boldsymbol{A} \prec \boldsymbol{B}$ means

$$(\boldsymbol{A}x, x) \leqq (\boldsymbol{B}x, x) \qquad (x \in H).$$

It is easy to see that orthogonal projections $\boldsymbol{P}_1$ and $\boldsymbol{P}_2$ on subspaces H_1, $H_2 \subset H$ satisfy the inequality $\boldsymbol{P}_1 \prec \boldsymbol{P}_2$ if and only if $H_1 \subset H_2$. In this case the operator $\boldsymbol{P}_3 = \boldsymbol{P}_2 - \boldsymbol{P}_1$ is the orthogonal projection on the subspace H_3 which is the orthogonal complement to H_1 in H_2 ($H_3 = H_2 \ominus H_1$).

An operator-function $\boldsymbol{P}_\lambda$ of real parameter λ ($-\infty < \lambda < \infty$) is called a *spectral function* if it possesses the following properties:

1°) The values of this function $\boldsymbol{P}_\lambda$ are orthogonal projections.

2°) $\boldsymbol{P}_{\lambda_1} \prec \boldsymbol{P}_{\lambda_2}$ if $\lambda_1 < \lambda_2$.

3°) For each element $x \in H$ the following relations are valid:

$$\lim_{\lambda \to -\infty} \|\boldsymbol{P}_\lambda x\| = 0 \text{ and } \lim_{\lambda \to \infty} \|\boldsymbol{P}_\lambda x - x\| = 0.$$

These relations permit us to write that $\boldsymbol{P}_{-\infty} = 0$ and $\boldsymbol{P}_\infty = \boldsymbol{I}$ (where $\boldsymbol{I}$ is the identity).

---

[1] The theory of selfadjoint operators can be found in the books by L. B. Kantorovic and G. P. Akilov [1], N. I. Akniezer and I. M. Glazman [1], and M. A. Naimark [1]. The theorems on fractional powers presented in this paragraph were established in weaker form by M. A. Krasnoselskii [5, 7] using other methods. Exact results were obtained by M. A. Krasnoselskii and E. I. Pustylnik [1].

4°) For each fixed $x \in H$ the function $\boldsymbol{P}_\lambda x$ is continuous from the right:

$$\lim_{\varepsilon \to 0+} \|\boldsymbol{P}_{\lambda+\varepsilon} x - \boldsymbol{P}_\lambda x\| = 0.$$

Denote by H_λ the subspace on which the operator $\boldsymbol{P}_\lambda$ projects $-\infty < \lambda < \infty$. It is easy to see that *the family of subspaces H_λ possesses the following properties: $H_{\lambda_1} \subset H_{\lambda_2}$ for $\lambda_1 < \lambda_2$, the intersection of all H_λ consists only of the zero element, the union of all H_λ is dense in H and, finally, $\bigcap_{\varepsilon>0} H_{\lambda+\varepsilon} = H_\lambda$ for each fixed λ.* The converse assertion is valid: *if a family of subspaces H_λ $(-\infty < \lambda < \infty)$ possesses the properties mentioned above, the orthogonal projections $\boldsymbol{P}_\lambda$ on these subspaces constitute a spectral function.*

Let us present two examples of spectral functions.

Let $e_1, e_2, \dots$ be a complete orthogonal system in H, and let $\lambda_1, \lambda_2, \dots$ be a non-decreasing sequence. Denote by H_λ the finite-dimensional subspace which is the linear span of all e_i such that $\lambda_i \leqq \lambda$. The orthogonal projection $\boldsymbol{P}_\lambda$ on H_λ is obviously defined by the relation

$$\boldsymbol{P}_\lambda = \sum_{\lambda_i \leqq \lambda} (x, e_i) e_i. \tag{10.1}$$

The operators (10.1) constitute a spectral function.

Consider for the second example, the space $L_{\frac{1}{2}}$ of functions square summable on the interval $[a, b]$. Put $\boldsymbol{P}_\lambda = 0$ for $-\infty < \lambda < a$, $\boldsymbol{P}_\lambda x(t) = x(t)\kappa(t; \lambda)$ for $a < \lambda \leqq b$, where $\kappa(t, \lambda)$ is the characteristic function of the interval $[a, \lambda]$ and, finally, $\boldsymbol{P}_\lambda x(t) = x(t)$ for $\lambda \geqq b$. The operator-function $\boldsymbol{P}_\lambda$ defined in this way is spectral.

Let $\boldsymbol{P}_\lambda$ be a spectral function. Let $\Delta = (\lambda_1, \lambda_2]$. As remarked already, the operator $\boldsymbol{P}_\Delta = \boldsymbol{P}_{\lambda_2} - \boldsymbol{P}_{\lambda_1}$ is an orthogonal projection on some subspace H_Δ. Suppose that the real axis is split into intervals $\Delta_1 = (-\infty, \lambda_1]$, $\Delta_2 = (\lambda_1, \lambda_2], \dots, \Delta_n = (\lambda_{n-1}, \infty]$. Then the subspaces $H_{\Delta_1}, H_{\Delta_2}, \dots$ are orthogonal to each other and their orthogonal sum coincides with the whole space

$$H = H_{\Delta_1} \oplus H_{\Delta_2} \oplus \dots \oplus H_n.$$

Correspondingly, the following formula holds:

$$\boldsymbol{I} = \boldsymbol{P}_{\Delta_1} + \boldsymbol{P}_{\Delta_2} + \dots + \boldsymbol{P}_{\Delta_n}, \tag{10.2}$$

decomposition of identity. Here $\boldsymbol{P}_{\Delta_j}\boldsymbol{P}_{\Delta_i} = 0$ for $i \neq j$.

10.2 *Fractional powers of bounded selfadjoint operators*

Let A be a bounded selfadjoint operator acting in a Hilbert space H. It can be shown that it determines uniquely a spectral function P_λ with the following two properties. First, each subspace H_Δ is invariant under the operator A. Second, on each subspace H_Δ, where $\Delta = (\lambda - \varepsilon, \lambda + \varepsilon]$, the operator A is 'almost' an operator of stretch multiplication by the factor λ:

$$\|Ax - \lambda x\| \leqq \varepsilon \|x\| \qquad (x \in H).$$

These properties imply, as is easily seen, that $P_\lambda = 0$ for $\lambda < -\|A\|$ and that $P_\lambda = I$ for $\lambda \geqq \|A\|$. By the second property the operator A can be approximated with any desired accuracy (in operator norm) by sums of the form $\sum \lambda_j P_{\Delta_j}$ where $\lambda_0 P_0 = -\|A\| P_{-\|A\|}$, and other λ_j are numbers in the intervals Δ_j $(j = 1, 2, \ldots, n)$ into which the interval $(-\|A\|, \|A\|)$ is split. The limit of these sums can be written in the form of Stieltjes integral relative to the spectral function; this limit coincides with the operator A.

Thus we have arrived at the functional formula[1]

$$A = \int_{-\|A\|-0}^{\|A\|} \lambda \, \mathrm{d}P_\lambda. \tag{10.3}$$

In the case when A is compact, formula (10.3) takes a simple form

$$Ax = \sum_{i=1}^{\infty} \lambda_i (x, e_i) e_i, \tag{10.4}$$

where the λ_i are the eigenvalues of the operator A and the e_i are the eigenelements associated with them (see formula (9.24)).

---

[1] Here

$$\int_{\alpha-0}^{\beta} \lambda \, \mathrm{d}P_\lambda = \lim_{\varepsilon \to 0+} \int_{\alpha-\varepsilon}^{\beta} \lambda \, \mathrm{d}P_\lambda.$$

In the sequel α, instead of $\alpha - 0$, will be used to denote the lower limit.

For a positive definite selfadjoint operator the spectral function $\boldsymbol{P}_\lambda$ is equal to zero for $\lambda < 0$. For such operators, formula (10.3) has the form

$$\boldsymbol{A} = \int_0^{\|\boldsymbol{A}\|} \lambda \mathrm{d}\boldsymbol{P}_\lambda; \tag{10.5}$$

and in formula (10.4) all eigenvalues are non-negative.

Fractional powers $\boldsymbol{A}^\tau$ $(0 < \tau < \infty)$ of a positive definite bounded selfadjoint operator $\boldsymbol{A}$ are defined by the formula

$$\boldsymbol{A}^\tau = \int_0^{\|\boldsymbol{A}\|} \lambda^\tau \mathrm{d}\boldsymbol{P}_\lambda. \tag{10.6}$$

The integral on the right hand side means, as usual, the limit of the sums $\sum_{i=1}^{n} \lambda_i^\tau \boldsymbol{P}_{\Delta_i}$. The fractional powers $\boldsymbol{A}^\tau$ are bounded operators. It is easy to see that $\|\boldsymbol{A}^\tau\| = \|\boldsymbol{A}\|^\tau$ $(0 < \tau < \infty)$. It is convenient to assume that $\boldsymbol{A}^0$ coincides with the identity operator $\boldsymbol{I}$.

The operators $\boldsymbol{A}^\tau$ form a semi-group in the sense that

$$\boldsymbol{A}^\tau \boldsymbol{A}^\sigma = \boldsymbol{A}^\sigma \boldsymbol{A}^\tau = \boldsymbol{A}^{\sigma+\tau} \qquad (\tau, \sigma \geqq 0). \tag{10.7}$$

If $\tau_0 > 0$, then the following relation is obvious:

$$\lim_{\tau \to \tau_0} \|\boldsymbol{A}^\tau - \boldsymbol{A}^{\tau_0}\| = 0. \tag{10.8}$$

At $\tau_0 = 0$ formula (10.8) is not generally true. In this case a weaker assertion is valid: for each fixed $x \in H$

$$\lim_{\tau \to 0} \|\boldsymbol{A}^\tau x - \boldsymbol{Q}x\| = 0 \tag{10.9}$$

where $\boldsymbol{Q}$ is the orthogonal projection on the closure of the range of the operator $\boldsymbol{A}$.

10.3 *The fundamental theorem*

Our subsequent constructions will concern a linear operator A which acts, on the one hand, in the space $L_{\frac{1}{2}}$ and is, on the other hand, a continuous operator acting from some space L_α ($\alpha \geqq \frac{1}{2}$) to a space L_β with $\beta < \frac{1}{2}$. Previously such operators were called constrictions. It will be shown that fractional powers of a constriction are also constrictions. Formulas will be developed which permit us to find the corresponding parts of the L-characteristics of the operators A^τ when the L-characteristic of the operator A is known.

Assume that an operator A acts in the space $L_{\frac{1}{2}}$. Then it is defined on a dense set D in each space L_α, where $\alpha > \frac{1}{2}$. It may turn out that A admits continuous extension onto the whole of L_α, whereupon it can be considered as an operator acting from L_α to some L_β. Usually the same notation A is used for the operator extended continuously in this way. In order that this notation be justified, it is necessary first, that the values of the extended operator do not depend on choice of a space L_β and second, that the values of the operator, extended onto different spaces L_{α_1} and L_{α_2}, coincide on the smaller of them. The reasoning will hereafter be left to the reader.

After the notion of the natural extension of an operator A given originally only on the space $L_{\frac{1}{2}}$ is introduced, it is possible to speak of its L-characteristic. If A is selfadjoint in $L_{\frac{1}{2}}$, its L-characteristic will be symmetric (Fig. 10.1) with respect to the line $\alpha + \beta = 1$. Here if $\{\alpha, \beta\} \in L(A;$ cont.), then

$$(Ax, y) = (x, Ay) \qquad (x \in L_\alpha,\ y \in L_{1-\beta}).$$

In the preceding section fractional powers A^τ of a positive definite self-

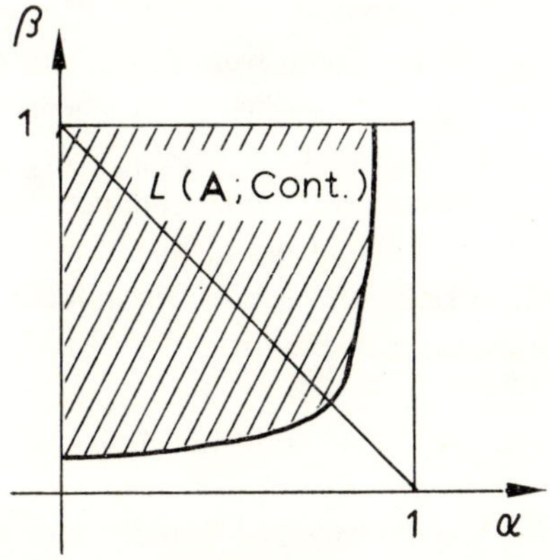

Figure 10.1

adjoint operator A were defined. The same symbols A^τ will denote their continuous extensions.

In this section operators in the complex spaces L_α are considered.

THEOREM 10.1: *Let a positive definite selfadjoint operator A, continuous in $L_{\frac{1}{2}}$, be, at the same time, a continuous operator acting from $L_{\frac{1}{2}}$ to L_β, where $0 \leqq \beta < \frac{1}{2}$, and suppose that*

$$\|A\|_{\frac{1}{2}\to\beta} \leqq a. \tag{10.10}$$

Then for each $\tau \in [0, 1]$ the operator A^τ is a continuous operator acting from $L_{\frac{1}{2}}$ to $L_{\beta(\tau)}$, where

$$\beta(\tau) = \tfrac{1}{2} + \tau(\beta - \tfrac{1}{2}); \tag{10.11}$$

and

$$\|A^\tau\|_{\frac{1}{2}\to\beta(\tau)} \leqq a^\tau. \tag{10.12}$$

PROOF: Assume that the assertion of the theorem is already proved for two values $\tau_1, \tau_2 \in [0, 1]$. Let us show that this implies its validity for $\tau_3 = (\tau_1 + \tau_2)/2$. The operator A^{τ_1} acts, by assumption, from $L_{\frac{1}{2}}$ to $L_{\beta(\tau_1)}$ and its norm does not exceed a^{τ_1}. By selfadjointness it acts from $L_{1-\beta(\tau_1)}$ to $L_{\frac{1}{2}}$ and its norm also does not exceed a^{τ_1}. Similarly the operator A^{τ_2} acts from $L_{\frac{1}{2}}$ to $L_{\beta(\tau_2)}$ and from $L_{1-\beta(\tau_2)}$ to $L_{\frac{1}{2}}$, and its norm in both cases does not exceed a^{τ_2}. This means that the operator $A^{2\tau_3} = A^{\tau_1+\tau_2}$ acts at the same time from $L_{1-\beta(\tau_1)}$ to $L_{\beta(\tau_2)}$ and from $L_{1-\beta(\tau_2)}$ to $L_{\beta(\tau_1)}$; here its norm does not exceed $a^{\tau_1+\tau_2}$. Applying interpolation Theorem 2.4, we have that $A^{2\tau_3}$ acts from $L_{1-\beta(\tau_3)}$ to $L_{\beta(\tau_3)}$ and $\|A^{2\tau_3}\|_{1-\beta(\tau_3)\to\beta(\tau_3)} \leqq a^{2\tau_3}$. The assertion of the theorem for the operator A^{τ_3} is now obtained as an immediate application of Theorem 9.1 on the splitting of operator.

The above result permits the use of an inductive method to prove the theorem for all numbers τ of the form

$$\tau = k/2^n \quad (k = 0, 1, \ldots, 2^n; n = 0, 1, \ldots). \tag{10.13}$$

Indeed, for $n = 0$ (i.e. for the case $\tau = 0$ and $\tau = 1$) the assertion of the theorem is obvious. If this assertion is true for all numbers (10.13) for

$n = n_0$ then, putting

$$\tau_1 = k/2^{n_0},\ \tau_2 = (k+1)/2^{n_0} \qquad (k = 0, 1, \ldots, 2^{n_0} - 1)$$

we obtain its validity for numbers

$$\tau_3 = \frac{2k+1}{2^{n_0+1}}$$

which gives all numbers (10.13) for $n = n_0 + 1$.

Let now τ be an arbitrary number of the interval (0, 1). Denote by τ_i ($i = 1, 2, \ldots$) a sequence of binary fractions of the form (10.13) such that

$$\tau - 1/i < \tau_i < \tau \qquad (i = 1, 2, \ldots).$$

It was already proved that for each fixed function $x \in L_{\frac{1}{2}}$ for any $i = 1, 2, \ldots$ the following inequality is valid:

$$\|A^\tau x\|_{\beta(\tau_i)} = \|A^{\tau_i} A^{\tau - \tau_i} x\|_{\beta(\tau_i)} \leqq a^{\tau_i} \|A^{\tau - \tau_i} x\|_{\frac{1}{2}}.$$

Passing to the limit in this inequality as $i \to \infty$, we obtain

$$\|A^\tau x\|_{\beta(\tau)} \leqq a^\tau \|x\|_{\frac{1}{2}}. \qquad (10.14)$$

The theorem has been proved.

10.4 *Operators in real spaces*

Theorem 10.1 is valid even in the case of real spaces L_α. Even the proof is not changed. It is justified by the fact that inequality (2.17) used in the proof, which followed from interpolation Theorem 2.4, remains valid for real spaces L_α, if the numbers $\alpha_0, \alpha_1, \beta_0, \beta_1$, satisfy the supplementary inequalities

$$\alpha_0 \geqq \beta_0,\ \alpha_1 \geqq \beta_1. \qquad (10.15)$$

To prove this assertion[1] it suffices to show (see the proof of Theorem 2.4;

[1] Here we follow M. Riesz [1].

in this section the notation in 2.3° is used) that the inequalities (2.14) and (2.15) imply the inequality

$$|(Ax, y)| \leqq M_0^{1-\tau} M_1^{\tau} \|x\|_{\alpha(\tau)} \|y\|_{1-\beta(\tau)} \tag{10.16}$$

for any pair of functions $x(s) \in L_{\alpha(\tau)}$ and $y(t) \in L_{1-\beta(\tau)}$.

Assume first that the operator A is finite dimensional. Consider the function

$$M(\tau) = \sup |(Ax, y)|, \tag{10.17}$$

where the supremum is taken over all functions $x(s)$ and $y(t)$ satisfying the condition $\|x\|_{\alpha(\tau)} \leqq 1$, $\|y\|_{1-\beta(\tau)} \leqq 1$.

The inequality (10.16) is equivalent to the inequality

$$M(\tau) \leqq M(0)^{1-\tau} M(1)^{\tau} \quad (0 < \tau < 1).$$

To prove this inequality, it suffices to show that the function $M(\tau)$ is logarithmically convex. Since $M(\tau)$ is continuous, it is even sufficient to show that in any interval $(\tau_0, \tau_1) \subset (0, 1)$ there exists a number $\tau = (1 - \lambda)\tau_0 + \lambda\tau_1$ which satisfies the inequality

$$M(\tau) \leqq M(\tau_0)^{1-\tau} \cdot M(\tau_1)^{\tau}. \tag{10.18}$$

This last fact will now be proved.

Given two arbitrary fixed numbers τ_0, τ_1 in the interval $(0, 1)$, it follows from the inequalities $\alpha_0 \geqq \beta_0$, $\alpha_1 \geqq \beta_1$ that the inequalities $\alpha(\tau_0) \geqq \beta(\tau_0)$ and $\alpha(\tau_1) \geqq \beta(\tau_1)$ hold. Hence there is a number $\lambda \in (0, 1)$, such that

$$\frac{\beta(\tau_1)}{1 - \beta(\tau_0)} \leqq \frac{1 - \lambda}{\lambda} \leqq \frac{\alpha(\tau_1)}{1 - \alpha(\tau_0)}. \tag{10.19}$$

Putting

$$\tau = (1 - \lambda)\tau_0 + \lambda\tau_1;$$

we will prove that for this τ the inequality (10.18) is satisfied.

By the finite dimensionality of the operator A there are two functions x_0 and y_0 satisfying the conditions $\|x_0\|_{\alpha(\tau)} = \|y_0\|_{1-\beta(\tau)} = 1$ and such that

$$M(\tau) = (Ax_0, y_0).$$

Hence

$$M(\tau) = \|Ax_0\|_{\beta(\tau)} = \|A^* y_0\|_{1-\alpha(\tau)},$$

where A^* is the adjoint operator of A. The above relation can hold only when[1]

$$\begin{aligned} |Ax_0(t)| &= M(\tau)|y_0(t)|^{\beta(\tau)/(1-\beta(\tau))}, \\ |A^* y_0(s)| &= M(\tau)|x_0(s)|^{(1-\alpha(\tau))/\alpha(\tau)}. \end{aligned} \tag{10.20}$$

Introduce for simplicity the following notations:

$$\mu(\tau) = \frac{1-\alpha(\tau)}{\alpha(\tau)}, \quad \nu(\tau) = \frac{1-\beta(\tau)}{\beta(\tau)}, \quad \delta = \frac{1-\lambda}{\lambda}.$$

It is easy to check that

$$\alpha(\tau_1) = [1 - \delta\mu(\tau)]\alpha(\tau) + \delta\mu(\tau)\,\frac{1-\alpha(\tau_0)}{\mu(\tau)},$$

$$1 - \beta(\tau_0) = \left[1 - \frac{1}{\delta\nu(\tau)}\right][1 - \beta(\tau)] + \frac{1}{\delta\nu(\tau)}\,\beta(\tau_1)\nu(\tau).$$

Moreover (by (10.19)),

$$0 \leqq \delta\mu(\tau) \leqq 1, \quad 0 \leqq \frac{1}{\delta\nu(\tau)} \leqq 1.$$

---

[1] See G. Hardy, D. Littlewood, G. Polya [1].

Hence (by (1.6))

$$\|x_0\|_{\alpha(\tau_1)} \leqq (\|x_0\|_{\alpha(\tau)})^{1-\delta\mu(\tau)}(\|x_0\|_{(1-\alpha(\tau_0))/\mu(\tau)})^{\delta\mu(\tau)}$$

$$= (\|x_0\|_{(1-\alpha(\tau_0))/\mu(\tau)})^{\delta\mu(\tau)}, \tag{10.21}$$

$$\|y_0\|_{1-\beta(\tau_0)} \leqq (\|y_0\|_{1-\beta(\tau)})^{1-1/\delta\nu(\tau)}(\|y_0\|_{\nu(\tau)\beta(\tau_1)})^{1/\delta\nu(\tau)}$$

$$= (\|y_0\|_{\nu(\tau)\beta(\tau_1)})^{1/\delta\nu(\tau)}. \tag{10.22}$$

The relation (10.20) implies that

$$M(\tau)(\|x_0\|_{(1-\alpha(\tau_0))/\mu(\tau)})^{\mu(\tau)} = \|A^*y_0\|_{1-\alpha(\tau_0)} \leqq M(\tau_0)\|y_0\|_{1-\beta(\tau_0)}, \tag{10.23}$$

$$M(\tau)(\|y_0\|_{\nu(\tau)\beta(\tau_1)})^{1/\nu(\tau)} = \|Ax_0\|_{\beta(\tau_1)} \leqq M(\tau_1)\|x_0\|_{\alpha(\tau_1)}. \tag{10.24}$$

Combine the inequalities (10.21) and (10.22) with the inequalities (10.23) and (10.24) to get

$$(\|x_0\|_{\alpha(\tau_1)})^{1/\delta} \leqq \frac{M(\tau_0)}{M(\tau)}\|y_0\|_{1-\beta(\tau_0)},$$

$$(\|y_0\|_{1-\beta(\tau_0)})^{\delta} \leqq \frac{M(\tau_1)}{M(\tau)}\|x_0\|_{\alpha(\tau_1)},$$

whence

$$\|x_0\|_{\alpha(\tau_1)} \leqq \left[\frac{M(\tau_0)}{M(\tau)}\right]^{\delta}\left[\frac{M(\tau_1)}{M(\tau)}\right]\|x_0\|_{\alpha(\tau_1)}.$$

Canceling $\|x_0\|_{\alpha(\tau_1)}$ we arrive at the expected inequality (10.18).

Now let $\boldsymbol{P}_n$ be an arbitrary sequence of finite dimensional projection operators, which is strongly convergent to the identity operator in each L_β ($\beta > 0$) with $\|\boldsymbol{P}_n\| = 1$[1]. For each n the operator $\boldsymbol{P}_n\boldsymbol{A}$ is finite dimen-

[1] Such sequences of operators were constructed in 1.5°.

sional; hence the following relation holds:

$$|(P_n Ax, y)| \leqq M_0^{1-\tau} M_1^{\tau} \|x\|_{\alpha(\tau)} \|y\|_{1-\beta(\tau)}.$$

Passing to the limit as $n \to \infty$ we get inequality (10.16) for an arbitrary operator A.

10.5 *Fractional powers of compact operators*

If a positive definite selfadjoint operator A is compact in $L_{\frac{1}{2}}$ all of its fractional powers A^{τ} ($\tau > 0$) are compact in $L_{\frac{1}{2}}$. Further the representation (10.4) for the operator A implies that the operator A^{τ} is written in the series

$$A^{\tau}x = \sum_{i=1}^{\infty} \lambda_i^{\tau}(x, e_i)e_i, \tag{10.25}$$

which converges uniformly in the ball $\|x\|_{\frac{1}{2}} \leqq 1$ of $L_{\frac{1}{2}}$.

THEOREM 10.2: *Let a positive definite selfadjoint operator A in $L_{\frac{1}{2}}$ be simultaneously a compact operator acting from $L_{\frac{1}{2}}$ to L_{β}, where $0 \leqq \beta < \frac{1}{2}$. Then for each $\tau \in (0, 1)$ the operator A^{τ} is a compact operator acting from $L_{\frac{1}{2}}$ to $L_{\beta(\tau)}$, where $\beta(\tau)$ is defined by the formula* (10.11).

PROOF: Consider the operators

$$A_n x = \sum_{i=1}^{n} \lambda_i(x, e_i)e_i, \quad (A - A_n)x = \sum_{i=n+1}^{\infty} \lambda_i(x, e_i)e_i.$$

Then $A^{\tau} = A_n^{\tau} + (A - A_n)^{\tau}$, because

$$A_n^{\tau}x = \sum_{i=1}^{n} \lambda_i^{\tau}(x, e_i)e_i, \quad (A - A_n)^{\tau}x = \sum_{i=n+1}^{\infty} \lambda_i^{\tau}(x, e_i)e_i \quad (x \in L_{\frac{1}{2}}).$$

By Theorem 9.3 the series (10.4) converges to the operator A in operator norm, as operators acting from $L_{\frac{1}{2}}$ to L_{β}, i.e. for any $\varepsilon > 0$ there is n_{ε} such that for all $n \geqq n_{\varepsilon}$

$$\|(A - A_n)x\|_{\beta} \leqq \varepsilon \|x\|_{\frac{1}{2}}.$$

Apply Theorem 10.1 to the operator $A - A_n$ $(n \geqq n_\varepsilon)$. This gives the inequality

$$\|(A^\tau - A_n^\tau)x\|_{\beta(\tau)} = \|(A - A_n)^\tau x\|_{\beta(\tau)} \leqq \varepsilon^\tau \|x\|_{\frac{1}{2}},$$

whence it follows that the operators A_n^τ converge to A^τ in operator norm, as operators acting from $L_{\frac{1}{2}}$ to $L_{\beta(\tau)}$. The operator A_n^τ, for each n, is finite dimensional; consequently the operator A^τ is compact. The theorem has been proved.

At the same time, it has been proved that the series

$$\sum_{i=1}^{\infty} \lambda_i^\tau (x, e_i) e_i$$

converges to A in operator norm, as operators acting from $L_{\frac{1}{2}}$ to $L_{\beta(\tau)}$.

10.6 *L-Characteristics of fractional powers of operators*

Theorems 10.1 and 10.2 permit us to draw parts of the L-characteristics of the operators A^τ by examining the L-characteristics $L(A; \text{cont.})$ and $L(A; \text{comp.})$. The same construction which is applied to determine a part of the L-characteristic $L(A^\tau; \text{cont.})$ from the L-characteristic $L(A; \text{cont.})$ can be used to determine a part of the L-characteristics $L(A^\tau; \text{comp.})$ from the L-characteristic $L(A; \text{comp.})$. Hence we confine ourselves here to the consideration of the L-characteristics $L(A; \text{cont.})$ and $L(A^\tau; \text{cont.})$.

An immediate application of Theorem 10.1 permits us to utilize those points $\{\frac{1}{2}, \beta_0\}$ in $L(A; \text{cont.})$ for which $\beta_0 < \frac{1}{2}$ (the point M in Fig. 10.2). To this end, it is necessary to find the point (M_τ in Fig. 10.2), which divides the segment joining the points $M = \{\frac{1}{2}, \beta_0\}$ and $M_0 = \{\frac{1}{2}, \frac{1}{2}\}$ in the ratio $\tau\colon 1-\tau$. Then we can use the symmetry and convexity of the L-characteristic $L(A^\tau; \text{cont.})$ (in Fig. 10.2 the corresponding part of $L(A^\tau; \text{cont.})$ is hatched).

Now suppose that two points of the form $M = \{\frac{1}{2}, \beta_0\}$ and $N = \{\alpha_0, 1 - \alpha_0\}$ are known to belong to the L-characteristic $L(A; \text{cont.})$. Assume that $\alpha_0 + \beta_0 \leqq 1$, $\alpha_0 \geqq \frac{3}{4} - \beta_0/2$ (Fig. 10.3). Denote by $M_{\frac{1}{2}}$ the point $\{\frac{1}{2}, 1 - \alpha_0\}$; by Theorem 9.1 this point belongs to $L(A^{\frac{1}{2}}; \text{cont.})$.

Suppose first that $0 < \tau < \frac{1}{2}$. Since $A^\tau = (A^{\frac{1}{2}})^{2\tau}$, Theorem 10.1 implies that the L-characteristic $L(A^\tau; \text{cont.})$ contains the point M_τ which divides the

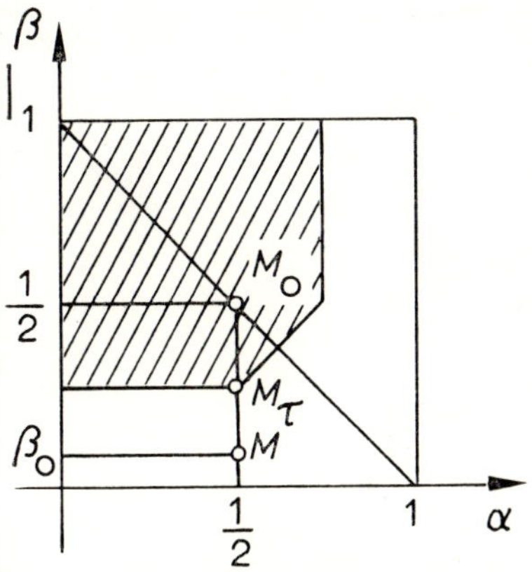

Figure 10.2

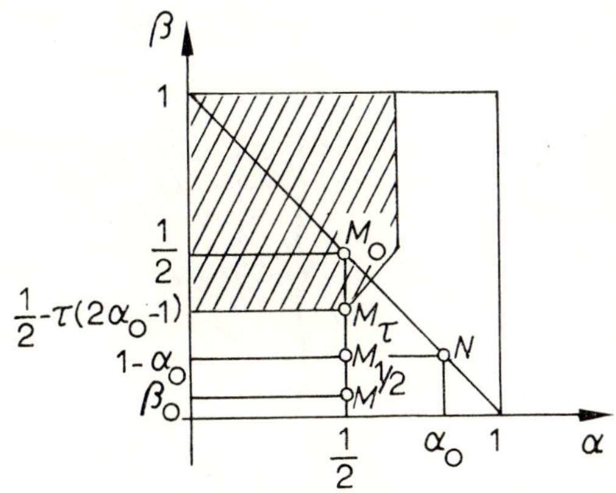

Figure 10.3

segment joining the point $M_{\frac{1}{2}}$ to the point $M_0 = \{\frac{1}{2}, \frac{1}{2}\}$ in the ratio 2τ: $1 - 2\tau$. It then follows that $L(A^\tau; \text{cont.})$ contains the hatched set in Fig. 10.3.

Consider next the case when $\tau = \frac{1}{2} + \sigma$, where $0 < \sigma < \frac{1}{2}$. In constructing points of the L-characteristic $L(A^\tau; \text{cont.})$ in this case, it is convenient to use the following auxiliary assertion, which is of independent interest.

Theorem 10.3: *Let* $\{\frac{1}{2}, \beta_0\} \in L(A^{\tau_0}; \text{cont.})$ *and* $\{\frac{1}{2}; \beta_1\} \in L(A^{\tau_1}; \text{cont.})$. *Put*

$$\beta_\lambda = (1 - \lambda)\beta_0 + \lambda\beta_1; \ \ \tau_\lambda = (1 - \lambda)\tau_0 + \lambda\tau_1.$$

Then for any $\lambda \in (0, 1)$ *the point* $\{\frac{1}{2}, \beta_\lambda\}$ *belongs to* $L(A^\tau; \text{cont.})$.

The proof is similar to that of Theorem 10.1. Details are left to the reader.

Now let us return to the construction of points of the L-characteristic $L(A^\tau; \text{cont.})$. Theorem 10.3 shows that the L-characteristic $L(A^{\frac{1}{2}+\sigma}; \text{cont.})$ contains the point M_τ, dividing the segment which joins the points $M_{\frac{1}{2}} = \{\frac{1}{2}, 1 - \alpha_0\}$ and $M = \{\frac{1}{2}, \beta_0\}$ in the ratio 2σ: $1 - 2\sigma$ (Fig. 10.4). The operator A^σ acts from $L_{\frac{1}{2}+\sigma(2\alpha_0-1)}$ to $L_{\frac{1}{2}}$, and the operator $A^{\frac{1}{2}}$ acts from $L_{\frac{1}{2}}$ to $L_{1-\alpha_0}$. Hence the operator $A^{\frac{1}{2}+\sigma} = A^{\frac{1}{2}}A^\sigma$ acts continuously from $L_{\frac{1}{2}+\sigma(2\alpha_0-1)}$ to $L_{1-\alpha_0}$. This means that the point $N_\tau = \{\frac{1}{2} + \sigma(2\alpha_0 - 1), 1 - \alpha_0\}$, which divides the segment $M_{\frac{1}{2}}N$ in the ratio 2σ: $1 - 2\sigma$ belongs to the L-characteristic $L(A^\tau; \text{cont.})$.

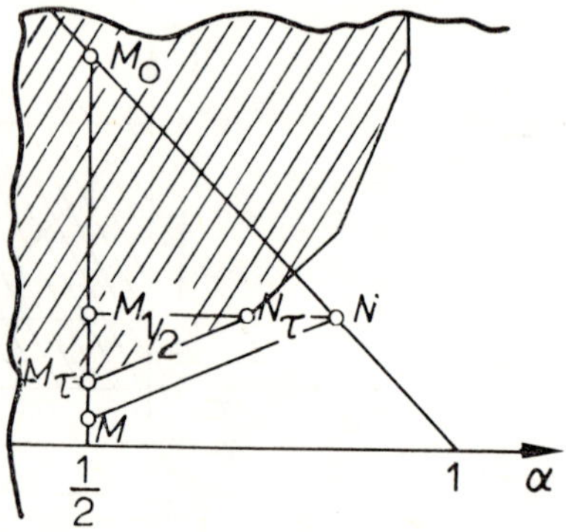

Figure 10.4

Thus for the case $\tau \in (\frac{1}{2}, 1)$ we have constructed two points M_τ and N_τ belonging to $L(A^\tau; \text{cont.})$. Consequently the L-characteristic $L(A^\tau; \text{cont.})$ contains the hatched polygon in Fig. 10.4.

It is not known whether it is possible, in general, to obtain supplementary information on the L-characteristic of an operator A^τ from consideration of points of the L-characteristic of the operator A not lying on the lines $\alpha = \frac{1}{2}$, $\beta = \frac{1}{2}$ and $\alpha + \beta = 1$. For a special class of operators more complete assertions will be obtained in what follows.

10.7 *Fractional powers of integral operators*

Let $\boldsymbol{K}$ be an integral operator with a symmetric positive definite kernel:

$$\boldsymbol{K}x(t) = \int_\Omega K(t, s)x(s)\mathrm{d}s. \tag{10.26}$$

Let $\boldsymbol{K}$ be continuous (or compact) as an operator in $L_{\frac{1}{2}}$. In this case Theorems 10.1 and 10.2 can be applied to $\boldsymbol{K}$ to permit one to obtain fractional powers $\boldsymbol{K}^\tau$ of the operator $\boldsymbol{K}$. These fractional powers, however, may not be integral operators.

Theorems 10.1 and 10.2 can also be used to obtain criteria for continuity and compactness of the integral operator (10.26), once some properties of operators with iterated kernels are developed.

Consider an iterate

$$K_r(t, s) = \int_\Omega \dots \int_\Omega K(t, s_1) \dots K(s_{r-1}, s) ds_1 \dots ds_{r-1} \tag{10.27}$$

of the kernel $K(t, s)$ and the corresponding integral operator

$$K_r x(t) = \int_\Omega K_r(t, s) x(s) ds. \tag{10.28}$$

If the operator $\boldsymbol{K}$ is regular in $L_{\frac{1}{2}}$, then $\boldsymbol{K}_r$ is bounded in $L_{\frac{1}{2}}$ and $\boldsymbol{K}_r = \boldsymbol{K}^r$ (see 4.6°), consequently $\boldsymbol{K} = (\boldsymbol{K}_r)^{1/r}$.

Theorems 10.1 and 10.2 imply

THEOREM 10.4: *Let the operator* $\boldsymbol{K}_r$ *in* (10.28) *act from* $L_{\frac{1}{2}}$ *to* L_β $(\beta < \frac{1}{2})$ *and suppose that the operator* (10.26) *is regular in* $L_{\frac{1}{2}}$. *Then the operator* $\boldsymbol{K}$ *in* (10.26) *acts from* $L_{\frac{1}{2}}$ *to* $L_{\beta(r)}$, *where*

$$\beta(r) = \tfrac{1}{2} + \frac{1}{r}(\beta - \tfrac{1}{2}), \tag{10.29}$$

and

$$\|\boldsymbol{K}\|_{\frac{1}{2}\to\beta(r)} \leqq \|\boldsymbol{K}_r\|_{\frac{1}{2}\to\beta}^{1/r}. \tag{10.30}$$

If the operator $\boldsymbol{K}_r$ *acts from* $L_{\frac{1}{2}}$ *to* L_β *and is, in addition, compact, then the operator* $\boldsymbol{K}$ *acts from* $L_{\frac{1}{2}}$ *to* $L_{\beta(\tau)}$ *and is also compact.*

§ 11 Unbounded selfadjoint operators[1]

11.1 *Closed operators*

Consider a linear operator A in a Hilbert space H, defined on a set $D(A) \subset H$.

The operator A is called *closed* if for each fundamental sequence $x_n \in D(A)$ ($\|x_n - x_m\| \to 0$ as $n, m \to \infty$) the convergence of the sequence

[1] See, for instance, N. I. Akhiezer and I. M. Glazman [1], M. A. Naimark [1], V. I. Smirnov [1].

Ax_n to some limit y_0 implies that the limit x_0 of the sequence x_n belongs to $D(A)$ and $Ax_0 = y_0$.

It is not difficult to see that each bounded linear operator is closed. The converse assertion is also valid: *an operator, defined on a closed subspace of a Hilbert space is bounded if it is closed.*

Denote by $\tilde{H}$ the Hilbert space, whose elements are ordered pairs $\{x, y\}$, where $x, y \in H$. Linear operations are given by the natural relation

$$\alpha\{x, y\} + \beta\{x_1, y_1\} = \{\alpha x + \beta x_1, \alpha y + \beta y_1\},$$

and a scalar product by the relation

$$(\{x, y\}, \{x_1, y_1\}) = (x, x_1) + (y, y_1).$$

In other words, $\tilde{H}$ is an orthogonal sum of two copies of the space H.

The *graph* Γ of an operator A is the totality of elements of the form $\{x, Ax\}$ in the space $\tilde{H}$, where x runs over $D(A)$. The graph is a linear set if the operator A is linear, because a linear combination of two elements $\{x_1, Ax_1\}$ and $\{x_2, Ax_2\}$ is an element of the same form.

It is easy to see that *an operator A is closed if and only if its graph Γ is a subspace, i.e. a closed linear set in $\tilde{H}$.*

Assume that a closed operator A has an inverse A^{-1}. The operator A^{-1} is also closed. This follows from the fact that the graph of the operator A^{-1} consists of elements of the form $\{Ax, x\}$, and the totality of these elements is closed if the totality of elements of the form $\{x, Ax\}$ is closed.

An operator A_1 with domain $D(A_1)$ is called an *extension* of an operator A with its domain $D(A)$ if $D(A) \subset D(A_1)$ and $A_1 x = Ax$ for $x \in D(A)$. It is obvious that *the graph of an extension A_1 contains the graph of the operator A as a subset.*

An operator A is said to *admit closure* if it has at least one extension A_1 which is a closed operator. The graph Γ of the operator A is then a subset of a closed subspace of Γ_1, the graph of the operator A_1. The closure $\bar{\Gamma}$ of the graph Γ is then also a subset of Γ_1. Hence $\bar{\Gamma}$ can be considered the graph of some operator $\bar{A}$. The operator $\bar{A}$ is also a closed extension of the operator A. This minimum closed extension (it is clear from the above considerations that any closed extension of the operator A is necessarily an extension of the operator $\bar{A}$) is called the *closure* of the operator A.

Consider all possible sequences $x_n \in D(A)$ converging to some element

x_0. Consider further the sequences Ax_n; some of these sequences will be fundamental. It is left to the reader to show that the operator A admits closure if and only if the limit of the above sequences Ax_n (those sequences for which this limit exists) is uniquely determined by the element x_0. The domain of the closure $\bar{A}$ of the operator A then coincides with the totality of all those $x_0 \in H$ for which at least one of the sequences Ax_n described above has a limit; this limit y_0 is the value of the closure $\bar{A}$ at the element x_0.

11.2 *Adjoint operators*

In the preceding paragraphs the notion of an operator which is adjoint to a continuous linear operator has been used repeatedly. Let us now introduce a generalization of this notion to the case of unbounded operators.

Denote by $D(A^*)$ the totality of those $y \in H$ for which an element y^* can be chosen such that the following relation is satisfied for all $x \in D(A)$:

$$(Ax, y) = (x, y^*). \tag{11.1}$$

The element y^* is defined uniquely by the element y because (11.1) and the relation

$$(Ax, y) = (x, y^{**}) \qquad (x \in D(A))$$

imply that $y^* - y^{**}$ is orthogonal to the set $D(A)$ which is assumed to be dense in H; hence it must be equal to zero. This permits us to take y^* as the value at y of some operator A^* which is called the operator *adjoint* to the operator A. The basic relation (11.1) for the adjoint operator can be written in the form

$$(Ax, y) = (x, A^*y) \qquad (x \in D(A),\ y \in D(A^*)).$$

The above definition immediately implies that the domain $D(A^*)$ of the adjoint operator is a linear set and that the operator A^* is linear.

Let Γ be the graph of the operator A in H. The definition of adjoint operator implies that $y \in D(A^*)$ if and only if there is a y^* such that $\{-y^*, y\}$ is orthogonal to the graph Γ. In other words, the orthogonal complement of Γ in $\tilde{H}$ coincides with the totality of all elements of the form $\{-A^*y, y\}$. Clearly, the totality of these elements is closed. Hence the totality of elements of the form $\{y, A^*y\}$, which is the graph Γ^* of the operator A^*, is also closed. Consequently *an adjoint operator is closed.*

It is left to readers to show that $A^{**} = (A^*)^*$ coincides with the closure of the operator A if this closure exists.

An operator A is called *selfadjoint* if it coincides with its adjoint. This definition implies that a selfadjoint operator is closed. A selfadjoint operator possesses the following symmetry property:

$$(Ax, y) = (x, Ay) \qquad (x, y \in D(A)).$$

Finally, the relation $(Ax, y) = (x, y^*)$ for all $x \in D(A)$ implies that $y \in D(A^*)$.

11.3 *Integration with respect to spectral functions*

The simplest integral involving a spectral function, the integral representation of a bounded selfadjoint operator, was treated in the preceding paragraph. In what follows, however, integrals of a more general form are necessary.

Let P_λ be a spectral function (see 10.1°) and, as usual, put $P_\Delta = P_{\lambda_2} - P_{\lambda_1}$, where $\Delta = (\lambda_1, \lambda_2]$. The integral

$$J_{ab}x = \int_a^b f(\lambda)\,\mathrm{d}P_\lambda x \tag{11.2}$$

of a continuous function $f(\lambda)$ is defined as the limit of the Riemann sums

$$Sx = \sum_{i=0}^{n} f(\xi_i) P_{\Delta_i} x,$$

where

$$\xi_0 = a, \; P_{\Delta_0} = \lim_{\varepsilon \to 0+} (P_a - P_{a-\varepsilon}),$$

and ξ_i are arbitrary points in the intervals $\Delta_i = (\lambda_{i-1}, \lambda_i]$, into which the interval $[a, b]$ is split. Passing to the limit in the obvious relation

$$\|Sx\|^2 = \sum_{i=0}^{n} |f(\xi_i)|^2 \cdot (P_{\Delta_i} x, x),$$

we obtain the important formula

$$\left\| \int_a^b f(\lambda)\,\mathrm{d}\boldsymbol{P}_\lambda x \right\|^2 = \int_a^b |f(\lambda)|^2 \cdot \mathrm{d}(\boldsymbol{P}_\lambda x, x). \tag{11.3}$$

We define the integral on the infinite interval by use of the formula

$$\int_{-\infty}^{\infty} f(\lambda)\,\mathrm{d}\boldsymbol{P}_\lambda x = \lim_{\substack{a \to -\infty \\ b \to \infty}} \int_a^b f(\lambda)\,\mathrm{d}\boldsymbol{P}_\lambda x. \tag{11.4}$$

Given a function $f(\lambda)$ this latter integral need not be defined for all elements x. For convergence of the integral (11.4) it is necessary that as $a, a_1 \to -\infty$ and $b, b_1 \to \infty$, the norms of the integrals

$$\int_{a_1}^{a} f(\lambda)\,\mathrm{d}\boldsymbol{P}_\lambda x, \quad \int_{b}^{b_1} f(\lambda)\,\mathrm{d}\boldsymbol{P}_\lambda x$$

converge to zero, i.e. that the integrals

$$\int_{a_1}^{a} |f(\lambda)|^2\,\mathrm{d}(\boldsymbol{P}_\lambda x, x), \quad \int_{b}^{b_1} |f(\lambda)|^2\,\mathrm{d}(\boldsymbol{P}_\lambda x, x)$$

converge to zero. In other words, the integral (11.4) is defined if and only if the following condition is fulfilled:

$$\left\| \int_{-\infty}^{\infty} f(\lambda)\,\mathrm{d}\boldsymbol{P}_\lambda x \right\|^2 = \int_{-\infty}^{\infty} |f(\lambda)|^2\,\mathrm{d}(\boldsymbol{P}_\lambda x, x) < \infty. \tag{11.5}$$

If $f(\lambda) \equiv 1$, we arrive at the relation

$$\int_{-\infty}^{\infty} \mathrm{d}\boldsymbol{P}_\lambda x = x, \quad \int_{-\infty}^{\infty} \mathrm{d}(\boldsymbol{P}_\lambda x, x) = \|x\|^2. \tag{11.6}$$

The formula

$$\boldsymbol{B}x = \int_{-\infty}^{\infty} f(\lambda)\,\mathrm{d}\boldsymbol{P}_\lambda x \tag{11.7}$$

defines a linear operator. If the function $f(\lambda)$ is bounded, then in view of (11.5) and (11.6) the operator $\boldsymbol{B}$ is defined for all $x \in H$ and

$$\|\boldsymbol{B}x\|^2 = \int_{-\infty}^{\infty} |f(\lambda)|^2 \mathrm{d}(\boldsymbol{P}_\lambda x, x)$$

$$\leqq \sup_\lambda |f(\lambda)|^2 \int_{-\infty}^{\infty} \mathrm{d}(\boldsymbol{P}_\lambda x, x) = \sup_\lambda |f(\lambda)|^2 \|x\|^2,$$

i.e.

$$\|\boldsymbol{B}\| \leqq \sup_\lambda |f(\lambda)|. \tag{11.8}$$

If, on the other hand, the function $|f(\lambda)|$ is bounded from below by a positive number, then the following inequality is valid:

$$\|\boldsymbol{B}x\|^2 \geqq \inf_\lambda |f(\lambda)|^2 \int_{-\infty}^{\infty} \mathrm{d}(\boldsymbol{P}_\lambda x, x) = \inf_\lambda |f(\lambda)|^2 \cdot \|x\|^2.$$

11.4 *The fundamental theorem on spectral representation of unbounded self-adjoint operators*

For each selfadjoint operator $\boldsymbol{A}$ there is one and only one spectral function $\boldsymbol{P}_\lambda$ with the following properties:

1°) $x \in D(\boldsymbol{A})$ *if and only if the following integral converges*:

$$\int_{-\infty}^{\infty} |\lambda|^2 \mathrm{d}(\boldsymbol{P}_\lambda x, x) < \infty. \tag{11.9}$$

2°) *The relation*:

$$\boldsymbol{A}x = \int_{-\infty}^{\infty} \lambda \mathrm{d}\boldsymbol{P}_\lambda x \tag{11.10}$$

holds, and

$$\|\boldsymbol{A}x\|^2 = \int_{-\infty}^{\infty} |\lambda|^2 \mathrm{d}(\boldsymbol{P}_\lambda x, x). \tag{11.11}$$

Formula (11.10) is a generalization of formula (10.3). The representation (11.10) implies that the scalar product (Ax, y) for $x \in D(A)$ and any $y \in H$ can be written in the form

$$(Ax, y) = \int_{-\infty}^{\infty} \lambda \mathrm{d}(\boldsymbol{P}_\lambda x, y). \tag{11.12}$$

A number λ is called a *regular point* of an operator A if there exists a continuous operator $\boldsymbol{R}(\lambda) = (\lambda \boldsymbol{I} - A)^{-1}$, defined on the whole space; this operator is called a *resolvent* of the operator A.

All non-regular points constitute the *spectrum* of the operator A. Clearly all eigenvalues of the operator A belong to the spectrum. We remark that *all eigenvalues of selfadjoint operators (if they exist) are real.*

Non-real numbers do not belong to the spectrum of a selfadjoint operator A. For the proof it suffices to observe that

$$\|(\lambda \boldsymbol{I} - A)x\| = \|\int_{-\infty}^{\infty} (\lambda - \mu) \mathrm{d}\boldsymbol{P}_\mu x\| \geqq |\mathrm{Im}\, \lambda| \cdot \|x\|. \tag{11.13}$$

This means that the operator $(\lambda \boldsymbol{I} - A)^{-1}$ is continuous on the range of the operator $\lambda \boldsymbol{I} - A$, and

$$\|(\lambda \boldsymbol{I} - A)^{-1}\| \leqq \frac{1}{|\mathrm{Im}\, \lambda|}. \tag{11.14}$$

Since $(\lambda \boldsymbol{I} - A)^{-1}$ is closed, the range of the operator $\lambda \boldsymbol{I} - A$ is closed. To complete the proof it suffices to show that this set is dense in H. If an element y is orthogonal to this range, i.e. $(\lambda x - Ax, y) = 0$ for all $x \in D(A)$, then it follows that $y \in D(A)$ and $(x, \bar{\lambda} y - Ay) = 0$. Hence since $D(A)$ is dense in H, the relation $Ay = \bar{\lambda} y$ holds. But a non-real number can not be an eigenvalue of the selfadjoint operator A, hence $y = 0$.

Real numbers can be regular points of a selfadjoint operator. It is possible to show that regularity of the real number λ is equivalent to the existence of an interval $(\lambda - \varepsilon, \lambda + \varepsilon)$ on which the spectral function $\boldsymbol{P}_\lambda$ is constant (whence, in particular, it follows that the spectrum is a closed set). Thus the fundamental formula of integral representation for the operator A can be written in the form

$$A = \int_E \lambda \mathrm{d}\boldsymbol{P}_\lambda, \tag{11.15}$$

where E is an arbitrary set containing the spectrum, or is the spectrum itself.

For instance, if the operator A is bounded, the entire spectrum of the operator A lies on the segment $[-\|A\|, \|A\|]$. Thus for a bounded operator the following formula is valid:

$$A = \int_{-\|A\|}^{\|A\|} \lambda \mathrm{d}\boldsymbol{P}_\lambda. \tag{11.16}$$

This fact has been used already; see the preceding paragraph.

An operator A is called *semi-bounded* if

$$(Ax, x) \geqq m(x, x) \qquad (x \in D(A)). \tag{11.17}$$

An immediate verification reveals that the spectrum of such an operator is contained in the interval $[m, \infty)$. Hence its spectral representation has the form

$$A = \int_m^\infty \lambda \mathrm{d}\boldsymbol{P}_\lambda. \tag{11.18}$$

In particular, if $m = 0$, i.e. $(Ax, x) \geqq 0$ then the operator A is called *positive definite*, while if $(Ax, x) \geqq a(x, x)$ where $a > 0$ then the operator A is called *strictly positive definite*.

11.5 *Functions of selfadjoint operators*

Let $f_1(\lambda)$ and $f_2(\lambda)$ be continuous functions on $[a, b]$. Define continuous operators A_1 and A_2 by the formulas

$$A_1 = \int_a^b f_1(\lambda) \mathrm{d}\boldsymbol{P}_\lambda, \quad A_2 = \int_a^b f_2(\lambda) \mathrm{d}\boldsymbol{P}_\lambda,$$

where $\boldsymbol{P}_\lambda$ is a spectral function. Then

$$A_1 A_2 = \int_a^b f_1(\lambda) f_2(\lambda) \mathrm{d}\boldsymbol{P}_\lambda. \tag{11.19}$$

In fact, the operators A_1 and A_2 can be approximated with any desired accuracy by the corresponding sums

$$\boldsymbol{S}_1 = \sum_{i=0}^{n} f_1(\xi_i)\boldsymbol{P}_{\Delta_i}, \quad \boldsymbol{S}_2 = \sum_{i=0}^{n} f_2(\xi_i)\boldsymbol{P}_{\Delta_i},$$

where Δ_i are intervals into which the interval $[a, b]$ has been split. Thus the operators $\boldsymbol{S}_1\boldsymbol{S}_2$ will approximate the operator $\boldsymbol{A}_1\boldsymbol{A}_2$. However by the properties of spectral functions,

$$\boldsymbol{S}_1\boldsymbol{S}_2 = \sum_{i=0}^{n} f_1(\xi_i)f_2(\xi_i)\boldsymbol{P}_{\Delta_i},$$

whence there follows formula (11.19).

In particular, if the selfadjoint operator $\boldsymbol{A}$ is bounded then

$$\boldsymbol{A}^n = \int_{-\|\boldsymbol{A}\|}^{\|\boldsymbol{A}\|} \lambda^n \mathrm{d}\boldsymbol{P}_\lambda. \tag{11.20}$$

This equation implies that for any polynomial of a bounded selfadjoint operator $\boldsymbol{A}$ the following representation holds:

$$c_n\boldsymbol{A}^n + c_{n-1}\boldsymbol{A}^{n-1} + \ldots + c_0\boldsymbol{I}$$

$$= \int_{-\|\boldsymbol{A}\|}^{\|\boldsymbol{A}\|} (c_n\lambda^n + c_{n-1}\lambda^{n-1} + \ldots + c_0)\mathrm{d}\boldsymbol{P}_\lambda. \tag{11.21}$$

Let the Taylor series of the analytic function $f(\lambda)$

$$f(\lambda) = c_0 + c_1\lambda + c_2\lambda + \ldots$$

converge in a disc whose radius is greater than $\|\boldsymbol{A}\|$. Then the series

$$c_0\boldsymbol{I} + c_1\boldsymbol{A} + c_2\boldsymbol{A}^2 + \ldots$$

converges in norm. The sum of this series is also a linear operator; denote it by $f(\boldsymbol{A})$. Passing to the limit in (11.21), we arrive at the formula

$$f(\boldsymbol{A}) = \int_{-\|\boldsymbol{A}\|}^{\|\boldsymbol{A}\|} f(\lambda)\mathrm{d}\boldsymbol{P}_\lambda. \tag{11.22}$$

For a non-analytic function $f(\lambda)$ the operator $f(A)$ is to be defined by the formula (11.22).

In the case of a compact operator A, which can be represented in the series

$$Ax = \sum_{i=1}^{\infty} \lambda_i(x, e_i)e_i$$

(where the e_i are the orthonormal sequence of eigenvectors and the λ_i are the corresponding eigenvalues), formula (11.22) takes the form

$$f(A)x = \sum_{i=1}^{\infty} f(\lambda_i)(x, e_i)e_i.$$

Let us now consider the case of unbounded operators. It is not difficult to show that the following formula holds:

$$A^n = \int_{-\infty}^{\infty} \lambda^n \mathrm{d}\boldsymbol{P}_\lambda, \tag{11.23}$$

where

$$A = \int_{-\infty}^{\infty} \lambda \mathrm{d}\boldsymbol{P}_\lambda.$$

By the fundamental spectral theorem, the domain of the operator A^n consists of those elements $x \in H$ for which

$$\int_{-\infty}^{\infty} |\lambda|^{2n} \mathrm{d}(\boldsymbol{P}_\lambda x, x) < \infty.$$

Hence it follows that the domain $D(A^n)$ diminishes as n increases.

Continuous functions of unbounded selfadjoint operators are defined by a formula analogous to (11.22):

$$f(A) = \int_{-\infty}^{\infty} f(\lambda) \mathrm{d}\boldsymbol{P}_\lambda. \tag{11.24}$$

Let $\sigma(A)$ be the spectrum of the operator A and let

$$\sup_{\sigma(A)} |f(\lambda)| = m < \infty.$$

Then the operator $f(A)$ is bounded and

$$\|f(A)\| = m. \tag{11.25}$$

In fact, the inequality $\|f(A)\| \leqq m$ results from the chain of relations

$$\|f(A)x\| = \|\int_{\sigma(A)} f(\lambda)\,\mathrm{d}\boldsymbol{P}_\lambda x\| \leqq m\,\|x\|.$$

Now fix $\lambda_0 \in \sigma(A)$ arbitrarily and let $x \in H_\Delta$, where $\Delta = (\lambda_0 - \varepsilon, \lambda_0 + \varepsilon)$. Then

$$\|f(A)x\|^2 = \int_{\lambda_0 - \varepsilon}^{\lambda_0 + \varepsilon} |f(\lambda)|^2\,\mathrm{d}(\boldsymbol{P}_\lambda x, x) \geqq \inf_{\Delta} |f(\lambda)|^2\,\|x\|^2,$$

whence it follows that

$$\|f(A)\| \geqq \inf_{\lambda_0 - \varepsilon \leqq \lambda \leqq \lambda_0 + \varepsilon} |f(\lambda)|.$$

As $\varepsilon \to 0$, we arrive at the inequality $\|f(A)\| \geqq |f(\lambda_0)|$, whence it follows that $\|f(A)\| \geqq m$.

In what follows we will be interested in functions of a positive definite operator $\boldsymbol{A}$. Let

$$A = \int_a^\infty \lambda\,\mathrm{d}\boldsymbol{P}_\lambda,$$

where $a \geqq 0$. Consider the following functions of this operator

$$(t\boldsymbol{I} + A)^{-1} = \int_a^\infty \frac{1}{t + \lambda}\,\mathrm{d}\boldsymbol{P}_\lambda, \quad A(t\boldsymbol{I} + A) = \int_a^\infty \frac{\lambda}{t + \lambda}\,\mathrm{d}\boldsymbol{P}_\lambda \qquad (t > 0). \tag{11.26}$$

It is clear that

$$\|(tI + A)^{-1}\| \leqq \frac{1}{a + t}, \ \|A(tI + A)^{-1}\| \leqq 1 \quad (t \geqq 0). \tag{11.27}$$

The functions

$$A^\alpha = \int_a^\infty \lambda^\alpha \mathrm{d}P_\lambda \quad (-\infty < \alpha < \infty) \tag{11.28}$$

will be analyzed in detail in what follows. If the operator A has discrete spectrum (with eigenvalues λ_i and corresponding eigenvectors e_i), it can be represented by the series

$$Ax = \sum_{i=1}^\infty \lambda_i (x, e_i) e_i;$$

in this case formula (11.28) takes the form

$$A^\alpha x = \sum_{i=1}^\infty \lambda_i^\alpha (x, e_i) e_i.$$

Here the eigenvalues λ_i can be arbitrary non-negative numbers.

11.6 *Commuting selfadjoint operators*

Consider two functions of the same selfadjoint operator A:

$$A_1 = \int_{-\infty}^\infty f_1(\lambda) \mathrm{d}P_\lambda, \ A_2 = \int_{-\infty}^\infty f_2(\lambda) \mathrm{d}P_\lambda.$$

In general the operators $A_1 A_2$ and $A_2 A_1$ have distinct domains if a product of operators is understood as the successive application of its factors. For instance, if A is an unbounded operator and λ is one of its regular points, then the operator $(\lambda I - A)(\lambda I - A)^{-1}$ is defined on the whole space H, while the operator $(\lambda I - A)^{-1}(\lambda I - A)$ makes sense only for $x \in D(A)$. It can be shown that on the common domain of the operators $A_1 A_2$ and $A_2 A_1$ their values are the same and will coincide with the value of the operator defined by the integral

$$\int_{-\infty}^{\infty} f_1(\lambda) f_2(\lambda) \mathrm{d}\boldsymbol{P}_\lambda. \tag{11.29}$$

In this sense, the operators $\boldsymbol{A}_1$ and $\boldsymbol{A}_2$ are said to *commute.*

The domain of the operator (11.29) is, in general, larger than the totality of elements to which $\boldsymbol{A}_1$ and $\boldsymbol{A}_2$ can be successively applied. In the sequel, the product $\boldsymbol{A}_1\boldsymbol{A}_2$ will be defined by the integral (11.29):

$$\boldsymbol{A}_1\boldsymbol{A}_2 = \boldsymbol{A}_2\boldsymbol{A}_1 = \int_{-\infty}^{\infty} f_1(\lambda) f_2(\lambda) \mathrm{d}\boldsymbol{P}_\lambda \tag{11.30}$$

with the natural domain.

The converse assertion is also valid. Suppose that the selfadjoint operators $\boldsymbol{A}_1$ and $\boldsymbol{A}_2$ commute in the sense that

$$\boldsymbol{A}_1\boldsymbol{A}_2 x = \boldsymbol{A}_2\boldsymbol{A}_1 x$$

for all x, for which both sides of the relation are defined. It can be shown that in this case the operators $\boldsymbol{A}_1$ and $\boldsymbol{A}_2$ are functions of some selfadjoint operator.

11.7 *Integrals of operator-functions*

In the preceding section, functions of selfadjoint operators were constructed by use of the spectral function. In many cases it is convenient to represent such functions by integrals of another type.

Let $\boldsymbol{B}(t)$ be an operator-valued function whose values are continuous linear operators. Three types of continuity of such operator-functions are distinguished. A function $\boldsymbol{B}(t)$ is called *uniformly continuous* (or continuous in operator norm) at a point t_0, if

$$\lim_{t \to t_0} \|\boldsymbol{B}(t) - \boldsymbol{B}(t_0)\| = 0. \tag{11.31}$$

A function $\boldsymbol{B}(t)$ is called *strongly continuous* at a point t_0, if for each fixed

$x \in H$ the following relation holds:

$$\lim_{t \to t_0} \|\boldsymbol{B}(t)x - \boldsymbol{B}(t_0)x\| = 0.$$

It is not difficult to see that strong continuity is a consequence of uniform continuity (but not conversely!). Finally, a function $\boldsymbol{B}(t)$ is called *weakly continuous* at a point t_0, if for any fixed pair of elements $x, y \in H$

$$\lim_{t \to t_0} |(\boldsymbol{B}(t)x, y) - (\boldsymbol{B}(t_0)x, y)| = 0.$$

Weak continuity follows from strong continuity.

Operator-integrals of continuous operator-functions on a finite interval $[a, b]$ are understood in the sense of Riemann. The definition of the Riemann integral depends, of course, on the type of continuity of the function $\boldsymbol{B}(t)$.

If $\boldsymbol{B}(t)$ is uniformly continuous, the integral

$$\boldsymbol{B} = \int_a^b \boldsymbol{B}(t)\,\mathrm{d}t \tag{11.32}$$

is understood as the limit, in operator norm, of integral sums

$$\sum_{i=1}^{n} \boldsymbol{B}(t_i)\Delta t_i$$

where t_i are arbitrary points in the intervals Δt_i into which the interval $[a, b]$ has been split. In the case of a strongly continuous function $\boldsymbol{B}(t)$ the values of the operator-integral (11.32) are defined for each fixed x as the limit, in the norm of the space H, of integral sums

$$\sum_{i=1}^{n} \boldsymbol{B}(t_i)x\,\Delta t_i.$$

Finally, if $\boldsymbol{B}(t)$ is weakly continuous one first defines the value of the scalar product $(\boldsymbol{B}x, y)$ as the limit of integral sums

$$\sum_{i=1}^{n} (\boldsymbol{B}(t_i)x, y)\Delta t_i;$$

then it is shown that these limits define a linear functional of y, hence the general representation of a functional on a Hilbert space implies that the element $\boldsymbol{B}x$ is defined.

These three types of integrals are called *uniform*, *strong* and *weak integrals*, respectively.

If an operator-function $\boldsymbol{B}(t)$ is strongly or uniformly continuous, it is clear that its weak integral coincides with its strong or uniform integral. This simple remark permits one to reduce the calculation of strong or uniform integrals of an operator-function to the calculation of integrals of scalar functions.

Integrals of functions of the following form are often encountered:

$$\boldsymbol{B}(t) = f(t, \boldsymbol{A}), \tag{11.33}$$

where $\boldsymbol{A}$ is a fixed positive definite selfadjoint operator. For all three types of integrals (11.32) of the function (11.33) the following representation holds under natural hypotheses:

$$\int_a^b f(t, \boldsymbol{A})\,\mathrm{d}t = \int_{-\infty}^{\infty} \left[\int_a^b f(t, \lambda)\,\mathrm{d}t\right] \mathrm{d}\boldsymbol{P}_\lambda. \tag{11.34}$$

This is proved through a change of order of integration. We do not here examine the legitimacy of such a change of order, except to remark that it suffices to prove the legitimacy of this change of order merely for weak integrals.

Integrals of operator-functions on infinite intervals or integrals of functions with singularities on a finite interval are defined in the usual manner as improper integrals. For instance, the integral on the interval $[0, \infty)$ of a function which is continuous on $(0, \infty)$ is defined as the limit of the integral (11.32) under $a \to 0$, $b \to \infty$. The limit is to be understood in the sense of operator norm in the case of the uniform integral, as the limit in norm of the values for each fixed element x in the case of the strong integral, and finally as the weak limit of the values in the case of the weak integral.

For integrals on infinite intervals, representations similar to (11.34) are valid. For instance,

$$\int_0^\infty f(t, \boldsymbol{A})\,\mathrm{d}t = \int_{-\infty}^{\infty} \left[\int_0^\infty f(t, \lambda)\,\mathrm{d}t\right] \mathrm{d}\boldsymbol{P}_\lambda. \tag{11.35}$$

Integrals of operator-functions satisfy the following inequalities, which can be proved in the usual manner:

$$\| \int_a^b \boldsymbol{B}(t)\,\mathrm{d}t \| \leqq \int_a^b \|\boldsymbol{B}(t)\|\,\mathrm{d}t, \tag{11.36}$$

$$\| \int_a^b \boldsymbol{B}(t)x\,\mathrm{d}t \| \leqq \int_a^b \|\boldsymbol{B}(t)x\|\,\mathrm{d}t. \tag{11.37}$$

On applying these inequalities together with Cauchy's criterion on the convergence of improper integrals, which can be formulated for integrals of operator-functions and vector-functions just as for improper integrals of scalar functions, it is possible to obtain conditions for the convergence of integrals of operator-functions.

For the uniform convergence of the improper integral (11.35), for instance, it is sufficient that the following inequality be satisfied:

$$\int_0^\infty \|\boldsymbol{B}(t)\|\,\mathrm{d}t < \infty. \tag{11.38}$$

For the strong convergence of the integral (11.35) it is sufficient that for each fixed $x \in H$ the following inequality holds:

$$\int_0^\infty \|\boldsymbol{B}(t)x\|\,\mathrm{d}t < \infty. \tag{11.39}$$

Let $\boldsymbol{F}$ be a closed operator. Suppose that for each t the range of the operator $\boldsymbol{B}(t)$ is contained in $D(\boldsymbol{F})$. Finally, suppose the following strong integrals exist as proper or improper integrals:

$$y = \int_a^b \boldsymbol{B}(t)x\,\mathrm{d}t, \quad z = \int_a^b \boldsymbol{F}\boldsymbol{B}(t)x\,\mathrm{d}t.$$

Then $y \in D(\boldsymbol{F})$ and $z = \boldsymbol{F}y$, i.e.

$$\boldsymbol{F}\int_a^b \boldsymbol{B}(t)x\,\mathrm{d}t = \int_a^b \boldsymbol{F}\boldsymbol{B}(t)x\,\mathrm{d}t.$$

To prove this formula, it suffices to remark that the element y can be obtained as the strong limit of a sequence $\sum_{i=1}^{n} \boldsymbol{B}(t_i)x\,\Delta t_i$ and that the sequence

$$\boldsymbol{F}\sum_{i=1}^{n} \boldsymbol{B}(t_i)x\,\Delta t_i = \sum_{i=1}^{n} \boldsymbol{F}\boldsymbol{B}(t_i)x\,\Delta t_i$$

also converges to the element z. Hence $y \in D(\boldsymbol{F})$ and $z = \boldsymbol{F}y$.

11.8 *Integral representation of fractional powers of an operator*

Let a selfadjoint operator $\boldsymbol{A}$ satisfy the condition

$$(\boldsymbol{A}x, x) \geqq a(x, x) \qquad (x \in D(\boldsymbol{A})), \tag{11.40}$$

where a is a positive number. Then the spectrum of the operator $\boldsymbol{A}$ is distributed in $[a, \infty)$. Hence

$$\boldsymbol{A} = \int_a^\infty \lambda\,\mathrm{d}\boldsymbol{P}_\lambda.$$

Consider the operator function

$$\boldsymbol{B}(t) = t^{-\alpha}(t\boldsymbol{I} + \boldsymbol{A})^{-1} \qquad (0 < t < \infty)$$

where α is a fixed number in $(0, 1)$. An immediate verification shows that this function is continuous in operator norm at each point $t > 0$. Moreover (11.27) implies

$$\|\boldsymbol{B}(t)\| \leqq \frac{t^{-\alpha}}{a + t}.$$

Hence it follows that the operator function $\boldsymbol{B}(t)$ is uniformly integrable on the interval $(0, \infty)$ and that its integral defines a bounded operator. The relation (11.35) now yields

$$\int_0^\infty t^{-\alpha}(t\boldsymbol{I} + A)^{-1}\,\mathrm{d}t = \int_a^\infty \left[\int_0^\infty \frac{t^{-\alpha}}{t+\lambda}\,\mathrm{d}t\right]\mathrm{d}\boldsymbol{P}_\lambda$$

$$= \int_a^\infty \lambda^{-\alpha}\left[\int_0^\infty \frac{s^{-\alpha}}{1+s}\,\mathrm{d}s\right]\mathrm{d}\boldsymbol{P}_\lambda.$$

Thus

$$A^{-\alpha} = k(\alpha)\int_0^\infty t^{-\alpha}(t\boldsymbol{I} + A)^{-1}\,\mathrm{d}t, \tag{11.41}$$

where

$$\frac{1}{k(\alpha)} = \int_0^\infty \frac{s^{-\alpha}}{1+s}\,\mathrm{d}s = \frac{\pi}{\sin\alpha\pi}. \tag{11.42}$$

§ 12 Properties of fractional powers of unbounded operators

12.1 *Problem setting*

In the preceding paragraph bounded linear operators A and their fractional powers A^τ were investigated. Attention was mainly focused on a detailed study of the L-characteristics $L(A;\text{cont.})$, $L(A^\tau;\text{cont.})$ and $L(A;\text{comp.})$, $L(A^\tau;\text{comp.})$. Many problems in mathematical physics require the study of unbounded operators (differential operators with boundary conditions). Here it is often the case that the operators $A^{-\tau}$ $(\tau > 0)$ are continuous.

The present section is devoted to the investigation of fractional powers of unbounded selfadjoint operators acting in a Hilbert space H. As it will turn out, these fractional powers satisfy important relations, which can be applied to establish properties of the operators A^{-1} and $A^{-\tau}$ (in particular, to study their L-characteristics) as well as to justify various constructions in which unbounded operators A take part.

An unbounded selfadjoint operator A in $L_{\frac{1}{2}}$ may turn out to be a continuous operator acting from $L_{\frac{1}{2}}$ to some space L_β, where $\beta > \frac{1}{2}$. As it will turn out at the end of the present paragraph, fractional powers of such operators become continuous operators acting from $L_{\frac{1}{2}}$ to some $L_{\beta(\tau)}$, where $\frac{1}{2} < \beta(\tau) \leqq 1$.

12.2 *The moment inequality for fractional powers*

In this section Hölder's inequality plays an essential role:

$$\left|\int_G f(\lambda)g(\lambda)\,\mathrm{d}\sigma(\lambda)\right| \leqq \{\int_G |f(\lambda)|^{1/\tau}\mathrm{d}\sigma(\lambda)\}^{\tau}\cdot\{\int_G |g(\lambda)|^{1/(1-\tau)}\mathrm{d}\sigma(\lambda)\}^{1-\tau}, \quad (12.1)$$

where G is an arbitrary set on which a positive measure σ is given, and $\tau \in (0, 1)$. In addition, Young's inequality is useful:

$$uv \leqq \tau|u|^{1/\tau} + (1-\tau)|v|^{1/(1-\tau)} \qquad (0 < \tau < 1). \qquad (12.2)$$

Inserting into Young's inequality the substitution

$$u = \varepsilon^{\tau}\tau^{-\tau}a, \quad v = \varepsilon^{-\tau}\tau^{\tau}b,$$

we arrive at the inequality

$$ab \leqq \varepsilon|a|^{1/\tau} + \left(\frac{\tau}{\varepsilon}\right)^{\tau/(1-\tau)}(1-\tau)|b|^{1/(1-\tau)}, \qquad (12.3)$$

valid for any non-negative a, b and any $\varepsilon > 0$.

THEOREM 12.1 *Let A be a positive definite selfadjoint operator. Then for any $\tau \in (0, 1)$ the following moment inequality*[1] *is fulfilled*:

$$\|A^{\tau}x\| \leqq \|Ax\|^{\tau}\cdot\|x\|^{1-\tau} \qquad (x \in D(A)). \qquad (12.4)$$

---

[1] Relations of the same type as inequality (12.4) appear in various branches of analysis. For instance, it is well known that there are similar inequalities which relate the norms of derivatives of different orders (here sharp results are due to G. E. Silov and A. N. Kolmogorov and others). It is appropriate to mention also multiplicative inequalities occurring in imbedding theorems (V. P. Glushko and S. G. Krein [1], V. P. Ilin [1–3], L. Nirenberg [1] and others). The moment inequality (12.4) was used to establish a general theory of fractional powers of operators by M. A. Krasnoselskii [5, 7]. The moment inequality plays an essential role in the theory of fractional powers of operators in Banach spaces (see § 14–16).

PROOF: For the proof we use the spectral representation of the operator $\boldsymbol{A}$:

$$\boldsymbol{A} = \int_0^\infty \lambda \, \mathrm{d}\boldsymbol{P}_\lambda.$$

This representation implies that

$$\|\boldsymbol{A}^\tau x\|^2 = \int_0^\infty \lambda^{2\tau} \mathrm{d}(\boldsymbol{P}_\lambda x, x) \qquad (x \in D(\boldsymbol{A})).$$

It follows by inequality (12.1) that

$$\|\boldsymbol{A}^\tau x\|^2 \leqq \{\int_0^\infty \lambda^2 \mathrm{d}(\boldsymbol{P}_\lambda x, x)\}^\tau \cdot \{\int_0^\infty \mathrm{d}(\boldsymbol{P}_\lambda x, x)\}^{1-\tau} = \|\boldsymbol{A}x\|^{2\tau} \|x\|^{2(1-\tau)}.$$

The theorem has been proved.

We remark that inequality (12.4) can be considered valid for all $x \in D(\boldsymbol{A}^\tau)$; here if $x \notin D(\boldsymbol{A})$, (12.4) should be understood in the sense that the left side is finite but the right side is infinite. If $x \notin D(\boldsymbol{A}^\tau)$, both sides of the inequality (12.4) are infinite.

Applying inequality (12.3) to the right side of inequality (12.4) we obtain the relation

$$\|\boldsymbol{A}^\tau x\| \leqq \varepsilon \|\boldsymbol{A}x\| + \left(\frac{\tau}{\varepsilon}\right)^{\tau/(1-\tau)} (1-\tau)\|x\| \qquad (x \in D(\boldsymbol{A})). \tag{12.5}$$

We remark that inequality (12.4) is, in turn, a consequence of inequality (12.5). This results from the fact that for each fixed $x \in D(\boldsymbol{A})$ the right side of inequality (12.5) takes its minimum value with respect to ε where the derivative with respect to ε of the right side vanishes, i.e. at

$$\varepsilon = \tau \frac{\|x\|^{1-\tau}}{\|\boldsymbol{A}x\|^{1-\tau}}.$$

Inserting this value for ε in the right side of inequality (12.5) we arrive at inequality (12.4).

It is sometimes convenient to write inequality (12.4) in the form of an inequality for quadratic forms:

$$(A^{\tau}x, x) \leqq (Ax, x)^{\tau}(x, x)^{1-\tau}. \tag{12.6}$$

From the spectral representation of fractional powers A^{τ} of the operator A it follows that

$$A^{\tau} = (A^{\tau_1})^{\tau/\tau_1} \quad (\tau, \tau_1 > 0). \tag{12.7}$$

Hence Theorem 12.1 implies that the important inequality

$$\|A^{\tau}x\| \leqq \|A^{\tau_0}x\|^{(\tau-\tau_1)/(\tau_0-\tau_1)} \cdot \|A^{\tau_1}x\|^{(\tau_0-\tau)/(\tau_0-\tau_1)} \tag{12.8}$$

is valid for $\tau_1 < \tau < \tau_0$.

It is possible to go from the multiplicative inequalities (12.6) and (12.8) to additive inequalities in the same manner as inequality (12.4) was transformed into the inequality (12.5).

12.3 *Subordinate operators*

Let us establish first an auxiliary assertion of general character.

LEMMA 12.1: *Let A and B be closed linear operators and let $D(A) \subset D(B)$. Then there is a constant M such that for all $x \in D(A)$ the following inequality holds:*

$$\|Bx\| \leqq M(\|Ax\| + \|x\|). \tag{12.9}$$

PROOF: Denote by Γ the graph of the operator A. Define the operator B_1 on the elements $\{x, Ax\}$ of the graph Γ by the relation

$$B_1\{x, Ax\} = Bx.$$

Suppose that sequences $\{x_n, Ax_n\}$ and Bx_n converge to elements $\{x_0, y_0\}$ and z_0 respectively. This means that $\|x_n - x_0\| \to 0$, $\|Ax_n - y_0\| \to 0$, $\|Bx_n - z_0\| \to 0$. Closedness of the operators A and B now implies that $x_0 \in D(A)$ and $x_0 \in D(B)$ with $y_0 = Ax_0$ and $z_0 = Bx_0$, whence it follows that the operator B_1 is closed. The graph is a subspace. As already remarked, a closed operator defined on a subspace is bounded, whence it follows that

$$\|B_1\{x, Ax\}\| \leqq M\,\|\{x, Ax\}\|,$$

i.e.

$$\|Bx\| \leqq M\sqrt{\|x\|^2 + \|Ax\|^2} \leqq M(\|x\| + \|Ax\|).$$

The lemma has been proved.

It is clear that the assertion of the lemma remains true if the operator B is not closed but admits closure.

An operator B is called *subordinate* to an operator A if $D(A) \subset D(B)$ and

$$\|Bx\| \leqq k_0\,\|Ax\| \qquad (x \in D(A)), \tag{12.10}$$

where k_0 is a positive constant.

If a closed operator A has a bounded inverse, then by Lemma 12.1, a closed operator B is, obviously, subordinate to the operator A whenever $D(A) \subset D(B)$.[1]

It is in some cases convenient to write inequality (12.10) in the following form:

$$\|BA^{-1}x\| \leqq k_0\,\|x\| \tag{12.11}$$

(assuming, of course, that the operator A^{-1} is defined).

Let an operator B be subordinate to an operator A which admits the closure A_1, and let $D(A) = D(B)$. Then the operator B has an extension B_1 on $D(A_1)$ which is subordinate to the operator A_1.

The proof is almost trivial. If $x_0 \in D(A_1)$ then there is a sequence $x_n \in D(A)$ $(n = 1, 2, \ldots)$ for which $\|x_n - x_0\| \to 0$ and $\|A_1x_n - Ax_0\| \to 0$. The inequalities

$$\|Bx_n - Bx_m\| \leqq k_0\,\|Ax_n - Ax_m\|$$

[1] Translator's note. A misleading sentence of the original text has been changed.

imply that the sequence $\boldsymbol{B}x_n$ is fundamental and that its limit y_0 is uniquely determined by the element $x_0 \in D(\boldsymbol{A}_1)$. To complete the proof, it suffices to define the operator $\boldsymbol{B}_1$ by the relation

$$\boldsymbol{B}_1 x_0 = y_0 \qquad (x_0 \in D(\boldsymbol{A}_1)).$$

Subordination of the operator $\boldsymbol{B}_1$ to the operator $\boldsymbol{A}_1$ follows from the inequality

$$\|\boldsymbol{B}_1 x_0\| \leqq k_0 \|\boldsymbol{A}_1 x_0\|,$$

which can be obtained by passing to the limit in the inequalities

$$\|\boldsymbol{B}x_n\| \leqq k_0 \|\boldsymbol{A}x_n\| \qquad (n = 1, 2, \ldots).$$

From this observation there results an important assertion.

LEMMA 12.2: *In order to prove the subordination of a closed operator* $\boldsymbol{B}$ *to a closed operator* $\boldsymbol{A}$*, it suffices to show that an inequality of the form* $\|\boldsymbol{B}x\| \leqq k_0 \|\boldsymbol{A}x\|$ *is satisfied for all elements* x *belonging to some linear set* $H_{\boldsymbol{A}}$ *(contained in* $D(\boldsymbol{A})$ *and in* $D(\boldsymbol{B})$*) with the property that* $\boldsymbol{A}$ *is the closure of the operator* $\boldsymbol{A}_1$ *with domain* $H_{\boldsymbol{A}}$ *and with* $\boldsymbol{A}_1 x = \boldsymbol{A}x$ $(x \in H_{\boldsymbol{A}})$.

If an operator $\boldsymbol{A}$ is selfadjoint:

$$\boldsymbol{A} = \int_{-\infty}^{\infty} \lambda \, \mathrm{d}\boldsymbol{P}_\lambda,$$

then it is possible to take as the set $H_{\boldsymbol{A}}$ the union of subspaces H_Δ (see 10.1°) on each of which the operator $\boldsymbol{A}$ is bounded.

In the sequel Lemma 12.2 will be used for positive definite operators $\boldsymbol{A}$. In the study of such operators the set $H_{\boldsymbol{A}}$ will be defined as the union of the subspaces H_Δ, where $\Delta = (t_1, t_2]$ and $t_1, t_2 > 0$. A simple verification shows that the linear sets $H_{\boldsymbol{A}}$, $H_{\boldsymbol{A}^\alpha}$, $H_{\boldsymbol{A}^{-\tau}}$, $H_{(t\boldsymbol{I}+\boldsymbol{A})^{-\tau}}$ and so on coincide.

The symbol $\boldsymbol{B} \ll \boldsymbol{C}$ means that the operator $\boldsymbol{B}$ is subordinate to the operator $\boldsymbol{C}$ and

$$\|\boldsymbol{B}x\| \leqq \|\boldsymbol{C}x\| \qquad (x \in D(\boldsymbol{C})). \tag{12.12}$$

Suppose that $\boldsymbol{B} \ll \boldsymbol{C}$ and both $\boldsymbol{B}$ and $\boldsymbol{C}$ are selfadjoint operators with inverses $\boldsymbol{B}^{-1}$ and $\boldsymbol{C}^{-1}$ (possibly unbounded). It follows from the substitution $\boldsymbol{C}x = y$ that (12.12) implies $\|\boldsymbol{BC}^{-1}y\| \leqq \|y\|$ for all $y \in D(\boldsymbol{C}^{-1})$. This means that for all $y \in D(\boldsymbol{C}^{-1})$ and each fixed $z \in D(\boldsymbol{B})$ the following inequality holds:

$$|(\boldsymbol{C}^{-1}y, \boldsymbol{B}z)| = |(\boldsymbol{BC}^{-1}y, z)| \leqq \|\boldsymbol{BC}^{-1}y\| \cdot \|z\| \leqq \|y\| \, \|z\|.$$

Consequently $|(\boldsymbol{C}^{-1}y, \boldsymbol{B}z)|$ can be extended as a continuous linear functional defined on all of H; that is,

$$(\boldsymbol{C}^{-1}y, \boldsymbol{B}z) = (y, y^*) \qquad (y \in D(\boldsymbol{C}^{-1})).$$

This implies that $\boldsymbol{B}z \in D(\boldsymbol{C}^{-1})$, i.e. $D(\boldsymbol{B}^{-1}) \subset D(\boldsymbol{C}^{-1})$. It is clear that

$$|(y, \boldsymbol{C}^{-1}\boldsymbol{B}z)| = |(\boldsymbol{BC}^{-1}y, z)| \leqq \|y\| \, \|z\| \qquad (y \in D(\boldsymbol{C}^{-1}), z \in D(\boldsymbol{B})),$$

hence

$$\|\boldsymbol{C}^{-1}\boldsymbol{B}z\| \leqq \|z\| \qquad (z \in D(\boldsymbol{B})).$$

Substituting $\boldsymbol{B}z = u$, we arrive at the inequality

$$\|\boldsymbol{C}^{-1}u\| \leqq \|\boldsymbol{B}^{-1}u\| \qquad (u \in D(\boldsymbol{B}^{-1})). \tag{12.13}$$

The following assertion has been proved.

LEMMA 12.3: *Let selfadjoint operators* $\boldsymbol{B}$ *and* $\boldsymbol{C}$, *with inverses* $\boldsymbol{B}^{-1}$ *and* $\boldsymbol{C}^{-1}$, *satisfy the relation* $\boldsymbol{B} \ll \boldsymbol{C}$. *Then* $\boldsymbol{C}^{-1} \ll \boldsymbol{B}^{-1}$.

It can be shown by argument of the same type that for the case of non-selfadjoint $\boldsymbol{B}$ and $\boldsymbol{C}$, $\boldsymbol{B} \ll \boldsymbol{C}$ implies

$$(\boldsymbol{C}^{-1})^* \ll (\boldsymbol{B}^*)^{-1},$$

provided, of course, that these inverses exist.

12.4 *Subordination of fractional powers*

Let A be a bounded positive definite selfadjoint operator. The relation $A^\sigma = A^{\sigma-\tau}A^\tau$ implies that

$$\|A^\sigma x\| \leqq \|A^{\sigma-\tau}\| \, \|A^\tau x\|.$$

This means that *an operator A^σ is subordinate to an operator A^τ whenever* $0 < \tau < \sigma$.

The reverse situation occurs in the case that A is a strictly positive definite unbounded selfadjoint operator:

$$A = \int_a^\infty \lambda \, \mathrm{d}P_\lambda \qquad (a > 0). \tag{12.14}$$

In this case, *A^σ is subordinate to an operator A^τ* if $0 < \sigma < \tau$. Lemma 12.3 could be used for this proof, but the reader can devise a direct proof without difficulty.

If a positive definite operator A is inverse to an unbounded operator then distinct fractional powers do not possess properties of subordination to each other.

Let us consider the following problem: to which fractional powers of a strictly positive definite operator A is another operator B subordinate?

Theorem 12.1 immediately implies:

THEOREM 12.2: *Let A be positive definite selfadjoint operator. The inequality*

$$\|Bx\| \leqq k \, \|Ax\|^{\tau_0} \, \|x\|^{1-\tau_0} \qquad (x \in D(A)) \tag{12.15}$$

is a necessary condition in order that a linear operator B be subordinate to the operator A^{τ_0} where $\tau_0 \in (0, 1)$.

It is possible to give an 'almost converse' to this theorem.

THEOREM 12.3:[1] *Let A be a strictly positive definite selfadjoint operator. Let B be a closed linear operator satisfying condition* (12.15) *for some* $\tau_0 \in (0, 1)$. *Then the operator B is subordinate to all operators A^τ, where $\tau > \tau_0$.*

PROOF: Let A be the operator (12.14). Then (12.15) implies $D(A) \subset D(B)$. By (11.27) the operator A is defined on elements of the form $(tI + A)^{-1}x$ $(t > 0)$; hence the vector function $t^{-\tau}B(tI + A)^{-1}x$ is defined for all $t > 0$. This vector function is continuous for $t > 0$ because the operator BA^{-1} is bounded and

$$t^{-\tau}B(tI + A)^{-1}x = BA^{-1}t^{-\tau}A(tI + A)^{-1}x,$$

and the continuity of the vector-function $A(tI + A)^{-1}x$ is easily verified.

Finally by (12.15) and (11.27)

$$\int_0^\infty t^{-\tau}\|B(tI + A)^{-1}x\|\,dt$$

$$\leqq k\int_0^\infty t^{-\tau}\|A(tI + A)^{-1}x\|^{\tau_0}\cdot\|(tI + A)^{-1}x\|^{1-\tau_0}dt$$

$$\leqq k\int_0^\infty \frac{t^{-\tau}dt}{(a + t)^{1-\tau_0}}\,\|x\| = k_1\|x\| < \infty.$$

This means that

$$\int_0^\infty t^{-\tau}(tI + A)^{-1}x\,dt \in D(B)$$

and

$$B\int_0^\infty t^{-\tau}(tI + A)^{-1}x\,dt = \int_0^\infty t^{-\tau}B(tI + A)^{-1}dt.$$

---

[1] This theorem was established by S. G. Krein and P. E. Sobolevskii [1]. Here we give a different proof.

From the integral representation (11.41) of a fractional power it follows that $D(A^\tau) \subset D(\boldsymbol{B})$ and

$$\|\boldsymbol{B}A^{-\tau}x\| \leqq k_1 \|x\|.$$

The theorem has been proved.

In 12.2° the equivalence of the conditions (12.4) and (12.5) was shown. Hence the assertions of the above two theorems can be expressed in the following form.

THEOREM 12.4: *In order that a closed operator $\boldsymbol{B}$ be subordinate to an operator A^τ $(0 < \tau < 1)$, where A is a strictly positive definite selfadjoint operator, it is necessary that the inequality*

$$\|\boldsymbol{B}x\| \leqq \varepsilon^{1-\tau_0}\|Ax\| + \frac{k_1}{\varepsilon^{\tau_0}}\|x\| \qquad (x \in D(A)) \tag{12.16}$$

be fulfilled for some k_1, for $\tau_0 = \tau$ and for all $\varepsilon > 0$, and it is sufficient that this inequality be fulfilled for some k_1, any one $\tau_0 < \tau$ and all $\varepsilon > 0$.

The infimum of those non-negative numbers τ, for which $\boldsymbol{B}$ is subordinate to A^τ is called the *order* of an operator with respect to a strictly positive definite selfadjoint operator A.

It is clear that each bounded operator is of order zero. Furthermore, a bounded operator $\boldsymbol{B}$ is subordinate to the 0-th power of the operator A. There also exist unbounded operators of order zero with respect to an operator A. For example, the operator

$$\boldsymbol{B} = \ln A = \int_a^\infty \ln \lambda \, \mathrm{d}\boldsymbol{P}_\lambda \qquad (a > 0)$$

has order zero. In fact, for any $\tau > 0$

$$|\ln \lambda|^2 \leqq c(\tau) + \lambda^{2\tau} \qquad (a \leqq \lambda < \infty),$$

where $c(\tau)$ is a positive function; hence for $x \in D(A^\tau)$ the following relation is valid:

$$\|Bx\|^2 = \int_a^\infty |\ln \lambda|^2 \mathrm{d}(P_\lambda x, x) \leqq \int_a^\infty [c(\tau) + \lambda^{2\tau}] \mathrm{d}(P_\lambda x, x)$$

$$= c(\tau)\|x\|^2 + \|A^\tau x\|^2 \leqq \left[\frac{c(\tau)}{a^{2\tau}} + 1\right] \|A^\tau x\|^2,$$

i.e. B is subordinate to the operators A^τ for all $\tau > 0$. Similarly the operator $B = A^{\tau_0} \ln A$ has order τ_0; in this case the operator B is subordinate to all operators A^τ where $\tau > \tau_0$, but not to the operator A^{τ_0}—in the condition of subordination

$$\|Bx\| \leqq k(\tau)\|A^\tau x\| \qquad (x \in D(A^\tau);\ \tau > \tau_0)$$

the constant $k(\tau)$ increases to infinity as $\tau \to \tau_0$.

Theorem 12.3 means that an operator B satisfying the inequality

$$\|Bx\| \leqq k\|Ax\|^\tau \|x\|^{1-\tau} \qquad (x \in D(A)), \tag{12.17}$$

has order (with respect to the operator A) not exceeding τ.

12.5 *Heinz' first inequality*

In this section we study a problem on the mutual subordination of fractional powers of two positive definite selfadjoint operators.

LEMMA 12.4: *Let A and B be bounded linear operators acting in a Hilbert space, and let AB be a selfadjoint operator. Then*

$$\|AB\| \leqq \|BA\|. \tag{12.18}$$

PROOF: From formula (11.25) it follows that for any selfadjoint operator C the following relation holds:

$$\|C^n\| = \|C\|^n \qquad (n = 1, 2, \ldots).$$

Hence

$$\|AB\| = \|(AB)^n\|^{1/n} = \|A(BA)\ldots(BA)B\|^{1/n}$$

$$\leqq [\|A\|\,\|BA\|^{n-1}\|B\|]^{1/n} = \|A\|^{1/n}\|BA\|^{1-1/n}\|B\|^{1/n}.$$

Taking the limit as n goes to infinity, we obtain (12.18). The lemma has been proved.

THEOREM 12.5:[1] *Let* $\boldsymbol{A}$ *be a positive definite selfadjoint operator, acting in a Hilbert space* H_1 *and let* $\boldsymbol{B}$ *be a positive definite selfadjoint operator, acting in a Hilbert space* H_2. *Let* $\boldsymbol{T}$ *be a bounded operator acting from the space* H_2 *to the space* H_1 *with* $\|\boldsymbol{T}\| = N$, *and finally, let*

$$\|\boldsymbol{A}\boldsymbol{T}x\|_{H_1} \leqq M \|\boldsymbol{B}x\|_{H_2} \quad (x \in D(\boldsymbol{B})). \tag{12.19}$$

Then for any τ, $0 \leqq \tau \leqq 1$

$$\boldsymbol{T}D(\boldsymbol{B}^\tau) \subset D(\boldsymbol{A}^\tau)$$

and for any $x \in D(\boldsymbol{B}^\tau)$ *the following inequality is valid*:

$$\|\boldsymbol{A}^\tau \boldsymbol{T}x\|_{H_1} \leqq M^\tau N^{1-\tau} \|\boldsymbol{B}^\tau x\|_{H_2}. \tag{12.20}$$

PROOF: Assume first that the operator $\boldsymbol{A}$ is continuous and that the operator $\boldsymbol{B}$ has a continuous inverse $\boldsymbol{B}^{-1}$. Then for any $\tau \in [0, 1]$ the operator $\boldsymbol{A}^\tau$ is bounded and the operator $\boldsymbol{B}^\tau$ has a bounded inverse $\boldsymbol{B}^{-\tau}$. The inequality (12.20) will be proved if it is shown that

$$\varphi(\tau) = \|\boldsymbol{A}^\tau \boldsymbol{T}\boldsymbol{B}^{-\tau}\|_{H_2 \to H_1} \leqq M^\tau N^{1-\tau}. \tag{12.21}$$

Let us prove that the function $\varphi(\tau)$ is logarithmically convex, i.e.

$$\varphi[(1-\lambda)\tau_0 + \lambda\tau_1] \leqq [\varphi(\tau_0)]^{1-\lambda}[\varphi(\tau_1)]^\lambda. \tag{12.22}$$

Inequality (12.21) results from this, because $\varphi(0) = N$ and $\varphi(1) \leqq M$. The function $\varphi(\tau)$ is continuous. Hence it suffices to prove that for any $\tau \in (0, 1)$ and $\varepsilon \in (0, \min(\tau, 1-\tau))$ the following inequality holds:

$$\varphi(\tau) \leqq \sqrt{\varphi(\tau-\varepsilon)\cdot\varphi(\tau+\varepsilon)}. \tag{12.23}$$

---

[1] This theorem was established by T. Kato [3]. The assertion at the end of this section was proved earlier by E. Heinz [1] by another method. The present proof of Theorem 12.6, which is also due to Heinz, is borrowed from the paper of Kato [3].

The identity

$$(A^{\tau}TB^{-\tau}x, A^{\tau}TB^{-\tau}x)_{H_1} = (B^{-\tau}T^*A^{2\tau}TB^{-\tau}x, x)_{H_2}$$

implies the inequality

$$[\varphi(\tau)]^2 \leqq \|B^{-\tau}T^*A^{2\tau}TB^{-\tau}\|_{H_2 \to H_2}.$$

Since the operator $B^{-\tau}T^*A^{2\tau}TB^{-\tau}$ is bounded and selfadjoint in H_2, it follows from the representation

$$B^{-\tau}T^*A^{2\tau}TB^{-\tau} = B^{-\varepsilon}(B^{-\tau+\varepsilon}T^*A^{2\tau}TB^{-\tau})$$

and Lemma 12.4 that

$$\|B^{-\tau}T^*A^{2\tau}TB^{-\tau}\|_{H_2 \to H_2} \leqq \|B^{-\tau+\varepsilon}T^*A^{2\tau}TB^{-\tau-\varepsilon}\|_{H_2 \to H_2}$$

$$\leqq \|B^{-\tau+\varepsilon}T^*A^{\tau-\varepsilon}\|_{H_1 \to H_2} \|A^{\tau+\varepsilon}TB^{-\tau-\varepsilon}\|_{H_2 \to H_1},$$

so by the equality of the norms of adjoint operators

$$[\varphi(\tau)]^2 \leqq \|A^{\tau-\varepsilon}TB^{-\tau+\varepsilon}\|_{H_2 \to H_1} \|A^{\tau+\varepsilon}TB^{-\tau-\varepsilon}\|_{H_2 \to H_1}.$$

The inequality (12.23) has been proved.

Now consider the general case. Let

$$A = \int_0^\infty \lambda \, \mathrm{d}P_\lambda,$$

and introduce the notation

$$A_n = \int_0^n \lambda \, \mathrm{d}P_\lambda, \quad B_n = B + \frac{1}{n} I \quad (n = 1, 2, \ldots).$$

The operators A_n $(n = 1, 2, \ldots)$ are obviously bounded and the operators B_n have bounded inverses. From the trivial inequalities

$$\|A_n y\|_{H_1} \leqq \|Ay\|_{H_1} \quad (y \in D(A)),$$

$$\|Bx\|_{H_2} \leqq \|B_n x\|_{H_2} \quad (x \in D(B))$$

and from (12.19) it follows that

$$\|A_n Tx\|_{H_1} \leqq M\,\|B_n x\| \quad (x \in D(B);\ n = 1, 2, \ldots).$$

By the result proved above, for each $\tau \in [0, 1]$ the following inequalities are valid:

$$\|A_n^\tau Tx\|_{H_1} \leqq M^\tau N^{1-\tau} \cdot \|B_n^\tau x\|_{H_2} \quad (x \in D(B^\tau),\ n = 1, 2, \ldots). \tag{12.24}$$

It is easy to see that $\|B_n^\tau x\|_{H_2} \to \|B^\tau x\|_{H_2}$ as $n \to \infty$; hence $\sup\limits_n \|A_n^\tau Tx\|_{H_1} < \infty$. This means, in turn, that $Tx \in D(A^\tau)$ and $\|A_n^\tau Tx\|_{H_1} \to \|A^\tau Tx\|_{H_1}$. Passing to the limit in the above inequality we obtain inequality (12.20). The theorem has been completely proved.

This theorem implies, in particular, that *the relation $A \ll B$ for positive definite selfadjoint operators A and B acting in a Hilbert space H implies the relation $A^\tau \ll B^\tau$* $(0 < \tau < 1)$.

12.6 *Heinz' second inequality*

THEOREM 12.6: *Let a positive definite selfadjoint operator A act in a Hilbert space H_1 and let a positive definite selfadjoint operator B act in a Hilbert space H_2. Let C be a closed linear operator with domain $D(C) \subset H_1$ and range $R(C) \subset H_2$. Let $D(A) \subset D(C)$, $D(B) \subset D(C^*)$ and suppose the following inequality holds for every $x \in D(A)$, $y \in D(B)$:*

$$\|Cx\|_{H_2} \leqq \|Ax\|_{H_1}, \quad \|C^* y\|_{H_1} \leqq \|By\|_{H_2}. \tag{12.25}$$

Then for every $\tau \in [0, 1]$, $x \in D(A)$ and $y \in D(B)$ the following inequality is valid:

$$|(Cx, y)_{H_2}| \leqq \|A^\tau x\|_{H_1} \cdot \|B^{1-\tau} y\|_{H_2}. \tag{12.26}$$

PROOF: Assume first that the operator A has a bounded inverse A^{-1}. Then from the first inequality of (12.25) it follows that the operator $T = CA^{-1}$

acts from H_1 to H_2, and

$$\|T\|_{H_1 \to H_2} \leqq 1. \tag{12.27}$$

Consider the operator $S = TA$ defined on $D(A) \subset H_1$ and acting from H_1 to H_2. It is clear that $D(S) \subset D(C)$ and

$$Sx = Cx \quad (x \in D(S)),$$

whence it follows that the operator T^* transforms $D(C^*)$ to $D(A)$ and $S^*y = AT^*y = C^*y$ for $y \in D(C^*)$. Then (12.25) implies that for $y \in D(B)$ one has the inequality

$$\|AT^*y\|_{H_1} \leqq \|By\|_{H_2}. \tag{12.28}$$

From (12.27), (12.28) and Theorem 12.5 it follows that $T^*D(B^\tau) \subset D(A^\tau)$ for every $\tau \in [0, 1]$ and

$$\|A^\tau T^*y\|_{H_1} \leqq \|B^\tau y\|_{H_2} \quad (y \in D(B^\tau)). \tag{12.29}$$

Now let $x \in D(A)$, $y \in D(B)$. Then

$$(Cx, y)_{H_2} = (TAx, y)_{H_2} = (Ax, T^*y)_{H_1} = (A^\tau x, A^{1-\tau}T^*y)_{H_1};$$

hence from inequality (12.29) there follows

$$|(Cx, y)_{H_2}| \leqq \|A^\tau x\|_{H_1} \|A^{1-\tau}T^*y\|_{H_1} \leqq \|A^\tau x\|_{H_1} \|B^{1-\tau}y\|_{H_2}.$$

The assertion of the theorem has been proved under the assumption that the operator A^{-1} is bounded. This restriction can be removed just as in the proof of Theorem 12.5. The theorem has been proved.

12.7 *Fractional powers of projected operators*[1]

Let a subspace H_0 be contained in the domain $D(C)$ of a closed linear operator C acting in a Hilbert space H. The operator C remains closed

[1] The results of this section were obtained by P. E. Sobolevskii [13].

when considered only on the subspace H_0; hence it is continuous on H_0, i.e.

$$\|Cx\| \leqq k\|x\| \quad (x \in H_0).$$

Now consider the operator PC acting in H_0, where P is the orthogonal projection on H_0; this operator is also continuous.

If C is selfadjoint, the operator PC is also selfadjoint. If C is positive definite, the operator PC is also positive definite. This last assertion results from the identity

$$(PCx, x) = (Cx, x) \quad (x \in H_0).$$

This identity can be written in the form

$$\|(PC)^{\frac{1}{2}}x\| = \|C^{\frac{1}{2}}x\| \quad (x \in H_0). \tag{12.30}$$

It is natural to raise the question of the relation between the norms of the elements $C^\tau x$ and $(PC)^\tau x$ $(x \in H_0)$ for various exponents τ.

THEOREM 12.7: *Let* C *be a positive definite selfadjoint operator and let* $\tau \in [0, \frac{1}{2}]$. *Then the following holds*:

$$\|C^\tau x\| \leqq \|(PC)^\tau x\| \quad (x \in H_0). \tag{12.31}$$

PROOF: To use Theorem 12.5, put

$$H_1 = H,\ H_2 = H_0,\ A = C^{\frac{1}{2}},\ B = (PC)^{\frac{1}{2}},\ Tx = x \quad (x \in H_0).$$

Then the relation (12.30) implies condition (12.19) with $M = 1$ and inequality (12.31) coincides with the estimate (12.20). The theorem has been proved.

THEOREM 12.8: *Let* C *be a positive definite selfadjoint operator and let*

$$\|Cx\| \leqq k\|PCx\| \quad (x \in H_0), \tag{12.32}$$

where k *is a constant. Let* $\tau \in (\frac{1}{2}, 1]$. *Then the following inequality holds*:

$$\|C^\tau x\| \leqq k^{2\tau-1}\|(PC)^\tau x\| \quad (x \in H_0). \tag{12.33}$$

PROOF: Assume first that $\boldsymbol{C}$ is a strictly positive definite operator. Then the operators $(\boldsymbol{PC})^{-\tau}$ $(\tau > 0)$ exist and are continuous in H_0. In the proof of Theorem 12.5 it was shown that the function (12.21) is logarithmically convex for $0 \leqq \tau \leqq 1$. Put

$$H_1 = H,\ H_2 = H_0,\ \boldsymbol{A} = \boldsymbol{C},\ \boldsymbol{B} = \boldsymbol{PC},\ \boldsymbol{T}x = x \quad (x \in H_0).$$

Then the function (12.21) take the form

$$\varphi(\tau) = \|\boldsymbol{C}^{\tau}(\boldsymbol{PC})^{-\tau}\|_{H_0 \to H}.$$

From logarithmic convexity of this function it follows that

$$\varphi(\tau) \leqq [\varphi(\tfrac{1}{2})]^{2(1-\tau)} \cdot [\varphi(1)]^{2\tau-1} \quad (\tfrac{1}{2} < \tau \leqq 1). \tag{12.34}$$

The relation (12.30) means that $\varphi(\frac{1}{2}) = 1$, while the inequality (12.32) means that $\varphi(1) \leqq k$. Hence (12.34) implies (12.33).

The supplementary assumption on strict positive definiteness of the operator $\boldsymbol{C}$ can be removed by considering first operators of the form $\boldsymbol{C} + (1/n)\boldsymbol{I}$ and then letting n tend to infinity (see the end of the proof of Theorem 12.5). The theorem has been proved.

Theorems 12.7 and 12.8 admit a generalization to the case where H_0 is not contained completely in $D(\boldsymbol{C})$, but the set $D_0 = H_0 \cap D(\boldsymbol{C})$ is dense in H_0. The operator $\boldsymbol{PC}$ considered on D_0, is symmetric and positive definite. It is known that each such operator admits a positive definite selfadjoint extension. The same notation $\boldsymbol{PC}$ will be used for this so-called Friedrichs extension.[1] It can be shown that for $\tau \in [0, \frac{1}{2}]$ the inclusion $D((\boldsymbol{PC})^{\tau}) \subset D(\boldsymbol{C}^{\tau})$ occurs and the following inequality holds:

$$\|\boldsymbol{C}^{\tau}x\| \leqq \|(\boldsymbol{PC})^{\tau}x\| \quad (x \in D(\boldsymbol{PC})). \tag{12.35}$$

If $D(\boldsymbol{PC}) = H_0 \cap D(\boldsymbol{C})$ and

$$\|\boldsymbol{C}x\| \leqq k\|\boldsymbol{PC}x\| \quad (x \in D(\boldsymbol{PC})). \tag{12.36}$$

---

[1] See, for instance, S. G. Mikhlin [2].

for $\tau \in (\frac{1}{2}, 1]$ the inclusion $D((\boldsymbol{PC})^\tau) \subset D(\boldsymbol{C}^\tau)$ occurs and the following inequality is valid:

$$\|\boldsymbol{C}^\tau x\| \leqq k^{2\tau-1} \cdot \|(\boldsymbol{PC})^\tau x\| \quad (x \in D((\boldsymbol{PC})^\tau)). \tag{12.37}$$

12.8 *On a special class of selfadjoint operators*

Let an operator $\boldsymbol{B}$ be given on a set G. When $F \subset G$ denote by $\boldsymbol{B}_F$ the operator which is defined on F and whose values on F coincide with the values of the operator $\boldsymbol{B}$. The operator $\boldsymbol{B}_F$ constructed in this way will be called the *restriction of the operator* $\boldsymbol{B}$ *on the set* F or simply the *restriction of the operator* $\boldsymbol{B}$.

Let a continuous linear operator A act from $L_{\frac{1}{2}}$ to L_β where $\frac{1}{2} < \beta \leqq 1$. Suppose that the restriction of the operator A on a linear set $D_0 \subset L_{\frac{1}{2}}$ is a positive definite selfadjoint operator acting in $L_{\frac{1}{2}}$. The operator A_{D_0} will not, of course, be bounded in $L_{\frac{1}{2}}$ if $D_0 \neq L_{\frac{1}{2}}$. It is possible to define fractional powers $A^\tau_{D_0}$ of the operator A_{D_0}; these fractional powers will be unbounded operators. By analogy to the theorems proved in § 10 it is reasonable to expect that operators $A^\tau_{D_0}$ admit extensions to continuous operators acting from $L_{\frac{1}{2}}$ to certain $L_{\beta(\tau)}$, where $\beta(\tau) > \frac{1}{2}$. The present and following sections are devoted to giving proofs of the corresponding theorems as well as to an analysis of the L-characteristics of continuous extensions A^τ of operators $A^\tau_{D_0}$.

Consider again a continuous linear operator A acting from $L_{\frac{1}{2}}$ to L_β; its adjoint operator A^* acts from $L_{1-\beta}$ to $L_{\frac{1}{2}}$. The operator A will be called *symmetric* if $A^*x = Ax$ $(x \in L_{1-\beta})$. Denote by D the totality of all elements $x \in L_{\frac{1}{2}}$ for which $Ax \in L_{\frac{1}{2}}$. Then $L_{1-\beta} \subset D$ and the following relation is valid:

$$(Ax, y) = (x, Ay) \quad (x \in L_{1-\beta}, y \in L_{\frac{1}{2}}). \tag{12.38}$$

Let y be an element of $L_{\frac{1}{2}}$ such that

$$(Ax, y) = (x, y^*) \quad (x \in L_{1-\beta})$$

where y^* also belongs to $L_{\frac{1}{2}}$. Then (12.38) implies that $y \in D$ and $y^* = A_D y$. This means that

$$(A_{L_{1-\beta}})^* = A_D. \tag{12.39}$$

The operator $A_{L_{1-\beta}}$ is symmetric in view of (12.38). *The relation* (12.39) *means that all its symmetric extensions (if they exist) are restrictions of the operator A.*

The operator A is called *positive definite* if

$$(Ax, x) \geqq 0 \quad (x \in L_{1-\beta}). \tag{12.40}$$

The operator A then admits a positive definite selfadjoint extension in $L_{\frac{1}{2}}$[1]. The above observations show that this selfadjoint extension is the restriction of the operator A_D to some set $D_0 \subset L_{\frac{1}{2}}$.

Thus the following has been proved:

THEOREM 12.9: *Let A be a continuous symmetric operator acting from $L_{\frac{1}{2}}$ to L_β $(\frac{1}{2} < \beta \leqq 1)$ which is positive definite in the sense of condition* (12.40). *Then there exists a restriction A_{D_0} $(D_0 \subset L_{\frac{1}{2}})$ of the operator A which is a positive definite selfadjoint operator in $L_{\frac{1}{2}}$.*

It is natural to raise the question of determining the cases for which the operator A_D is selfadjoint, where D is the set of all $x \in L_{\frac{1}{2}}$ such that $Ax \in L_{\frac{1}{2}}$.

THEOREM 12.10: *Let a positive*[2] *operator A satisfy the conditions of Theorem* 12.9. *Let the set D possess the property that $x \in D$ implies $|x| \in D$. Then the operator A_D is selfadjoint.*

PROOF: It suffices to show that the operator A_D is the closure of the operator $A_{L_{1-\beta}}$. To this end, it is necessary to show that for any $x \in D$ there is a sequence $x_n \in L_{1-\beta}$ such that $\|x_n - x\|_{\frac{1}{2}} \to 0$ and, in addition, $\|Ax_n - Ax\|_{\frac{1}{2}} \to 0$.

Put

$$x_n(t) = \min\{|x(t)|, n\} \operatorname{sgn} x(t) \quad (n = 1, 2, \dots);$$

then $\|x_n - x\|_{\frac{1}{2}} \to 0$. The relation $\|Ax_n - Ax\|_\beta \to 0$ implies that the sequence Ax_n converges to Ax in measure. Moreover, it implies that the

---

[1] See N. I. Akhiezer and I. M. Glazman [1]. A complete determination of such extensions was given by M. G. Krein.

[2] Recall that an operator A is called positive if it transforms non-negative functions to non-negative functions.

functions $Ax_n(t)$ have equi-absolutely continuous norms, since by positivity of the operator A

$$|Ax_n(t)| \leqq \varphi(t) \in L_{\frac{1}{2}},$$

where $\varphi = A|x|$. Hence $\|Ax_n - Ax\|_{\frac{1}{2}} \to 0$. The theorem has been proved.

12.9 *Theorems on splitting*[1]

Let a regular triple of spaces $E^* \subset H \subset E$ be given (see 9.2°).

THEOREM 12.11: *Let A be a positive definite selfadjoint operator in a Hilbert space H. Let the set $D = D(A) \cap E^*$ be dense in the space E^*. Finally, let the operator A admit an extension to a continuous operator $\tilde{A}$ acting from E^* to E:*

$$\|Ax\|_E \leqq a\|x\|_{E^*} \quad (x \in D). \tag{12.41}$$

Then the operator $A^{\frac{1}{2}}$ admits an extension to a continuous operator acting from H to E, and

$$\|A^{\frac{1}{2}}x\|_E \leqq \sqrt{a}\,\|x\|_H \quad (x \in D). \tag{12.42}$$

PROOF: Let $x \in D$, then

$$\|A^{\frac{1}{2}}x\|_H = \sqrt{(A^{\frac{1}{2}}x, A^{\frac{1}{2}}x)} = \sqrt{(Ax, x)}$$

$$\leqq \sqrt{\|Ax\|_E \cdot \|x\|_{E^*}} \leqq \sqrt{a}\,\|x\|_{E^*}.$$

This means that $A^{\frac{1}{2}}$ can be extended to an operator $\boldsymbol{B}$ acting from E^* to H. The adjoint operator $\boldsymbol{B}^*$ is continuous as an operator from H to E^{**}. The range of the operator $\boldsymbol{B}^*$ turns out to belong to E. To establish this fact, it suffices to prove that $\boldsymbol{B}^*y = A^{\frac{1}{2}}y$ for $y \in D$, since D is dense in E^* by assumption and consequently dense in H. The relation $\boldsymbol{B}^*y = A^{\frac{1}{2}}y$ $(y \in D)$

[1] The results in this section and the next were established by E. J. Pustylnik [2].

follows from the relation

$$(x, \boldsymbol{B}^*y - A^{\frac{1}{2}}y) = (\boldsymbol{B}x - A^{\frac{1}{2}}x, y) = 0 \qquad (x, y \in D).$$

Inequality (12.42) is now clear:

$$\|A^{\frac{1}{2}}x\|_E = \|\boldsymbol{B}^*x\|_E \leqq \|\boldsymbol{B}^*\|_{H\to E}\|x\|$$

$$= \|\boldsymbol{B}\|_{E^*\to H}\|x\|_H \leqq \sqrt{a}\|x\|_H \qquad (x \in D).$$

The theorem has been proved.

THEOREM 12.12: *Let the conditions of Theorem* 12.11 *be fulfilled. Let the extension* $\tilde{A}$ *of the operator* A *be a compact operator acting from* E^* *to* E. *Then the operator* $A^{\frac{1}{2}}$ *admits an extension to a compact operator acting from* H *to* E.

PROOF: Let $\boldsymbol{B}$ and $\boldsymbol{B}^*$ be the operators introduced in the proof of Theorem 12.11. The assertion of the theorem is equivalent to stating that the operators $\boldsymbol{B}$ and $\boldsymbol{B}^*$ are compact (the first as an operator from E^* to H, the second as an operator from H to E). It suffices[1] to prove that the operator $\boldsymbol{B}$ is compact. To this end, it suffices to establish the compactness of the range of the operator on the intersection of D with the unit ball $\|x\|_{E^*} \leqq 1$.

Let $x_n \in D$, $\|x_n\|_{E^*} \leqq 1$ $(n = 1, 2, \ldots)$. On account of the compactness of the operator $\tilde{A}$ it is possible to choose a subsequence x_{n_k} such that the elements Ax_{n_k} converge in E. But then

$$\lim_{k,l\to\infty} \|\boldsymbol{B}x_{n_k} - \boldsymbol{B}x_{n_l}\|_H = \lim_{k,l\to\infty} \sqrt{(A(x_{n_k} - x_{n_l}), x_{n_k} - x_{n_l})} = 0.$$

The theorem has been proved.

Let the operator A satisfy the conditions of Theorem 12.11 and let $\tilde{A}$ be its continuous extension to an operator acting from E^* to E. Now Theorem 12.11 can be treated as asserting the possibility of splitting the operator $\tilde{A}$ into a superposition

$$\tilde{A} = \boldsymbol{B}^*\boldsymbol{B} \tag{12.43}$$

[1] See L. V. Kantorovic and G. P. Akilov [1].

where $\boldsymbol{B}$ acts from E^* to H and $\boldsymbol{B}^*$ from H to E. Theorem 12.12 means that the operators $\boldsymbol{B}$ and $\boldsymbol{B}^*$ are compact if the operator $\tilde{\boldsymbol{A}}$ is compact.

Theorems 12.11 and 12.12 are analogous to Theorems 9.1 and 9.2. The assertions of all these theorems can be restated for the case of a triple of spaces of the type $E \subset H \subset E^*$.

Theorems 12.11 and 12.12 would be more useful, if it were possible to replace the assumption on the denseness of the set $D = D(A) \cap E^*$ by a less restrictive assumption involving denseness of the set $D(A^{\frac{1}{2}}) \cap E^*$ in E^*.

12.10 *Theorems on fractional powers*

Here we begin an investigation of such operators in the scale of spaces L_α. We will study fractional powers A^τ of unbounded positive definite selfadjoint operators A which act in $L_{\frac{1}{2}}$ and admit extensions to continuous operators acting from $L_{\frac{1}{2}}$ to certain spaces L_β ($\frac{1}{2} < \beta \leqq 1$). It will be established that the operators A^τ admit extensions to continuous operators acting from $L_{\frac{1}{2}}$ to certain spaces $L_{\beta(\tau)}$; it will be convenient to use the same notations A^τ for these extended operators.

THEOREM 12.13: *Let a positive definite selfadjoint operator A defined in $L_{\frac{1}{2}}$ be, at the same time, a continuous operator acting from $L_{\frac{1}{2}}$ to L_β ($\frac{1}{2} < \beta \leqq 1$), with*

$$\|A\|_{\frac{1}{2}\to\beta} \leqq a. \tag{12.44}$$

Further let the set $D = D(A^2) \cap L_{1-\beta}$ be dense[1] *in the space $L_{1-\beta}$. Then for each $\tau \in (0, 1]$ the operator A^τ is a continuous operator acting from $L_{\frac{1}{2}}$ to $L_{\beta(\tau)}$, where*

$$\beta(\tau) = \tfrac{1}{2} + \tau(\beta - \tfrac{1}{2}), \tag{12.45}$$

and

$$\|A^\tau\|_{\frac{1}{2}\to\beta(\tau)} \leqq a^\tau; \tag{12.46}$$

[1] Translator's note. This assumption seems unnecessary.

in the case of real spaces the above inequality is replaced by

$$\|A^\tau\|_{\frac{1}{2}\to\beta(\tau)} \leqq 2^\tau a^\tau. \tag{12.47}$$

PROOF: In the case of complex spaces L_α, the proof is almost a direct repetition of the proof of Theorem 10.1.

Consider next the case of real spaces L_α. Extend the operator A to the space $\tilde{L}_\alpha$ of complex functions $z = x + iy$ by the relation

$$\tilde{A}(x + iy) = Ax + iAy.$$

The operator $\tilde{A}$ is also selfadjoint and positive definite. Its fractional powers $\tilde{A}^\tau$ are connected with the fractional powers A^τ of the operator A through the relation

$$\tilde{A}^\tau(x + iy) = A^\tau x + iA^\tau y. \tag{12.48}$$

The operator $\tilde{A}$ satisfies the conditions of the theorem, hence

$$\|\tilde{A}^\tau\|_{\frac{1}{2}\to\beta(\tau)} \leqq \tilde{a}^\tau, \tag{12.49}$$

where $\tilde{a}$ is the norm of $\tilde{A}$ as an operator from $\tilde{L}_{\frac{1}{2}}$ to $\tilde{L}_\beta$; it is clear that $\tilde{a} \leqq 2a$. Inequality (12.47) now follows from (12.49) because

$$\|A^\tau\|_{\frac{1}{2}\to\beta(\tau)} = \sup_{\|x\|_{\frac{1}{2}}\leqq 1} \|A^\tau x\|_{\beta(\tau)} \leqq \sup_{\|z\|_{\frac{1}{2}}\leqq 1} \|\tilde{A}^\tau z\|_{\beta(\tau)} \leqq \|\tilde{A}^\tau\|_{\frac{1}{2}\to\beta(\tau)}.$$

The theorem has been proved.

If $\tau \in (0, \frac{1}{2}]$ the assertion of Theorem 12.13 remains true without the assumption that the set $D(A^2) \cap L_{1-\beta}$ is dense in $L_{1-\beta}$.

THEOREM 12.14: *Let the conditions of Theorem* 12.13 *be fulfilled. Let A be a compact operator acting from $L_{\frac{1}{2}}$ to L_β ($\frac{1}{2} < \beta \leqq 1$). Then for each $\tau \in (0, 1)$ the operator A^τ is a compact operator acting from $L_{\frac{1}{2}}$ to $L_{\beta(\tau)}$, where $\beta(\tau)$ is defined by formula* (12.45).

PROOF: The proof of Theorem 10.2 cannot be applied here because of the unboundedness of the operator A in $L_{\frac{1}{2}}$. The following proof uses another idea.

First let $\tau \in (\frac{1}{2}, 1)$, then $2\tau - 1 = \sigma \in (0, 1)$. Consequently by Theorem 12.13 the operator A^σ acts from $L_{\frac{1}{2}}$ to $L_{\beta(\sigma)}$ and is continuous. The operator A (by selfadjointness) acts from $L_{1-\beta}$ to $L_{\frac{1}{2}}$ and is compact. This means that the operator $A^{2\tau} = A^\sigma A$ acts from $L_{1-\beta}$ to $L_{\beta(\sigma)}$ and is compact. Since the operator $A^{2\tau}$ is also selfadjoint it is also compact as an operator from $L_{1-\beta(\sigma)}$ to L_β. Applying Theorem 3.10 we deduce that $A^{2\tau}$ is compact as an operator from $L_{1-\beta(\tau)}$ to $L_{\beta(\tau)}$. By Theorem 12.12 (on splitting of a linear operator) A^τ is compact as an operator from $L_{\frac{1}{2}}$ to $L_{\beta(\tau)}$, as was asserted.

The relation (12.7) now implies that the theorem is valid for all exponents of the form

$$\tau = \sigma^n \quad (\tfrac{1}{2} < \sigma < 1;\, n = 1, 2, \ldots). \tag{12.50}$$

But any number τ in the interval $(0, \frac{1}{2}]$ can be written in the form (12.50). The theorem has been proved.

12.11 *L-Characteristics of fractional powers*

The present section is related to 10.6°; just as in that paragraph we will confine ourselves to an investigation of the L-characteristics $L(A^\tau; \text{cont.})$. In contrast to 10.6°, we will here examine the case in which certain known points of the L-characteristic $L(A; \text{cont.})$ lie on the line $\beta = \alpha$.

Let an operator A satisfy the conditions of Theorem 12.13 for $\beta = \beta_0$. Then the L-characteristic $L(A^\tau; \text{cont.})$ is symmetric and contains the point $M_\tau = \{\frac{1}{2}, \frac{1}{2} + \tau(\beta_0 - \frac{1}{2})\}$ dividing the segment joining the points $M = \{\frac{1}{2}, \beta_0\}$ and $M_0 = \{\frac{1}{2}, \frac{1}{2}\}$ in the ratio $\tau : (1 - \tau)$ (Fig. 12.1). In this way an immediate application of Theorem 12.13 permits us to derive that part of the L-characteristic $L(A^\tau; \text{cont.})$ hatched in Fig. 12.1.

Assume next that two points $M = \{\frac{1}{2}, \beta_0\}$ and $N = \{\alpha_0, 1 - \alpha_0\}$ are known to belong to the L-characteristic $L(A; \text{cont.})$, and that $\alpha_0 \leqq \frac{1}{2}$, $\beta_0 \geqq \frac{1}{2}$, $\alpha_0 \geqq \frac{3}{4} - \beta_0/2$ (Fig. 12.2). Theorem 12.11 on the splitting of a selfadjoint operator implies that $L(A^{\frac{1}{2}}; \text{cont.})$ contains the point $M_{\frac{1}{2}} = \{\frac{1}{2}, 1 - \alpha_0\}$.

Assume first that $0 < \tau < \frac{1}{2}$. Then $A^\tau = (A^{\frac{1}{2}})^{2\tau}$ and the L-characteristic $L(A^\tau; \text{cont.})$ contains the point M_τ dividing the segment joining the points $M_{\frac{1}{2}}$ and M_0 in the ratio $2\tau : (1 - 2\tau)$ (see Fig. 12.2); the corresponding portion of the L-characteristic $L(A^\tau; \text{cont.})$ is hatched in Fig. 12.2.

Now examine the case $\tau \in (\frac{1}{2}, 1)$. In this case, just as in 10.6°, it is

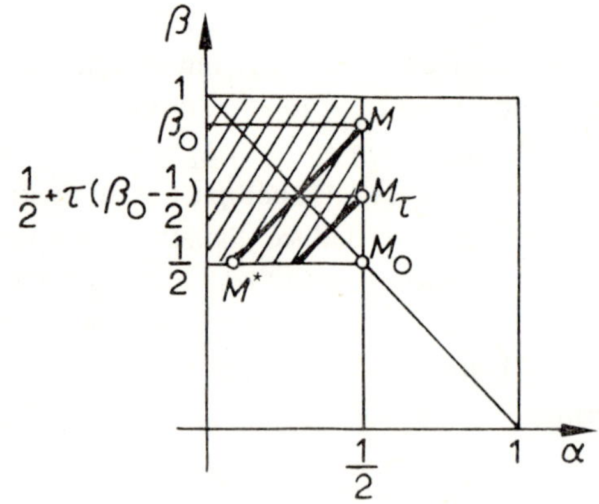

Figure 12.1

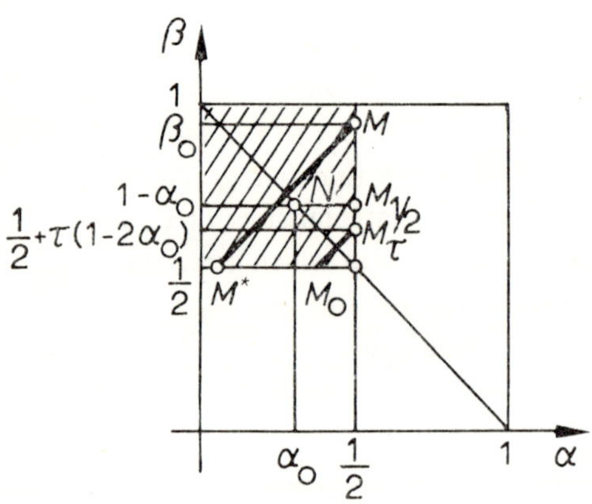

Figure 12.2

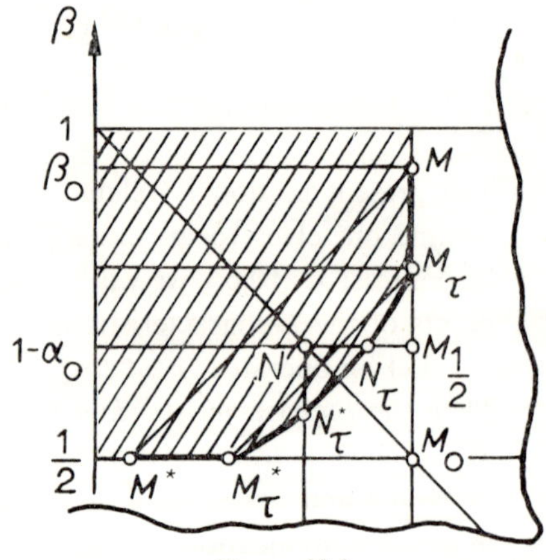

Figure 12.3

convenient to use Theorem 10.3 which is valid for an operator A satisfying the conditions of Theorem 12.13. Theorem 10.3 implies that $L(A^\tau; \text{cont.})$ contains the point M_τ dividing the segment joining the points $M_{\frac{1}{2}} = \{\frac{1}{2}, 1 - \alpha_0\}$ and $M = \{\frac{1}{2}, \beta_0\}$ in the ratio $\lambda: (1 - \lambda)$ where $\lambda = 2\tau - 1$ (see Fig. 12.3). Repeating the arguments in 10.6°, we deduce that for $\tau = \frac{1}{2} + \sigma/2$ with $0 < \sigma < 1$ the L-characteristic $L(A^\tau; \text{cont.})$ also contains the point $N_\tau = \{\frac{1}{2} + \sigma(\alpha_0 - \frac{1}{2}), 1 - \alpha_0\}$ which divides the segment $M_{\frac{1}{2}}N$ in the ratio $\sigma: (1 - \sigma)$. In this way, the L-characteristic $L(A^\tau; \text{cont.})$ contains the polygon hatched in Fig. 12.3.

It is not known (as was true for the cases examined in 10.6°) whether it is possible to utilize points of $L(A; \text{cont.})$ not lying on the lines $\alpha = \frac{1}{2}$, $\beta = \frac{1}{2}$ and $\alpha + \beta = 1$ to obtain additional information about the structure of $L(A^{\tau}; \text{cont.})$.

The discussion presented in this section can be applied (see 10.7°) to construct the L-characteristics of fractional powers of integral operators as well as to construct the L-characteristics of the integral operators

$$\boldsymbol{K}x(t) = \int_{\Omega} K(t, s)x(s)\,\mathrm{d}s$$

themselves, provided that properties of the operators formed with iterates of the kernel are known. We will not describe the appropriate observations and results, except to remark that the greatest difficulties arise in the study of the domain of integral operators which are studied as unbounded operators in L_{α}. Here the following simple remark may be useful: if $K(s, t) \geqq 0$ and $\boldsymbol{K}x_0(t) \in L_{\alpha}$ for some function $x_0(t)$ which is positive almost everywhere, then $\boldsymbol{K}x(t) \in L_{\alpha}$ for all functions lying in some dense set in L_{α}.

4

Fractional powers of operators of positive type

§ 13 Semi-groups of operators[1]

13.1 *Vector-functions and operator-functions*

In the sequel we need some results from the theory of functions of a real or complex variable with values in a Banach space. Some of the corresponding assertions for the case of Hilbert space were presented in Chapter 3.

Let a function $x(t)$ with values in a Banach space E be defined on an interval $\{a, b\}$ (finite or infinite; open, closed or half-open). The function $x(t)$ is *continuous* on $\{a, b\}$ if

$$\lim_{t \to t_0} \|x(t) - x(t_0)\| = 0 \qquad (t_0 \in \{a, b\}). \tag{13.1}$$

The function $x(t)$ is called *weakly continuous* if for any continuous linear functional $f \in E^*$ the scalar function $f[x(t)]$ is continuous. Observe that a function $x(t)$ which is weakly continuous on a bounded closed interval is bounded: $\|x(t)\| \leqq C$.

The notions of differentiability and weak differentiability of $x(t)$ are

---

[1] The theory of semi-groups of bounded operators (see E. Hille and R. Phillips [1], N. Dunford and J. T. Schwartz [1]) is one of the important branches of functional analysis. It plays a decisive role in various parts of mathematical physics, in probability theory and elsewhere.

defined analogously. A weakly differentiable function is continuous and differentiable if its weak derivative is continuous.

Let a function $x(t)$ be defined on a bounded interval $[a, b]$. Its Riemann integral $\int_a^b x(t)\,dt$ is the limit, in the norm of the space E, of the integral sums

$$\sigma = \sum_{i=1}^{n-1} x(\xi_i)(t_{i+1} - t_i), \tag{13.2}$$

as $\lambda \to 0$, where $a = t_0 < t_1 < \ldots < t_n = b$ is an arbitrary partition of the interval $[a, b]$, $\lambda = \max(t_{i+1} - t_i)$ and ξ_i is an arbitrary point in $[t_i, t_{i+1}]$. It is easy to see that continuous functions are integrable. This Riemann integral possesses the usual properties of the scalar integral. In particular, if $x(t)$ is continuous, the function

$$y(t) = \int_a^t x(\tau)\,d\tau \tag{13.3}$$

is differentiable and $y'(t) = x(t)$.

The weak Riemann integral is analogously defined as the weak limit of the integral sums (13.2) as $\lambda \to 0$.

Let A be a continuous linear operator, acting in E. Then the following relation holds for any integrable function $x(t)$:

$$A \int_a^b x(\tau)\,d\tau = \int_a^b Ax(\tau)\,d\tau. \tag{13.4}$$

If $x(t)$ is weakly integrable then the above relation holds if the integrals in it are taken in the weak sense. The relation (13.4) also remains valid for certain classes of unbounded operators. We note that one always has the inequality

$$\left\| \int_a^b x(\tau)\,d\tau \right\| \leqq \int_a^b \|x(\tau)\|\,d\tau. \tag{13.5}$$

Let a function $x(t)$ be defined on $[a, \infty)$ and integrable (weakly integrable) on each finite interval $[a, b]$. The improper (or weak improper) integral

$\int_a^\infty x(\tau)\,d\tau$ is defined by the formula

$$\int_a^\infty x(\tau)\,d\tau = \lim_{b\to\infty} \int_a^b x(\tau)\,d\tau, \tag{13.6}$$

where the limit is taken in the corresponding sense. Improper integrals on intervals of the form $(-\infty, b]$, $(-\infty, \infty)$ and improper integrals of unbounded functions are defined analogously. Formulas (13.4) and (13.5) are also valid for improper integrals.

Integrals depending on a parameter will often occur. Theorems on differentiability or integrability with respect to a parameter are formulated for such integrals just as for integrals of numerical functions. For example, differentiability is proved if, after differentiation, the resulting integral converges uniformly; a verification of the convergence or uniform convergence of the integral of a function $f(t, \mu)$, depending on a parameter μ, is usually reduced to a verification of the uniform convergence of the integral of the numerical function $\varphi(t, \mu) = \|f(t, \mu)\|$.

Operator-functions $A(t)$ (i.e. functions, whose values are bounded operators) are special examples of functions with values in a Banach space (of bounded operators acting from E_1 to E_2). For operator-functions we have the notions of continuity in operator norm ($\|A(t) - A(t_0)\| \to 0$ as $t \to t_0$), of strong continuity ($\|A(t)x - A(t_0)x\| \to 0$ as $t \to t_0$ for each fixed $x \in E_1$) and of weak continuity ($(A(t)x, l) \to (A(t_0)x, l)$ as $t \to t_0$ for each fixed $x \in E$, $l \in E^*$). Analogously there are distinct notions of differentiability (differentiability in operator norm), of strong differentiability (differentiability of all functions $A(t)x$, $x \in E_1$) and of weak differentiability.

The well-known theorem of Banach-Steinhaus shows that the norms of the values of a weakly continuous operator function $A(t)$ on a compact interval are bounded.

It is not difficult to show that an operator-function $A(t)$ is strongly continuous for the entire space E_1, if its norm is a bounded function and the functions $A(t)x$ are continuous for all elements x in some dense set in E_1.

The following assertion, which will be used in the sequel, is obvious: *if an operator-function $A(t)$ is strongly continuous and a function $x(t)$ with values in E_1 is continuous in norm, then the function $A(t)x(t)$ is continuous in norm.*

It is possible to introduce the notions of Riemann integral, strong

Riemann integral and weak Riemann integral for operator-functions. For these integrals, analogues of the relations (13.4) and (13.5) are valid.

All the above definitions and properties can be carried over without difficulty to the case of vector-functions and operator-functions of a complex variable (here, of course, curvilinear integrals appear).

Additional features arise in the theory of analytic vector-functions and operator-functions. A function $x(\lambda)$ (an operator-function $A(\lambda)$) is called *analytic* in a domain Λ of the complex plane if it has a derivative at each point $\lambda \in \Lambda$. It turns out that *the existence of a weak derivative in the domain Λ implies the existence of a derivative in norm.*

The basic facts of the theory of scalar analytic functions are also valid for analytic vector-functions (operator-functions). In particular, Cauchy's theorem is valid: *if Γ is a rectifiable contour lying completely in a simply connected domain of analyticity Λ of the function $x(\lambda)$ ($A(\lambda)$), then*

$$\int_\Gamma x(\lambda)\,\mathrm{d}\lambda = 0 \qquad \left(\int_\Gamma A(\lambda)\,\mathrm{d}\lambda = 0\right). \tag{13.7}$$

As in the scalar case, Cauchy's theorem is useful in the calculation of integrals: it permits replacement of a path of integration by a more convenient one without changing the value of the integral.

For instance, let an operator-function $A(\lambda)$ be analytic in the strip $\sigma_1 < \operatorname{Re}\lambda < \sigma_2$ and let

$$\lim_{|\lambda|\to\infty} \max_{\sigma_1' \leqq \operatorname{Re}\lambda \leqq \sigma_2'} \|A(\lambda)\| = 0$$

for any σ_1', σ_2' satisfying the inequality $\sigma_1 < \sigma_1' < \sigma_2' < \sigma_2$. Then the improper integral

$$J(\sigma) = \int_{\sigma - i\infty}^{\sigma + i\infty} A(\lambda)\,\mathrm{d}\lambda$$

exists for all $\sigma \in (\sigma_1, \sigma_2)$, if it exists for some $\sigma_0 \in (\sigma_1, \sigma_2)$ and its value does not depend on σ.

Analogously, if an operator-function $A(\lambda)$ is analytic in a sector $\varphi_1 < \arg\lambda < \varphi_2$ and if

$$\lim_{|\lambda|\to\infty} \max_{\varphi_1' \leqq \arg\lambda \leqq \varphi_2'} \|\lambda A(\lambda)\| = 0$$

for any φ_1', φ_2', satisfying the inequality $\varphi_1 < \varphi_1' < \varphi_2' < \varphi_2$, then the improper integral

$$J(\varphi) = \int_{\Gamma(\varphi)} A(\lambda)\,\mathrm{d}\lambda \quad (\Gamma(\varphi) = \{\lambda;\ \lambda = \rho e^{i\varphi},\ 0 \leqq \rho < \infty\})$$

exists for all $\varphi \in (\varphi_1, \varphi_2)$ if it exists for some $\varphi_0 \in (\varphi_1, \varphi_2)$ and its value does not depend on φ.

Just as for scalar analytic functions, Cauchy's formula is valid for analytic functions with values in a Banach space. For an operator-function it has the form

$$A(\lambda) = \frac{1}{2\pi i} \int_{\Gamma} \frac{A(z)}{z - \lambda}\,\mathrm{d}z. \tag{13.8}$$

From this formula follow the usual Cauchy estimates for the derivatives of a function which is analytic in a disc $|\lambda - \lambda_0| \leqq r$:

$$\|A^{(n)}(\lambda_0)\| \leqq M(r) r^{-n} n! \quad (n = 1, 2, \ldots), \tag{13.9}$$

where

$$M(r) = \max_{|\lambda - \lambda_0| = r} \|A(\lambda)\|.$$

Finally, we remark that a uniqueness theorem holds for analytic functions with values in Banach spaces: if a function $x(\lambda)$ (an operator function $A(\lambda)$) analytic in a domain Λ is equal to zero on some set with a limit point in Λ, then it is identically equal to zero.

13.2 *Unbounded operators*

Let E be a Banach space and A be a linear operator defined on some linear set $D(A) \subset E$ and taking values in E.

An operator A is called *closed* (see 11.1°) if $x_n \in D(A)$, $\|x_n - x_0\| \to 0$, and $\|Ax_n - y_0\| \to 0$ imply that $x_0 \in D(A)$ and $Ax_0 = y_0$. In other words, closedness of an operator A means closedness of the graph $\Gamma(A)$ of the operator A (the notion of graph is introduced in the same way as was done in § 11 for operators in Hilbert spaces).

If an operator A has a bounded inverse A^{-1}, defined on all of E, then the operator A is closed. More generally, if for some λ the operator $\lambda \boldsymbol{I} + A$ has a bounded inverse, then the operator A is closed.

Let the domain of an operator A be dense in E and let the operator A have a bounded inverse defined on E. Then the domains of the operators A^n are also dense in E.

To prove this, let us first show that the range of any bounded operator $\boldsymbol{B}$ on a set D which is dense in E, will itself be dense in E if $\boldsymbol{B}E$ is dense in E. In fact, for each $x \in E$ a sequence $x_n \in D$ $(n = 1, 2, \ldots)$ can be taken such that $\|x_n - x\| \to 0$, and consequently $\|\boldsymbol{B}x_n - \boldsymbol{B}x\| \to 0$. This means that $\boldsymbol{B}D$ is dense in $\boldsymbol{B}E$ and, consequently, in E.

From this observation it follows that the ranges of the operators A^{-n} are dense in E. It is only necessary to observe that the range of the operator A^{-n} coincides with the domain of the operator A^n.

Similarly it can be shown that the domains of the operators A^n are dense in E if the domain $D(A)$ of the operator A is dense in E and if for some λ the operator $\lambda \boldsymbol{I} + A$ has a bounded inverse.

Again let an operator A be continuously invertible. Denote by $A_{[n]}$ the operator defined on $D(A^n)$ by the relation $A_{[n]}x = Ax$. It turns out that the closure[1] of the operator $A_{[n]}$ coincides with the operator A. For the proof it suffices to show that for any $x_0 \in D(A)$ there is a sequence of elements $x_i \in D(A^n)$ $(i = 1, 2, \ldots)$ such that $\|x_i - x_0\| \to 0$ and $\|Ax_i - Ax_0\| \to 0$. Now since $D(A^{n-1})$ is dense in E it is possible to construct a sequence $y_i \in D(A^{n-1})$ such that $\|y_i - Ax_0\| \to 0$. Define the sequence x_i by the relation

$$x_i = A^{-1}y_i \qquad (i = 1, 2, \ldots).$$

It is clear that $x_i \in D(A^n)$, while

$$\|x_i - x_0\| = \|A^{-1}y_i - x_0\| \leqq \|A^{-1}\| \, \|y_i - Ax_0\| \to 0$$

and

$$\|Ax_i - Ax_0\| = \|y_i - Ax_0\| \to 0.$$

[1] The closure A of an operator A in the case of Banach spaces is defined exactly as in the case of Hilbert spaces (see § 11); the graph of $\bar{A}$ is the closure of the graph of A.

Our assertion has been proved. We point out again that this result too remains valid whenever for some λ the operator $\lambda \boldsymbol{I} + \boldsymbol{A}$ is continuously invertible.

Note the following fact which will be important in the sequel. Let $\boldsymbol{A}$ be a closed operator and let $x(t)$ be a function integrable on a finite interval $\{a, b\}$. Suppose that the value of the function $x(t)$ at each $t \in \{a, b\}$ belong to $D(\boldsymbol{A})$ and that the function $\boldsymbol{A}x(t)$ is integrable on $\{a, b\}$. Then the following relation holds:

$$\boldsymbol{A} \int_a^b x(\tau)\,\mathrm{d}\tau = \int_a^b \boldsymbol{A}x(\tau)\,\mathrm{d}\tau. \tag{13.10}$$

In fact, let $\bar{x}_k = \sum x(\tau_i)(t_{i+1} - t_i)$ be a sequence of integral sums for the function $x(t)$, and let $\bar{y}_k = \boldsymbol{A}\bar{x}_k = \sum \boldsymbol{A}x(\tau_i)(t_{i+1} - t_i)$ be the corresponding integral sums for the function $\boldsymbol{A}x(t)$. As $\lambda \to 0$ the elements $\bar{x}_k$ converge to $\int_a^b x(\tau)\,\mathrm{d}\tau$ and the elements $\bar{y}_k$ converge to $\int_a^b \boldsymbol{A}x(\tau)\,\mathrm{d}\tau$. From the closedness of $\boldsymbol{A}$ it follows that the element $\int_a^b x(\tau)\,\mathrm{d}\tau$ belongs to $D(\boldsymbol{A})$ and that relation (13.10) is valid.

13.3 *Resolvents*

The *resolvent set* $\rho(\boldsymbol{A})$ of an operator $\boldsymbol{A}$ is the set of all complex[1] numbers λ, for which the operator

$$\boldsymbol{R}(\lambda; \boldsymbol{A}) = (\lambda \boldsymbol{I} - \boldsymbol{A})^{-1} \tag{13.11}$$

is defined on all of E and is continuous. The operator $\boldsymbol{R}(\lambda; \boldsymbol{A})$ is called the *resolvent* of the operator $\boldsymbol{A}$. The resolvent $\boldsymbol{R}(\lambda; \boldsymbol{A})$ is by definition an operator such that

$$(\lambda \boldsymbol{I} - \boldsymbol{A})\boldsymbol{R}(\lambda; \boldsymbol{A})x = x \qquad (x \in E) \tag{13.12}$$

[1] If the operator $\boldsymbol{A}$ is considered in a real space, then in order to construct the resolvent set the space is imbedded to its complex extension $\tilde{E}$, consisting of elements of the form $x + iy$, where $x, y \in E$; the norm in $\tilde{E}$ is defined, for instance, by the relation $\|x + iy\| = \max \|x \cos \theta + y \sin \theta\|$. The value of $\boldsymbol{A}$ on $\tilde{E}$ is given by the relation $\boldsymbol{A}(x + iy) = \boldsymbol{A}x + i\boldsymbol{A}y$.

and

$$\boldsymbol{R}(\lambda; A)(\lambda \boldsymbol{I} - A)x = x \qquad (x \in D(A)). \tag{13.13}$$

If the resolvent set $\rho(A)$ is non-empty then A is a closed operator. Hence it suffices to check relation (13.12) only for elements $x \in D(A)$.

If an operator A is bounded, then $\lambda \in \rho(A)$ whenever $|\lambda| > \|A\|$. This follows from the fact that in this case the resolvent is given by the following series, which is convergent in operator norm:

$$\boldsymbol{R}(\lambda; A) = \boldsymbol{I}/\lambda + A/\lambda^2 + \ldots + A^n/\lambda^{n+1} + \ldots \tag{13.14}$$

The convergence of this series results from the fact that the norms of the terms of the series are dominated by the terms of a decreasing geometric series. The operator $(\lambda \boldsymbol{I} - A)^{-1}$ is equal to the series in (13.14) because

$$(\lambda \boldsymbol{I} - A)(\boldsymbol{I}/\lambda + A/\lambda^2 + \ldots + A^n/\lambda^{n+1} + \ldots)$$

$$= (\boldsymbol{I}/\lambda + A/\lambda^2 + \ldots + A^n/\lambda^{n+1} + \ldots)(\lambda \boldsymbol{I} - A) = \boldsymbol{I}.$$

For selfadjoint operators A in a Hilbert space H the resolvent set $\rho(A)$ contains all non-real λ. The resolvent set of a positive definite operator contains the entire left half-plane of the complex plane.

The resolvent set is always an open set. This fact is a consequence of the following easy lemma:

LEMMA 13.1: *Let* $\lambda_0 \in \rho(A)$. *Then all points* λ *in the disc* $|\lambda - \lambda_0| < \|\boldsymbol{R}(\lambda_0; A)\|^{-1}$ *also belong to* $\rho(A)$ *and*

$$\|\boldsymbol{R}(\lambda; A)\|$$

$$\leqq \frac{1}{\|\boldsymbol{R}(\lambda_0; A)\|^{-1} - |\lambda - \lambda_0|} \qquad (|\lambda - \lambda_0| < \|\boldsymbol{R}(\lambda_0; A)\|^{-1}). \tag{13.15}$$

Note that for $|\lambda - \lambda_0| < \|\boldsymbol{R}(\lambda_0; A)\|^{-1}$ the operators $\boldsymbol{R}(\lambda; A)$ can be defined by the following series which is convergent in operator norm:

$$\boldsymbol{R}(\lambda; A) = \boldsymbol{R}(\lambda_0; A) - (\lambda - \lambda_0)\boldsymbol{R}^2(\lambda_0; A) + \ldots$$

$$\ldots + (-1)^n(\lambda - \lambda_0)^n \boldsymbol{R}^{n+1}(\lambda_0; A) + \ldots \tag{13.16}$$

Next note an important identity which the resolvent satisfies. Let $\lambda, \mu \in \rho(A)$. Since $\boldsymbol{R}(\lambda; A)x$, $\boldsymbol{R}(\mu; A)x \in D(A)$ for any $x \in E$,

$$(\lambda \boldsymbol{I} - A)\boldsymbol{R}(\mu; A)x = (\mu \boldsymbol{I} - A)\boldsymbol{R}(\mu; A)x - (\mu - \lambda)\boldsymbol{R}(\mu; A)x,$$

and on account of (13.12)

$$(\lambda \boldsymbol{I} - A)\boldsymbol{R}(\mu; A)x = x - (\mu - \lambda)\boldsymbol{R}(\mu; A)x.$$

Hence on account of (13.13)

$$\boldsymbol{R}(\mu; A)x = \boldsymbol{R}(\lambda; A)x - (\mu - \lambda)\boldsymbol{R}(\lambda; A)\boldsymbol{R}(\mu; A)x,$$

i.e.

$$\boldsymbol{R}(\lambda; A) - \boldsymbol{R}(\mu; A) = (\mu - \lambda)\boldsymbol{R}(\lambda; A)\boldsymbol{R}(\mu; A). \tag{13.17}$$

Equation (13.17) implies, in particular, that the operators $\boldsymbol{R}(\lambda; A)$ and $\boldsymbol{R}(\mu; A)$ commute with each other. It is not difficult to see that for elements in $D(A)$ the operator $\boldsymbol{R}(\lambda; A)$ commutes with the operator A:

$$\boldsymbol{R}(\lambda; A)Ax = -\boldsymbol{R}(\lambda; A)(\lambda \boldsymbol{I} - A)x + \lambda \boldsymbol{R}(\lambda; A)x = -x + \lambda \boldsymbol{R}(\lambda; A)x$$

and

$$A\boldsymbol{R}(\lambda; A)x = -(\lambda \boldsymbol{I} - A)\boldsymbol{R}(\lambda; A)x + \lambda \boldsymbol{R}(\lambda; A)x$$

$$= -x + \lambda \boldsymbol{R}(\lambda; A)x. \tag{13.18}$$

We remark that formula (13.18) is actually valid for all $x \in E$.

Now (13.16) implies that the function $\boldsymbol{R}(\lambda; A)$ is differentiable at each point $\lambda_0 \in \rho(A)$. Hence $\boldsymbol{R}(\lambda; A)$ is analytic in the domain $\rho(A)$. From (13.16) it also follows that

$$\frac{\mathrm{d}^k}{\mathrm{d}\lambda^k} \boldsymbol{R}(\lambda_0; A) = (-1)^k k! [\boldsymbol{R}(\lambda_0; A)]^{k+1}. \tag{13.19}$$

13.4 *Definition of a semi-group*

Given a family of bounded operators $T(t)$ $(t \geqq 0)$ in a Banach space E, this family is said to form a *strongly continuous semi-group* if the following conditions are fulfilled:

1°) $T(t+s) = T(t)T(s) = T(s)T(t)$ for $t, s \geqq 0$ and $T(0) = I$.

2°) The function $T(t)x$ is continuous in norm on $[0, \infty)$ for each fixed $x \in E$.

Of course it is assumed that $T(t)x$ is continuous from the right at the point $t = 0$.

In the case of a one dimensional space the continuous semi-groups are the functions $T(t) = e^{\alpha t}$, where α is an arbitrary complex number.

The simplest example of a strongly continuous semi-group $T(t) = e^{-At}$ in a Banach space E is given by the formula:

$$T(t) = I - tA + \frac{t^2}{2!} A^2 + \ldots + (-1)^n \frac{t^n}{n!} A^n + \ldots \tag{13.20}$$

where A is a bounded operator acting in E. The series (13.20) converges in operator norm for all t since its terms are clearly dominated by the terms of the power series expansion of the function $e^{\|A\|t}$; hence $\|T(t)\| \leqq e^{\|A\|t}$.

Power series can be differentiated term by term (this is proved just as in the case of series with numerical coefficients); hence

$$\frac{\mathrm{d}T(t)}{\mathrm{d}t} + AT(t) = 0. \tag{13.21}$$

Thus formula (13.20) defines a solution of the differential equation

$$\frac{\mathrm{d}X}{\mathrm{d}t} + AX = 0, \tag{13.22}$$

satisfying the initial condition $X(0) = I$ (equation (13.22) is considered in the space of bounded operators acting in E).

A uniqueness theorem is valid for equation (13.22) (both this and a corresponding existence theorem can be proved for instance by successive approximations). Hence the operator-functions $\boldsymbol{T}(t+s)$ and $\boldsymbol{T}(t)\boldsymbol{T}(s)$ coincide, because both of them satisfy equation (13.22) for fixed s and both equal $\boldsymbol{T}(s)$ at $t = 0$. Thus the operator-function (13.20) satisfies condition 1°. Condition 2° is clearly fulfilled, since the function (13.20) is differentiable.

The operator $-\boldsymbol{A}$ is called the *generator of the semi-group* (13.20).

Let $\boldsymbol{A}$ and $\boldsymbol{B}$ be two commuting bounded operators. An immediate calculation shows that the functions $e^{-(\boldsymbol{A}+\boldsymbol{B})t}$ and $e^{-\boldsymbol{A}t}e^{-\boldsymbol{B}t}$ satisfy the same differential equation

$$\frac{\mathrm{d}X}{\mathrm{d}t} + (\boldsymbol{A}+\boldsymbol{B})X = 0,$$

for which a uniqueness theorem is valid. Both of these functions equal the identity operator at $t = 0$. Hence

$$e^{-(\boldsymbol{A}+\boldsymbol{B})t} = e^{-\boldsymbol{A}t}e^{-\boldsymbol{B}t} = e^{-\boldsymbol{B}t}e^{-\boldsymbol{A}t}.$$

In particular, for any bounded operator $\boldsymbol{A}$ and any complex number α

$$e^{(\alpha \boldsymbol{I}-\boldsymbol{A})t} = e^{\alpha \boldsymbol{I}t}e^{-\boldsymbol{A}t} = e^{\alpha t}e^{-\boldsymbol{A}t}.$$

We remark also that an operator $\boldsymbol{B}$ commutes with all operators $e^{-\boldsymbol{A}t}$ if $\boldsymbol{B}$ commutes with $\boldsymbol{A}$; this results from the representation (13.20) of the operator $e^{-\boldsymbol{A}t}$. As is noted above, $\|e^{-\boldsymbol{A}t}\| \leqq e^{\|\boldsymbol{A}\|t}$. Thus the preceding identities imply that

$$\|e^{(\alpha \boldsymbol{I}-\boldsymbol{A})t}\| \leqq e^{(\mathrm{Re}\,\alpha + \|\boldsymbol{A}\|)t}.$$

Let us consider one more example. Let $\boldsymbol{A}$ be a positive definite selfadjoint operator acting in a Hilbert space H, i.e. a selfadjoint operator such that $(\boldsymbol{A}x, x) \geqq 0$ for $x \in D(\boldsymbol{A})$. Then the spectral representation of the operator $\boldsymbol{A}$ has the form

$$\boldsymbol{A} = \int_0^\infty \lambda \,\mathrm{d}\boldsymbol{P}_\lambda. \tag{13.23}$$

Consider the operators

$$\boldsymbol{T}(t) = e^{-\boldsymbol{A}t} = \int_0^\infty e^{-\lambda t}\,\mathrm{d}\boldsymbol{P}_\lambda, \tag{13.24}$$

where the numbers t are non-negative. The operators (13.24) are selfadjoint. They are also bounded, because by (11.8)

$$\|e^{-\boldsymbol{A}t}\| \leqq \sup_{\lambda \geqq 0} e^{-\lambda t} = 1. \tag{13.25}$$

When the operator $\boldsymbol{A}$ is bounded the operators (13.24) and (13.20) coincide.

The operators (13.24) satisfy condition 1°—this follows from the formula (see § 11) by which functions of operators are calculated. They satisfy condition 2°, as well. In fact, for each $x \in H$ and any $t,\ t + \Delta t \geqq 0$

$$\|e^{-\boldsymbol{A}(t+\Delta t)}x - e^{-\boldsymbol{A}t}x\|^2 = \int_0^\infty [e^{-\lambda(t+\Delta t)} - e^{-\lambda t}]^2\,\mathrm{d}(\boldsymbol{P}_\lambda x, x)$$

$$\leqq \int_0^\infty [e^{-\lambda|\Delta t|} - 1]^2 \cdot \mathrm{d}(\boldsymbol{P}_\lambda x, x),$$

whence we obtain, after passing to the limit under the integral sign,

$$\lim_{\Delta t \to 0} \|e^{-\boldsymbol{A}(t+\Delta t)}x - e^{-\boldsymbol{A}t}x\| = 0.$$

Consequently formula (13.24) defines a strongly continuous semi-group of bounded operators. The operator $-\boldsymbol{A}$ is called the generator of this semi-group.

It is interesting to note that (as in case of the semi-group (13.20)) for $t > 0$ the function $x(t) = \boldsymbol{T}(t)x$ is a solution of the differential equation

$$\frac{\mathrm{d}x}{\mathrm{d}t} = -\boldsymbol{A}x \tag{13.26}$$

considered in the space H. In the case of $x \in D(\boldsymbol{A})$, the function $x(t) = \boldsymbol{T}(t)x$ also satisfies equation (13.26) at $t = 0$.

LEMMA 13.2: *Let $T(t)$ be a strongly continuous semi-group of operators. Then*

$$\|T(t)\| \leqq c(\omega) e^{\omega t}, \tag{13.27}$$

where ω and $c(\omega)$ are certain constants.

PROOF: Put

$$c_1 = \sup_{0 \leqq t \leqq 1} \|T(t)\|.$$

For each $t \geqq 0$ the following identity holds:

$$T(t) = T([t])\, T(t - [t]) = \{T(1)\}^{[t]} \cdot T(t - [t]),$$

where $[t]$ is the integral part of the number t. Hence it follows that

$$\|T(t)\| \leqq c_1 e^{\omega [t]},$$

where $\omega = \ln \|T(1)\|$. This inequality implies that

$$\|T(t)\| \leqq c e^{\omega t} \quad (t \geqq 0),$$

where $c = c_1$ if $\omega \geqq 0$ and $c = c_1 e^{-\omega}$ if $\omega < 0$. The lemma has been proved.

Denote by ω_0 the infimum of the set of numbers ω, for which inequality (13.27) is fulfilled; this number is called the *order* of *growth* of the semi-group. It can be shown (see, for instance, E. Hille and R. S. Phillips [1]) that

$$\omega_0 = \lim_{t \to \infty} \frac{\ln \|T(t)\|}{t}. \tag{13.28}$$

13.5 *Generator of a semi-group*

Let $T(t)$ be a strongly continuous semi-group of operators acting in E. Denote by A the linear operator which is defined as the strong limit

$$Ax = -\lim_{\Delta t \to 0} \frac{1}{\Delta t} [T(\Delta t) - I]x \tag{13.29}$$

on those x for which this limit exists. The operator $-\boldsymbol{A}$ is called the *generator* of the semi-group $\boldsymbol{T}(t)$. In other words, the values of the generator are defined to be the values of the right derivative of the functions $\boldsymbol{T}(t)x$ at the point $t = 0$. The term 'generator' has already been used by us previously in certain special cases (the semi-groups (13.20) and (13.24)).

Let us now show that the generator $-\boldsymbol{A}$ of any strongly continuous semi-group has a domain $D(A)$ which is dense in E. To show this, let us prove that the domain contains all elements x of the form

$$x_\varepsilon = \frac{1}{\varepsilon}\int_0^\varepsilon \boldsymbol{T}(t)x_0\,\mathrm{d}t, \tag{13.30}$$

where $\varepsilon > 0$ and x_0 is an element in E. The elements of the form (13.30) yield a set which is dense in E, since for any $x_0 \in E$

$$\lim_{\varepsilon\to 0}\|x_\varepsilon - x_0\| \leqq \lim_{\varepsilon\to 0}\frac{1}{\varepsilon}\int_0^\varepsilon \|\boldsymbol{T}(\tau)x_0 - x_0\|\,\mathrm{d}\tau = 0.$$

It is clear that

$$\frac{\boldsymbol{T}(\Delta t) - \boldsymbol{I}}{\Delta t}x_\varepsilon = \frac{1}{\varepsilon\Delta t}\int_0^\varepsilon [\boldsymbol{T}(\tau + \Delta t)x_0 - \boldsymbol{T}(\tau)x_0]\,\mathrm{d}\tau$$

$$= \frac{1}{\varepsilon\Delta t}\left\{\int_{\Delta t}^{\varepsilon+\Delta t} \boldsymbol{T}(\tau)x_0\,\mathrm{d}\tau - \int_0^\varepsilon \boldsymbol{T}(\tau)x_0\,\mathrm{d}\tau\right\}$$

$$= \frac{1}{\varepsilon\Delta t}\left\{\int_\varepsilon^{\varepsilon+\Delta t} \boldsymbol{T}(\tau)x_0\,\mathrm{d}\tau - \int_0^{\Delta t} \boldsymbol{T}(\tau)x_0\,\mathrm{d}\tau\right\},$$

whence

$$\lim_{\Delta t\to 0}\left\|\frac{\boldsymbol{T}(\Delta t) - \boldsymbol{I}}{\Delta t}x_\varepsilon - \frac{\boldsymbol{T}(\varepsilon) - \boldsymbol{I}}{\varepsilon}x_0\right\| = 0.$$

We have proved that $x_\varepsilon \in D(A)$ and

$$-Ax_\varepsilon = \frac{1}{\varepsilon}\,[T(\varepsilon)x_0 - x_0]. \tag{13.31}$$

For each $x_0 \in D(A)$ and any $t \geqq 0$ the relation

$$AT(t)x_0 = T(t)Ax_0 \tag{13.32}$$

holds, as follows in an obvious way from the identity

$$\frac{T(\Delta t) - I}{\Delta t}\,T(t)x_0 = T(t)\,\frac{T(\Delta t) - I}{\Delta t}\,x_0. \tag{13.33}$$

This means that the operator-function $T(t)$ transforms the domain $D(A)$ of the operator A into itself. From (13.33) it follows also that

$$\frac{\mathrm{d}}{\mathrm{d}t}\{T(t)x_0\} = -AT(t)x_0 \qquad (x_0 \in D(A)) \tag{13.34}$$

(for $t = 0$ the symbol $\mathrm{d}/\mathrm{d}t$ means right derivative).

Let us prove next that *the generator* $-A$ *is closed.* Let $x_n \in D(A)$ $(n = 1, 2, \ldots)$ be such that the sequences x_n and $-Ax_n$ converge to x_0 and y_0 respectively. Equations (13.34) and (13.32) imply that for $\Delta t > 0$

$$\frac{1}{\Delta t}\,[T(\Delta t)x_n - x_n]$$

$$= \frac{1}{\Delta t}\int_0^{\Delta t} T'(\tau)x_n\,\mathrm{d}\tau = -\frac{1}{\Delta t}\int_0^{\Delta t} T(\tau)Ax_n\,\mathrm{d}\tau.$$

Passing to limit in this relation as $n \to \infty$ (with Δt fixed), we obtain

$$\frac{1}{\Delta t}\,[T(\Delta t)x_0 - x_0] = \frac{1}{\Delta t}\int_0^{\Delta t} T(\tau)y_0\,\mathrm{d}\tau.$$

Now let Δt approach zero. The limit of the right side exists and is equal to y_0; hence $x_0 \in D(A)$ and $-Ax_0 = y_0$. The closedness of the operator $-A$ has been proved.

Now let us examine the problem of determining those numbers λ for which the operator $\lambda I + A$ has a bounded inverse $R(\lambda; -A) = (\lambda I + A)^{-1}$, i.e. the problem of determining the numbers λ belonging to the resolvent set $\rho(-A)$ of the generator $-A$ of a strongly continuous semi-group $T(t)$.

LEMMA 13.3: *Let ω_0 be the order of growth of a semi-group $T(t)$. Then all points λ in the half-plane* $\operatorname{Re} \lambda > \omega_0$ *belong to the resolvent set of the generator $-A$, and for such λ the following relation holds:*

$$R(\lambda; -A)x = \int_0^\infty e^{-\lambda t} T(t)x\,dt. \tag{13.35}$$

PROOF: Let $\operatorname{Re} \lambda > \omega > \omega_0$. Then for some $c(\omega)$ inequality (13.27) is satisfied. Hence on this set the integral

$$J(\lambda)x = \int_0^\infty e^{-\lambda t} T(t)x\,dt$$

converges uniformly and absolutely, and defines a bounded operator. Since the operator A is closed, it suffices in order to prove the lemma to verify that

$$(\lambda I + A)J(\lambda)x = J(\lambda)(\lambda I + A)x = x. \qquad (x \in D(A)).$$

Let λ ($\operatorname{Re} \lambda > \omega$) be fixed and select $x \in D(A)$. From the relation

$$(\lambda I + A)T(t)x = T(t)(\lambda I + A)x \qquad (x \in D(A)) \tag{13.36}$$

it follows that the function $e^{-\lambda t}(\lambda I + A)T(t)x$ is integrable on $[0, \infty)$; further by the closedness of the operator $\lambda I + A$,

$$(\lambda I + A)J(\lambda)x = \int_0^\infty e^{-\lambda t}(\lambda I + A)T(t)x\,dt$$

$$= \lambda J(\lambda)x - \int_0^\infty e^{-\lambda t}\frac{dT(t)}{dt}x\,dt.$$

Integrating by parts, we obtain

$$(\lambda \boldsymbol{I} + A)\boldsymbol{J}(\lambda)x$$

$$= \lambda \boldsymbol{J}(\lambda)x + x - \lambda \int_0^\infty e^{-\lambda t}\boldsymbol{T}(t)x\,\mathrm{d}t$$

$$= \lambda \boldsymbol{J}(\lambda)x + x - \lambda \boldsymbol{J}(\lambda)x = x.$$

Finally, it also follows from (13.36) that

$$\boldsymbol{J}(\lambda)(\lambda \boldsymbol{I} + A)x = \int_0^\infty e^{-\lambda t}\boldsymbol{T}(t)(\lambda \boldsymbol{I} + A)x\,\mathrm{d}t$$

$$= \int_0^\infty e^{-\lambda t}(\lambda \boldsymbol{I} + A)\boldsymbol{T}(t)x\,\mathrm{d}t = x.$$

The lemma has been proved.

Equation (13.35) implies the following formulas for successive derivatives of the resolvent $\boldsymbol{R}(\lambda; -\boldsymbol{A})$:

$$\frac{d^k \boldsymbol{R}(\lambda; -A)}{d\lambda^k} = (-1)^k \int_0^\infty t^k e^{-\lambda t}\boldsymbol{T}(t)x\,\mathrm{d}t$$

$$(\mathrm{Re}\,\lambda > \omega_0,\ k = 1, 2, \ldots). \qquad (13.37)$$

To justify these formulas, it is only necessary to remark that all the integrals in (13.37) converge absolutely.

Formula (13.35) gives an explicit expression for the resolvent $\boldsymbol{R}(\lambda; -A)$ of the generator $-\boldsymbol{A}$ in terms of the semi-group $\boldsymbol{T}(t)$. Conversely, it is possible to express a semi-group in terms of the resolvents of its generator. Namely, it can be proved that for any $\sigma > \omega_0$

$$\boldsymbol{T}(t)x = \frac{1}{2\pi i}\lim_{c\to\infty} \int_{\sigma - ic}^{\sigma + ic} e^{\lambda t}\boldsymbol{R}(\lambda; -A)x\,\mathrm{d}\lambda \qquad (x \in D(A)).$$

13.6 *Theorem of Hille-Phillips-Miyadera*

We are here interested in the problem of which operators $-\boldsymbol{A}$ are generators of strongly continuous semi-groups. A general answer to this problem is given by the following theorem.

THEOREM 13.1[1]: *An operator $-\boldsymbol{A}$ is the generator of a strongly continuous semi-group $\boldsymbol{T}(t)$ if and only if the resolvent set $\rho(-\boldsymbol{A})$ contains all λ for which* $\operatorname{Re}\lambda > \sigma_0$, *where σ_0 is some real number, and if for such λ and any $k = 1, 2, \ldots$ the resolvent $\boldsymbol{R}(\lambda) = \boldsymbol{R}(\lambda; -\boldsymbol{A})$ satisfies the inequality*

$$\|\boldsymbol{R}(\lambda)^k\| \leqq \frac{c}{(\operatorname{Re}\lambda - \sigma_0)^k}. \tag{13.38}$$

A direct verification of condition (13.38) is complicated. Hence simpler sufficient conditions are of interest. For instance, if the resolvent $\boldsymbol{R}(\lambda)$ satisfies the condition

$$\|\boldsymbol{R}(\lambda)\| \leqq \frac{1}{\operatorname{Re}\lambda - \sigma_0} \qquad (\operatorname{Re}\lambda > \sigma_0), \tag{13.39}$$

then the conditions (13.38) are fulfilled for all $k = 1, 2, \ldots$ and $-\boldsymbol{A}$ is the generator of some semi-group. Note that condition (13.39) is satisfied by any positive definite selfadjoint operator in a Hilbert space.

PROOF of Theorem 13.1: Let $-\boldsymbol{A}$ be the generator of the strongly continuous semi-group $\boldsymbol{T}(t)$. Then from Lemma 13.3 it follows that $\lambda \in \rho(-\boldsymbol{A})$ if $\operatorname{Re}(\lambda) > \omega_0$, where ω_0 is the order of growth of the semi-group $\boldsymbol{T}(t)$. From (13.19) and (13.37) it follows that for any k

$$\boldsymbol{R}(\lambda)^k x = \frac{1}{(k-1)!}\int_0^\infty t^{k-1} e^{-\lambda t}\boldsymbol{T}(t)x\,\mathrm{d}t.$$

[1] Translator's note. See E. Hille and R. S. Phillips [1], M. Miyadera [1].

Hence from (13.27) it follows that for any fixed $\omega > \omega_0$ and for $\operatorname{Re}\lambda > \omega$ the following inequality holds:

$$\|\boldsymbol{R}(\lambda)^k\| \leqq \frac{c(\omega)}{(k-1)!}\int_0^\infty t^{k-1}e^{(\omega-\operatorname{Re}\lambda)t}\,\mathrm{d}t = \frac{c(\omega)}{(\operatorname{Re}\lambda-\omega)^k}.$$

The necessity has been proved.

Let us turn to the proof of sufficiency. Let an operator A satisfy the conditions of Theorem 13.1. We shall construct a semi-group whose generator is $-A$.

Consider the operators

$$A_n = nA\boldsymbol{R}(n) = nA(n\boldsymbol{I} + A)^{-1}, \tag{13.40}$$

where n assumes integer values greater than σ_0. It is clear that the operators A_n are bounded.

Let us show that on elements $x \in D(A)$ the operators A_n converge to A:

$$\lim_{n\to\infty} \|A_n x - Ax\| = 0. \tag{13.41}$$

For this, we observe first that for any $x \in E$

$$\lim_{n\to\infty} \|n(n\boldsymbol{I} + A)^{-1}x - x\| = 0. \tag{13.42}$$

This relation is obvious for elements $x \in D(A)$, because in this case

$$n(n\boldsymbol{I} + A)^{-1}x - x = -(n\boldsymbol{I} + A)^{-1}Ax = -\boldsymbol{R}(n)Ax$$

while by (13.38)

$$\|n(n\boldsymbol{I} + A)^{-1}x - x\| \leqq \frac{c}{n-\sigma_0}\|Ax\|.$$

The validity of the relation (13.42) for other elements $x \in E$ results from the fact that $D(A)$ is dense in E while the norms of the operators

$n(n\boldsymbol{I} + A)^{-1}$ are uniformly bounded:

$$\|n(n\boldsymbol{I} + A)^{-1}\| = n\|\boldsymbol{R}(n)\| \leqq \frac{cn}{n - \sigma_0}.$$

Now relation (13.41) is proved without difficulty, because for $x \in D(A)$

$$A_n x - Ax = n(n\boldsymbol{I} + A)^{-1} Ax - Ax.$$

Define the operators $e^{-A_n t}$ according to formula (13.20). Since

$$A_n = n\boldsymbol{I} - n^2(n\boldsymbol{I} + A)^{-1},$$

$$e^{-A_n t} = e^{-nt} e^{n^2(n\boldsymbol{I}+A)^{-1}t} = e^{-nt} \sum_{k=0}^{\infty} \frac{t^k n^{2k}}{k!} \boldsymbol{R}(n)^k.$$

The estimate (13.38) implies that

$$\|e^{-A_n t}\| \leqq e^{-nt} \sum_{k=0}^{\infty} \frac{t^k n^{2k}}{k!} \|\boldsymbol{R}(n)^k\|$$

$$\leqq ce^{-nt} \sum_{k=0}^{\infty} \frac{1}{k!} \left(\frac{tn^2}{n - \sigma_0} \right)^k \leqq ce^{(-n + n^2/(n - \sigma_0))t},$$

i.e. for sufficiently large n

$$\|e^{-A_n t}\| \leqq ce^{n\sigma_0 t/(n - \sigma_0)} \leqq ce^{(\sigma_0 + \varepsilon)t} \qquad (t \geqq 0), \tag{13.43}$$

where ε is any small fixed positive number.

Now let us show that the sequence of bounded operators $e^{-A_n t}$ converges strongly on all of E and uniformly with respect to t in each finite interval $[0, t_0]$. By virtue of (13.43) it suffices to establish this fact for elements x in the dense subset $D(A)$ of E.

Let $x \in D(A)$. Then for sufficiently large m and n

$$e^{-A_m t}x - e^{-A_n t}x = \int_0^t \frac{\mathrm{d}}{\mathrm{d}s} e^{-A_n(t-s)-A_m s}x\,\mathrm{d}s$$

$$= \int_0^t e^{-A_n(t-s)}e^{-A_m s}(A_n - A_m)x\,\mathrm{d}s,$$

whence by (13.43)

$$\|e^{-A_m t}x - e^{-A_n t}x\| \leqq c^2 e^{(\sigma_0+\varepsilon)t}t\,\|A_n x - A_m x\|$$

$$\leqq c_1\|A_n x - A_m x\|,$$

so by (13.41)

$$\lim_{n,m\to\infty} \|e^{-A_m t}x - e^{-A_n t}x\| = 0.$$

Thus the operators $e^{-A_n t}$ converge to a limit which is denoted by $\boldsymbol{T}(t)$.

Now (13.43) implies that for each $x \in E$

$$\|e^{-A_n t}x\| \leqq c e^{n\sigma_0 t/(n-\sigma_0)}\|x\|,$$

whence we obtain after passing to limit (as $n \to \infty$),

$$\|\boldsymbol{T}(t)x\| \leqq c e^{\sigma_0 t}\|x\| \qquad (x \in E),$$

i.e.

$$\|\boldsymbol{T}(t)\| \leqq c e^{\sigma_0 t}. \tag{13.44}$$

Since the operator-functions $e^{-A_n t}$ are strongly continuous and the functions $e^{-A_n t}x$ converge to $\boldsymbol{T}(t)x$ uniformly in t, the operator-function $\boldsymbol{T}(t)$ is also strongly continuous. It is clear that $\boldsymbol{T}(0) = \boldsymbol{I}$. Passing to the limit in the relation

$$e^{-A_n(t+s)} = e^{-A_n t}e^{-A_n s},$$

we obtain the relation $\boldsymbol{T}(t+s) = \boldsymbol{T}(t)\boldsymbol{T}(s)$. Thus $\boldsymbol{T}(t)$ is a strongly continuous semi-group of bounded operators.

Now let us show that the operator $-A$ is the generator of the semi-group $T(t)$.

Let $x \in D(A)$. Passing to limit in the relation

$$e^{-A_n t}x - x = -\int_0^t e^{-A_n s}A_n x\,\mathrm{d}s,$$

we obtain the relation

$$T(t)x - x = -\int_0^t T(s)Ax\,\mathrm{d}s, \tag{13.45}$$

which implies that x belongs to the domain of the generator $-\tilde{A}$ of the semi-group $T(t)$ and that $\tilde{A}x = Ax$ for $x \in D(A)$. The range of the operator $\lambda I + \tilde{A}$ for sufficiently large $\operatorname{Re}\lambda$ coincides (by Lemma 13.3) with the entire space E, i.e. with the range of the operator $\lambda I + A$. Hence it follows that the operators A and $\tilde{A}$ coincide. Theorem 13.1 has been completely proved.

It is useful to note that the proof of sufficiency is based only on the validity of inequality (13.38) for sufficiently large integers $\lambda = n$.

13.7 *Analytic semi-groups*

Let A be a positive definite selfadjoint operator in a Hilbert space H and let

$$T(t) = e^{-At} = \int_0^\infty e^{-\lambda t}\,\mathrm{d}P_\lambda \tag{13.46}$$

be the strongly continuous semi-group generated by this operator (see 13.3°). It turns out that for $t > 0$ this semi-group is analytic.

In fact, consider for $\operatorname{Re}(z) \geqq 0$ the operator-function

$$T(z) = \int_0^\infty e^{-\lambda z}\,\mathrm{d}P_\lambda. \tag{13.47}$$

Since the function $e^{-\lambda z}$ is continuous in z while for any $x \in E$ the integral

$$(T(z)x, x) = \int_0^\infty e^{-\lambda z}\,\mathrm{d}(P_\lambda x, x) \quad (x \in H)$$

converges uniformly in z for $\operatorname{Re} z \geqq 0$, the operator-function $\boldsymbol{T}(z)$ is strongly continuous in z for $\operatorname{Re} z \geqq 0$. Finally, since the function $e^{-\lambda z}$ is analytic in z, the operator-function $\boldsymbol{T}(z)$ is analytic in z for $\operatorname{Re} z \geqq 0$. This operator-function is the analytic continuation of the operator-function $\boldsymbol{T}(t)$ to the right halfplane.

Consider now the general case of semi-groups in a Banach space. A semi-group $\boldsymbol{T}(t)$ is called *analytic* if it is possible to continue it from the semi-axis as an operator-function analytic in some sector

$$S_\alpha = \{z\colon |\arg z| < \alpha,\ 0 < |z| < \infty\},$$

where $0 < \alpha < \pi/2$, and strongly continuous on the closure $\bar{S}_\alpha$ of this sector. This continuation will possess the semi-group property:

$$\boldsymbol{T}(z_1 + z_2) = \boldsymbol{T}(z_1)\boldsymbol{T}(z_2) \qquad (z_1, z_2 \in \bar{S}_\alpha), \tag{13.48}$$

as follows from the semi-group identity for the operator-function $\boldsymbol{T}(z)$ on the real semi-axis $0 < t < \infty$ and from the analyticity of the operator-function $\boldsymbol{T}(z)$ in the sector S_α containing this semi-axis.

THEOREM 13.2:[1] *In order that an operator* $-\boldsymbol{A}$ *be the generator of an analytic semi-group* $\boldsymbol{T}(t)$ *it is necessary and sufficient that the resolvent set* $\rho(-\boldsymbol{A})$ *of this operator contain some halfplane* $\operatorname{Re}\lambda \geqq \sigma_0$ *and that for* $\operatorname{Re}\lambda \geqq \sigma_0$ *the following inequality holds*:

$$\|\boldsymbol{R}(\lambda; -\boldsymbol{A})\| \leqq \frac{c}{1 + |\lambda|}. \tag{13.49}$$

PROOF OF NECESSITY: Let $\boldsymbol{T}(t)$ be an analytic semi-group and let $\boldsymbol{T}(z)$ be its analytic continuation to some sector S_α, such that it is strongly continuous on the closure of this sector. From (13.48) and the strong continuity of $\boldsymbol{T}(z)$ in $\bar{S}_\alpha$ it follows that for some ω and $c = c(\omega)$ the inequality

$$\|\boldsymbol{T}(z)\| \leqq c(\omega)e^{\omega|z|} \qquad (z \in \bar{S}_\alpha) \tag{13.50}$$

---

[1] M. Z. Solomjak [1], K. Yosida [1].

holds (this inequality can be proved by the same arguments as were applied in 13.4° to derive the inequality (13.27)). It is clear that $\omega > \omega_0$, where ω_0 is the order of growth of the semi-group.

From the estimate (13.50) it follows that the analytic operator-function defined by $\boldsymbol{A}(z) = e^{-\lambda z}\boldsymbol{T}(z)$ for all $z \in S_\alpha$, satisfies the inequality

$$\|\boldsymbol{A}(z)\| \leqq c(\omega)e^{-\rho(\sigma\cos\varphi - \tau\sin\varphi - \omega)} \qquad (\lambda = \sigma + i\tau,\ z = \rho e^{i\varphi}). \tag{13.51}$$

Let Γ_1 be a ray $z = \rho e^{i\varphi_0}$ $(0 \leqq \rho < \infty)$, where φ_0 is some fixed number in $(0, \alpha)$. The estimate (13.51) means, in particular, that $\max_{0 \leqq \arg z \leqq \varphi_0} \|z\boldsymbol{A}(z)\| \to 0$ as $z \to \infty$ when $\sigma > \omega/\cos\varphi_0$ and $\tau < 0$. Consequently for these values of λ the following relation is satisfied:

$$\int_0^\infty e^{-\lambda t}\boldsymbol{T}(t)\,\mathrm{d}t = \int_{\Gamma_1} e^{-\lambda z}\boldsymbol{T}(z)\,\mathrm{d}z.$$

On the other hand by Lemma 13.3 the operator $\lambda\boldsymbol{I} + \boldsymbol{A}$ has a bounded inverse $\boldsymbol{R}(\lambda; -\boldsymbol{A})$ for $\operatorname{Re}\lambda > \omega_0$ and formula (13.35) is valid. Hence for any $\lambda = \sigma + i\tau$ where $\sigma > \omega/\cos\varphi_0$, $\tau < 0$

$$\boldsymbol{R}(\lambda; -\boldsymbol{A}) = \int_{\Gamma_1} e^{-\lambda z}\boldsymbol{T}(z)\,\mathrm{d}z.$$

It is proved similarly that for any $\lambda = \sigma + i\tau$, where $\sigma > \omega/\cos\varphi_0$ and $\tau > 0$,

$$\boldsymbol{R}(\lambda; -\boldsymbol{A}) = \int_{\Gamma_2} e^{-\lambda z}\boldsymbol{T}(z)\,\mathrm{d}z,$$

where Γ_2 is the ray $z = \rho e^{-i\varphi_0}$ $(0 \leqq \rho < \infty)$.

From these two formulas for $\boldsymbol{R}(\lambda; -\boldsymbol{A})$ and inequality (13.51) it follows that for $\lambda = \sigma + i\tau$, $\operatorname{Re}\lambda \geqq \sigma_0 > \omega/\cos\varphi_0$ the following inequality holds:

$$\|\boldsymbol{R}(\lambda; -\boldsymbol{A})\| \leqq \frac{c(\omega)}{\sigma\cos\varphi_0 + |\tau|\sin\varphi_0 - \omega} \leqq \frac{c}{1 + |\lambda|}. \tag{13.52}$$

The necessity of the condition has been proved. We leave it to the reader to prove that inequality (13.49) is fulfilled in any halfplane $\operatorname{Re}\lambda \geqq \sigma_0$ with $\sigma_0 > \omega_0$ (ω_0 being the order of growth of the semi-group $\boldsymbol{T}(t)$).

To prove sufficiency let us give a formula defining the semi-group $\boldsymbol{T}(t)$. Assume $\sigma_0 > 0$.

Inequality (13.49) and Lemma 13.1 imply that the resolvent $\boldsymbol{R}(\lambda; -\boldsymbol{A})$ of the operator $-\boldsymbol{A}$ is defined not only for $\operatorname{Re}\lambda \geqq \sigma_0$, but also for all $\lambda = \sigma_0 + \rho e^{i\varphi}$, where $0 \leqq \rho < \infty$, $|\sin(\varphi - \pi/2)| < 1/c$. A simple calculation shows that the norm of the resolvent $\boldsymbol{R}(\lambda; -\boldsymbol{A})$ satisfies the inequality

$$\|\boldsymbol{R}(\lambda; -\boldsymbol{A})\| \leqq \frac{c(\alpha)}{1 + |\lambda|} \qquad (\lambda = \sigma_0 + \rho e^{i\varphi}, 0 \leqq \varphi \leqq \alpha), \tag{13.53}$$

where α is any fixed number in the interval $(\pi/2, \pi/2 + \arcsin(1/c))$.

Denote by $\Pi_1(\alpha, \sigma_0) + \Pi_2(\alpha, \sigma_0)$ the broken line consisting of the two rays $\lambda = \sigma_0 + \rho e^{-i\alpha}$ $(0 \leqq \rho < \infty)$ and $\lambda = \sigma_0 + \rho e^{i\alpha}$ $(0 \leqq \rho < \infty)$; this is to be oriented as indicated in Fig. 13.1.

Put

$$\boldsymbol{T}(z) = \frac{1}{2\pi i} \int\limits_{\Pi_1(\alpha, \sigma_0) + \Pi_2(\alpha, \sigma_0)} e^{\lambda z} \boldsymbol{R}(\lambda; -\boldsymbol{A}) \, d\lambda. \tag{13.54}$$

The estimate (13.53) implies that the integral in (13.54) converges for z in the sector $S_{\alpha - \pi/2}$ and defines an analytic semi-group in this sector.

Now α was fixed in the definition of the semi-group $\boldsymbol{T}(z)$. From analyticity of the resolvent $\boldsymbol{R}(\lambda; -\boldsymbol{A})$ and the estimate (13.53) it follows

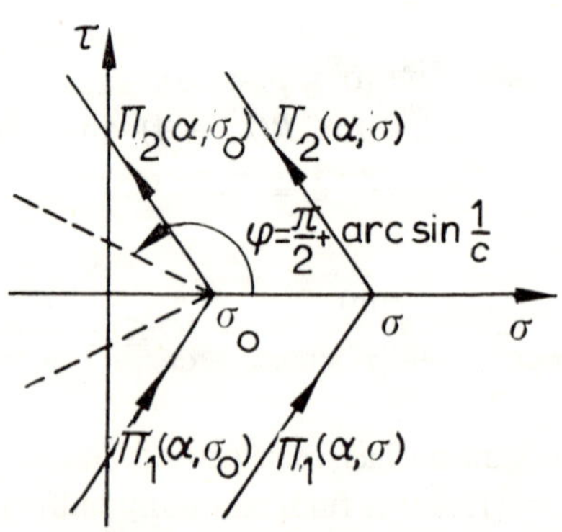

Figure 13.1

(details are left to the reader) that the value of the integral does not depend on α, provided of course that $\arg z$ is sufficiently small. Hence in the following it is permissible to assume that the operator-function $\boldsymbol{T}(z)$ is defined in the open sector $S_{\arcsin 1/c}$.

In the above we gave an integral representation of the operator-function $\boldsymbol{T}(z)$ in which path of integration consisted of a broken line with vertex at the point σ_0. Analyticity of the resolvent together with the estimate (13.53) permit us to go to a representation of the form:

$$\boldsymbol{T}(z) = \frac{1}{2\pi i} \int\limits_{\Pi_1(\alpha,\sigma)+\Pi_2(\alpha,\sigma)} e^{\lambda z} \boldsymbol{R}(\lambda; -\boldsymbol{A})\,\mathrm{d}\lambda, \tag{13.55}$$

where $\Pi_1(\alpha, \sigma) + \Pi_2(\alpha, \sigma)$ is the contour consisting of the rays $\lambda = \sigma + \rho e^{-i\alpha}$ $(0 \leqq \rho < \infty)$ and $\lambda = \sigma + \rho e^{i\alpha}$ $(0 \leqq \rho < \infty)$, for some $\sigma \geqq \sigma_0$, $\pi/2 < \alpha < \pi/2 + \arcsin(1/c)$ (see Fig. 13.1).

Formula (13.55) permits us to obtain simple estimates for norms of the operators $\boldsymbol{T}(z)$. To obtain these estimates for a fixed $z = |z| e^{i\theta}$ formula (13.55) is used, where the path of integration $\Pi_1(\alpha, \sigma) + \Pi_2(\alpha, \sigma)$ is chosen so that $\alpha - \pi/2 > |\theta|$ and

$$\sigma = \sigma(z) = \begin{cases} \dfrac{1}{|z|}, & \text{if } |z| < \dfrac{1}{\sigma_0} \\[2ex] \sigma_0, & \text{if } |z| \geqq \dfrac{1}{\sigma_0}. \end{cases} \tag{13.56}$$

Then (13.53) implies that

$$\|\boldsymbol{T}(z)\| \leqq \frac{c(\alpha)|e^{\sigma(z)z}|}{2\pi}$$

$$\times \left\{ \int\limits_0^\infty \frac{e^{\rho|z|\cos(\theta-\alpha)}}{1 + |\sigma(z) + \rho e^{-i\alpha}|}\,\mathrm{d}\rho + \int\limits_0^\infty \frac{e^{\rho|z|\cos(\theta+\alpha)}}{1 + |\sigma(z) + \rho e^{i\alpha}|}\,\mathrm{d}\rho \right\}$$

$$\leqq \frac{c(\alpha)e^{\sigma(z)\operatorname{Re} z}}{2\pi[1 + \sigma(z)\sin\alpha]} \left\{ \int\limits_0^\infty e^{\rho|z|\cos(\theta-\alpha)}\,\mathrm{d}\rho + \int\limits_0^\infty e^{\rho|z|\cos(\theta+\alpha)}\,\mathrm{d}\rho \right\}$$

$$= \frac{c(\alpha) e^{\sigma(z) \operatorname{Re} z}}{2\pi |z| [1 + \sigma(z) \sin \alpha]} \left[\frac{1}{|\cos(\theta - \alpha)|} + \frac{1}{|\cos(\theta + \alpha)|} \right],$$

whence follows the inequality

$$\| \boldsymbol{T}(z) \| \leqq k(\alpha) e^{\sigma_0 \operatorname{Re} z} \qquad (z \in S_{\alpha - \pi/2}). \tag{13.57}$$

The operator-function $\boldsymbol{T}(z)$ is strongly continuous in the closure $\bar{S}_{\alpha_0}$ of each sector S_{α_0} ($\alpha_0 < \operatorname{arc} \sin(1/c)$) if it is defined at zero by the relation

$$\boldsymbol{T}(0) = I.$$

Since by (13.57) the norms of the $\boldsymbol{T}(z)$ are uniformly bounded for small $z \in S_{\alpha_0}$, it suffices to show that

$$\lim_{z \in S_\alpha, z \to 0} \| \boldsymbol{T}(z) x - x \| = 0$$

for elements x in the domain $D(A)$ of the operator, which is dense in E.

Let $x \in D(A)$. From the formula

$$(\lambda \boldsymbol{I} + A)^{-1} x = \frac{1}{\lambda} x - \frac{1}{\lambda} (\lambda \boldsymbol{I} + A)^{-1} A x$$

it follows that

$$\boldsymbol{T}(z) x = \frac{1}{2\pi i} \int_{\Pi_1(\alpha, \sigma) + \Pi_2(\alpha, \sigma)} \left[\frac{e^{\lambda z}}{\lambda} x - \frac{e^{\lambda z}}{\lambda} \boldsymbol{R}(\lambda; -A) A x \right] \mathrm{d}\lambda,$$

and since

$$\int_{\Pi_1(\alpha, \sigma) + \Pi_2(\alpha, \sigma)} \frac{e^{\lambda z}}{\lambda} \mathrm{d}\lambda = 2\pi i,$$

we have

$$\boldsymbol{T}(z) x - x = -\frac{1}{2\pi i} \int_{\Pi_1(\alpha, \sigma) + \Pi_2(\alpha, \sigma)} \frac{e^{\lambda z}}{\lambda} \boldsymbol{R}(\lambda; -A) A x \, \mathrm{d}\lambda.$$

From (13.53) it follows that

$$\left\| -\frac{1}{2\pi i} \int\limits_{\Pi_1(\alpha,\sigma)+\Pi_2(\alpha,\sigma)} \frac{e^{\lambda z}}{\lambda} \boldsymbol{R}(\lambda; -\boldsymbol{A})\boldsymbol{A}x\,\mathrm{d}\lambda \right\|$$

$$\leqq \frac{c(\alpha)e^{\sigma \operatorname{Re} z}}{2\pi|z|(1+\sigma\sin\alpha)\sigma\sin\alpha}\left[\frac{1}{|\cos(\theta-\alpha)|}+\frac{1}{|\cos(\theta+\alpha)|}\right].$$

Choosing as σ the number defined by the relation (13.56), we obtain the inequality

$$\left\| -\frac{1}{2\pi i} \int\limits_{\Pi_1(\alpha,\sigma)+\Pi_2(\alpha,\sigma)} \frac{e^{\lambda z}}{\lambda} \boldsymbol{R}(\lambda; -\boldsymbol{A})\boldsymbol{A}x\,\mathrm{d}\lambda \right\|$$

$$\leqq |z|k(\alpha)e^{\sigma_0 \operatorname{Re}(z)} \qquad (z \in S_{\alpha_0}),$$

which implies that $\boldsymbol{T}(z)x - x \to 0$ as $z \to 0$ and $z \in S_{\alpha_0}$.

To complete the proof it remains only to show that $\boldsymbol{T}(z)$ possesses the semi-group property

$$\boldsymbol{T}(z_1 + z_2) = \boldsymbol{T}(z_1)\boldsymbol{T}(z_2) \tag{13.58}$$

and that the operator $-\boldsymbol{A}$ is the generator of the semi-group $\boldsymbol{T}(t)$.

Let z_1 and z_2 be fixed numbers in the sector S_{α_0}. The superposition of the operators $\boldsymbol{T}(z_1)$ and $\boldsymbol{T}(z_2)$ can be written in the form

$$\boldsymbol{T}(z_1)\boldsymbol{T}(z_2) = \frac{1}{(2\pi i)^2} \times$$

$$\times \int\limits_{\Pi_1(\alpha,\sigma_1)+\Pi_2(\alpha,\sigma_1)} \int\limits_{\Pi_1(\alpha,\sigma_2)+\Pi_2(\alpha,\sigma_2)} e^{\lambda z_1} e^{\nu z_2} \boldsymbol{R}(\lambda; -\boldsymbol{A})\boldsymbol{R}(\nu; -\boldsymbol{A})\,\mathrm{d}\lambda\,\mathrm{d}\nu.$$

Assume here that the numbers σ_1, σ_2 are distinct and, for definiteness, that $\sigma_0 < \sigma_1 < \sigma_2$. The identity

$$\boldsymbol{R}(\lambda; -\boldsymbol{A})\boldsymbol{R}(\nu; -\boldsymbol{A}) = \frac{\boldsymbol{R}(\lambda; -\boldsymbol{A}) - \boldsymbol{R}(\nu; -\boldsymbol{A})}{\nu - \lambda}$$

implies that

$$T(z_1)T(z_2) =$$

$$= \frac{1}{(2\pi i)^2} \int\limits_{\Pi_1(\alpha,\sigma_1)+\Pi_2(\alpha,\sigma_1)} e^{\lambda z_1}\left[- \int\limits_{\Pi_1(\alpha,\sigma_2)+\Pi_2(\alpha,\sigma_2)} \frac{e^{\nu z_2}R(\nu; -A)}{\nu - \lambda}\,\mathrm{d}\nu\right]\mathrm{d}\lambda$$

$$+ \frac{1}{(2\pi i)^2} \int\limits_{\Pi_1(\alpha,\sigma_2)+\Pi_2(\alpha,\sigma_2)} e^{\nu z_2}\left[- \int\limits_{\Pi_1(\alpha,\sigma_1)+\Pi_2(\alpha,\sigma_1)} \frac{e^{\lambda z_1}R(\lambda; -A)}{\lambda - \nu}\,\mathrm{d}\lambda\right]\mathrm{d}\nu.$$

The inner integrals on the right side are easily calculated (it is necessary to apply Cauchy's integral formula for the case of unbounded contours); the first of them is equal to zero, while the second is equal to $-2\pi i e^{\nu z_1}R(\nu; -A)$. Hence

$$T(z_1)T(z_2) = \frac{1}{2\pi i} \int\limits_{\Pi_1(\alpha,\sigma_2)+\Pi_2(\alpha,\sigma_2)} e^{\nu(z_1+z_2)}R(\nu; -A)\,\mathrm{d}\nu = T(z_1 + z_2)$$

and the relation (13.58) has been proved.

Since the operator-function $T(z)$ is analytic it is, in particular, differentiable for positive real t, and

$$T'(t) = \frac{1}{2\pi i} \int\limits_{\Pi_1(\alpha,\sigma)+\Pi_2(\alpha,\sigma)} e^{\lambda t}\lambda R(\lambda; -A)\,\mathrm{d}\lambda. \tag{13.59}$$

Let $x \in D(A)$. Then

$$\lambda R(\lambda; -A)x = x - R(\lambda; -A)Ax$$

and by (13.59),

$$T'(t)x = \frac{1}{2\pi i} \int\limits_{\Pi_1(\alpha,\sigma)+\Pi_2(\alpha,\sigma)} e^{\lambda t}\,\mathrm{d}\lambda x - \frac{1}{2\pi i} \int\limits_{\Pi_1(\alpha,\sigma)+\Pi_2(\alpha,\sigma)} e^{\lambda t}R(\lambda; -A)\,\mathrm{d}\lambda Ax,$$

i.e.

$$T'(t)x = -T(t)Ax \qquad (t > 0,\ x \in D(A)). \tag{13.60}$$

This relation implies that

$$\frac{T(t)x - x}{t} + Ax = \frac{1}{t}\int_0^t T'(\tau)x\,\mathrm{d}\tau + Ax = \frac{1}{t}\int_0^t [-T(\tau) + I]Ax\,\mathrm{d}\tau,$$

whence

$$\left\| \frac{T(t)x - x}{t} + Ax \right\| \leqq \sup_{0 \leqq \tau \leqq t} \|[T(\tau) - I]Ax\|.$$

The strong continuity of the operator-function $T(z)$ implies the relation

$$\lim_{t \to 0} \frac{T(t)x - x}{t} = -Ax \qquad (x \in D(A)).$$

This means that the operator $-A$ is the generator of the semi-group $T(t)$. The theorem has been proved.

13.8 *Estimates for the operators* $A^n T(t)$

Let us continue our investigation of an analytic semi-group $T(t)$. Assume that the following inequality holds:

$$\|R(\lambda; -A)\| \leqq \frac{c}{1 + |\lambda|} \qquad (\operatorname{Re} \lambda \geqq \sigma_0).$$

As was shown in the proof of Theorem 13.2, the operator function $T(z)$ is then the analytic continuation of the semi-group $T(t)$ to the sector $S_{\arcsin 1/c}$.

Denote by $K(\alpha, t)$ the circle in the complex plane with center at a point t ($t > 0$) and radius $R = t \sin \alpha$; each such circle lies in the sector S_α. If $\alpha < \arcsin(1/c)$, then inequality (13.50) implies the estimate

$$\|T(z)\| \leqq c(\alpha, \omega) e^{\omega(\alpha)(1 + \sin \alpha)t} \qquad (z \in K(\alpha, t)).$$

Hence the estimate (13.9) for derivatives of analytic functions implies the inequalities

$$\|T^{(n)}(t)\| \leqq \frac{n!\,c(\alpha, \omega)\,e^{\omega(\alpha)(1+\sin\alpha)t}}{t^n(\sin\alpha)^n} \qquad (n = 1, 2, \ldots). \tag{13.61}$$

Let $x \in D(A)$. Then as shown in the preceding section, the function $T(t)x$ is continuously differentiable for all $t \geqq 0$ and

$$T'(t)x = -AT(t)x = -T(t)Ax. \tag{13.62}$$

Using the fact that for $t > 0$ the operator $T'(t)$ is bounded while the set $D(A)$ is dense in E, we see that the operator $AT(t)$ can be continuously extended to all of E. Since the operator A is closed, this extension coincides with the operator $AT(t)$ itself. Thus, the following relation is valid:

$$T'(t) = -AT(t) \qquad (t > 0).$$

Similarly one can prove the relations,

$$T^{(n)}(t) = (-1)^n A^n T(t) \qquad (t > 0;\ n = 1, 2, \ldots). \tag{13.63}$$

These relations mean, in particular, that operators $A^nT(t)$ $(t > 0)$ are bounded; (13.61) implies the estimates

$$\|A^nT(t)\| \leqq \frac{n!\,c(\alpha, \omega)\,e^{\omega(\alpha)(1+\sin\alpha)t}}{t^n(\sin\alpha)^n} \qquad (n = 1, 2, \ldots). \tag{13.64}$$

In the next paragraph we will introduce the notion of fractional powers of the operator A, and estimates similar to (13.64) will be established for these operators.

Assume further that the operator A^{-1} is compact. Then the range of each operator $A^nT(t)$ $(t > 0)$ on the ball $\|x\| \leqq 1$ will be compact, because it is contained in the range of the operator A^{-1} on the ball $\|x\| \leqq \|A^{n+1}T(t)\|$. This means that *all the operators* $A^nT(t)$ $(t > 0, n = 1, 2, \ldots)$ *are compact.*

§ 14 Fractional powers of positive-type operators[1]

14.1 *Positive-type operators*

Throughout this paragraph A denotes a closed linear operator acting in a Banach space E and having a domain $D(A)$ dense in E. The operator A is called *of positive type* if the operators $(t\boldsymbol{I} + A)^{-1}$ exist for all $t \geqq 0$ and if

$$\|(t\boldsymbol{I} + A)^{-1}\| \leqq \frac{c}{1+t} \qquad (t \geqq 0). \tag{14.1}$$

The simplest examples of positive-type operators are positive definite self-adjoint operators acting in Hilbert spaces. As a second example one can

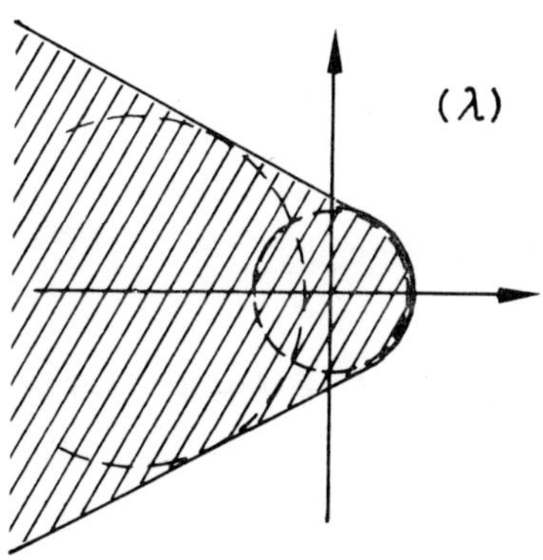

Figure 14.1

[1] This paragraph is written essentially along the lines of the paper of M. A. Krasnoselskii and P. E. Sobolevskii [1]. In addition many results of A. V. Balakrishnan [1, 2] have been used. Fractional powers of various classes of operators have been investigated by many authors: M. Riesz, A. N. Kolmogorov, P. Levy, W. Feller, N. Dunford, E. Hille, R. Phillips, T. Kato, S. G. Krein and others.

take operators of the form $A = A_1 + \sigma_0 I$, where $-A_1$ is the generator of a strongly continuous semi-group and σ_0 is a sufficiently large positive number.

Inequality (14.1) and Lemma 13.1 together imply that the resolvent set $\rho(A)$ of the operator A contains all circles of the form $|\lambda + t| < (1 + t)/c$ $(t > 0)$ (in Fig. 14.1 the union of these circles is hatched); in particular, $\rho(A)$ contains the sector

$$|\arg \lambda - \pi| < \arcsin(1/c).$$

As can be shown by examples, positive-type operators are not necessarily the generators of strongly continuous semi-groups.

14.2 *Negative fractional powers*

Let A be a positive-type operator. Define the operators A^{-z}, where the complex number z satisfies the inequality $0 < \operatorname{Re}(z) < 1$, by the formula

$$A^{-z} = \frac{\sin \pi z}{\pi} \int_0^\infty t^{-z}(tI + A)^{-1} \mathrm{d}t. \tag{14.2}$$

Equation (14.1) implies the inequality

$$\left\| \int_0^\infty t^{-z}(tI + A)^{-1} \mathrm{d}t \right\| \leqq \int_0^\infty t^{-\operatorname{Re} z} \|(tI + A)^{-1}\| \, \mathrm{d}t$$

$$\leqq c \int_0^\infty \frac{t^{-\operatorname{Re} z}}{1 + t} \mathrm{d}t = \frac{c\pi}{\sin(\pi \operatorname{Re} z)},$$

whence follows the boundedness of the operators A^{-z} and the corresponding estimate

$$\|A^{-z}\| \leqq c \frac{|\sin \pi z|}{|\sin(\pi \operatorname{Re} z)|} \qquad (0 < \operatorname{Re} z < 1) \tag{14.3}$$

for their norms. In particular, for real $z = \tau$,

$$\|A^{-\tau}\| \leqq c \quad (0 < \tau < 1).$$

It is possible to obtain integral representations different from (14.2) for operators A^{-z}. The derivation is based on the formulas

$$\frac{\mathrm{d}^n}{\mathrm{d}t^n}(tI + A)^{-1} = (-1)^n n!(tI + A)^{-n-1} \quad (n = 1, 2, \ldots). \tag{14.4}$$

Integrating the right side of formula (14.2) by parts, we obtain

$$A^{-z} = \frac{\sin \pi z}{\pi(1-z)} t^{1-z}(tI + A)^{-1}\Big|_0^\infty$$

$$+ \frac{\sin \pi z}{\pi} \frac{1}{1-z} \int_0^\infty t^{1-z}(tI + A)^{-2}\,\mathrm{d}t.$$

Here the first term vanishes by (14.1), hence

$$A^{-z} = \frac{\sin \pi z}{\pi(1-z)} \int_0^\infty t^{1-z}(tI + A)^{-2}\,\mathrm{d}t. \tag{14.5}$$

The integral in (14.5) actually converges absolutely in operator norm for $0 < \mathrm{Re}(z) < 2$; moreover coefficients in this integral are entire functions. Hence formula (14.5) gives an analytic continuation of the operator-function (14.2) to the strip $0 < \mathrm{Re}\, z < 2$.

It is possible by successive integration by parts to obtain new formulas for the operators A^{-z} of the form:

$$A^{-z} = \frac{\sin \pi z}{\pi} \frac{n!}{(1-z)(2-z)\ldots(n-z)} \int_0^\infty t^{n-z}(tI + A)^{-n-1}\,\mathrm{d}t. \tag{14.6}$$

Each of these formulas defines an analytic continuation of the operator-function (14.2) to the strip $0 < \mathrm{Re}(z) < n+1$. Thus we may consider the operators A^{-z} to be defined for all z in the right halfplane.

We define the operator A^0 by the relation

$$A^0 = I.$$

The notation A^{-k} is ordinarily used to denote the k-th power of the inverse operator A^{-1}. Such powers of the inverse operator coincide with the operator in (14.6), for $z = k$. To prove this, it suffices to remark that for $z = k$ the right side of formula (14.6) (for $n > k$) is easily calculated by making use of the following formula:

$$\frac{\mathrm{d}}{\mathrm{d}t}(tI + A)^{-n} = -n(tI + A)^{-n-1} \qquad (n = 1, 2, \ldots).$$

In the case of a positive definite selfadjoint operator A acting in a Hilbert space (see § 11), formula (14.2) yields the usual fractional powers. It turns out that in the case of arbitrary positive-type operators A, the operators A^{-z} can also be considered as fractional powers. This is a consequence of the following assertion.

THEOREM 14.1: *Let A be a positive-type operator. Then the semi-group identity*

$$A^{-z_1}A^{-z_2} = A^{-z_1-z_2} \qquad (0 < \mathrm{Re}(z_1), \mathrm{Re}(z_2) < \infty) \tag{14.7}$$

is valid and the semi-group A^{-t} $(t \geqq 0)$ is analytic.

PROOF: To establish identity (14.7) it suffices to check its validity for real z_1, z_2, satisfying the inequality $0 < z_1, z_2 < \frac{1}{2}$. For such numbers z_1 and z_2 (14.2) implies the relation

$$A^{-z_1}A^{-z_2} = \frac{\sin \pi z_1}{\pi}\,\frac{\sin \pi z_2}{\pi}\int_0^\infty\!\!\int_0^\infty \lambda^{-z_1}\mu^{-z_2}(\lambda I + A)^{-1}(\mu I + A)^{-1}\,\mathrm{d}\lambda\,\mathrm{d}\mu.$$

The double integral on the right side is taken over the quadrant $\lambda, \mu \geqq 0$. Split it into two integrals J_1 and J_2, the one on the domain $\lambda \leqq \mu$ and the other on the domain $\mu \leqq \lambda$, and perform the change of variables $\lambda = \mu u$ in the first and $\mu = \lambda u$ in the second. This yields

$$A^{-z_1}A^{-z_2} = \frac{\sin \pi z_1}{\pi} \frac{\sin \pi z_2}{\pi}$$

$$\times \left[\int_0^\infty \mu^{1-z_1-z_2} \boldsymbol{R}(\mu; -\boldsymbol{A}) \int_0^1 u^{-z_1} \boldsymbol{R}(\mu u; -\boldsymbol{A}) \mathrm{d}u \mathrm{d}\mu\right.$$

$$\left. + \int_0^\infty \lambda^{1-z_1-z_2} \boldsymbol{R}(\lambda; -\boldsymbol{A}) \int_0^1 u^{-z_2} \boldsymbol{R}(\lambda u; -\boldsymbol{A}) \mathrm{d}u \mathrm{d}\lambda\right],$$

whence it follows that

$$A^{-z_1}A^{-z_2} = \frac{\sin \pi z_1}{\pi} \frac{\sin \pi z_2}{\pi}$$

$$\times \int_0^\infty \lambda^{1-z_1-z_2} \left[\int_0^1 (u^{-z_1} + u^{-z_2})(\lambda \boldsymbol{I} + \boldsymbol{A})^{-1}(\lambda u \boldsymbol{I} + \boldsymbol{A})^{-1} \mathrm{d}u\right] \mathrm{d}\lambda.$$

Now the identity

$$(\lambda \boldsymbol{I} + \boldsymbol{A})^{-1}(\lambda u \boldsymbol{I} + \boldsymbol{A})^{-1} = \frac{(\lambda u \boldsymbol{I} + \boldsymbol{A})^{-1} - (\lambda \boldsymbol{I} + \boldsymbol{A})^{-1}}{\lambda(1 - u)}$$

implies that

$$A^{-z_1}A^{-z_2} = \frac{\sin \pi z_1}{\pi} \frac{\sin \pi z_2}{\pi}$$

$$\times \int_0^\infty \lambda^{-z_1-z_2} \left\{ \int_0^1 \frac{u^{-z_1} + u^{-z_2}}{1 - u} [(\lambda u \boldsymbol{I} + \boldsymbol{A})^{-1} - (\lambda \boldsymbol{I} + \boldsymbol{A})^{-1}] \mathrm{d}u \right\} \mathrm{d}\lambda$$

$$= \frac{\sin \pi z_1}{\pi} \frac{\sin \pi z_2}{\pi}$$

$$\times \int_0^1 \frac{u^{-z_1} + u^{-z_2}}{1 - u} \left\{ \int_0^\infty \lambda^{-z_1-z_2} [(\lambda u \boldsymbol{I} + \boldsymbol{A})^{-1} - (\lambda \boldsymbol{I} + \boldsymbol{A})^{-1}] \mathrm{d}\lambda \right\} \mathrm{d}u.$$

Since

$$\int_0^\infty \lambda^{-z_1-z_2}[(\lambda u\boldsymbol{I} + A)^{-1} - (\lambda\boldsymbol{I} + A)^{-1}]\mathrm{d}\lambda$$

$$= \int_0^\infty \lambda^{-z_1-z_2}(u^{z_1+z_2-1} - 1)(\lambda\boldsymbol{I} + A)^{-1}\mathrm{d}\lambda,$$

we arrive at the following representation for the operator $A^{-z_1}A^{-z_2}$:

$$A^{-z_1}A^{-z_2} = \left\{\frac{\sin \pi z_1}{\pi}\ \frac{\sin \pi z_2}{\pi}\right.$$

$$\times \left.\int_0^1 \frac{(u^{-z_1} + u^{-z_2})(1 - u^{1-z_1-z_2})}{(1-u)\cdot u^{1-z_1-z_2}}\,\mathrm{d}u\right\} \int_0^\infty \lambda^{-z_1-z_2}(\lambda\boldsymbol{I} + A)^{-1}\mathrm{d}\lambda. \qquad (14.8)$$

This relation holds, in particular, if A is the multiplication operator by a number a; in this case A^{-z} is the multiplication operator by a^{-z}. Hence for this operator the semi-group identity $a^{-z_1}a^{-z_2} = a^{-z_1-z_2}$ is fulfilled and consequently the coefficient in the above bracket is equal to

$$\{\sin \pi(z_1 + z_2)\}/\pi.$$

Thus

$$A^{-z_1}A^{-z_2} = \frac{\sin \pi(z_1 + z_2)}{\pi}\int_0^\infty \lambda^{-z_1-z_2}\boldsymbol{R}(\lambda; -A)\mathrm{d}\lambda = A^{-z_1-z_2}.$$

It remains for us to verify that the operator-function A^{-z} is strongly continuous at zero when it is considered in some sector $S_\alpha = \{z;\ z = |z|\cdot e^{i\theta}, |z| > 0, |\theta| \leqq \alpha\}$, where $\alpha < \pi/2$. In each such sector, the norms of the operators A^{-z} for small z are uniformly bounded—this follows from the estimate (14.3). Hence to prove strong continuity it suffices to show that

$$\lim_{z\in S_\alpha,\, z\to 0} \|A^{-z}x - x\| = 0 \qquad (14.9)$$

for the elements x in some dense subset of E. We take the domain $D(A)$ of the operator A as an appropriate subset of this kind.

Let $x \in D(A)$. Then $x = A^{-1}y$ where $y \in E$, and

$$A^{-z}x - x = \frac{\sin \pi z}{\pi} \int_0^\infty t^{-z}(tI + A)^{-1}A^{-1}y \,\mathrm{d}t$$

$$- \frac{\sin \pi z}{\pi} \int_0^\infty \frac{t^{-z}}{1+t} A^{-1}y \,\mathrm{d}t$$

$$= \frac{\sin \pi z}{\pi} \int_0^\infty \frac{t^{-z}}{1+t} (tI + A)^{-1} \mathrm{d}t(A^{-1}y - y).$$

The relation (14.1) then implies the estimate

$$\|A^{-z}x - x\| \leqq \frac{|\sin \pi z|}{\pi} \int_0^\infty \frac{ct^{-\operatorname{Re} z}}{(1+t)^2} \mathrm{d}t \, \|A^{-1}y - y\|$$

$$\leqq \frac{c}{\pi} \left[\frac{1}{1 - \operatorname{Re} z} + \frac{1}{2} \right] |\sin \pi z| \, \|A^{-1}y - y\|,$$

from which (14.9) follows. The theorem has been proved.

14.3 *Positive fractional powers*

Let A again be a positive-type operator. The relation (14.1) implies, in particular, the existence of the inverse operator of A. Hence all the operators A^{-n}, where $n = 1, 2, \ldots$, vanish only on the zero element. Let z be an arbitrary number in the right half plane and choose $n > |\operatorname{Re} z|$. Then the identity $A^{-n} = A^{-(n-z)}A^{-z}$ implies that the operator A^{-z} vanishes only on the zero element, that is the operator A^{-z} has an inverse, which will be denoted by A^{z}:

$$A^{z} = (A^{-z})^{-1}. \tag{14.10}$$

The operators A^z for $\mathrm{Re}(z) > 0$ will be unbounded if the operator A is unbounded. The boundedness of the operators A^{-z} implies that the operators A^z are closed. The domain $D(A^z)$ of each operator A^z $(\mathrm{Re}(z) > 0)$ coincides with the range $R(A^{-z})$ of the operator A^{-z}. From the relation

$$A^{-z_1} = A^{-z_2}A^{-(z_1-z_2)} \qquad (0 < \mathrm{Re}(z_2) < \mathrm{Re}(z_1))$$

it follows that $R(A^{-z_1}) \subset R(A^{-z_2})$; hence

$$D(A^{z_1}) \subset D(A^{z_2}) \qquad (\mathrm{Re}(z_1) > \mathrm{Re}(z_2)). \tag{14.11}$$

These inclusions imply that the domains of all the operators A^z $(\mathrm{Re}(z) > 0)$ are dense in E because the domains of the operators A^n for positive integer n are dense in E (see 13.2°).

Again let $\mathrm{Re}(z_1) > \mathrm{Re}(z_2)$ and denote by $\boldsymbol{B}$ the operator defined on $D(A^{z_1})$ by the relation $\boldsymbol{B}x = A^{z_2}x$. It turns out that the closure of the operator $\boldsymbol{B}$ coincides with the operator A^{z_2}. Let us prove this simple but important fact.

Let x_0 be any element in $D(A^{z_2})$. Choose in $D(A^{z_1-z_2})$ a sequence of elements y_n converging to $A^{z_2}x_0$. Put $x_n = A^{-z_2}y_n$; then $x_n \in D(A^{z_1})$ and $\boldsymbol{B}x_n = y_n$. It is clear that

$$\|x_n - x_0\| = \|A^{-z_2}y_n - A^{-z_2}A^{z_2}x_0\| \leqq \|A^{-z_2}\| \, \|y_n - A^{z_2}x_0\| \to 0$$

and that $\boldsymbol{B}x_n$ converges to $A^{z_0}x_0$. That is, x_0 belongs to the domain of the closure $\overline{\boldsymbol{B}}$ of the operator $\boldsymbol{B}$ and $\overline{\boldsymbol{B}}x_0 = A^{z_2}x_0$. The assertion has been proved.

We have already established the identity

$$A^{z_1}A^{z_2} = A^{z_2}A^{z_1} = A^{z_1+z_2} \tag{14.12}$$

for z_1 and z_2 in the left half plane. It is clear that it remains true for any z whose real part is different from zero. Here in the case of unbounded operators relation (14.12) means that $A^{z_1}A^{z_2}x = A^{z_2}A^{z_1}x = A^{z_1+z_2}x$ for all those elements x for which $A^{z_1}x$, $A^{z_2}x$, $A^{z_1+z_2}x$ are defined.

It can be shown without difficulty that the operators A^z for any z $(\mathrm{Re}\, z \neq 0)$ commute with the resolvents $(\lambda \boldsymbol{I} - A)^{-1}$. In particular if

$\operatorname{Re} z > 0$ then for any $x \in D(A^z)$

$$A^z(\lambda \boldsymbol{I} - A)^{-1}x = (\lambda \boldsymbol{I} - A)^{-1}A^z x. \tag{14.13}$$

Note further that the operators $A^z(t\boldsymbol{I} + A)^{-n}$ are uniformly bounded for $t \geqq 0$ when z is fixed and n satisfies $0 < \operatorname{Re} z < n$:

$$\|A^z(t\boldsymbol{I} + A)^{-n}\| \leqq C(z, n) \quad (0 \leqq t < \infty). \tag{14.14}$$

The proof results from the trivial inequality

$$\|A^z(t\boldsymbol{I} + A)^{-n}\| \leqq \|A^z A^{-n}[A(t\boldsymbol{I} + A)^{-1}]^n\|$$

$$\leqq \|A^{z-n}\| \, \|A(t\boldsymbol{I} + A)^{-1}\|^n \leqq \|A^{z-n}\| \, \|\boldsymbol{I} - t(t\boldsymbol{I} + A)^{-1}\|^n$$

$$\leqq \|A^{z-n}\| \cdot (1 + c)^n.$$

Assume that $0 < \operatorname{Re} z < n$. Then by (14.6)

$$A^z x = \frac{\sin \pi z}{\pi} \, \frac{n!}{z(z-1)\dots(z-n+1)} \int_0^\infty t^z (t\boldsymbol{I} + A)^{-n-1} A^n x \, \mathrm{d}t$$

$$(x \in D(A^n)). \quad (14.15)$$

This formula can be considered as defining the values of the operator A^z on $D(A^n)$. The closure of the operator defined by the right side permits one to determine its values on the whole domain $D(A^z)$, as has already been shown.

In some cases it is convenient to replace formula (14.15) by an equivalent one:

$$A^z x = \frac{\sin \pi z}{\pi} \int_0^\infty \lambda^z \left[(\lambda \boldsymbol{I} + A)^{-1} x - \frac{\theta(\lambda) x}{\lambda} + \dots \right.$$

$$\left. \dots + (-1)^n \frac{\theta(\lambda) A^{n-1} x}{\lambda^n} \right] \mathrm{d}\lambda + \frac{\sin \pi z}{\pi}$$

$$\times \left[\frac{x}{z} - \frac{Ax}{z-1} + \dots + (-1)^{n-1} \frac{A^{n-1} x}{z-n+1} \right] \quad (x \in D(A^n)), \quad (14.16)$$

where

$$\theta(\lambda) = \begin{cases} 0 & \text{if } 0 \leqq \lambda \leqq 1 \\ 1 & \text{if } 1 < \lambda < \infty; \end{cases} \tag{14.17}$$

here we assume that the right side is equal to $A^k x$ for real integers $z = k$. The integrals on the right side of formulas (14.15) and (14.16) can converge even for non-positive-type operators A—it is sufficient that the following inequality, which is less restrictive than (14.1) be valid:

$$\|t(tI + A)^{-1}\| < c \qquad (t > 0). \tag{14.18}$$

If formula (14.16) is taken as the definition of fractional powers, then the operators A^z are defined first on sets dense in E and are then extended by means of the closure procedure. The semi-group identity is also first proved on certain sets dense in E, and then extended by a limit procedure.

14.4 *A moment inequality*

In 12.2° an important moment inequality was established for positive definite selfadjoint operators A

$$\|A^\tau x\| \leqq \|Ax\|^\tau \|x\|^{1-\tau} \qquad (x \in D(A)). \tag{14.19}$$

In this section we indicate the appropriate generalization to positive-type operators acting in Banach spaces as well as Hilbert spaces.

LEMMA 14.1: *Let A be a positive-type operator:*

$$\|(tI + A)^{-1}\| \leqq \frac{c}{1 + t} \qquad (t \geqq 0). \tag{14.20}$$

Then the fractional powers A^δ $(\delta > 0)$ satisfy the inequality

$$\|A^\delta (tI + A)^{-n}\| \leqq \frac{C(n, \delta)}{(1 + t)^{n-\delta}} \qquad (t \geqq 0) \tag{14.21}$$

for each $n > \delta$.

PROOF: As was noted above, the operators $A^n(tI + A)^{-n}$ are bounded; in particular the range of the operator $(tI + A)^{-n}$ is contained in $D(A^n)$. Consequently, (14.15) implies that

$$A^\delta(tI + A)^{-n} = \frac{\sin \pi\delta}{\pi} \frac{n!}{\delta(\delta - 1)\dots(\delta - n + 1)}$$

$$\times \int_0^\infty s^\delta(sI + A)^{-n-1}A^n(tI + A)^{-n}\mathrm{d}s. \tag{14.22}$$

To estimate the norm of the operators $A^\delta(tI + A)^{-n}$, split the integral on the right side of (14.22) into two integrals—an integral on the interval $[0, N]$ and an integral on the interval $[N, \infty)$, where N is a number which is not fixed in advance. Let us estimate each of these two terms separately. The estimates obtained will depend, of course, on N. Finally we find, by the usual method of calculus, that N at which the right side of the estimate takes on its minimum value.

Let

$$A^\delta(tI + A)^{-n} = \frac{\sin \pi\delta}{\pi} \frac{n!}{\delta(\delta - 1)\dots(\delta - n + 1)} [J_1 + J_2],$$

where

$$J_1 = \int_0^N s^\delta(sI + A)^{-n-1}A^n(tI + A)^{-n}\mathrm{d}s,$$

$$J_2 = \int_N^\infty s^\delta(sI + A)^{-n-1}A^n(tI + A)^{-n}\mathrm{d}s.$$

To estimate both these integrals we use the trivial inequality

$$\|A^n(uI + A)^{-n}\| \leqq (c + 1)^n \quad (0 \leqq u < \infty),$$

which follows from (14.20):

$$\|A^n(uI + A)^{-n}\| \leqq \|A(uI + A)^{-1}\|^n$$

$$= \|I - u(uI + A)^{-1}\|^n \leqq (c + 1)^n.$$

From this inequality we obtain for the first integral J_1 the estimate:

$$\|J_1\| \leqq c(c+1)^n \int_0^N s^{\delta-1}\,ds\,\|(tI + A)^{-n}\| \leqq \frac{c^{n+1}(c+1)^{n+1}}{\delta(1+t)^n} N^\delta,$$

and for the second the estimate:

$$\|J_2\| \leqq c^{n+1}(c+1)^n \int_N^\infty s^{\delta-n-1}\,ds = \frac{c^{n+1}(c+1)^{n+1}}{n-\delta} N^{\delta-N}.$$

Thus

$$\|A^\delta(tI + A)^{-n}\| \leqq \frac{|\sin \pi\delta|}{\pi} \frac{n!}{|\delta(\delta-1)\dots(\delta-n+1)|}$$

$$\times\, c^{n+1}(c+1)^n \left[\frac{N^\delta}{\delta(1+t)^n} + \frac{N^{\delta-n}}{n-\delta}\right]. \tag{14.23}$$

The right side takes its minimum value at

$$N = 1 + t.$$

Substituting this value for N into the right side of formula (14.23), we obtain the estimate (14.21) where

$$C(n, \delta) = \frac{|\sin \pi\delta|}{\pi} \frac{n!}{|\delta(\delta-1)\dots(\delta-n+1)|} \frac{nc^{n+1}(c+1)^n}{\delta(n-\delta)}.$$

The lemma has been proved.

THEOREM 14.2: *Let A be a positive-type operator. Let the real numbers α and β have the same sign with $0 < |\alpha| < |\beta|$. Then the following moment inequality is valid:*

$$\|A^\alpha x\| \leqq k(\alpha, \beta)\,\|A^\beta x\|^{\alpha/\beta}\,\|x\|^{1-\alpha/\beta} \qquad (x \in D(A^\beta)). \tag{14.24}$$

PROOF: Consider first the case in which the numbers α and β are negative. Represent the operator A by means of the relation (14.6):

$$A^\alpha x = -\frac{\sin \pi\alpha}{\pi} \frac{n!}{(1+\alpha)\dots(n+\alpha)} \int_0^\infty t^{n+\alpha}(tI + A)^{-n-1} x \, dt,$$

where $n < |\beta|$. As in the proof of Lemma 14.1, we split the integral on the right side into integrals over the intervals $[0, N]$ and $[N, \infty)$, estimate each integral separately and then choose that N at which the right side of the estimate takes on its minimum value.

Let

$$A^\alpha x = -\frac{\sin \pi\alpha}{\pi} \frac{n!}{(1+\alpha)\dots(n+\alpha)} [J_1 + J_2],$$

where

$$J_1 = \int_0^N t^{n+\alpha}(tI + A)^{-n-1} x \, dt,$$

$$J_2 = \int_N^\infty t^{n+\alpha}(tI + A)^{-n-1} x \, dt.$$

Lemma 14.1 implies that

$$\|J_1\| \leqq \| \int_0^N t^{n+\alpha} A^{-\beta}(tI + A)^{-n-1} A^\beta x \, dt \|$$

$$\leqq C(n, |\beta|) \int_0^N t^{\alpha-\beta-1} dt \, \|A^\beta x\| \leqq \frac{C(n, |\beta|)}{|\alpha - \beta|} \|A^\beta x\| \, N^{\alpha-\beta}.$$

The estimate for the second integral J_2 follows from (14.20):

$$\|J_2\| = \|\int_N^\infty t^{n+\alpha}(tI + A)^{-n-1}x\,dt\|$$

$$\leqq c^{n+1}\int_N^\infty t^{\alpha-1}\,dt\,\|x\| = \frac{c^{n+1}}{|\alpha|}\|x\|\,N^\alpha.$$

From these estimates it follows that

$$\|A^\alpha x\| \leqq \frac{|\sin \pi\alpha|}{\pi}\,\frac{n!}{|(1+\alpha)\dots(n+\alpha)|}$$

$$\times\left\{\frac{C(n, |\beta|)}{|\beta-\alpha|}\|A^\beta x\|\,N^{\alpha-\beta} + \frac{c^{n+1}}{|\alpha|}\|x\|\,N^\alpha\right\}. \tag{14.25}$$

The expression in the bracket takes on its minimum value at

$$N = \left[\frac{c^{n+1}}{C(n, |\beta|)}\,\frac{\|x\|}{\|A^\beta x\|}\right]^{1/\beta}.$$

With this value of N inequality (14.25) takes the form

$$\|A^\alpha x\| \leqq \frac{|\sin \pi\alpha|}{\pi}\,\frac{n!}{|(1+\alpha)\dots(n+\alpha)|}$$

$$\times\left[\frac{C(n, |\beta|)^{\alpha/\beta}c^{(n+1)(1-\alpha/\beta)}|\beta|}{|(\beta-\alpha)\alpha|}\|A^\beta x\|^{\alpha/\beta}\|x\|^{1-\alpha/\beta}\right].$$

The assertion of the theorem for negative α, β has been proved.

Now let α, β be positive. The part of the theorem proved above, implies the inequality

$$\|A^{-\beta+\alpha}y\| \leqq k(-\beta+\alpha, -\beta)\|A^{-\beta}y\|^{1-\alpha/\beta}\|y\|^{\alpha/\beta} \quad (y \in E). \tag{14.26}$$

On putting $y = A^\beta x$ ($x \in D(A^\beta)$), this leads to the inequality (14.24). The theorem has been completely proved.

14.5 *Operators subordinate to fractional powers of a positive-type operator*

The notion of subordination of operators introduced in 12.3° for operators in a Hilbert space, can be extended without change to operators acting in Banach spaces: an operator $\boldsymbol{B}$ is *subordinate* to an operator $\boldsymbol{A}$ if $D(\boldsymbol{A}) \subset D(\boldsymbol{B})$ and

$$\|\boldsymbol{B}x\| \leqq k_0 \|\boldsymbol{A}x\| \qquad (x \in D(\boldsymbol{A})), \tag{14.27}$$

where k_0 is some positive number.

Lemma 12.1 (together with its proof) remains true under the transfer to the Banach space situation. This implies that the subordination of an operator $\boldsymbol{B}$ to a closed operator $\boldsymbol{A}$ follows from $D(\boldsymbol{A}) \subset D(\boldsymbol{B})$, provided that $\boldsymbol{A}^{-1}$ is bounded. Note further that Lemma 12.2 also remains true under the transfer to Banach spaces.

Theorem 14.2 implies:

THEOREM 14.3:[1] *Let an operator $\boldsymbol{B}$ be subordinate to a fractional power $\boldsymbol{A}^{\tau}$ ($\tau > 0$) of a positive-type operator $\boldsymbol{A}$. Then for any $\tau_0 > \tau$ the following inequality holds:*

$$\|\boldsymbol{B}x\| \leqq k(\tau_0) \|\boldsymbol{A}^{\tau_0}x\|^{\tau/\tau_0} \|x\|^{1-\tau/\tau_0} \qquad (x \in D(\boldsymbol{A}^{\tau_0})). \tag{14.28}$$

As in the case of strictly positive definite selfadjoint operators acting in a Hilbert space (see Theorem 12.3), it is possible to provide an 'almost converse' to this theorem.

THEOREM 14.4: *Let $\boldsymbol{A}$ be a positive-type operator. Let $\boldsymbol{B}$ be a closed linear operator satisfying condition* (14.28) *for some τ and τ_0, $0 < \tau < \tau_0$. Then the operator $\boldsymbol{B}$ is subordinate to all operators $\boldsymbol{A}^{\tau+\varepsilon}$, where $\varepsilon > 0$.*

PROOF: The assertion of the theorem is equivalent to claiming boundedness of all operators $\boldsymbol{B}\boldsymbol{A}^{-\tau-\varepsilon}$.

---

[1] Translator's note. The original assertion of this theorem was incorrect and is modified.

Consider the integral

$$Jx = \int_0^\infty t^{n-\tau-\varepsilon} B(tI + A)^{-n-1} x \, dt, \tag{14.29}$$

where n is an integer and $n > \tau_0$. Inequality (14.28) implies the estimate

$$\begin{aligned} &\|B(tI + A)^{-n-1}x\| \\ &\leqq k\|A^{\tau_0}(tI + A)^{-n-1}x\|^{\tau/\tau_0} \|(tI + A)^{-n-1}x\|^{1-\tau/\tau_0}, \end{aligned}$$

whence by Lemma 14.1

$$\|B(tI + A)^{-n-1}x\| \leqq \frac{k_1\|x\|}{(1 + t)^{n-\tau+1}}.$$

Hence the operator-function which is the integrand in (14.29), is absolutely integrable on $[0, \infty)$. This means that formula (14.29) determines a continuous operator defined on all of E.

The operator $A^{-\tau-\varepsilon}$ can be represented (see (14.6)) by the integral

$$\begin{aligned} A^{-\tau-\varepsilon}x = &\frac{\sin \pi(\tau + \varepsilon)}{\pi} \frac{n!}{(1 - \tau - \varepsilon)\dots(n - \tau - \varepsilon)} \\ &\times \int_0^\infty t^{n-\tau-\varepsilon}(tI + A)^{-n-1}x \, dt \quad (x \in E). \end{aligned}$$

From the closedness of the operator B it follows that

$$\begin{aligned} BA^{-\tau-\varepsilon}x = &\frac{\sin \pi(\tau + \varepsilon)}{\pi} \frac{n!}{(1 - \tau - \varepsilon)\dots(n - \tau - \varepsilon)} \\ &\times \int_0^\infty t^{n-\tau-\varepsilon}B(tI + A)^{-n-1}x \, dt \\ = &\frac{\sin \pi(\tau + \varepsilon)}{\pi} \frac{n!}{(1 - \tau - \varepsilon)\dots(n - \tau - \varepsilon)} Jx. \end{aligned}$$

Hence the operator $BA^{-\tau-\varepsilon}$ is bounded. The theorem has been proved.

As in the case of operators in a Hilbert space (see 12.4°), the *order* of an operator $\boldsymbol{B}$ with respect to a positive-type operator $\boldsymbol{A}$ is the infimum of the non-negative numbers τ such that $\boldsymbol{B}$ is subordinate to $\boldsymbol{A}^{\tau}$. Theorem 14.4 implies that the order of an operator $\boldsymbol{B}$ with respect to a positive-type operator $\boldsymbol{A}$ is not greater than τ, provided that inequality (14.28) is fulfilled for some τ_0, $0 < \tau < \tau_0$.

In § 16 the following assertion will be used. It is a generalization of Theorem 14.4 (and is proved similarly).

THEOREM 14.5: *Let E and E_1 be two Banach spaces. Let $\boldsymbol{A}$ be a positive-type operator in the space E and $\boldsymbol{B}$ be a closed linear operator acting from E to E_1. Let the following inequality be fulfilled*:

$$\|\boldsymbol{B}x\|_{E_1} \leqq k \|\boldsymbol{A}^{\tau_0}x\|_E^{\tau/\tau_0} \|x\|_E^{1-\tau/\tau_0} \qquad (x \in D(\boldsymbol{A}^{\tau_0})),$$

where $0 < \tau < \tau_0$. Then for each $\varepsilon > 0$ the operator $\boldsymbol{B}\boldsymbol{A}^{-\tau-\varepsilon}$ acts from E to E_1 and is continuous.

The possibility and importance of modifying assertions of the kind in Theorem 14.4 to assertions of the kind in Theorem 14.5 were pointed out by V. P. Glushko.

14.6 *General theorems on subordination*

Let us now present more exact theorems on the subordination of operators to fractional powers of positive type operators.

THEOREM 14.6: *Let E and E_0 be two Banach spaces. Let $\boldsymbol{A}$ be a positive-type operator in the space E and let $\boldsymbol{B}$ be a closed linear operator acting from E to E_0. Let the following inequality hold*:

$$\|\boldsymbol{B}x\|_{E_0} \leqq c \|\boldsymbol{A}^{\tau_0}x\|_E^{\tau/\tau_0} \|x\|_E^{1-\tau/\tau_0} \qquad (x \in D(\boldsymbol{A}^{\tau_0})), \tag{14.30}$$

where $0 < \tau < \tau_0$. Finally, suppose $0 < \varepsilon_1 \leqq \tau_0 - \tau$, $0 < \varepsilon_2 \leqq \tau$. Then the following inequality is satisfied:

$$\|\boldsymbol{B}x\|_{E_0} \leqq \frac{c_1(\varepsilon_1 + \varepsilon_2)}{\varepsilon_1 \varepsilon_2} \|\boldsymbol{A}^{\tau+\varepsilon_1}x\|_E^{\varepsilon_2/(\varepsilon_1+\varepsilon_2)} \|\boldsymbol{A}^{\tau-\varepsilon_2}x\|_E^{\varepsilon_1/(\varepsilon_1+\varepsilon_2)} \qquad (x \in D(\boldsymbol{A}^{\tau+\varepsilon_1})). \tag{14.31}$$

PROOF: From formula (14.6) for $z = \tau + \varepsilon_1$ and $x \in D(A^{\tau_0})$ there follows the identity

$$x = \frac{\sin \pi(\tau + \varepsilon_1)}{\pi} \frac{m!}{(1 - \tau - \varepsilon_1)\dots(m - \tau - \varepsilon_1)}$$

$$\times \int_0^\infty t^{m-\tau-\varepsilon_1} (tI + A)^{-m-1} A^{\tau+\varepsilon_1} x \, dt \qquad (m > \tau_0).$$

The element x obviously belongs to the domain $D(B)$ of the operator B. Let us estimate $\|Bx\|_0$.

To this end, consider the integral

$$J_1 x = \frac{\sin \pi(\tau + \varepsilon_1)}{\pi} \frac{(n + k)!}{(1 - \tau - \varepsilon_1)\dots(n + k - \tau - \varepsilon_1)}$$

$$\times \int_0^\infty t^{n+k-\tau-\varepsilon_1} B(tI + A)^{-n-k-1} A^{\tau+\varepsilon_1} x \, dt,$$

where n and k are integers satisfying the inequalities $n > \tau_0$, $k \geqq \varepsilon_1 + \varepsilon_0$. Inequality (14.30) and Lemma 14.1 imply that

$$\|B(tI + A)^{-n-k-1} A^{\tau+\varepsilon_1} x\|_{\varepsilon_0}$$

$$\leqq \|B(tI + A)^{-n}\|_{E\to E_0} \|A^{\tau+\varepsilon_1-\delta}(tI + A)^{-k-1} A^{\delta} x\|_E$$

$$\leqq c\|A^{\tau_0}(tI + A)^{-n}\|_{E\to E}^{\tau/\tau_0} \|(tI + A)^{-n}\|_{E\to E}^{1-\tau/\tau_0}$$

$$\times \|A^{\tau+\varepsilon_1-\delta}(tI + A)^{-k-1}\|_{E\to E} \|A^{\delta} x\|_E$$

$$\leqq \frac{c(\delta)}{t^{n+k-2\tau-\varepsilon_1+\delta+1}} \|A^{\delta} x\|_E. \qquad (14.32)$$

Here δ is any number in the interval $[\tau - \varepsilon_2, \tau + \varepsilon_1]$. This estimate with $\delta = \tau + \varepsilon_1$ implies the convergence of the integral $J_1 x$ at infinity, and with $\delta = \tau - \varepsilon_2$ it implies the convergence of this integral at zero. Hence

$$Bx = \frac{\sin \pi(\tau + \varepsilon_1)}{\pi} \frac{(n+k)!}{(1-\tau-\varepsilon_1)\dots(n+k-\tau-\varepsilon_1)}$$

$$\times \int_0^\infty t^{n+k-\tau-\varepsilon_1} B(tI + A)^{-n-k-1} A^{\tau+\varepsilon_1} x \, \mathrm{d}t. \tag{14.33}$$

From (14.33) there follows the inequality

$$\|Bx\|_{E_0} \leqq c_2 \{ \int_0^N t^{n+k-\tau-\varepsilon_1} \|B(tI + A)^{-n-k-1} A^{\tau+\varepsilon_1} x\|_{E_0} \mathrm{d}t$$

$$+ \int_N^\infty t^{n+k-\tau-\varepsilon_1} \|B(tI + A)^{-n-k-1} A^{\tau+\varepsilon_1} x\|_{E_0} \mathrm{d}t \}.$$

By (14.32) with $\delta = \tau - \varepsilon_2$

$$\int_0^N t^{n+k-\tau-\varepsilon_1} \|B(tI + A)^{-n-k-1} A^{\tau+\varepsilon_1} x\|_{E_0} \mathrm{d}t \leqq \frac{c_3}{\varepsilon_2} \|A^{\tau-\varepsilon_2} x\|_E N^{\varepsilon_2}$$

and by the same inequality with $\delta = \tau + \varepsilon_1$

$$\int_N^\infty t^{n+k-\tau-\varepsilon_1} \|B(tI + A)^{-n-k-1} A^{\tau+\varepsilon_1} x\|_{E_0} \mathrm{d}t \leqq \frac{c_4}{\varepsilon_1} \|A^{\tau+\varepsilon_1} x\|_E N^{-\varepsilon_1}.$$

Hence

$$\|Bx\|_{E_0} \leqq c_5 \left\{ \|A^{\tau+\varepsilon_1} x\|_E \frac{N^{-\varepsilon_1}}{\varepsilon_1} + \|A^{\tau-\varepsilon_2} x\|_E \frac{N^{\varepsilon_2}}{\varepsilon_2} \right\}.$$

In particular, with

$$N = \left[\frac{\|A^{\tau+\varepsilon_1} x\|_E}{\|A^{\tau-\varepsilon_2} x\|_E} \right]^{1/(\varepsilon_1+\varepsilon_2)}$$

we obtain the estimate (14.31).

Thus inequality (14.31) is valid for all $x \in D(A^{\tau_0})$. By the closedness of the operator $\boldsymbol{B}$ it follows that inequality (14.31) is also valid for all $x \in D(A^{\tau+\varepsilon_1})$. The theorem has been proved.

Theorem 14.6 has a valid converse: *if for any* $\varepsilon_1 \in (0, \tau_0 - \tau]$ *and* $\varepsilon_2 \in (0, \tau]$ *inequality* (14.31) *holds, then the moment inequality* (14.24) *implies inequality* (14.30).

Inequality (14.30) means, roughly speaking, that the operator $\boldsymbol{B}$ 'consists of a τ/τ_0 part of the operator A^{τ_0} and of a $1 - \tau/\tau_0$ part of the identity operator'. From Theorem 14.6 it follows that in this case the operator $\boldsymbol{B}$ 'consists of an $\varepsilon_2/(\varepsilon_1 + \varepsilon_2)$ part of the operator $A^{\tau+\varepsilon_1}$ and of an $\varepsilon_1/(\varepsilon_1 + \varepsilon_2)$ part of the operator $A^{\tau-\varepsilon_2}$'.

14.7 *Estimates for elements of the form* $\boldsymbol{B}A^{-\tau}x$

In this section some special estimates for the E_0 norm of elements of the form $\boldsymbol{B}A^{-\tau}x$ $(x \in E)$ will be established. These estimates play a basic role in the study of fractional powers of elliptic operators (see § 16).

Let $\boldsymbol{B}$ be a closed linear operator acting from a Banach space E to a Banach space E_0 and let the inequality (14.30) hold. Then by Theorem 14.6 for any $\varepsilon_1 \in (0, \tau_0 - \tau]$, $\varepsilon_2 \in (0, \tau]$ one has the inequality:

$$\|\boldsymbol{B}A^{-\tau-\varepsilon_1}x\|_{E_0} \leqq c_1 \|A^{-\varepsilon_1-\varepsilon_2}x\|_E^{\varepsilon_1/(\varepsilon_1+\varepsilon_2)} \|x\|_E^{\varepsilon_2/(\varepsilon_1+\varepsilon_2)} \qquad (x \in E). \tag{14.34}$$

Hence the estimation of the norm of elements of the form $\boldsymbol{B}A^{-\tau-\varepsilon_1}x$ in E_0 is reduced to the estimation of the norms of elements of the form $A^{-\varepsilon_1-\varepsilon_2}x$.

LEMMA 14.2: *Let A be a positive-type operator in a space E and let k be an integer satisfying the inequality* $0 < \varepsilon < k$. *Let* $0 \leqq \mu < \varepsilon < \nu \leqq k$. *Then*

$$\|A^{-\varepsilon}x\|_E \leqq \frac{c(\nu - \mu)}{(\nu - \varepsilon)(\varepsilon - \mu)} [\varphi_{k,k-\mu}(x)]^{(\nu-\varepsilon)/(\nu-\mu)} [\varphi_{k,k-\nu}(x)]^{(\varepsilon-\mu)/(\nu-\mu)}, \tag{14.35}$$

where

$$\varphi_{k,\delta}(x) = \sup_{0<t<\infty} \|t^\delta(t\boldsymbol{I} + A)^{-k}x\|_E.$$

The proof of this lemma is left to the reader.

LEMMA 14.3: *Let a Banach space E be imbedded continuously in a space* $\tilde{E}$, *and let A be a positive-type operator in the space* $\tilde{E}$ *with* $D(A^k) \subset E$. *Let the following inequality hold*:

$$\|x\|_E \leqq \tilde{c}\,\|A^k x\|_{\tilde{E}}^{\delta}\,\|x\|_{\tilde{E}}^{1-\delta} \qquad (x \in D(A^k)). \tag{14.36}$$

Then

$$\varphi_{k,k(1-\delta)}(x) \leqq l\tilde{c}\,\|x\|_{\tilde{E}}, \tag{14.37}$$

where l is some constant depending only on the operator A and the number k.

The assertion of the lemma results from the string of inequalities

$$\|(tI + A)^{-k}x\|_{\tilde{E}} \leqq \tilde{c}\,\|A^k(tI + A)^{-k}x\|_{\tilde{E}}^{\delta}\,\|(tI + A)^{-k}x\|_{\tilde{E}}^{1-\delta}$$

$$\leqq \tilde{c}\,\|A^k(tI + A)^{-k}\|_{\tilde{E}\to\tilde{E}}^{\delta}\,\|(tI + A)^{-k}\|_{\tilde{E}\to\tilde{E}}^{1-\delta}\,\|x\|_{\tilde{E}}$$

$$\leqq l\tilde{c}t^{k(\delta-1)}\,\|x\|_{\tilde{E}}.$$

THEOREM 14.7: *Let* E_0, E, E_1, E_2 *be Banach spaces, and let the space E be imbedded continuously in each of the spaces* E_1, E_2. *Let an operator A be of positive-type in each of spaces* E_1, E_2 *and let* D_i $(i = 1, 2)$ *the domain of A as an operator in* E_i, *be contained in E, and suppose the following inequality holds*:

$$\|x\|_E \leqq c_1\,\|A^k x\|_{E_i}^{\delta_i}\,\|x\|_{E_i}^{1-\delta_i} \qquad (x \in D(A^k),\ 0 \leqq \delta_1 < \delta_2 \leqq 1). \tag{14.38}$$

Finally, let B be a closed linear operator acting from E to E_0 *and satisfying the inequality* (14.30). *Then for any numbers* $\varepsilon_1, \varepsilon_2$ *satisfying the inequality* $k\delta_1 < \varepsilon_1 + \varepsilon_2 < k\delta_2$, *the inequality*

$$\|BA^{-\tau-\varepsilon_1}x\|_{E_0} \leqq c_0\,\|x\|_{E_1}^{[\varepsilon_1/(\varepsilon_1+\varepsilon_2)][(k\delta_2-\varepsilon_1-\varepsilon_2)/(k(\delta_2-\delta_1))]}$$

$$\times\ \|x\|_{E_2}^{[\varepsilon_1/(\varepsilon_1+\varepsilon_2)]\{(\varepsilon_1+\varepsilon_2-k\delta_1)/[k(\delta_2-\delta_1)]\}}\,\|x\|_E^{\varepsilon_2/(\varepsilon_1+\varepsilon_2)} \tag{14.39}$$

is satisfied.

This assertion follows in an obvious way from inequality (14.34) and Lemmas 14.2 and 14.3.

The theorems in sections 14.6 and 14.7 were originally obtained by P. E. Sobolevskii [14].

14.8 *Comparison of fractional powers of two operators*

Let A and B be positive-type operators acting in a Banach space E.

THEOREM 14.8: *Let an operator A^{τ_1} be subordinate to an operator B^{τ_2}, where τ_1, τ_2 are positive numbers. Then for $0 \leqq \varepsilon_1 < \varepsilon_2 \leqq 1$ the operator $A^{\varepsilon_1\tau_1}$ is subordinate to the operator $B^{\varepsilon_2\tau_2}$.*

PROOF: The moment inequality for the operator A and the subordination of the operator A^{τ_1} to the operator B^{τ_2} imply that

$$\|A^{\varepsilon_1\tau_1}x\| \leqq k\|A^{\tau_1}x\|^{\varepsilon_1}\|x\|^{1-\varepsilon_1} \leqq k_1\|B^{\tau_2}x\|^{\varepsilon_1}\|x\|^{1-\varepsilon_1}. \tag{14.40}$$

Hence by Theorem 14.4 the operator $A^{\varepsilon_1\tau_1}$ is subordinate to all the operators B^{τ} for $\tau > \varepsilon_1\tau_2$. In particular, $A^{\varepsilon_1\tau_1}$ is subordinate to the operator $B^{\varepsilon_2\tau_2}$. The theorem has been proved.

Theorem 14.8 is a generalization of Theorem 12.5 to the case of operators in Banach spaces. Note that in the case of positive definite selfadjoint operators acting in a Hilbert space, the operators $A^{\varepsilon_1\tau_1}$ are subordinate to the operators $B^{\varepsilon_2\tau_2}$ with $\varepsilon_1 = \varepsilon_2$. In the general case this is not true. It would be interesting to find other broad classes of operators for which the subordination of an operator A^{τ_1} to an operator B^{τ_2} implies the subordination of the operator $A^{\varepsilon\tau_1}$ to the operator $B^{\varepsilon\tau_2}$ $(0 < \varepsilon < 1)$. In this direction certain important results in the theory of dissipative operators were obtained by T. Kato.

Previously we used the notion of an operator adjoint to a bounded operator T acting from one Banach space to another. In a similar way one can introduce the notion of an operator adjoint to an unbounded operator. Here we need only consider the case in which both the domain $D(T)$ and the range $R(T)$ of an operator T belong to the same Banach space E. Let us give the corresponding definitions.

Let $D(T)$ be dense in E. For each fixed $y \in E^*$ the formula $m(x) = (Tx, y)$ yields an additive homogeneous functional defined on $D(T)$. Suppose that the following inequality is satisfied:

$$|m(x)| = |(Tx, y)| \leqq k\|x\| \qquad (x \in D(T)).$$

Then the functional $m(x)$ admits a continuous extension to all of E. Using the same notation for this extension, it follows that

$$(\boldsymbol{T}x, y) = (x, m) \qquad (x \in D(\boldsymbol{T})). \tag{14.41}$$

In this case y is said to belong to the domain $D(\boldsymbol{T}^*) \subset E^*$ of the operator $\boldsymbol{T}^*$ adjoint to $\boldsymbol{T}$, and $\boldsymbol{T}^*$ is defined by the relation $\boldsymbol{T}^*y = m$. Relation (14.41) can be written in the form

$$(\boldsymbol{T}x, y) = (x, \boldsymbol{T}^*y) \qquad (x \in D(\boldsymbol{T}),\ y \in D(\boldsymbol{T}^*)). \tag{14.42}$$

This relation defines the operator $\boldsymbol{T}^*$ unambiguously because the set $D(\boldsymbol{T})$ is dense in E.

THEOREM 14.9: *Let a positive-type operator* $\boldsymbol{A}$ *act in* E, *and a positive-type operator* $\boldsymbol{B}$ *act in* E^*. *Suppose a closed operator* $\boldsymbol{C}$ *satisfies the conditions*

$$\|\boldsymbol{C}x\| \leqq M_1\|\boldsymbol{A}^{\gamma_1}x\| \qquad (x \in D(\boldsymbol{A}^{\gamma_1}) \subset D(\boldsymbol{C}) \subset E) \tag{14.43}$$

and

$$\|\boldsymbol{C}^*y\| \leqq M_2\|\boldsymbol{B}^{\gamma_2}y\| \qquad (y \in D(\boldsymbol{B}^{\gamma_2}) \subset D(\boldsymbol{C}^*) \subset E^*), \tag{14.44}$$

where the numbers γ_1 *and* γ_2 *are positive. Then*

$$|(\boldsymbol{C}x, y)| \leqq k(\varepsilon_1, \varepsilon_2)\|\boldsymbol{A}^{\varepsilon_1\gamma_1}x\|_E \cdot \|\boldsymbol{B}^{\varepsilon_2\gamma_2}y\|_{E^*}$$
$$(x \in D(\boldsymbol{A}^{\gamma_1}),\ y \in D(\boldsymbol{B}^{\gamma_2})), \tag{14.45}$$

where ε_1 *and* ε_2 *are positive numbers for which* $0 < \varepsilon_1, \varepsilon_2 < 1$, $\varepsilon_1 + \varepsilon_2 > 1$.

PROOF: In the proof we shall need to use, in addition to fractional powers of the operators $\boldsymbol{A}$ and $\boldsymbol{B}$, fractional powers of the operator $\boldsymbol{A}^*$ acting in the space E^*. Fractional powers of the operator $\boldsymbol{A}^*$ can be defined in the usual way[1] described in 14.2°, because the fact that $\boldsymbol{A}^*$ is of positive-type

[1] We do not bother to discuss the supplementary difficulties which arise in the case where the domain of this operator is not dense in E^*.

follows from the fact that A is of positive-type. We leave it to the reader to check the validity of the relation

$$[A^\alpha]^* = [A^*]^\alpha.$$

Theorem 14.2 implies the inequality

$$\|[A^*]^{-\varepsilon_1\gamma_1}y\|_{E^*} \leqq k_1\|[A^*]^{-\gamma_1}y\|_{E^*}^{\varepsilon_1}\|y\|_{E^*}^{1-\varepsilon_1} \qquad (y \in E^*).$$

Let $y = \boldsymbol{C}^*z$ $(z \in D(\boldsymbol{C}^*) \subset E^*)$. Then the above inequality means that

$$\|[A^*]^{-\varepsilon_1\gamma_1}\boldsymbol{C}^*z\|_{E^*} \leqq k_1\|[A^*]^{-\gamma_1}\boldsymbol{C}^*z\|_{E^*}^{\varepsilon_1}\|\boldsymbol{C}^*z\|_{E^*}^{1-\varepsilon_1},$$

and since the operator $[A^*]^{-\gamma_1}\boldsymbol{C}^*$ is bounded simultaneously with $\boldsymbol{C}A^{-\gamma_1}$, condition (14.43) implies

$$\|[A^*]^{-\varepsilon_1\gamma_1}\boldsymbol{C}^*z\|_{E^*} \leqq k_1M_1\|\boldsymbol{C}^*z\|_{E^*}^{1-\varepsilon_1}\cdot\|z\|_{E^*}^{\varepsilon_1} \qquad (z \in D(\boldsymbol{C}^*)).$$

Then (14.44) implies the further inequality

$$\|[A^*]^{-\varepsilon_1\gamma_1}\boldsymbol{C}^*z\|_{E^*} \leqq k_1M_1M_2\|\boldsymbol{B}^{\gamma_2}z\|_{E^*}^{1-\varepsilon_1}\cdot\|z\|_{E^*}^{\varepsilon_1} \qquad (z \in D(\boldsymbol{B}^{\gamma_2})).$$

This inequality and Theorem 14.4 imply that the operator $[A^*]^{-\varepsilon_1\gamma_1}\boldsymbol{C}^*$ is subordinate to all the operators $\boldsymbol{B}^\gamma$, where $\gamma > (1-\varepsilon_1)\gamma_2$. In particular the operator $[A^*]^{-\varepsilon_1\gamma_1}\boldsymbol{C}^*$ is subordinate to the operator $\boldsymbol{B}^{\varepsilon_2\gamma_2}$ i.e.

$$\|[A^*]^{-\varepsilon_1\gamma_1}\cdot\boldsymbol{C}^*z\|_{E^*} \leqq k_2\|\boldsymbol{B}^{\varepsilon_2\gamma_2}z\|_{E^*} \qquad (z \in D(\boldsymbol{B}^{\varepsilon_2\gamma_2})).$$

Now inequality (14.45) is almost obvious:

$$|(\boldsymbol{C}x, y)| = |(x, \boldsymbol{C}^*y)| = |(A^{\varepsilon_1\gamma_1}x, [A^{-\varepsilon_1\gamma_1}]^*\boldsymbol{C}^*y)|$$

$$\leqq \|A^{\varepsilon_1\gamma_1}x\|_E\cdot\|[A^*]^{-\varepsilon_1\gamma_1}\boldsymbol{C}^*y\|_{E^*} \leqq k_2\|A^{\varepsilon_1\gamma_1}x\|_E\|\boldsymbol{B}^{\varepsilon_2\gamma_2}y\|_{E^*}$$

$$(x \in D(A^{\gamma_1}),\ y \in D(\boldsymbol{B}^{\gamma_2})).$$

The theorem has been proved.

This theorem is a generalization of Theorem 12.6.

14.9 *Fractional powers of positive-type generators*

Let $\boldsymbol{T}(t)$ $(0 \leqq t < \infty)$ be a strongly continuous semi-group satisfying an exponential decay condition

$$\|\boldsymbol{T}(t)\| \leqq ce^{-\sigma_0 t} \qquad (0 \leqq t < \infty), \tag{14.46}$$

where $\sigma_0 > 0$. Let A be the generator of this semi-group. Theorem 13.1 implies that

$$\|(t\boldsymbol{I} + A)^{-1}\| \leqq \frac{c}{\sigma_0 + t} \qquad (0 \leqq t < \infty). \tag{14.47}$$

This inequality implies that the operator A is of positive type. Hence its fractional powers $A^{-\tau}$ can be defined. It is natural to raise the problem of representating these fractional powers in terms of the semi-group $\boldsymbol{T}(t)$.

THEOREM 14.10: *Let an operator A be the generator of a strongly continuous semi-group satisfying the condition* (14.46). *Then the negative fractional powers $A^{-\tau}$ admit the representation*

$$A^{-\tau} = \frac{1}{\Gamma(\tau)} \int_0^\infty s^{\tau-1} \boldsymbol{T}(s)\, \mathrm{d}s. \tag{14.48}$$

PROOF: We will establish the following more general relation:

$$A^{-z} = \frac{1}{\Gamma(z)} \int_0^\infty s^{z-1} \boldsymbol{T}(s)\, \mathrm{d}s, \tag{14.49}$$

valid for all z in the right half plane.

It follows from (14.46) that the right side of formula (14.49) is analytic for z in the right halfplane; the operator-function A^{-z} is also analytic in the right half plane. Hence it suffices to prove relation (14.49) for real z in the interval $(0, 1)$. In other words, it suffices to prove that the relation (14.48) holds for $\tau \in (0, 1)$.

To prove this relation we use representation (14.2) for the fractional powers $A^{-\tau}$, in which the operator $(tI + A)^{-1}$ is replaced by the right side of relation (13.35). Thus we obtain the formula

$$A^{-\tau} = \frac{\sin \pi\tau}{\pi} \int_0^\infty t^{-\tau} \left[\int_0^\infty e^{-ts} T(s) \mathrm{d}s \right] \mathrm{d}t$$

or, after a change in the order of integration,

$$A^{-\tau} = \int_0^\infty \left[\frac{\sin \pi\tau}{\pi} \int_0^\infty t^{-\tau} e^{-ts} \mathrm{d}t \right] T(s) \mathrm{d}s.$$

Calculating the inner integral we obtain formula (14.48). The theorem has been proved.

An operator A is called of *strongly positive type* if it satisfies the following inequality which is stronger than condition (14.1):

$$\|(\lambda I + A)^{-1}\| \leqq \frac{c}{1 + |\lambda|} \qquad (\mathrm{Re}\, \lambda \geqq 0). \tag{14.50}$$

By Theorem 13.2 a strongly positive-type operator is the generator of an analytic semi-group $T(t)$. This semi-group decreases exponentially. Hence the fractional powers of the operator A admit the representation (14.48).

We need one more relation between fractional powers and the semi-group $T(t)$ which generalizes the estimate (13.64).

THEOREM 14.11: *Let A be a strongly positive-type operator. Let $\tau > 0$. Then the following inequality is valid:*

$$\|A^\tau T(t)\| \leqq \frac{c(\tau)}{t^\tau} \qquad (0 < t < \infty), \tag{14.51}$$

where $T(t)$ is the semi-group with generator $-A$.

PROOF: Let τ be fixed and select an integer n greater than τ. Theorem 14.2 implies that

$$\|A^\tau T(t)\| \leqq c(\tau, n)\,\|A^n T(t)\|^{\tau/n} \cdot \|T(t)\|^{1-\tau/n}$$

so that by the inequality (13.64)

$$\|A^\tau T(t)\| \leqq \frac{c(\tau)}{t^\tau}\,\|T(t)\|^{1-\tau/n} \qquad (0 < t < \infty).$$

It remains only to use the uniform boundedness of the semi-group $T(t)$. The theorem has been proved.

14.10 *Compactness of fractional powers*

THEOREM 14.12: *Let an operator A be of positive type and suppose the operator A^{-1} is compact. Then the operators A^{-z}* ($\operatorname{Re} z > 0$) *are compact.*

PROOF: Let $n > \operatorname{Re} z$. Use the representation (14.6) for the operator A^{-z}:

$$A^{-z} = \frac{\sin \pi z}{z}\,\frac{n!}{(1-z)\dots(n-z)}\int_0^\infty t^{z-n}(tI + A)^{-n-1}\,\mathrm{d}t,$$

where the integral converges in operator norm. This representation implies that it suffices to establish compactness of the operators $(tI + A)^{-1}$. It remains only to note that

$$(tI + A)^{-1} = A^{-1}(tA^{-1} + I)^{-1}$$

and to use the compactness of the operator A^{-1}. The theorem has been proved.

14.11 *Supplementary remarks*

Let us present, without proof, some theorems of E. Hille, A. Balakrishnan and T. Kato on fractional powers of positive-type operators.

Note first that an operator A being of positive type implies the same property for all operators A^τ, $0 < \tau < 1$. Similarly an operator A being of strongly positive-type implies that the operators A^τ are of strongly positive type, for the same set of τ values. The most interesting and important fact in this direction is that an operator A being of positive-type implies that the operators A^τ are of strongly positive-type, $0 < \tau \leqq \frac{1}{2}$.

These results lead quite simply to the natural relation

$$(A^\tau)^z = A^{\tau z}, \tag{14.52}$$

which holds for instance for $0 < \tau < 1$ and $0 < \mathrm{Re}(z) < 1$.

Finally note the formula

$$(\lambda \boldsymbol{I} + A^\tau)^{-1} = \frac{\sin \pi\tau}{\pi} \int_0^\infty \frac{\mu^\tau(\mu \boldsymbol{I} + A)^{-1}}{\lambda^2 + 2\lambda\mu^\tau \cos \pi\tau + \mu^{2\tau}} \, \mathrm{d}\mu \qquad (0 < \tau < 1), \tag{14.53}$$

by means of which the resolvents of fractional powers of an operator A are expressed in terms of the resolvent of the operator A itself.

§ 15 Moment inequalities and *L*-characteristics of fractional powers[1]

15.1 *Lorentz spaces*

In this paragraph we investigate operators in L_α spaces by utilizing the Lorentz spaces Λ_α which are 'close to' the L_α spaces.

$\lambda(x; h)$ denotes, as usual, the measure of the set of points $s \in \Omega$ such that $|x(s)| \geqq h$. Λ_α $(0 < \alpha \leqq 1)$ denotes the totality of functions $x(s)$ for which the following norm is finite:

$$\|x(s)\|_{\Lambda_\alpha} = \int_0^\infty \lambda^\alpha(x; h)\,\mathrm{d}h. \tag{15.1}$$

[1] This paragraph is based on the paper of P. P. Zabreiko, M. A. Krasnoselskii, and E. I. Pustylnik [1].

It can be easily verified that Λ_α is a normed space. In many cases it will be convenient to denote by Λ_0 the space L_0 of essentially bounded functions on Ω.

A simple calculation shows that

$$\|\kappa(s; D)\|_{\Lambda_\alpha} = (\text{mes}\,(D))^\alpha \tag{15.2}$$

for characteristic functions $\kappa(s; D)$ of sets $D \subset \Omega$. For each function $x(s) \in \Lambda_\alpha$ the norm $\|x\|_{\Lambda_\alpha}$ coincides with the norm of its absolute value $|x(s)|$. Hence to calculate the norm of a step function $x(s)$ it is first necessary to go over to its absolute value $|x(s)|$, which can be written in the form

$$|x(s)| = \sum_{i=1}^{n} c_i \cdot \kappa(s; D_i) \tag{15.3}$$

where the numbers c_i decrease: $c_1 > c_2 > \ldots > c_n > c_{n+1} = 0$ and the sets D_i are mutually disjoint. Then

$$\|x(s)\|_{\Lambda_\alpha} = \sum_{i=1}^{n} (c_i - c_{i+1}) \cdot [\text{mes}(\bigcup_{k=1}^{i} D_k)]^\alpha. \tag{15.4}$$

Note that the set of step functions is dense in each space Λ_α $(0 < \alpha \leqq 1)$.

Denote by T_α the set of functions $x(s)$ of the form

$$x(s) = (\text{mes}\, D_1 + \text{mes}\, D_2)^{-\alpha} \times \{\kappa(s; D_1) - \kappa(s; D_2)\}, \tag{15.5}$$

where D_1 and D_2 are disjoint subsets of Ω. Equation (15.2) implies that the norm of each function of the form (15.5) is equal to 1.

LEMMA 15.1: *The unit ball $\|x\|_{\Lambda_\alpha} \leqq 1$ is the closed convex hull of the set T_α.*

PROOF: It is necessary to show that each function $z(s)$ in the unit ball of the space Λ_α belongs to the closure of the convex hull of the set T_α. It will suffice to consider functions $z(s)$ such that $\|z\|_{\Lambda_\alpha} = 1$.

For each such function $z(s)$ it is possible to construct a sequence of step functions $x_n(s)$ such that $\|x_n\|_{\Lambda_\alpha} = 1$ and

$$\lim_{n\to\infty} \|x_n(s) - z(s)\|_{\Lambda_\alpha} = 0.$$

Hence to prove the lemma it suffices to show that each step function $x(s)$ whose norm is equal to one, belongs to the convex hull of the set T_α.

Let $x(s)$ be a step function and $\|x\|_{\Lambda_\alpha} = 1$. The function $x(s)$ can be written in the form

$$x(s) = \sum_{i=1}^{n} c_i[\kappa(s;\, D_i^+) - \kappa(s;\, D_i^-)],$$

where D_i^+ is the set on which the function $x(s)$ takes the value c_i and D_i^- is the set on which the function $x(s)$ takes the value $-c_i$. Hence

$$x(s) = \sum_{i=1}^{n} \alpha_i x_i(s),$$

where

$$\alpha_i = (c_i - c_{i+1})[\operatorname{mes}(\bigcup_{k=1}^{i} D_k^+) + \operatorname{mes}(\bigcup_{k=1}^{i} D_k^-)]^\alpha,$$

and the functions

$$x_i(s) = [\operatorname{mes}(\bigcup_{k=1}^{i} D_k^+) + \operatorname{mes}(\bigcup_{k=1}^{i} D_k^-)]^{-\alpha}\{\kappa(s;\, \bigcup_{k=1}^{i} D_k^+) - \kappa(s;\, \bigcup_{k=1}^{i} D_k^-)\}$$

belong to T_α. Formula (15.4) implies that

$$\sum_{i=1}^{n} \alpha_i = \sum_{i=1}^{n} (c_i - c_{i+1})\cdot[\operatorname{mes}(\bigcup_{k=1}^{i} D_k^+) + \operatorname{mes}(\bigcup_{k=1}^{i} D_k^-)]^\alpha = \|x\|_{\Lambda_\alpha} = 1.$$

The lemma has been proved.

Note that the closure of the convex hull of the set T_α in the norm of the space Λ_α coincides with the closure of the convex hull in measure.

The Lorentz spaces Λ_α are closely connected with the Marcinkiewicz spaces M_α introduced in 2.7° and 8.3°. Recall that M_α $(0 < \alpha < 1)$ is the space of functions $x(s)$ for which the following norm is finite:

$$\|x\|_{M_\alpha} = \sup_{D\subset\Omega} \{(\operatorname{mes} D)^{\alpha-1} \int_D |x(s)|\, ds\}. \tag{15.6}$$

The norms in the spaces Λ_α, L_α, M_α are connected through the inequalities

$$\|x\|_{M_\alpha} \leqq \|x\|_{L_\alpha} \leqq \|x\|_{\Lambda_\alpha} \quad (x \in \Lambda_\alpha; 0 < \alpha < 1). \tag{15.7}$$

The first of the inequalities results from Hölder's inequality and the second from Lemma 15.1. It is easy to see that for characteristic functions $\kappa(s; D)$ one has the relations:

$$\|\kappa(s; D)\|_{M_\alpha} = \|\kappa(s; D)\|_{L_\alpha} = \|\kappa(s; D)\|_{\Lambda_\alpha}.$$

Now let us show that for any $\varepsilon > 0$ the space $\Lambda_{\alpha+\varepsilon}$ contains the space M_α with

$$\|x\|_{\Lambda_{\alpha+\varepsilon}} \leqq 2\left(\frac{\alpha}{\varepsilon}\right)^{\alpha/(\alpha+\varepsilon)} \times (\text{mes } \Omega)^{\varepsilon} \|x\|_{M_\alpha}. \tag{15.8}$$

For the proof write $\|x\|_{\Lambda_{\alpha+\varepsilon}}$ in the form

$$\|x\|_{\Lambda_{\alpha+\varepsilon}} = \boldsymbol{J}_1 + \boldsymbol{J}_2,$$

where

$$\boldsymbol{J}_1 = \int_0^c \lambda^{\alpha+\varepsilon}(x; h)\,\mathrm{d}h, \quad \boldsymbol{J}_2 = \int_c^\infty \lambda^{\alpha+\varepsilon}(x; h)\,\mathrm{d}h.$$

It is clear that

$$\boldsymbol{J}_1 \leqq c(\text{mes } \Omega)^{\alpha+\varepsilon}.$$

Equations (8.17) and (8.12) now imply that

$$h \cdot \lambda^\alpha(x; h) \leqq \|x\|_{M_\alpha},$$

hence

$$\boldsymbol{J}_2 \leqq \frac{\alpha}{\varepsilon} \|x\|_{M_\alpha}^{(\alpha+\varepsilon)/\alpha} c^{-\varepsilon/\alpha}.$$

Thus

$$\|x\|_{\Lambda_{\alpha+\varepsilon}} \leqq c(\text{mes }\Omega)^{\alpha+\varepsilon} + \frac{\alpha}{\varepsilon} \|x\|_{M_\alpha}^{(\alpha+\varepsilon)/\alpha} c^{-\varepsilon/\alpha},$$

whence (15.8) follows, provided that we put

$$c = \left(\frac{\alpha}{\varepsilon}\right)^{\alpha/(\alpha+\varepsilon)} (\text{mes }\Omega)^{-\alpha} \|x\|_{M_\alpha}.$$

The general form of a continuous linear functional $f(x)$ on a space Λ_α $(0 < \alpha \leqq 1)$ is given by the formula

$$f(x) = \int_\Omega x(s)y(s)\,\mathrm{d}s \qquad (x \in \Lambda_\alpha), \tag{15.9}$$

where $y(s) \in M_{1-\alpha}$. Thus the Marcinkiewicz space $M_{1-\alpha}$ can be considered to be the dual to the Lorentz space Λ_α. Note further that a Lorentz space is not reflexive—this fact follows, for instance, from the non-separability of Marcinkiewicz spaces.

A number of facts on the theory of Lorentz and Marcinkiewicz spaces is presented, for instance, in the papers of G. G. Lorentz [1], S. G. Krein and E. M. Semenov [1], E. M. Semenov [1].

15.2 *Linear operators*

Let an additive homogeneous operator $\boldsymbol{B}$ be defined on the characteristic functions $\kappa(s; D)$ of the measurable sets $D \subset \Omega$ and take its values in the set of summable functions on a set Ω^*. Then each characteristic function $\kappa(t; D^*)$ of a measurable set $D^* \subset \Omega^*$ defines a set function

$$\mu(D) = (\boldsymbol{B}\kappa(t; D), \kappa(t; D^*)) = \int_{D^*} \boldsymbol{B}\kappa(t; D)\,\mathrm{d}t \qquad (D \subset \Omega). \tag{15.10}$$

Assume that the set functions (15.10) are absolutely continuous with respect to Lebesgue measure on Ω. Then the Radon-Nikodym theorem (see N. Dunford and J. T. Schwartz [1]) implies that the measures (15.10) are uniquely representable in the form

$$\mu(D) = \int_D y^*(s)\,\mathrm{d}s, \tag{15.11}$$

where the $y^*(s)$ denote summable functions on Ω.

The relation (15.11) permits one to study the operator

$$\boldsymbol{B}^*y = y^*, \tag{15.12}$$

which is defined on the characteristic functions $y = \kappa(t; D^*)$ and takes values which belong to $L_1(\Omega)$. The operator $\boldsymbol{B}^*$ can be extended as an additive homogeneous operator (which will be denoted by the same symbol $\boldsymbol{B}^*$) defined on all step functions. This extended operator is completely determined by the relation

$$(\boldsymbol{B}\kappa(t; D), y) = (\kappa(t; D), \boldsymbol{B}^*y) \quad (D \subset \Omega). \tag{15.13}$$

Moreover this relation implies that

$$(\boldsymbol{B}x, y) = (x, \boldsymbol{B}^*y), \tag{15.14}$$

where $x(s)$ and $y(t)$ are arbitrary step functions.

The operator $\boldsymbol{B}^*$ will be called the *adjoint* to the operator $\boldsymbol{B}$. We emphasize that this notion of adjoint operator is closely related to the usual notion of adjoint operator in Banach spaces but does not coincide with it.

Assume that $\boldsymbol{B}$ is a continuous operator acting from L_α to L_β, where $\alpha \in (0, 1]$, $\beta \in (0, 1]$. Tenh the set functions (15.10) are obviously absolutely continuous with respect to Lebesgue measure; hence the operator (15.12) is defined—it coincides with the usual adjoint operator. For operators $\boldsymbol{B}$ acting from L_0 to L_β the absolute continuity of the measures (15.10) is an additional assumption.

Similar considerations show that the operator $\boldsymbol{B}^*$ acts from $\Lambda_{1-\beta}$ to $L_{1-\alpha}$ $(0 < \alpha \leqq 1,\ 0 \leqq \beta < 1)$ and is continuous when $\boldsymbol{B}$ acts from L_α to M_β and is continuous; likewise, $\boldsymbol{B}^*$ acts from $L_{1-\beta}$ to $M_{1-\alpha}$ $(0 < \alpha \leqq 1,$ $0 \leqq \beta \leqq 1)$ and is continuous when $\boldsymbol{B}$ acts from Λ_α to L_β and is continuous; finally, $\boldsymbol{B}^*$ acts from $\Lambda_{1-\beta}$ to $M_{1-\alpha}$ $(0 < \alpha \leqq 1,\ 0 \leqq \beta < 1)$ and is continuous when $\boldsymbol{B}$ acts from Λ_α to M_β and is continuous.

If the operator $\boldsymbol{B}$ acts from L_α to M_β $(0 < \alpha \leqq 1,\ 0 \leqq \beta < 1)$ and is continuous, then for each measurable set $D^* \subset \Omega^*$ the following inequality holds:

$$\int_{D^*} |\boldsymbol{B}x(t)|\,dt \leqq c\,\|x\|_{L_\alpha}\cdot(\text{mes } D^*)^{1-\beta} \tag{15.15}$$

where c is some constant. Such operators have already been studied in 2.7° and 8.3°; they were said to satisfy the Marcinkiewicz condition $LM(\alpha, \beta)$.

Let us take up the study of operators acting from Λ_α to L_β and from Λ_α to M_β. We need a general auxiliary assertion.

LEMMA 15.2: *Let a linear operator* $\boldsymbol{B}$ *be defined on step functions and satisfy the condition*

$$\|\boldsymbol{B}\kappa(s; D)\|_E \leqq k(\text{mes } D)^\alpha \qquad (D \subset \Omega), \tag{15.16}$$

where $0 < \alpha \leqq 1$ *and* E *is some Banach space. Then* $\boldsymbol{B}$ *can be extended to a continuous operator, acting from* Λ_α *to* E.

PROOF: Let $x(s)$ be a function of the form (15.5). Then by (15.16)

$$\|\boldsymbol{B}x(s)\|_E \leqq \frac{1}{(\text{mes } D_1 + \text{mes } D_2)^\alpha}\{\|\boldsymbol{B}\kappa(s; D_1)\|_E + \|\boldsymbol{B}\kappa(s; D_2)\|_E\}$$

$$\leqq k\,\frac{(\text{mes } D_1)^\alpha + (\text{mes } D_2)^\alpha}{(\text{mes } D_1 + \text{mes } D_2)^\alpha} \leqq 2^{1-\alpha}k.$$

Hence it follows that $\boldsymbol{B}$ is bounded on the convex hull of the set T_α, which is dense in the unit ball of the space Λ_α by Lemma 15.1. The lemma has been proved[1].

When applied to the case $E = L_\beta$, Lemma 15.2 means that $\boldsymbol{B}$ admits an extension to a continuous operator acting from Λ_α $(0 < \alpha \leqq 1)$ to L_β $(0 \leqq \beta \leqq 1)$ if it satisfies the condition

$$\|\boldsymbol{B}\kappa(s; D)\|_{L_\beta} \leqq c_1(\text{mes } D)^\alpha \qquad (D \subset \Omega).$$

---

[1] S. G. Krein and E. M. Semenov [1].

If an operator $\boldsymbol{B}$ satisfies this inequality it will be said to satisfy the condition $\Lambda L(\alpha, \beta)$.

In the case $E = M_\beta$, Lemma 15.2 means that $\boldsymbol{B}$ admits an extension to a continuous operator acting from Λ_α $(0 < \alpha \leqq 1)$ to M_β $(0 \leqq \beta \leqq 1)$, if it satisfies

$$\int_{D^*} |\boldsymbol{B}\kappa(s; D)|\,\mathrm{d}t \leqq c_2(\text{mes } D)^\alpha(\text{mes } D^*)^{1-\beta} \qquad (D \subset \Omega,\ D^* \subset \Omega^*).$$

Such operators have already appeared in 8.3°—they are the operators which satisfy the condition $\Lambda M(\alpha, \beta)$.

LEMMA 15.3: *Let an operator $\boldsymbol{B}$ satisfy the condition $LM(\alpha_0, \beta_0)$. Then the operator $\boldsymbol{B}$ acts from $L_{\alpha 0}$ to any L_β where $\beta > \beta_0$, and is continuous.*

LEMMA 15.4: *Let an operator $\boldsymbol{B}$ satisfy the condition $\Lambda L(\alpha_0, \beta_0)$. Then the operator $\boldsymbol{B}$ acts from any L_α where $\alpha < \alpha_0$, to L_{β_0}, and is continuous.*

LEMMA 15.5: *Let an operator $\boldsymbol{B}$ satisfy the condition $\Lambda M(\alpha_0, \beta_0)$. Then the operator $\boldsymbol{B}$ acts from any L_α where $\alpha < \alpha_0$ to any L_β where $\beta > \beta_0$, and is continuous.*

The assertion in Lemma 15.3 follows from the inequalities (15.7) and (15.8). Lemma 15.4 results from Lemma 15.3 on going over to the adjoint operator. The proof of Lemma 15.5 is left to the reader.

15.3 *Interpolation theorems*

In the previous chapters a number of theorems on the interpolation properties of continuous linear operators were established. We need one more analogous result.

THEOREM 15.1: *Let a linear operator $\boldsymbol{B}$ satisfy the conditions $\Lambda L(\alpha_0, \beta_0)$ and $\Lambda L(\alpha_1, \beta_1)$*

$$\|\boldsymbol{B}x\|_{L_{\beta_0}} \leqq k_0\|x\|_{\Lambda_{\alpha 0}} \qquad (x \in \Lambda_{\alpha 0}), \tag{15.17}$$

$$\|\boldsymbol{B}x\|_{L_{\beta_1}} \leqq k_1\|x\|_{\Lambda_{\alpha_1}} \qquad (x \in \Lambda_{\alpha_1}). \tag{15.18}$$

Let the inequalities

$$0 \leqq \beta_0 \leqq \alpha_0 \leqq 1, \ 0 \leqq \beta_1 \leqq \alpha_1 \leqq 1 \tag{15.19}$$

hold, with

$$\alpha_0 \neq \alpha_1. \tag{15.20}$$

Then for any $\tau \in (0, 1)$ *the operator* $\boldsymbol{B}$ *acts from* $L_{\alpha(\tau)}$ *into* $L_{\beta(\tau)}$, *where*

$$\alpha(\tau) = (1 - \tau)\alpha_0 + \tau\alpha_1; \ \beta(\tau) = (1 - \tau)\beta_0 + \tau\beta_1,$$

and is continuous, with

$$\|\boldsymbol{B}\|_{L_{\alpha(\tau)} \to L_{\beta(\tau)}} \leqq \frac{2}{|\alpha_0 - \alpha_1|\tau(1 - \tau)} k_0^{1-\tau} k_1^{\tau}. \tag{15.21}$$

PROOF: In the case of α_0 and α_1 both being positive, the assertion results from Theorem 2.9 by going over to the adjoint operator.

Suppose that $\alpha_1 > 0$, $\alpha_0 = 0$. Let $x(s)$ be a step function assuming only the three values: 0, 1, -1. Hölder's inequality and conditions (15.17) and (15.18) imply, in view of (15.2), that for each $\varepsilon \in (0, 1)$

$$\|\boldsymbol{B}x\|_{L_{\beta(\varepsilon)}} \leqq \|\boldsymbol{B}x\|_{L_{\beta_0}}^{1-\varepsilon} \|\boldsymbol{B}x\|_{L_{\beta_1}}^{\varepsilon}$$

$$\leqq k_0^{1-\varepsilon} \cdot k_1^{\varepsilon} \|x\|_{\Lambda_{\alpha_0}}^{1-\varepsilon} \|x\|_{\Lambda_{\alpha_1}}^{\varepsilon} = k_0^{1-\varepsilon} k_1^{\varepsilon} \cdot (\text{mes } D)^{\alpha(\varepsilon)},$$

where D is the support of the function $x(s)$ (the set of those s, at which $x(s) \neq 0$) and $\beta(\varepsilon)$, $\alpha(\varepsilon)$ are defined by

$$\beta(\varepsilon) = (1 - \varepsilon)\beta_0 + \varepsilon\beta_1, \ \alpha(\varepsilon) = (1 - \varepsilon)\alpha_0 + \varepsilon\alpha_1 = \varepsilon\alpha_1.$$

This means that the operator $\boldsymbol{B}$ satisfies the condition $\Lambda L(\alpha(\varepsilon), \beta(\varepsilon))$ with

$$\|\boldsymbol{B}x\|_{L_{\beta(\varepsilon)}} \leqq k_0^{1-\varepsilon} k_1^{\varepsilon} \|x\|_{\Lambda_{\alpha(\varepsilon)}} \quad (x \in \Lambda_{\alpha(\varepsilon)}).$$

The part of the theorem already proved implies that for each $\sigma \in (0, 1)$

the operator $\boldsymbol{B}$ acts from $L_{\alpha(\sigma;\varepsilon)}$ to $L_{\beta(\sigma;\varepsilon)}$ where

$$\alpha(\sigma;\varepsilon) = (1-\sigma)\alpha(\varepsilon) + \sigma\alpha_1;\ \beta(\sigma;\varepsilon) = (1-\sigma)\beta(\varepsilon) + \sigma\beta_1,$$

and

$$\|\boldsymbol{B}\|_{L_{\alpha(\sigma;\varepsilon)}\to L_{\beta(\sigma;\varepsilon)}} \leqq \frac{2}{|\alpha(\varepsilon)-\alpha_1|\sigma(1-\sigma)} k_0^{(1-\varepsilon)(1-\sigma)} k_1^{\varepsilon(1-\sigma)+\sigma}.$$

Let $\tau \in (0, 1)$ and $0 < \varepsilon < \tau$, and put

$$\sigma = \frac{\tau-\varepsilon}{1-\varepsilon}.$$

Then the following relations hold:

$$\alpha(\sigma;\varepsilon) = \alpha(\tau),\ \ \beta(\sigma;\varepsilon) = \beta(\tau).$$

Hence the operator $\boldsymbol{B}$ acts from $L_{\alpha(\tau)}$ to $L_{\beta(\tau)}$ and

$$\|\boldsymbol{B}\|_{L_{\alpha(\tau)}\to L_{\beta(\tau)}} \leqq \frac{2(1-\varepsilon)}{\alpha_1(\tau-\varepsilon)(1-\tau)} k_0^{1-\tau} k_1^{\tau}.$$

This implies inequality (15.21) because ε can be taken as small as desired. The theorem has been proved.

Let us now study the question of interpolation of the compactness property. Previously we proved theorems on the interpolation of this property (Theorems 3.10 and 3.11) which corresponded to the fundamental interpolation theorem of M. Riesz (Theorem 2.4). Here we give results on the interpolation of the compactness property which correspond to other interpolation theorems (Theorems 2.7, 8.2 and 15.1) connected with Lorentz and Marcinkiewicz spaces.

THEOREM 15.2: *Let $\boldsymbol{B}$ be a continuous linear operator, acting from Λ_{α_0} to L_{β_0} and from Λ_{α_1} to L_{β_1}, where*

$$0 \leqq \beta_0 \leqq \alpha_0 \leqq 1,\ 0 \leqq \beta_1 \leqq \alpha_1 \leqq 1,\ \alpha_0 \neq \alpha_1.$$

Let $\boldsymbol{B}$ *be compact as an operator from* Λ_{α_0} *to* L_{β_0}. *Then for any* $\tau \in (0, 1)$ $\boldsymbol{B}$ *is compact as an operator from* $L_{\alpha(\tau)}$ *to* $L_{\beta(\tau)}$, *where*

$$\alpha(\tau) = (1 - \tau)\alpha_0 + \tau\alpha_1, \ \beta(\tau) = (1 - \tau)\beta_0 + \tau\beta_1.$$

PROOF: Let $\boldsymbol{P}_n$ be a regular sequence of projection operators as in (1.12). Then (1.14) implies that $\|\boldsymbol{P}_n\|_{L_\beta \to L_\beta} = 1$ $(0 \leqq \beta \leqq 1)$. Hence it follows from Lemma 1.4 that the compact operators $\boldsymbol{P}_n\boldsymbol{B}$ converge to the operator $\boldsymbol{B}$ in operator norm, as operators acting from Λ_{α_0} to L_{β_0}:

$$\lim_{n\to\infty} \|\boldsymbol{P}_n\boldsymbol{B} - \boldsymbol{B}\|_{\Lambda_{\alpha_0} \to L_{\beta_0}} = 0. \tag{15.22}$$

According to Theorem 15.1 the operators $\boldsymbol{P}_n\boldsymbol{B} - \boldsymbol{B}$ which act from $L_{\alpha(\tau)}$ to $L_{\beta(\tau)}$ $(0 < \tau < 1)$ are continuous while (15.21) and (15.22) imply that

$$\lim_{n\to\infty} \|\boldsymbol{P}_n\boldsymbol{B} - \boldsymbol{B}\|_{L_{\alpha(\tau)} \to L_{\beta(\tau)}} = 0.$$

This means that the operator $\boldsymbol{B}$ can be approximated arbitrarily closely by the compact operators $\boldsymbol{P}_n\boldsymbol{B}$. The theorem has been proved.

Let us present two more assertions without proof.

THEOREM 15.3: *Let* $\boldsymbol{B}$ *be a continuous linear operator acting from* L_{α_0} *to* M_{β_0} *and from* L_{α_1} *to* M_{β_1}, *where*

$$0 < \alpha_0 < 1, \ 0 \leqq \beta_0 \leqq \alpha_0, \ 0 \leqq \beta_1 \leqq \alpha_1 \leqq 1$$

and

$$\beta_0 \neq \beta_1.$$

Let $\boldsymbol{B}$ *be compact as an operator from* L_{α_0} *to* M_{β_0}. *Then for each* $\tau \in (0, 1)$ $\boldsymbol{B}$ *is compact as an operator from* $L_{\alpha(\tau)}$ *to* $L_{\beta(\tau)}$, *where*

$$\alpha(\tau) = (1 - \tau)\alpha_0 + \tau\alpha_1, \ \beta(\tau) = (1 - \tau)\beta_0 + \tau\beta_1.$$

THEOREM 15.4: *Let **B** be a continuous linear operator acting from Λ_{α_0} to M_{β_0} and from Λ_{α_1} to M_{β_1} where*

$$0 \leqq \beta_0 < 1, \ \beta_0 \leqq \alpha_0 \leqq 1, \ 0 \leqq \beta_1 \leqq \alpha_1 \leqq 1$$

and

$$\alpha_0 \neq \alpha_1, \ \beta_0 \neq \beta_1.$$

*Let **B** be compact as an operator from Λ_{α_0} to M_{β_0} and let the values of **B** on Λ_{α_0} lie in the space $M^0_{\beta_0}$ where $M^0_{\beta_0}$ is the closure of step functions in norm of M_{β_0}. Then for each $\tau \in (0, 1)$ **B** is compact as an operator from $L_{\alpha(\tau)}$ to $L_{\beta(\tau)}$, where*

$$\alpha(\tau) = (1 - \tau)\alpha_0 + \tau\alpha_1, \ \beta(\tau) = (1 - \tau)\beta_0 + \tau\beta_1.$$

15.4 *Fundamental theorems*

In the following E_1 and E_2 denote arbitrary Banach spaces and τ denotes a number in the interval (0, 1].

THEOREM 15.5: *Suppose that a linear operator **B** satisfies the inequality*

$$\|B\kappa\|_{E_2} \leqq k \, \|A\kappa\|^{\tau}_{E_1} \|\kappa\|^{1-\tau}_{L_\gamma} \tag{15.23}$$

*on characteristic functions $\kappa(s; D)$ where $0 \leqq \gamma \leqq 1$ and **A** is a continuous operator acting from Λ_δ to E_1. Then the operator **B** acts from the Lorentz space $\Lambda_{\alpha(\tau, \gamma, \delta)}$, where*

$$\alpha(\tau, \gamma, \delta) = \tau\delta + (1 - \tau)\gamma, \tag{15.24}$$

to the space E_2 and is continuous.

PROOF: The inequality (15.23) implies that

$$\|B\kappa(s; D)\|_{E_2} \leqq k \, \|A\|^{\tau}_{\Lambda_\delta \to E_1} (\text{mes } D)^{\tau\delta + (1-\tau)\gamma}.$$

It only remains for us to apply Lemma 15.2. The theorem has been proved.

Lemma 15.4 implies that under the conditions of Theorem 15.5 $\boldsymbol{B}$ is a continuous operator acting from any space $L_{\alpha(\tau,\gamma,\delta)-\varepsilon}$ to the space E_2, where $0 < \varepsilon < \alpha(\tau, \gamma, \delta)$.

THEOREM 15.6: *Suppose that a linear operator* $\boldsymbol{B}$ *satisfies the inequality*

$$\|\boldsymbol{B}x\|_{E_2} \leqq k \|\boldsymbol{A}x\|_{E_1}^{\tau} \|x\|_{L_\gamma}^{1-\tau}, \tag{15.25}$$

on all step functions $x(s)$ *where* $0 \leqq \gamma \leqq 1$ *and* $\boldsymbol{A}$ *is a compact operator acting from* L_δ *to* E_1, *with* $0 < \delta < 1$. *Then the operator* $\boldsymbol{B}$ *is compact as an operator from* $\Lambda_{\alpha(\tau,\gamma,\delta)}$ *to the space* E_2, *where*

$$\alpha(\tau, \gamma, \delta) = \tau\delta + (1 - \tau)\gamma.$$

PROOF: Let $\boldsymbol{P}_n$ be a regular sequence of projections. Then (see 3.5°)

$$\lim_{n\to\infty} \|\boldsymbol{A} - \boldsymbol{A}\boldsymbol{P}_n\|_{L_\delta \to E_1} = 0. \tag{15.26}$$

Let $x \in T_{\alpha(\tau,\gamma,\delta)}$, where T_α is the set of functions (15.5). Then (15.25) implies that

$$\|\boldsymbol{B}(\boldsymbol{I} - \boldsymbol{P}_n)x\|_{E_2} \leqq k \|\boldsymbol{A}(\boldsymbol{I} - \boldsymbol{P}_n)\|_{L_\delta \to E_1}^{\tau} \|\boldsymbol{I} - \boldsymbol{P}_n\|_{L_\gamma \to L_\gamma}^{1-\tau},$$

because by (15.2) and (15.24)

$$\|x\|_{L_\delta}^{\tau} \|x\|_{L_\gamma}^{1-\tau} = 1.$$

Hence Lemma 15.1 implies that

$$\|\boldsymbol{B}(\boldsymbol{I} - \boldsymbol{P}_n)\|_{\Lambda_{\alpha(\tau,\gamma,\delta)} \to E_2} \leqq k \|\boldsymbol{A}(\boldsymbol{I} - \boldsymbol{P}_n)\|_{L_\beta \to E_1}^{\tau}. \tag{15.27}$$

The relations (15.26) and (15.27) mean that $\boldsymbol{B}$ can be approximated by finite dimensional operators $\boldsymbol{B}\boldsymbol{P}_n$ with arbitrary accuracy in operator norm, as operators acting from $\Lambda_{\alpha(\tau,\gamma,\delta)}$ to E_2. Consequently the operator $\boldsymbol{B}$ is compact. The theorem has been proved.

The assumption in Theorem 15.6 of the compactness of $\boldsymbol{A}$ as an operator from L_δ to E_1 can be replaced by the less restrictive assumption of compactness of A as an operator from Λ_δ to E_1, provided that the values of the operator A^* on B^* belong to the closure of the set of bounded functions in the Marcinkiewicz space $M_{1-\delta}$.

To investigate concrete operators B, it is often convenient to apply Theorems 15.5 and 15.6 in combination with interpolation theorems.

Let the conditions

$$\|B\kappa(s; D)\|_{L_{\beta_0}} \leqq k_0 \|A_0\kappa(s; D)\|_{E_0}^{\tau_0} \|\kappa(s; D)\|_{L_{\gamma_0}}^{1-\tau_0} \quad (D \subset \Omega), \tag{15.28}$$

and

$$\|B\kappa(s; D)\|_{L_{\beta_1}} \leqq k_1 \|A\kappa(s; D)\|_{E_1}^{\tau_1} \|\kappa(s; D)\|_{L_{\gamma_1}}^{1-\tau_1} \quad (D \subset \Omega) \tag{15.29}$$

hold, where the operators A_0 and A_1 act continuously from the spaces L_{δ_0} and L_{δ_1} to the spaces E_0 and E_1 respectively. Then it is possible to deduce from Theorem 15.5 that the operator B satisfies both condition $\Lambda L[\alpha(\tau_0, \gamma_0, \delta_0); \beta_0]$ and condition $\Lambda L[\alpha(\tau_1, \gamma_1, \delta_1); \beta_1]$. If the following inequalities all hold:

$$\left.\begin{aligned} &\alpha(\tau_0, \gamma_0, \delta_0) \geqq \beta_0, \quad \alpha(\tau_1, \gamma_1, \delta_1) \geqq \beta_1 \\ &\alpha(\tau_0, \gamma_0, \delta_0) \neq \alpha(\tau_1, \gamma_1, \delta_1), \end{aligned}\right\}, \tag{15.30}$$

it follows from Theorem 15.1 that the L-characteristic $L(B;\text{cont.})$ contains all interior points of the segment joining the points $\{\alpha(\tau_0, \gamma_0, \delta_0); \beta_0\}$ and $\{\alpha(\tau_1, \gamma_1, \delta_1); \beta_1\}$. If one of the relations (15.30) is not fulfilled, then Theorem 15.1 cannot be applied; by utilizing Theorem 2.4 it is possible to assert only that $L(B;\text{cont.})$ contains all the points situated above the segment in question. If $\alpha(\tau_0, \gamma_0, \delta_0) > \beta_0$, but $\alpha(\tau_1, \gamma_1, \delta_1) < \beta_1$ and $\alpha(\tau_0, \gamma_0, \delta_0) \neq \alpha(\tau_1, \gamma_1, \delta_1)$, then $L(B;\text{cont.})$ contains the portion, lying under the line $\beta = \alpha$, of the segment joining the points $\{\alpha(\tau_0, \gamma_0, \delta_0); \beta_0\}$ and $\{\alpha(\tau_1, \gamma_1, \delta_1); \beta_1\}$; for a proof it is possible, for instance, to use inequality (2.52) and Theorem 8.2.

Suppose that the following conditions, which are less restrictive than (15.28) and (15.29), are satisfied

$$\|B\kappa(s; D)\|_{M_{\beta_0}} \leqq k_0 \|A_0\kappa(s; D)\|_{E_0}^{\tau_0} \|\kappa(s; D)\|_{L_{\gamma_0}}^{1-\tau_0} \quad (D \subset \Omega) \tag{15.31}$$

and

$$\|B\kappa(s;\,D)\|_{M_1} \leqq k_1\|A_1\kappa(s;\,D)\|_{E_1}^{\tau_1}\,\|\kappa(s;\,D)\|_{L_{\gamma_1}}^{1-\tau_1} \qquad (D \subset \Omega). \tag{15.32}$$

Let the operators A_0 and A_1 act continuously from the spaces Λ_{δ_0} and Λ_{δ_1} to the spaces E_0 and E_1, respectively. Assume that the segment joining the points

$$\{\alpha(\tau_0, \gamma_0, \delta_0);\, \beta_0\} \text{ and } \{\alpha(\tau_1, \gamma_1, \delta_1);\, \beta_1\},$$

is situated under the line $\beta = \alpha$ and is not parallel to either coordinate axis. Then to investigate the operator B, which by Theorem 15.5 acts continuously from $\Lambda_{\alpha(\tau_0,\gamma_0,\delta_0)}$ to M_{β_0} and from $\Lambda_{\alpha(\tau_1,\gamma_1,\delta_1)}$ to M_{β_1} it is possible to apply interpolation Theorem 8.2. This interpolation theorem implies that all interior points of the segments in question belong to $L(B;\text{cont.})$.

Similarly, in order to construct the L-characteristic $L(B;\text{comp.})$ one can combine Theorem 15.6 with Theorems 3.10, 15.2–15.4 on the interpolation of the compactness property.

It can happen that inequalities of the type (15.23) or (15.25) are established more easily for the operator B^* adjoint to the operator B than for the operator B itself. Then using Theorems 15.5 and 15.6 it is also possible to obtain information about which spaces the operator B maps into which other spaces.

In particular, if $E_2 = L_\beta$ and $B^* = B$ then Theorem 15.5 implies that the operator B satisfies both the conditions $\Lambda L[\alpha(\tau, \gamma, \delta);\, \beta]$ and $LM[1 - \beta;\, 1 - \alpha(\tau, \gamma, \delta)]$. Let

$$\alpha(\tau, \gamma, \delta) \geqq \beta,\ \ \alpha(\tau, \gamma, \delta) \neq 1 - \beta.$$

Then it is possible to apply Theorem 8.2 which implies that the operator B is continuous as an operator from each $L_{r(\lambda)}$ to the corresponding space $L_{q(\lambda)}$, $0 < \lambda < 1$, where

$$r(\lambda) = (1 - \lambda)\alpha(\tau, \gamma, \delta) + \lambda(1 - \beta)$$

and

$$q(\lambda) = (1 - \lambda)\beta + \lambda[1 - \alpha(\tau, \gamma, \delta)].$$

15.5 *L-Characteristics of fractional powers*

In this section we study operators $\boldsymbol{B}$ for which inequalities (15.25) of the following special form are satisfied:

$$\|\boldsymbol{B}x\|_\beta \leqq k(\beta)\,\|\boldsymbol{A}x\|_\beta^\tau\,\|x\|_\beta^{1-\tau} \qquad (x \in L_\beta). \tag{15.33}$$

Suppose that the inequalities (15.33) are fulfilled for all values of β in some interval (β_0, β_1). Let the L-characteristic $L(\boldsymbol{A}$; cont.) or the L-characteristic $L(\boldsymbol{A}$; comp.) contain the curve

$$\alpha = \eta(\beta) \qquad (\beta_0 < \beta < \beta_1),$$

where $\eta(\beta)$ is a non-decreasing function (Fig. 15.1). The theorems proved in the preceding section imply that the corresponding L-characteristic $L(\boldsymbol{B}$; cont.) or $L(\boldsymbol{B}$; comp.) contains all points $\{\alpha, \beta\}$ situated to the left of the curve

$$\alpha = \eta_1(\beta) \qquad (\beta_0 < \beta < \beta_1)$$

where

$$\eta_1(\beta) = (1 - \tau)\beta + \tau\eta(\beta). \tag{15.34}$$

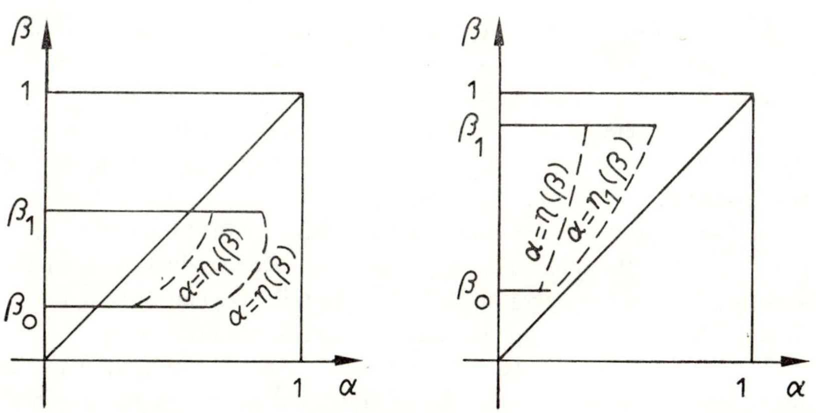

Figure 15.1

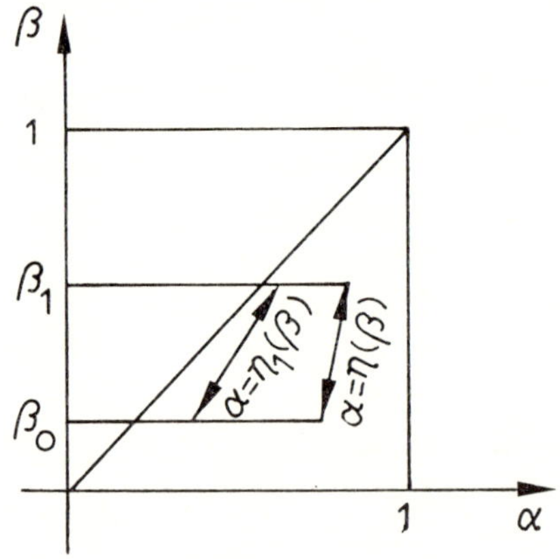

Figure 15.2

If the curve $\alpha = \eta(\beta)$ $(\beta_0 < \beta < \beta_1)$ is a segment lying under the line $\alpha = \beta$ (Fig. 15.2) then all points of the segment $\alpha = \eta_1(\beta)$ $(\beta_0 < \beta < \beta_1)$ belong to the corresponding L-characteristic of the operator $\boldsymbol{B}$. Let us make one further remark; if the inequality is fulfilled with $\beta = \beta_0$ and if one of the L-characteristics $L(\boldsymbol{A}$; cont.) and $L(\boldsymbol{A}$; comp.) contains the point $\alpha_0 = \eta(\beta_0)$, then the corresponding L-characteristic of the operator $\boldsymbol{B}$ contains all points of the line $\beta = \beta_0$ whose abscissa is less than $\eta_1(\beta_0) = (1 - \tau)\beta + \tau\eta(\beta_0)$.

Sometimes inequalities of the form (15.33) are established not for the operator $\boldsymbol{B}$ but for its adjoint operator $\boldsymbol{B}^*$:

$$\|\boldsymbol{B}^* x\|_\beta \leqq k_1(\beta) \|\boldsymbol{A}_1 x\|_\beta^{\tau_1} \|x\|_\beta^{1-\tau_1} \quad (x \in L_\beta). \tag{15.35}$$

In this case the above scheme permits one to construct part of the L-characteristic of the operator $\boldsymbol{B}^*$. The set which is symmetric to this part with respect to the line $\alpha + \beta = 1$, will then belong to the L-characteristic of the operator $\boldsymbol{B}$.

For some operators both inequalities (15.33) and (15.35) can be established. Each of these inequalities permits one to determine a part of the L-characteristic of the operator $\boldsymbol{B}$. Following this, one can analyse the L-characteristic further by applying interpolation theorems.

The discussion presented above finds its principal application in the construction of the L-characteristics of fractional powers of positive type operators.

Let C be an operator which is of positive type (see § 14) in each of the spaces L_β, $\beta_0 < \beta < \beta_1$ (note that C being of positive-type in $L_{\beta'}$ and $L_{\beta''}$ implies that it is also of positive type in every space L_β where $\beta \in (\beta', \beta'')$). Then by Theorem 14.2 the operator $C^{-\tau}$ $(0 < \tau < 1)$ satisfies the inequalities

$$\|C^{-\tau}x\|_\beta \leqq k(\beta)\|C^{-1}x\|_\beta^\tau \|x\|_\beta^{1-\tau} \quad (\beta_0 < \beta < \beta_1). \tag{15.36}$$

These inequalities can be considered as inequalities of the form (15.33). Hence using the L-characteristic of the operator C^{-1}, parts of the L-characteristic of the operator $C^{-\tau}$ can be constructed.

The operator C being of positive type in a space L_β, where $\beta_0 < \beta < \beta_1$, implies that C^* is of positive type in the space $L_{1-\beta}$. The fractional powers $(C^*)^{-\tau}$ of the operator C^* are connected with the operators $C^{-\tau}$ by the relation

$$(C^*)^{-\tau} = (C^{-\tau})^*, \tag{15.37}$$

which follows immediately from formula (14.2). This means that

$$\|(C^{-\tau})^*x\|_{1-\beta} \leqq k_1(\beta) \cdot \|(C^{-1})^*x\|_{1-\beta}^\tau \|x\|_{1-\beta}^{1-\tau} \quad (\beta_0 < \beta < \beta_1). \tag{15.38}$$

Hence it is also possible to obtain a description of part of the L-characteristic of the operator $C^{-\tau}$. Thus there are many cases in which one can obtain parts of the sought for L-characteristics which are different from those which can be obtained directly from the inequalities (15.36).

15.6 *One more theorem on compactness*

Let a continuous linear operator $\boldsymbol{B}$ act from the Lorentz space Λ_α to some Banach space E. Lemma 15.1 implies that the operator $\boldsymbol{B}$ is compact if its values on the set T_α of functions (15.5) form a compact set in E.

Denote by T_α^+ the set of non-negative functions in T_α. In other words, T_α^+ consists of functions of the form

$$x(s) = (\text{mes } D)^{-\alpha} x(s; D) \quad (D \subset \Omega). \tag{15.39}$$

It is easy to see that each function $x(s)$ in T_α can be written in the form

$$x(s) = a_1 x_1(s) - a_2 x_2(s),$$

where $x_1(s)$, $x_2(s)$ are functions in T_α^+ and $0 \leqq a_1, a_2 \leqq 1$. Here the following result is valid.

LEMMA 15.6: *Let an operator* $\boldsymbol{B}$ *transform* T_α^+ *to a compact set in E. Then* $\boldsymbol{B}$ *is compact as an operator from* Λ_α *to E.*

Suppose that $\boldsymbol{B}$ is continuous as an operator from Λ_{α_0} to E and compact as an operator from Λ_{α_1} to E. A simple calculation shows that $\boldsymbol{B}$ is compact as an operator acting from each space Λ_α, where $\alpha \in (\alpha_0, \alpha_1)$ to the space E. This remark and Lemma 15.6 imply the following supplement to Theorem 15.6.

THEOREM 15.7: *Let a linear operator* $\boldsymbol{B}$ *satisfy the inequality* $\|\boldsymbol{B}\chi\|_{E_2} \leqq \leqq k\|\boldsymbol{A}\chi\|_{E_1}^{\tau}\|\chi\|_{L_\gamma}^{1-\tau}$ $(0 < \tau \leqq 1)$, *on all functions of the form*

$$\chi(s) = \frac{\kappa(s; D_1)}{(\text{mes } D_1)^\delta} - \frac{\kappa(s; D_2)}{(\text{mes } D_2)^\delta} \qquad (D_1, D_2 \subset \Omega), \tag{15.40}$$

where $\boldsymbol{A}$ *is a compact operator acting from* Λ_δ *to* E_1, *with* $0 < \delta < \gamma \leqq 1$. *Then* $\boldsymbol{B}$ *is compact as an operator from each space* L_α, *with*

$$\delta < \alpha < \tau\delta + (1 - \tau)\gamma \tag{15.41}$$

to the space E_2.

§ 16 Fractional powers of elliptic operators

16.1 *Elliptic differential expressions*[1]

In the sequel $\boldsymbol{D}^r u$, where $r = \{r_1, r_2, \ldots, r_n\}$ is a vector of non-negative integer components, denotes the derivative

[1] There is a large number of papers concerning the general theory of elliptic operators. Important results in this field were obtained by S. N. Bernstein, J. Schauder, I. G. Petrovskii, S. L. Sobolev, I. N. Vekua, O. A. Oleinik, O. A. Ladyzenskaya, A. I. Koshelev, S. Agmon, A. Douglas, L. Nirenberg, S. G. Mikhlin, L. N. Slobodeckii, Y. B. Lopatinskii, Z. Y. Shapiro, F. Browder, N. Aronszajn, L. Hörmander, M. Schechter, V. A. Solonikov, M. I. Visik, G. I. Eskin and many others (a reasonably complete bibliography, up to 1962, appears in the article of S. Agmon, A. Douglas and L. Nirenberg [1]).

$$D^r u = \frac{\partial^{r_1+r_2+\ldots+r_n} u}{\partial x_1^{r_1} \ldots \partial x_n^{r_n}}; \tag{16.1}$$

the order $r_1 + \ldots + r_n$ of this derivative is denoted by $|r|$. To each differential expression[1]

$$\mathfrak{C}u = \sum_{0 \leq |r| \leq m} c_r(x) D^r u \tag{16.2}$$

there is associated its characteristic polynomial

$$\varphi(\xi; \mathfrak{C}, x) = \sum_{|r|=m} c_r(x)(i\xi)^r, \tag{16.3}$$

where $\xi = \{\xi_1, \ldots, \xi_n\}$ and

$$(i\xi)^r = (i\xi_1)^{r_1} \ldots (i\xi_n)^{r_n}. \tag{16.4}$$

In the sequel Ω will denote a bounded closed domain with sufficiently smooth boundary Γ. A differential expression

$$\mathfrak{A}u = \sum_{0 \leq |r| \leq 2k} a_r(x) D^r u \qquad (x \in \Omega) \tag{16.5}$$

is called *elliptic at a point* $x_0 \in \Omega$, if its characteristic polynomial

$$\varphi(\xi_1, \ldots, \xi_n; \mathfrak{A}, x_0) = \sum_{|r|=2k} a_r(x_0)(i\xi)^r \tag{16.6}$$

is positive for each non-zero real vector ξ. Ellipticity at a point x_0 of the differential expression (16.5) is unaffected by any orthogonal change of coordinate systems.

The differential expression (16.5) is called *elliptic in a domain* Ω if it is elliptic at each point $x \in \Omega$.

In the sequel we will analyze the differential expression (16.5) in the case of coefficients $a_r(x)$ which are smooth on the domain Ω. For a differ-

---

[1] For simplicity only differential expressions with real coefficients are considered.

ential expression of second order

$$\mathfrak{A}u(x) = \sum_{i,j=1}^{n} a_{ij}(x) \frac{\partial^2 u}{\partial x_i \partial x_j} + \sum_{k=1}^{n} b_k(x) \frac{\partial u}{\partial x_k} + c(x)u \tag{16.7}$$

the condition of ellipticity is equivalent to negative definiteness of the quadratic form:

$$\boldsymbol{P}(\xi_1, \ldots, \xi_n) = \sum_{i,j=1}^{n} a_{ij}(x)\xi_i\xi_j \quad (x \in \Omega). \tag{16.8}$$

Usually differential expressions are studied in conjunction with boundary conditions of the form

$$\mathfrak{B}_j u(x) = \sum_{0 \leqq |r| \leqq m_j} b_{rj}(x) \boldsymbol{D}^r u = 0 \quad (x \in \Gamma), \tag{16.9}$$

where $0 \leqq m_j \leqq 2k-1$, $j = 1, 2, \ldots, k$ and the coefficients $b_{rj}(x)$ are continuous.

Fix a point $x_0 \in \Gamma$ and assume that a coordinate system is chosen so that the inner normal to Γ at the point x_0 is directed along the axis of the argument x_n. The boundary conditions (16.9) are said to *cover* the differential expression (16.5) at the point x_0 if for any non-zero vector $\{\xi_1, \ldots, \xi_{n-1}, 0\}$ the system of k polynomials

$$\varphi_j(z) = \varphi(\xi_1, \ldots, \xi_{n-1}, z; \mathfrak{B}_j, x_0) \quad (j = 1, 2, \ldots, k) \tag{16.10}$$

is linearly independent modulo a polynomial of degree k $\varphi^+(z)$ whose roots coincide with those roots of the polynomial $\varphi(\xi_1, \ldots, \xi_{n-1}, z; \mathfrak{A}, x_0)$ lying in the upper halfplane.

In what follows it will be assumed that the boundary conditions (16.9) cover the differential expression (16.3) at each point $x_0 \in \Gamma$.

It is easy to check that the boundary condition

$$u(x) = 0 \quad (x \in \Gamma) \tag{16.11}$$

covers each elliptic differential expression of second order. A differential expression of second order is also covered by the boundary condition

$$\frac{\partial u(x)}{\partial \nu} + \sigma(x)u(x) = 0 \qquad (x \in \Gamma), \tag{16.12}$$

where $\partial/\partial\nu$ is the conormal derivative, and $\sigma(x)$ is a continuous positive function.

16.2 *Elliptic operators*

Denote by $\mathfrak{D}$ the totality of functions $u(x)$ $(x \in \Omega)$ possessing on Ω (classical) derivatives $\boldsymbol{D}^r u(x)$ of all orders $|r| \leqq 2k$, and satisfying on Γ the boundary conditions (16.9). Consider on this family $\mathfrak{D}$ the operator A defined by the differential expression:

$$Au(x) = \mathfrak{A}u(x). \tag{16.13}$$

The operator A depends on the elliptic differential expression (16.5) as well as on the boundary conditions (16.9). It is called an *elliptic operator*.

To investigate elliptic operators A it is convenient to take them to be unbounded linear operators acting in some function space E. As is customary we will consider elliptic operators in the spaces L_α.

When an elliptic operator A is considered in a function space E it could be assumed to be defined on the maximal linear subset $D(A)$ of the set $\mathfrak{D}$ which is contained in E and on which the values of A also belong to E. This construction unfortunately leads to an operator with 'bad' properties. In particular, these operators are not closed.

In view of this, elliptic operators are redefined so as to operate also on some functions not belonging to $\mathfrak{D}$. The terminology 'elliptic operator' will be kept for the redefined operators.

This redefinition of elliptic operators can be attained by the usual method of closure. However, we use another method, which is connected with a generalized notion of derivative and with a corresponding construction for the domain of the redefined operator. This natural method was developed by S. L. Sobolev and his successors.

Recall that the support of a function $u(x)$ is the set of points, at which $u(x)$ takes a non-zero value. A function $u(x)$ defined on Ω is said to be of *finite-type* if its support lies in Ω and has positive distance from Γ.

If a function $u(x)$ has derivatives of sufficiently high orders it satisfies the identities

$$\int_\Omega u(x)\boldsymbol{D}^r\varphi(x)\,\mathrm{d}x = (-1)^{|r|}\int_\Omega \boldsymbol{D}^r u(x)\varphi(x)\,\mathrm{d}x, \tag{16.14}$$

where $\varphi(x)$ is an arbitrary infinitely differentiable finite-type function. The identities (16.14) are proved simply by integration by parts. These identities suggest a natural way to generalize the notion of derivative.

Suppose that for some function $u(x)$ there can be found a summable function $v(x)$ such that

$$\int_\Omega u(x)\boldsymbol{D}^r\varphi(x)\,\mathrm{d}x = (-1)^{|r|}\int_\Omega v(x)\varphi(x)\,\mathrm{d}x, \tag{16.15}$$

where $\varphi(x)$ is any infinitely differentiable finite-type function and $r = \{r_1, r_2, \dots, r_n\}$. Then $u(x)$ is said to have the *generalized* (*in the sense of* S. L. *Sobolev*) *derivative* $\boldsymbol{D}^r u(x)$, which is defined by the relation

$$\boldsymbol{D}^r u(x) = v(x). \tag{16.16}$$

A generalized derivative is defined uniquely (up to values on a set of measure zero). The identities (16.14) mean that for a smooth function generalized derivatives coincide with the usual derivatives. In what follows the word 'generalized' will often be omitted.

Denote by $W_\alpha^l = W_\alpha^l(\Omega)$, where $0 < \alpha \leqq 1$, $l = 0, 1, 2, \dots$ the totality of all functions defined on Ω for which the following norm is finite:

$$\|u(x)\|_{W_\alpha{}^l} = \{\int_\Omega \sum_{0\leqq|r|\leqq l} |\boldsymbol{D}^r u(x)|^{1/\alpha}\mathrm{d}x\}^\alpha. \tag{16.17}$$

The spaces W_α^l are complete normed spaces; the space W_α^0 obviously coincides with the space L_α.

In the sequel, we will work not only with the spaces W_α^l, but also with the spaces $C^l = C^l(\Omega)$ of continuously differentiable functions $u(x)$ on Ω with norm

$$\|u(x)\|_{C^l} = \sum_{0\leqq|r|\leqq l} \max_{x\in\Omega} |\boldsymbol{D}^r u(x)|. \tag{16.18}$$

As mentioned earlier, generalized derivatives are only defined up to values on a set of measure zero. Hence at first glance it appears that it is impossible to speak about boundary values of functions which belong to the spaces W_α^l.

It turns out, however (see, for instance, S. L. Sobolev [1], L. V. Kantorovic and G. P. Akilov [1], V. I. Smirnov [1]), that each function $u(x)$ in $W_\alpha^l(\Omega)$ can be represented together with its generalized derivatives up to order $l - 1$ by means of linear integral operators applied to the generalized derivatives of order l of the function $u(x)$. The values taken on by these integral representations should then be considered to be the values of the function $u(x)$ and of its generalized derivatives up to order $l - 1$. An analysis of the corresponding integral operators shows that the function $u(x)$ and its generalized derivatives up to order $l - 1$ are thus defined uniquely on the boundary Γ of the domain Ω (up to values on a set of surface measure zero).

Hence for each function $u(x) \in W_\alpha^{2k}(\Omega)$ it is possible to speak about the values of boundary operators of the form

$$\mathfrak{B}u(x) = \sum_{0<|r|\leqq 2k-1} b_r(x)\boldsymbol{D}^r u \qquad (x \in \Gamma). \tag{16.19}$$

It turns out that those functions $u(x) \in W_\alpha^{2k}(\Omega)$, satisfying the boundary conditions (16.9) form a closed set in $W_\alpha^{2k}(\Omega)$ which is denoted by $W_\alpha^{2k}(\Omega; \mathfrak{B})$.

In the sequel we will often refer to an elliptic operator A, which is generated by the elliptic differential expression (16.5) and the boundary conditions (16.9) in the space L_α $(0 < \alpha < 1)$. This will mean the operator with domain $D(A) = W_\alpha^{2k}(\Omega; \mathfrak{B})$ whose values are calculated by the formula

$$Au(x) = \sum_{0<|r|\leqq 2k} a_r(x)\boldsymbol{D}^r u(x); \tag{16.20}$$

here the $\boldsymbol{D}^r u(x)$ are generalized derivatives. The operator A given by this definition possesses many 'good' properties.

16.3 *Positive-type elliptic operators*

Assume that at each point $x_0 \in \Gamma$ the boundary conditions (16.9) cover the differential expression (16.5) in the strong sense. This means that the polynomials (16.10) are linearly independent modulo each polynomial of degree k $\varphi^+(z; t)$ (for $t \in [0, \infty)$), whose roots coincide with those roots of the polynomial

$$\varphi(\xi_1, \ldots, \xi_{n-1}, z; \mathfrak{A}, x_0) + t$$

lying in the upper half plane.

It turns out (see S. Agmon [1]) that in this case when t is large the operator $\boldsymbol{A} + t\boldsymbol{I}$ has a bounded inverse in each space L_α $(0 < \alpha < 1)$, and for such values of t

$$\|(t\boldsymbol{I} + \boldsymbol{A})^{-1}\|_{L_\alpha \to L_\alpha} \leqq \frac{c(\alpha)}{1 + t}. \tag{16.21}$$

The elliptic operator $\boldsymbol{A}$ will be of positive type (see § 14) if the inequality (16.21) is fulfilled for all $t \geqq 0$. For an elliptic operator of second order, being of positive-type is guaranteed if the coefficient $c(x)$ in (16.7) is positive and sufficiently large. Let the elliptic operator $\boldsymbol{A}_1$ be defined by the elliptic differential expression (16.5) and the boundary conditions (16.9), which are assumed to cover it in the strong sense. Then for a sufficiently large positive number t_0 the operator $\boldsymbol{A} = \boldsymbol{A}_1 + t_0\boldsymbol{I}$ is of positive type.

The following constructions concern positive-type elliptic operators.

Many problems lead to the study of the operator $\boldsymbol{A}^{-1}$ which is inverse to an elliptic operator $\boldsymbol{A}$. For instance, the formula

$$u(x) = \boldsymbol{A}^{-1}f(x)$$

gives the solution of the differential equation

$$\sum_{0 \leqq |r| \leqq 2k} a_r(x)\boldsymbol{D}^r u(x) = f(x) \tag{16.22}$$

satisfying the boundary conditions (16.9), provided that the elliptic operator $\boldsymbol{A}$ is defined by the left side of equation (16.22) in conjunction with the boundary conditions (16.9).

We are interested in the properties of the operator $\boldsymbol{A}^{-1}$. This operator which is continuous in L_α, actually admits an integral representation. However, a relatively complete analysis of it can be obtained immediately, just on the basis of ellipticity.

Let E_1 and E_2 be two Banach spaces such that $E_1 \subset E_2$. Denote by J the operator which assigns to each element $u \in E_1$ the same element considered as a point in the space E_2. If the operator J is continuous, E_1 is said to *be continuously imbedded* in E_2; if J is compact, E is said to *be compactly imbedded* in E_2. Continuous imbedding of E_1 in E_2 is equivalent to the relation

$$\|u(x)\|_{E_2} \leqq M\,\|u(x)\|_{E_1} \qquad (u(x) \in E_1). \tag{16.23}$$

THEOREM 16.1: *Let a space W_α^{2k} $(0 < \alpha < 1)$ be continuously imbedded in a space E. Then the operator A^{-1} acts from L_α to E and is continuous.*

THEOREM 16.2: *Let a space W_α^{2k} $(0 < \alpha < 1)$ be compactly imbedded in E. Then the operator A^{-1} acts from L_α to E and is compact.*

Both of these theorems result immediately from the so-called *coerciveness inequality*

$$\|Au(x)\|_{L_\alpha} \geqq c_0\|u(x)\|_{W_\alpha{}^{2k}} \qquad (u \in D(A)), \tag{16.24}$$

which is one of the basic facts in the theory of elliptic operators. The coerciveness inequality follows from the fact of an elliptic operator being of positive type. In fact, let the operator A^{-1} exist. Then A establishes a one-to-one correspondence between the elements of the complete normed spaces W_α^{2k} and L_α and is continuous. A famous theorem of Banach now implies the continuity of the operator A^{-1} as an operator from L_α to $W_\alpha^{2k}(\Omega; \mathfrak{B})$; this means that coerciveness inequality is satisfied.

To apply Theorems 16.1 and 16.2, it is necessary to know in which space E a space W_α^{2k} is imbedded.

Put

$$\xi(\alpha) = \alpha - \frac{2k}{n} \qquad \left(\frac{2k}{n} < \alpha < 1\right), \tag{16.25}$$

where n is the dimension of the domain Ω. The famous imbedding theorem of S. L. Sobolev and V. I. Kondrasov implies that for $0 < \alpha < 2k/n$ the space W_α^{2k} is compactly imbedded in the space C; for $\alpha = 2k/n$ the space W_α^{2k} is compactly imbedded in all spaces L_β, where $\beta > 0$; for $2k/n < \alpha < 1$

the space W_α^{2k} is continuously imbedded in all spaces $L_{\xi(\alpha)}$ and is compactly imbedded in all spaces L_β where $\beta > \xi(\alpha)$; finally W_1^{2k} is compactly imbedded in all spaces L_β where $\beta > 1 - 2k/n$.

These imbedding theorems and Theorems 16.1 and 16.2 permit us to draw the L-characteristic of A^{-1} the inverse to the elliptic operator. If $2k > n$, A^{-1} acts from any L_α to C and is compact. If $2k < n$, the L-characteristics $L(A^{-1}$; cont.) and $L(A^{-1}$; comp.) are polygons. These polygons are hatched in Fig. 16.1. If $2k = n$, the L-characteristics $L(A^{-1}$; comp.) and $L(A^{-1}$; cont.) contain the half-strip $0 \leqq \alpha < 1$, $0 \leqq \beta < \infty$.

Theorems 16.1 and 16.2 can also be applied to construct the L-characteristics of the operators $\boldsymbol{D}^r A^{-1}$, where $\boldsymbol{D}^r$ is a fixed derivative of order $|r| < 2k$. To this end it is necessary to take the space $W_\alpha^{|r|}$ as the space E. The imbedding theorem of S. L. Sobolev and V. I. Kondrasov implies that the space W_α^{2k} for $0 < \alpha < (2k - |r|)/n$ is compactly imbedded in the space $C^{|r|}$; for $\alpha = (2k - |r|)/n$ it is compactly imbedded in all space $W_\beta^{|r|}$, where $\beta > 0$; for $(2k - |r|)/n < \alpha < 1$ the space W_α^{2k} is continuously imbedded in the space $W_{\xi_{|r|}(\alpha)}^{|r|}$, with

$$\xi_{|r|}(\alpha) = \alpha - \frac{2k - |r|}{n} \qquad \left(\frac{2k - |r|}{n} < \alpha < 1\right), \tag{16.26}$$

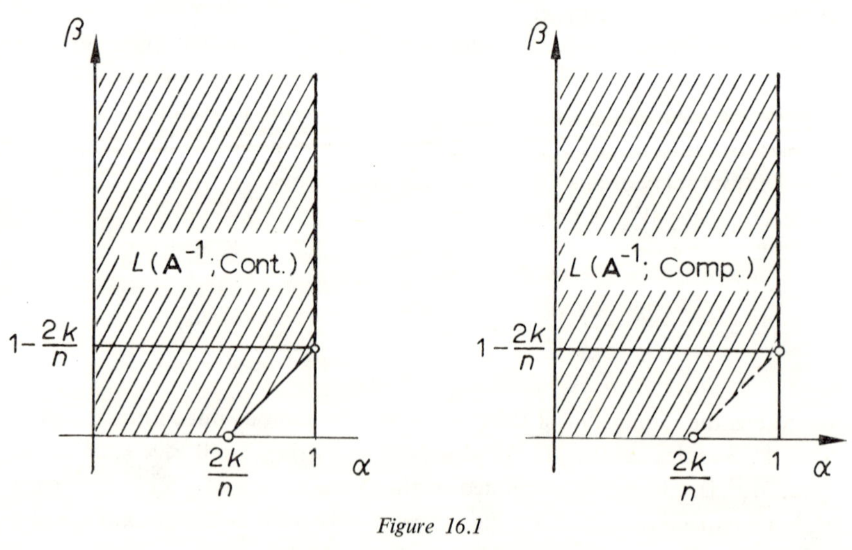

Figure 16.1

and is compactly imbedded in all spaces $W_\beta^{|r|}$, where $\beta > \xi_{|r|}(\alpha)$; finally, W_1^{2k} is compactly imbedded in all spaces $W_\beta^{|r|}$, where $\beta > 1 - (2k - |r|)/n$.

These theorems permit us to describe the L-characteristics of the operators $\boldsymbol{D}^r A^{-1}$: if $(2k - |r|)/n \geqq 1$, then the L-characteristics $L(\boldsymbol{D}^r A^{-1}; \text{cont.})$ and $L(\boldsymbol{D}^r A^{-1}; \text{comp.})$ contain the half-strip $0 \leqq \alpha < 1$, $0 \leqq \beta < \infty$; if $(2k - |r|)/n < 1$, the L-characteristics $L(\boldsymbol{D}^r A^{-1}; \text{cont.})$ and $L(\boldsymbol{D}^r A^{-1}; \text{comp.})$ are polygons—the L-characteristic $L(\boldsymbol{D}^r A^{-1}; \text{cont.})$ contains the set of points $\{\alpha, \beta\}$

$$0 \leqq \alpha < 1, \; 0 \leqq \beta < \infty, \; \beta \geqq \xi_{|r|}(\alpha),$$

from which the point $\{(2k - |r|)/n, 0\}$ is excluded, and the L-characteristic $L(\boldsymbol{D}^r A^{-1}; \text{comp.})$ contains the set of points $\{\alpha, \beta\}$, for which

$$0 \leqq \alpha < 1, \; 0 \leqq \beta < \infty, \; \beta > \xi_{|r|}(\alpha).$$

The coerciveness inequality (16.24) permits us to obtain more complete information about operators $\boldsymbol{D}^r A^{-1}$. It implies that

$$\|\boldsymbol{D}^r A^{-1} u(x)\|_{W_\alpha^{2k-|r|}} \leqq c_1 \|u(x)\|_{L_\alpha}. \tag{16.27}$$

Hence $\boldsymbol{D}^r A^{-1}$ is a continuous operator acting from L_α to each space E in which the space $W_\alpha^{2k-|r|}$ is continuously imbedded. Similarly $\boldsymbol{D}^r A^{-1}$ is a compact operator acting from L_α to each space E in which the space $W_\alpha^{2k-|r|}$ is compactly imbedded.

In conclusion we note that the application of the above scheme of analysis permits us to study A^{-1} and $\boldsymbol{D}^r A^{-1}$ as operators acting from $L_\alpha(\Omega)$ to various spaces E of functions defined on an m-dimensional manifold $S \subset \Omega$ (for instance, on Γ or on a part of Γ). For this purpose it is necessary to use theorems on the imbedding of the spaces $W_\alpha^l(\Omega)$ into spaces E of functions which are defined on S.

16.4 *Multiplicative inequalities and fractional powers of elliptic operators*[1]

Let us continue the study of positive-type elliptic operators A. In this section we are interested in the operators $A^{-\tau}$ and $\boldsymbol{D}^r A^{-\tau}$. Let us determine

[1] In this section we present certain results of V. P. Glushko [1].

into which spaces E they act from the spaces L_α. The results of this section will be obtained by combining some of the theorems in § 14, with the coerciveness and multiplicative inequalities and with the interrelations of the norms of certain spaces.

Let us first formulate a general theorem concerning abstract operators.

THEOREM 16.3: *Let E and E_1 be two Banach spaces. Let $\boldsymbol{A}$ be a positive-type operator in the space E and let $\boldsymbol{B}$ be a closed linear operator acting from E to E_1. Suppose the following inequality holds:*

$$\|\boldsymbol{B}u\|_{E_1} \leqq k\,\|\boldsymbol{A}u\|_E^\tau \|u\|_E^{1-\tau} \qquad (u \in E), \tag{16.28}$$

where $\tau \in (0, 1)$. *Then for each $\varepsilon > 0$ the operator $\boldsymbol{B}\boldsymbol{A}^{-\tau-\varepsilon}$ acts from E to E_1 and is continuous. If the operator $\boldsymbol{A}^{-1}$ is compact in E then $\boldsymbol{B}\boldsymbol{A}^{-\tau-\varepsilon}$ acts from E to E_1 and is compact.*

PROOF: The first assertion is a special case of Theorem 14.5 with $\tau_0 = 1$. To prove the second assertion write the operator $\boldsymbol{B}\boldsymbol{A}^{-\tau-\varepsilon}$ in the form

$$\boldsymbol{B}\boldsymbol{A}^{-\tau-\varepsilon} = \boldsymbol{B}\boldsymbol{A}^{-\tau-\varepsilon/2}\boldsymbol{A}^{-\varepsilon/2}. \tag{16.29}$$

By Theorem 14.12 the operator $\boldsymbol{A}^{-\varepsilon/2}$ is compact in E, while by the first assertion the operator $\boldsymbol{B}\boldsymbol{A}^{-\tau-\varepsilon/2}$ acts from E to E_1 and is continuous. Hence (16.29) implies the compactness of $\boldsymbol{B}\boldsymbol{A}^{-\tau-\varepsilon}$ as an operator from E to E_1. The theorem has been proved.

Let E_0 be a Banach space of functions defined either on Ω or on some m-dimensional manifold $S \subset \Omega$, and let the following multiplicative inequality be satisfied:

$$\|u(x)\|_{E_0} \leqq M_0 \|u(x)\|_{W_\alpha^{2k}}^\tau \|u(x)\|_{L_\alpha}^{1-\tau}. \tag{16.30}$$

Then the coerciveness inequality (16.24) implies that

$$\|u(x)\|_{E_0} \leqq M_0 \|\boldsymbol{A}u(x)\|_{L_\alpha}^\tau \|u(x)\|_{L_\alpha}^{1-\tau}, \tag{16.31}$$

where $\boldsymbol{A}$ is a positive-type elliptic operator of order $2k$. The inequality (16.31) can be taken as condition (16.28) of Theorem 16.3. This theorem then implies

THEOREM 16.4: *Let A be a positive-type elliptic operator of order $2k$. Let the inequality* (16.30) *be satisfied. Then for each $\varepsilon > 0$ the operator $A^{-\tau-\varepsilon}$ acts from L_α to E_0 and is compact.*

The applications of this theorem are connected with concrete inequalities of the form (16.30). Let the numbers α, β, l, l_0 be connected through the inequalities

$$0 \leqq l \leqq l_0, \; n\alpha - l_0 \leqq n\beta - l < n\alpha, \; \alpha < 1. \tag{16.32}$$

Then (L. Nirenberg [1], V. P. Glushko and S. G. Krein [1], V. P. Ilin [1–3]) the following assertions are valid: *if $\beta > 0$, then*

$$\|u(x)\|_{W_\beta{}^l} \leqq c \|u(x)\|^{\tau}_{W_\alpha{}^{l_0}} \|u(x)\|^{1-\tau}_{L_\alpha} \tag{16.33}$$

where

$$\tau = \frac{l}{l_0} + \frac{n}{l_0}(\alpha - \beta), \tag{16.34}$$

while if $\beta = 0$, then

$$\|u(x)\|_{C^l} \leqq c \|u(x)\|^{\tau}_{W_\alpha{}^{l_0}} \|u(x)\|^{1-\tau}_{L_\alpha} \tag{16.35}$$

where

$$\tau = \frac{l}{l_0} + \frac{n\alpha}{l_0}. \tag{16.36}$$

These assertions and Theorem 16.4 imply:

THEOREM 16.5: *Let*

$$l/2k < \tau < 1. \tag{16.37}$$

Then for $0 \leqq \alpha < (2k\tau - l)/n$, $\alpha < 1$, the operator $A^{-\tau}$ acts from L_α to C^l

and is compact; for $(2k\tau - l)/n \leqq \alpha < 1$ *it acts from* L_α *to any* W^l_β, *with*

$$\beta > \alpha - (2k\tau - l)/n \tag{16.38}$$

and is compact.

Theorem 16.5 permits us to draw certain parts of the L-characteristic of the operator $\boldsymbol{A}^{-\tau}$ and $\boldsymbol{D}^r\boldsymbol{A}^{-\tau}$. It implies that all points $\{\alpha, \beta\}$, with

$$0 \leqq \alpha < 1, \ \beta > 0, \ \beta > \alpha - \frac{2k\tau - |r|}{n},$$

belong to the L-characteristic $L(\boldsymbol{D}^r\boldsymbol{A}^{-\tau}; \text{cont.})$ and even to $L(\boldsymbol{D}^r\boldsymbol{A}^{-\tau}; \text{comp.})$ for

$$\frac{|r|}{2k} < \tau < 1.$$

In Fig. 16.2 the set of these points (for the case $2k\tau - |r| < n$) is hatched.

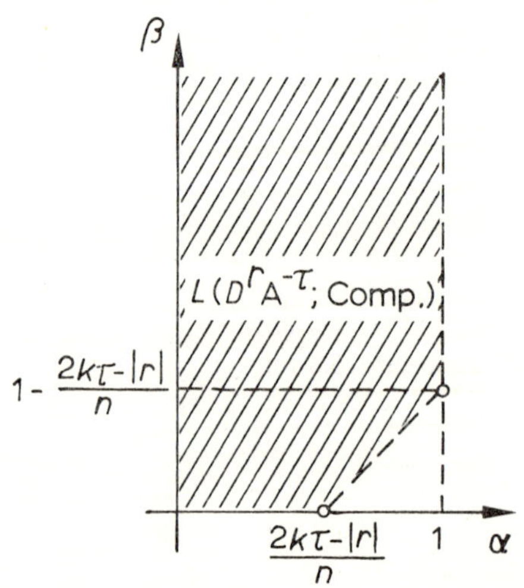

Figure 16.2

16.5 *L-characteristics of negative fractional powers of elliptic operators*[1]

In the preceding section we obtained some parts of the half-strip $0 \leqq \alpha \leqq 1$, $0 \leqq \beta < \infty$ which lie in the L-characteristic $L(A^{-\tau}; \text{cont.})$ (and in the L-characteristic $L(A^{-\tau}; \text{comp.})$) of a fractional power $A^{-\tau}$ of a positive-type elliptic operator A. In this section additional information concerning the L-characteristic $L(A^{-\tau}; \text{cont.})$ will be obtained. This additional information will be obtained by other methods—we use results, established in § 15.

THEOREM 16.6: *Let A be a positive-type elliptic operator of order $2k$. Let $2k < n$, where n is the dimension of the domain Ω. Then for each $\tau \in (0, 1)$ the L-characteristic* $L(A^{-\tau}; \text{cont.})$ *contains all points on the segment*

$$\beta = \alpha - \frac{2k\tau}{n} \quad \left(\frac{2k\tau}{n} < \alpha < 1\right). \tag{16.39}$$

PROOF: As was shown in 16.3°, the L-characteristic $L(A^{-1}; \text{cont.})$ of the operator A^{-1} contains the segment

$$\beta = \alpha - \frac{2k}{n} \quad \left(\frac{2k}{n} < \alpha < 1\right). \tag{16.40}$$

This means that for each fixed $\beta \in (0, 1 - 2k/n)$ the operator A^{-1} acts from $L_{\beta+2k/n}$ to L_β and is continuous. The fact that A is of positive-type and Theorem 14.2 imply that

$$\|A^{-\tau}u\|_\beta \leqq c(\beta)\,\|A^{-1}u\|_\beta^\tau\,\|u\|_\beta^{1-\tau}. \tag{16.41}$$

Hence Theorem 15.5 implies that the operator $A^{-\tau}$ acts from the Lorentz space $\Lambda_{\beta+2k\tau/n}$ to the space L_β and is continuous. In other words, the operator $A^{-\tau}$ acts from each space Λ_α where $2k\tau/n < \alpha < 1 - 2k(1 - \tau)/n$ to the corresponding space $L_{\alpha-2k\tau/n}$. Theorem 15.1 then implies that the operator $A^{-\tau}$ acts continuously from each space L_α where $2k\tau/n < \alpha < 1 - 2k(1 - \tau)/n$ to the corresponding space $L_{\alpha-2k\tau/n}$.

---

[1] P. P. Zabreiko, M. A. Krasnoselskii, E. I. Pustylnik [1].

We have shown that the L-characteristic $L(A^{-\tau}; \text{cont.})$ contains the segment

$$\beta = \alpha - \frac{2k\tau}{n} \quad \left(\frac{2k\tau}{n} < \alpha < 1 - \frac{2k(1-\tau)}{n}\right). \tag{16.42}$$

Similar arguments show that the L-characteristic $L((A^*)^{-\tau}; \text{cont.})$ of the operator $(A^*)^{-\tau}$ adjoint to the operator $A^{-\tau}$ also contains the segment (16.42). Then the relation

$$(A^*)^{-\tau} = (A^{-\tau})^* \tag{16.43}$$

implies that the L-characteristic $L(A^{-\tau}; \text{cont.})$ contains the segment which is located symmetrically to the segment (16.42) with respect to the line $\alpha + \beta = 1$. In other words, $L(A^{-\tau}; \text{cont.})$ contains the segment

$$\beta = \alpha - \frac{2k\tau}{n} \quad \left(\frac{2k}{n} < \alpha < 1\right). \tag{16.44}$$

If

$$\frac{4k}{n} < 1 + \frac{2k\tau}{n},$$

then the segments (16.42) and (16.44) cover the segment (16.39) and the theorem is proved. If

$$\frac{4k}{n} \geqq 1 + \frac{2k\tau}{n},$$

then in order to complete the proof it suffices to use M. Riesz' interpolation theorem 2.4. The theorem has been proved.

The segment (16.39) lies on the boundary of the part of the L-characteristic $L(A^{-\tau}; \text{cont.})$ which was drawn in the preceding section. It is possible to show that the segment (16.39) does not belong to the L-characteristic $L(A^{-\tau}; \text{comp.})$.

16.6 *Further theorems*

Theorem 16.6 is contained in the following more general assertion.

THEOREM 16.7: *Let $\boldsymbol{A}$ be a positive-type elliptic operator of order $2k$. Then for*

$$0 < \tau < \min\left\{1, \frac{n}{2k}\right\}$$

the L-characteristic $L(\boldsymbol{A}^{-\tau};\ \text{cont.})$ contains the segment

$$\beta = \alpha - \frac{2k\tau}{n} \quad \left(\frac{2k\tau}{n} < \alpha < 1\right).$$

Let us now analyse operators of the form $\boldsymbol{D}^r\boldsymbol{A}^{-\tau}$, where $\boldsymbol{A}$ is a positive-type elliptic operator of order $2k$ and $\boldsymbol{D}^r$ is a concrete operator of differentiation of order $|r| < 2k$.

THEOREM 16.8: *Let*

$$\frac{|r|}{2k} < \tau < \min\left\{1, \frac{n+|r|}{2k}\right\}.$$

Then the L-characteristic $L(\boldsymbol{D}^r\boldsymbol{A}^{-\tau};\ \text{cont.})$ contains the segment

$$\beta = \alpha - \frac{2k\tau - |r|}{n} \quad \left(\frac{2k\tau - |r|}{n} < \alpha < 1\right).$$

Here we present a proof due to P. E. Sobolevskii [14].

Let the number τ satisfy the conditions of one of the Theorems 16.7 or 16.8. We will show that the operator $\boldsymbol{D}^r\boldsymbol{A}^{-\tau}$ acts from each Lorentz space Λ_α $((2k\tau - |r|)/n < \alpha < 1)$ to the corresponding space L_β where $\beta = \alpha - (2k\tau - |r|)/n$, and that the following estimate is valid:

$$\|\boldsymbol{D}^r\boldsymbol{A}^{-\tau}\kappa_D\|_\beta \leqq c(\alpha)(\text{mes } D)^\alpha \quad (D \subset \Omega). \tag{16.45}$$

Then to complete the proof, it remains only to use interpolation theorem 15.1.

To prove (16.45) let us use Theorem 14.7. Put $E_0 = E = L_\beta$, $E_1 = L_{\gamma_1}$ and $E_2 = L_{\gamma_2}$, where γ_1 and γ_2 are numbers satisfying the inequalities

$$\beta + \frac{2k\tau - |r|}{n} < \gamma_1 < \gamma_2 < \min\left\{1, \beta + \frac{2k}{n}\right\}.$$

Then (16.33) (with $l = 0$, $l_0 = 2k$) and the coerciveness inequality (16.24) imply the estimates

$$\|u(x)\|_{L_\beta} \leqq c_i \|Au(x)\|_{L_{\gamma_i}}^{\delta_i} \|u(x)\|_{L_{\gamma_i}}^{1-\delta_i} \qquad (u \in D(A),\ i = 1, 2), \tag{16.46}$$

where

$$\delta_i = \frac{n}{2k}(\gamma_i - \beta) \qquad (i = 1, 2).$$

Next, put $\boldsymbol{B} = \boldsymbol{D}^r$; then again by (16.33) (with $l = |r|$, $l_0 = 2k$) and the coerciveness inequality (16.24) it follows that the estimate

$$\|\boldsymbol{D}^r u(x)\|_{L_\beta} \leqq c \|Au(x)\|_{L_\beta}^{|r|/2k} \|u(x)\|_{L_\beta}^{1-|r|/2k} \qquad (u \in D(A)) \tag{16.47}$$

is valid. The inequalities (16.47) and (16.48) can be considered to be the condition (14.38) and (14.30) of Theorem 14.7. Assume that the numbers ε_1 and ε_2 in the formulation of this theorem are defined by the relations

$$\varepsilon_1 = \tau - \frac{|r|}{2k}, \quad \varepsilon_2 = \frac{\delta_1 + \delta_2}{2} - \varepsilon_1.$$

Then inequality (16.45) follows:

$$\|\boldsymbol{B}A^{-\tau}\kappa_D\|_{L_\beta} \leqq c_0 (\text{mes } D)^{\gamma_1\varepsilon_1(\delta_2 - \varepsilon_1 - \varepsilon_2)/(\varepsilon_1 + \varepsilon_2)(\delta_2 - \delta_1)}$$

$$\times (\text{mes } D)^{\gamma_2\varepsilon_1(\varepsilon_1 + \varepsilon_2 - \delta_1)/(\varepsilon_1 + \varepsilon_2)(\delta_2 - \delta_1)} (\text{mes } D)^{\beta\varepsilon_2/(\varepsilon_1 + \varepsilon_2)}$$

$$= c_0 (\text{mes } D)^\alpha \qquad (D \subset \Omega).$$

If the operator A is of positive type not only in the spaces L_α but also in the space W_α^l for integer l, then the proof of Theorems 16.7 and 16.8 can be accomplished by means of a scheme similar to the proof of Theorem 16.6. Here it is convenient to use the following assertion (proved by P. P. Zabreiko) concerning abstract operators.

THEOREM 16.9: *Let a positive-type operator A in a Banach space E satisfy the conditions*

$$\|(t\boldsymbol{I}+A)^{-1}Ax\|_{E_1} \leqq M_1\|x\|_{E_2}\cdot t^{-\delta_1} \quad (0<t<\infty;\, x\in E_2\subset E) \tag{16.48}$$

and

$$\|(t\boldsymbol{I}+A)^{-1}x\|_{E_1} \leqq M_2\|x\|_{E}\cdot t^{-\delta_2} \quad (0<t<\infty;\, x\in E), \tag{16.49}$$

where E_1 and E_2 are Banach spaces, $0<\delta_1<\delta_2<1$, and M_1 and M_2 are constants. Then for any $\tau\in(1-\delta_2, 1-\delta_1)$ the following inequality holds:

$$\|A^{-\tau}x\|_{E_1} \leqq \frac{\sin\pi\tau}{\pi}\,\frac{(\delta_2-\delta_1)M_1^qM_2^{1-q}}{(1-\tau-\delta_1)(\tau+\delta_1-1)}\,\|A^{-1}x\|_{E_2}^q\,\|x\|_E^{1-q} \quad (A^{-1}x\in E_2), \tag{16.50}$$

where

$$q=\frac{\tau+\delta_2-1}{\delta_2-\delta_1}.$$

For the proof we use formula (14.2):

$$A^{-\tau}x=\frac{\sin\pi\tau}{\pi}\int_0^\infty t^{-\tau}(t\boldsymbol{I}+A)^{-1}x\,dt \qquad (x\in E).$$

Split the integral on the right side into two parts:

$$J_1=\int_0^N t^{-\tau}(t\boldsymbol{I}+A)^{-1}x\,dt,$$

$$J_2=\int_N^\infty t^{-\tau}(t\boldsymbol{I}+A)^{-1}x\,dt.$$

Now (16.48) implies the estimate

$$\|J_1\|_{E_1} \leqq \int_0^N t^{-\tau}\|(tI + A)^{-1}AA^{-1}x\|_{E_1}\,\mathrm{d}t$$

$$\leqq M_1 \int_0^N t^{-\tau-\delta_1}\mathrm{d}t\,\|A^{-1}x\|_{E_2} = \frac{M_1}{1-\tau-\delta_1} N^{1-\tau-\delta_1}\|A^{-1}x\|_{E_2}.$$

Similarly (16.49) implies that

$$\|J_2\|_{E_1} \leqq \int_N^\infty t^{-\tau}\|(tI + A)^{-1}x\|_{E_1}\,\mathrm{d}t$$

$$\leqq M_2' \int_N^\infty t^{-\tau-\delta_2}\mathrm{d}t\,\|x\|_E = \frac{M_2'}{\tau+\delta_2-1} N^{1-\tau-\delta_2}\|x\|_E.$$

Thus

$$\|A^{-\tau}x\|_{E_1} \leqq \frac{\sin \pi\tau}{\pi}$$

$$\times \left\{\frac{M_1\|A^{-1}x\|_{E_2}}{1-\tau-\delta_1} N^{1-\tau-\delta_1} + \frac{M_2\|x\|_E}{\tau+\delta_2-1} N^{1-\tau-\delta_2}\right\},$$

which implies (16.50) on putting

$$N = \left\{\frac{M_1}{M_2}\,\frac{\|x\|_E}{\|A^{-1}x\|_{E_2}}\right\}^{1/(\delta_2-\delta_1)}.$$

16.7 *On integral representations of fractional powers of elliptic operators*

Let A be a positive-type elliptic operator. As already mentioned, the operator A^{-1} admits an integral representation

$$A^{-1}u(x) = \int_\Omega G(x, y)u(y)\,\mathrm{d}y. \tag{16.51}$$

The kernel $G(x, y)$ is called the *Green's function* of the elliptic operator $\boldsymbol{A}$; the Green's function is determined by the differential expression (16.5) and the boundary conditions (16.9).

It is natural to raise the question of whether the operator $\boldsymbol{A}^{-\tau}$ can be written in the form

$$A^{-\tau}u(x) = \int_{\Omega} G_{\tau}(x, y)u(y)\,\mathrm{d}y. \tag{16.52}$$

The answer to this question is, generally speaking, affirmative, since every operator $\boldsymbol{B}$ acting from a space L_{α}, where $0 < \alpha < 1$, to the space L_0 or from L_1 to a space L_{β}, where $0 < \beta < 1$, is an integral operator (see 6.3° and 6.8°). But the construction and analysis of the kernel $G_{\tau}(x, y)$ meet with essential difficulties.

The various schemes for construction of the kernel $G_{\tau}(x, y)$ are connected with the various schemes for constructing fractional powers of abstract operators.

First let $\boldsymbol{A}$ be a positive definite selfadjoint elliptic operator in $L_{\frac{1}{2}}$. Denote by $e_1(x), e_2(x), \ldots$, the complete orthonormal system of its eigenfunctions and by $\lambda_1, \lambda_2, \ldots$ the corresponding eigenvalues. We remark that all the eigenfunctions $e_1(x), e_2(x), \ldots$ are bounded. The Green's function admits, as is well known, a bilinear expansion

$$G(x, y) = \sum_{j=1}^{\infty} \frac{e_j(x)e_j(y)}{\lambda_j} \qquad (x, y \in \Omega), \tag{16.53}$$

which however does not always converge, even in the mean. Let $\tau \in (0, 1)$. Consider the series

$$H_{\tau}(x, y) = \sum_{j=1}^{\infty} \frac{e_j(x)e_j(y)}{\lambda_j^{\tau}} \tag{16.54}$$

and assume that it converges in the norm of the space L_1, i.e. that

$$\int_{\Omega}\int_{\Omega} |H_{\tau}(x, y)|\,\mathrm{d}x\,\mathrm{d}y < \infty$$

and that

$$\lim_{m\to\infty} \int_\Omega \int_\Omega \left| H_\tau(x, y) - \sum_{j=1}^{m} \frac{e_j(x) e_j(y)}{\lambda_j^\tau} \right| \mathrm{d}x \mathrm{d}y = 0. \tag{16.55}$$

By Theorem 6.1 the operator

$$\boldsymbol{B}_\tau u(x) = \int_\Omega H_\tau(x, y) u(y) \mathrm{d}y$$

acts from L_0 to L_1. Equation (16.55) and the relation

$$\int_\Omega e_i(x) e_j(x) \mathrm{d}x = \delta_{ij} \quad (i, j = 1, 2, \ldots)$$

imply that

$$\boldsymbol{B}_\tau e_j(x) = \frac{1}{\lambda_j^\tau} e_j(x) \quad (j = 1, 2, \ldots).$$

Hence the operators $\boldsymbol{B}_\tau$ and $\boldsymbol{A}^{-\tau}$ assume the same values on the linear span of the eigenfunctions of the operator $\boldsymbol{A}$. This fact can be considered as a 'proof' that the kernel (16.54) coincides with the kernel of the integral operator (16.52). Thus, the 'Green's function' $G_\tau(x, y)$ of the operator $\boldsymbol{A}^{-\tau}$ can be sought for in the form of the series (16.54). It is necessary to study this series and to prove that the operators $\boldsymbol{B}_\tau$ and $\boldsymbol{A}^{-\tau}$ take identical values on the corresponding function spaces. Thereafter it is possible to derive from the properties of the function $G_\tau(x, y)$ various theorems on the fractional powers $\boldsymbol{A}^{-\tau}$ of the operator $\boldsymbol{A}$ (among them assertions already proved in this book).

An investigation of the series (16.54) for general elliptic operators has apparently not been made. For second order operators these series were investigated in an important series of papers by V. A. Ilin (see, for instance, [1, 2]).

Now let us describe several methods for construction of the kernel $G_\tau(x, y)$ which are applicable to cases in which the operator $\boldsymbol{A}$ is not self-adjoint. These methods are based on the integral representations (14.2) and (14.48) of the operators $\boldsymbol{A}^{-\tau}$.

Let $\boldsymbol{A}$ be a positive-type elliptic operator with

$$(t\boldsymbol{I} + \boldsymbol{A})^{-1} u(x) = \int_\Omega H(t, x, y) u(y) \mathrm{d}y \quad (0 < t < \infty). \tag{16.56}$$

Then (14.2) implies the relation

$$A^{-\tau}u(x) = \frac{\sin \pi\tau}{\pi} \int_0^\infty t^{-\tau} \left[\int_\Omega H(t, x, y) u(y) \mathrm{d}y \right] \mathrm{d}t.$$

If it is possible to change the order of integration in this relation, then we obtain the integral representation (16.52) for the operator $A^{-\tau}$ with

$$G_\tau(x, y) = \frac{\sin \pi\tau}{\pi} \int_0^\infty t^{-\tau} H(t, x, y) \mathrm{d}t. \tag{16.57}$$

as kernel. If the properties of the kernels $H(t, x, y)$ are known it is possible to establish the properties of the kernel $G_\tau(x, y)$ from them. Thereafter it is possible to go into the proofs of various theorems on fractional powers of the operator $\boldsymbol{A}$.

As far as we know, the representation (16.57) of the Green's functions of fractional powers of elliptic operators have not been studied.

Consider a problem of the form

$$\frac{\mathrm{d}u}{\mathrm{d}t} + Au = 0 \qquad u(0) = u_0, \tag{16.58}$$

where A is a strongly positive-type (for instance, selfadjoint) elliptic operator. By a solution of the problem (16.58) is meant a function taking values in some space L_α which is continuous on $[0, \infty)$ and differentiable on $(0, \infty)$ and which satisfies the equation for $t > 0$ and the initial condition for $t = 0$. Theorem 13.2 implies that the solution of the problem (16.58) is given by the formula

$$u(t) = \boldsymbol{T}(t)u \qquad (0 < t \leqq \infty), \tag{16.59}$$

where $\boldsymbol{T}(t)$ is the semi-group with generator $-\boldsymbol{A}$. Problem (16.58) is the equation of parabolic type

$$\frac{\partial u(t, x)}{\partial t} + \sum_{0 \leqq |r| \leqq 2k} a_r(x) \boldsymbol{D}^r u(t, x) = 0, \tag{16.60}$$

which we solve under the boundary conditions (16.9) and the initial condition

$$u(0, x) = u_0(x) \quad (x \in \Omega). \tag{16.61}$$

The solution of a parabolic equation can often be sought for in the form

$$u(t, x) = \int_\Omega G(t, x, y) u_0(y) \mathrm{d}y. \tag{16.62}$$

Here the properties of the Green's function $G(t, x, y)$ are investigated simultaneously with the equation (16.60). If formula (16.62) gives a solution of the parabolic equation (16.60) under the boundary conditions (16.9) for any initial value $u_0(x) \in L_\alpha$, then formula (14.48) implies that

$$A^{-\tau} u(x) = \frac{1}{\Gamma(\tau)} \int_0^\infty t^{\tau-1} \left[\int_\Omega G(t, x, y) u(y) \mathrm{d}y \right] \mathrm{d}t. \tag{16.63}$$

Hence it follows that the kernel $G_\tau(x, y)$ of the integral representation (16.52) of the fractional power $A^{-\tau}$ of the operator A can be sought for in the form

$$G_\tau(x, y) = \frac{1}{\Gamma(\tau)} \int_0^\infty t^{\tau-1} G(t, x, y) \mathrm{d}t. \tag{16.64}$$

These methods were studied in detail for selfadjoint elliptic operators of second order by P. E. Sobolevskii [7, 8]. He showed that in this case the Green's functions $G_\tau(x, y)$ $(0 < \tau < \infty)$ and their derivatives $(\partial/\partial x_i) G_\tau(x, y)$ $(\frac{1}{2} < \tau < \infty)$ are kernels of potential type. More precisely, the functions $G_\tau(x, y)$ are non-negative, symmetric and satisfy the inequalities

$$G_\tau(x, y) \leqq \begin{cases} c|x - y|^{2\tau - n}, & \text{if } 0 < \tau < n/2 \\ c_1 |\ln|x - y|| + c_2, & \text{if } \tau = n/2 \\ c, & \text{if } n/2 < \tau < \infty. \end{cases} \tag{16.65}$$

The functions $G_\tau(x, y)$ for $\tau > n/2$ are jointly continuous for $x, y \in \Omega$, and for $\tau \leqq n/2$ are jointly continuous in x, y for $x \neq y$. Note further that the kernels $G_\tau(x, y)$ satisfy the inequalities

$$|x - y|^{-\gamma}|G_\tau(x, z) - G_\tau(y, z)| \leqq \begin{cases} cr^{2\tau-\gamma-n}, & \text{if } \tau - \gamma/2 < n/2 \\ c_1|\ln r| + c_2, & \text{if } \tau - \gamma/2 = n/2 \\ c, & \text{if } \tau - \gamma/2 > n/2, \end{cases} \tag{16.66}$$

where $0 < \gamma < \min\{1, 2\tau\}$ and $r = \min\{|x - z|, |y - z|\}$. The following estimates, which are similar to (16.65) and (16.66) are valid for the functions $(\partial/\partial x_i)G_\tau(x, y)$:

$$\left|\frac{\partial}{\partial x_i} G_\tau(x, y)\right| \leqq \begin{cases} c|x - y|^{2\tau-n-1}, & \text{if } \frac{1}{2} < \tau < \dfrac{n+1}{2} \\ c_1|\ln|x - y|| + c_2, & \text{if } \tau = \dfrac{n+1}{2} \\ c, & \text{if } \dfrac{n+1}{2} < \tau < \infty \end{cases} \tag{16.67}$$

and

$$|x - y|^{-\gamma}\left|\frac{\partial}{\partial x_i} G_\tau(x, z) - \frac{\partial}{\partial x_i} G_\tau(y, z)\right| \leqq \begin{cases} cr^{2\tau-n-1-\gamma}, & \text{if } \dfrac{1+\gamma}{2} < \tau < \dfrac{n+1+\gamma}{2} \\ c_1|\ln r| + c_2, & \text{if } \tau = \dfrac{n+1+\gamma}{2} \\ c, & \text{if } \dfrac{n+1+\gamma}{2}. \end{cases} \tag{16.68}$$

We note that in the papers of P. E. Sobolevskii [7, 8] the notion of elliptic operator is introduced in a special way: first a mixed problem for a parabolic equation is studied and it is shown that the generalized solutions of this mixed problem define a semi-group of bounded operators, then the generator of this semi-group turns out to be an elliptic operator. Such a construction permits one to weaken the restrictions on the smoothness of boundaries and smoothness of the coefficients appearing in the differential expression and in the boundary conditions.

The assertions contained in Theorems 16.5–16.8 on the L-characteristics of negative fractional powers of elliptic operators of second order can, of course, be obtained as consequences of the estimates (16.65) and (16.67)—it suffices to apply the theory of operators of potential type presented in § 8.

We note in conclusion that it would be interesting to obtain estimates similar to (16.65)–(16.68) for Green's functions of fractional powers of elliptic operators of higher order.

5

Non-linear integral operators

§ 17 The superposition operators[1]

17.1 *On functions which are continuous in one variable*

In this section we study real functions $f(s, u)$, where s belongs to a set Ω with finite measure and $-\infty < u < \infty$.

If for each fixed u $f(s, u)$ is measurable in s and for almost all s it is continuous in u, then it is said to satisfy the *Caratheodory condition.*

In the sequel we often study superpositions of the form $f[s, x(s)]$. Such superpositions define the simplest non-linear operator

$$\mathfrak{f}x(s) = f[s, x(s)], \tag{17.1}$$

which is naturally called a *superposition operator*. It is not difficult to see that the operator (17.1) transforms measurable functions to measurable functions if the function $f(s, u)$ satisfies the Caratheodory condition. For the proof it suffices to remark that for almost all s

$$f[s, x(s)] = \lim_n f[s, x_n(s)],$$

where $x_n(s)$ is a sequence of step functions converging to $x(s)$. The measurability of the functions $f[s, x_n(s)]$ is obvious.

---

[1] The superposition operator has been studied by K. Caratheodory, V. V. Nemickii, M. M. Vainberg, M. A. Krasnoselskii, L. A. Ladyzenskii, Y. B. Rutickii and other authors. We follow M. M. Krasnoselskii [3, 7] in the basic constructions of this paragraph.

In this section we will prove two lemmas [1] on functions of two variables which will often be utilized in what follows.

LEMMA 17.1: *Let the function* $f(s, u)$ $(s \in \Omega, \ -\infty < u < \infty)$ *satisfy the Caratheodory condition. Then to each* $\delta > 0$ *there corresponds a set* $\Omega_\delta \subset \Omega$ *such that* $\operatorname{mes}(\Omega - \Omega_\delta) < \delta$ *and the function* $f(s, u)$ *is jointly continuous on* $\Omega_\delta \times (-\infty, \infty)$.

PROOF: Let a number $\delta > 0$ be given and let n_0 be a natural number.

Choose a sequence $u^{(1)}, u^{(2)}, \ldots, u^{(j)}, \ldots$, which is dense in the interval $[-n_0, n_0]$. Denote by $v(s)$ $(s \in \Omega)$ the supremum of those numbers v such that $|u^{(j_1)} - u^{(j_2)}| < v$ implies the inequality

$$|f(s, u^{(j_1)}) - f(s, u^{(j_2)})| \leqq 1/n_0.$$

Obviously the function $v(s)$ is positive for almost all $s \in \Omega$.

Consider the function

$$F_\beta(s) = \sup_{|u^{(j_1)} - u^{(j_2)}| \leqq \beta} |f(s, u^{(j_1)}) - f(s, u^{(j_2)})|.$$

These functions are measurable because each of them is the supremum of a countable set of measurable functions. Hence the set $N(\beta)$ of points $s \in \Omega$ for which $F_\beta(s) \leqq 1/n_0$ is measurable. But $N(\beta)$ coincides with the set of points, for which $v(s) \geqq \beta$. Consequently, the function $v(s)$ is measurable.

Denote by G_k $(k = 1, 2, \ldots)$ the set of points $s \in \Omega$ for which $v(s) \geqq 1/k$. It is clear that $G_k \subset G_{k+1}$ $(k = 1, 2, \ldots)$ and that $\operatorname{mes}(G_k) \to$ $\to \operatorname{mes} \Omega$ as $k \to \infty$. Hence there exists a number k_0 such that

$$\operatorname{mes}(\Omega - G_{k_0}) < 2^{-n_0 - 1}\delta. \tag{17.2}$$

If $|u_1 - u_2| < 1/k_0$ $(|u_1|, |u_2| \leqq n_0)$ then the inequality:

$$|f(s, u_1) - f(s, u_2)| \leqq 1/n_0 \qquad (s \in G_{k_0}). \tag{17.3}$$

---

[1] See M. A. Krasnoselskii and L. A. Ladyzenskii [1].

holds. In fact, there exist sequences $u^{(j_1)}$ and $u^{(j_2)}$, converging to u_1 and u_2 and satisfying the inequality $|u^{(j_1)} - u^{(j_2)}| < 1/k_0$. The inequality (17.3) follows from the inequality

$$|f(s, u^{(j_1)}) - f(s, u^{(j_2)})| \leqq 1/n_0 \quad (s \in G_{k_0}).$$

Now partition the interval $[-n_0, n_0]$ by points $-n_0 = u_1 < \ldots < u_q = n_0$ to parts of length less than $1/k_0$. By Lusin's theorem there exists for each function $f(s, u_i)$ $(i = 1, 2, \ldots, q)$ a closed set $G_{n_0}^{(q)}$ whose measure satisfies the inequality

$$\operatorname{mes}(G_{k_0} - G_{n_0}^{(i)}) < \frac{\delta}{2^{n_0+1}q} \quad (i = 1, \ldots, q)$$

and on which the function $f(s, u_i)$ is continuous.

Put $\Omega_{n_0} = \bigcap_i G_{n_0}^{(i)}$. It is clear that

$$\operatorname{mes}(G_{k_0} - \Omega_{n_0}) = \operatorname{mes}(G_{k_0} - \bigcap_i G_{n_0}^{(i)})$$

$$\leqq \sum_{i=1}^{q} \operatorname{mes}(G_{k_0} - G_{n_0}^{(i)}) \leqq 2^{-n_0-1}\delta$$

and since $\operatorname{mes}(\Omega - G_{k_0}) < 2^{-n_0-1}\delta$,

$$\operatorname{mes}(\Omega - \Omega_{n_0}) < 2^{-n_0}\delta.$$

Consider the function $f_{n_0}(s, u)$ $(s \in \Omega_{n_0}, -n_0 \leqq u \leqq n_0)$, which is defined by its value for $u = (1 - \tau)u_i + \tau u_{i+1}$, $0 \leqq \tau \leqq 1$ to be given by the relation

$$f_{n_0}(s, u) = (1 - \tau)f(s, u_i) + \tau f(s, u_{i+1}).$$

Obviously the function $f_{n_0}(s, u)$ is continuous and satisfies the inequality

$$|f(s, u) - f_{n_0}(s, u)| < 1/n_0 \quad (s \in \Omega_{n_0}, -n_0 \leqq u \leqq n_0).$$

The sequence of continuous functions $f_{n_0}(s, u)$ $(n_0 = 1, 2, \ldots)$ converges uniformly to the function $f(s, u)$ for u in any finite interval and for $s \in \Omega_\delta$

where $\Omega_\delta = \bigcap_{n_0} \Omega_{n_0}$. Consequently, the function $f(s, u)$ is continuous on $\Omega_\delta \times (-\infty, \infty)$. It is clear that $\operatorname{mes}(\Omega - \Omega_\delta) \leqq \delta$. The lemma has been proved.

The converse to Lemma 17.1 is also valid.

LEMMA 17.2: *Let the function $f(s, u)$ $(s \in \Omega, -\infty < u < \infty)$ satisfy the Caratheodory condition. Let $v_1(s)$ and $v_2(s)$ be measurable functions with $v_1(s) \leqq v_2(s)$ $(s \in \Omega)$. Then there exists a measurable function $u^*(s)$ such that*

$$v_1(s) \leqq u^*(s) \leqq v_2(s)$$

and

$$\sup_{v_1(s) \leqq u \leqq v_2(s)} f(s, u) = f[s, u^*(s)].$$

PROOF: Put

$$w^*(s) = \sup_{v_1(s) \leqq u \leqq v_2(s)} f(s, u)$$

and denote by $u^*(s)$ the smallest of those values $u \in [v_1(s), v_2(s)]$ for which $f(s, u)$ is equal to $w^*(s)$. We will show that the function $u^*(s)$ is measurable.

Lemma 17.1 implies the measurability of the function $w^*(s)$ and measurability of all functions

$$w_h(s) = \begin{cases} \sup\limits_{v_1(s) \leqq u \leqq \min\{h, v_2(s)\}} f(s, u), & \text{if } v_1(s) \leqq h \\ f[s, v_1(s)], & \text{if } v_1(s) \geqq h. \end{cases}$$

Hence the set $N(h)$ of points for which $w^*(s) \geqq w_h(s)$ is measurable. The set $N(h)$, $-\infty < h < \infty$, coincides as is easily seen, with the set $M(h)$ of the function $u^*(s)$, i.e. with the set of those points s for which $u^*(s) \geqq h$. The lemma has been proved.

17.2 *Simplest properties of the superposition operator*

Recall that the *support* of a function $x(s)$ is the set of those values of the argument for which $x(s) \neq 0$. In investigating the superposition operator $\mathfrak{f}$

we often use its 'partial additivity', which is expressed in the statement that for functions $x_1(s), x_2(s), \ldots, x_n(s)$ with disjoint supports the following relation is valid:

$$\mathfrak{f}(x_1 + \ldots + x_n) = \mathfrak{f}x_1 + \ldots + \mathfrak{f}x_n - (n-1)\mathfrak{f}\theta, \tag{17.4}$$

where θ is the function identically equal to zero. This property implies various simple corollaries.

First, if the superposition operator $\mathfrak{f}$ transforms some ball of a space L_α, where $\alpha > 0$ to a space L_β then it transforms each function in L_α to a function in L_β. Second, if $\mathfrak{f}$, as an operator from L_α, $\alpha > 0$ to L_β is continuous at all points of some ball, then it is continuous at each point of the space L_α. Third, if the norms of the values of $\mathfrak{f}$ on some ball of a space L_α, $\alpha > 0$, are bounded, then the norms of its values on any ball are bounded (of course, by different constants). The proofs of these assertions are left to the reader. We only point out that these assertions do not use the Caratheodory condition.

LEMMA 17.3: *Let a superposition operator $\mathfrak{f}$ act from L_α to L_β, with $\alpha, \beta > 0$. Then the operator $\mathfrak{f}$ transforms each set of functions with equi-absolutely continuous norms in L_α to a set of functions with equi-absolutely continuous norms in L_β.*

PROOF: Let $\mathfrak{M}$ be a set of functions with equi-absolutely continuous norms in L_α. Suppose that $\mathfrak{f}\mathfrak{M}$ does not possess the property of equi-absolute continuity. This means that there is a sequence of functions $y_k(s) \in \mathfrak{M}$ and a sequence of sets $F_k \subset \Omega$ such that $\operatorname{mes}(F_k) \to 0$ and

$$\int_{F_k} |f[s, y_k(s)]|^{1/\beta} ds > \gamma \quad (k = 1, 2, \ldots).$$

Here it can be assumed without loss of generality that

$$\sum_{k=1}^{\infty} \operatorname{mes} F_k < \infty.$$

Introduce the notation

$$D_k = \bigcup_{i=k}^{\infty} F_i \quad (k = 1, 2, \ldots).$$

Since mes $D_k \to 0$ as $k \to \infty$, there exists an integer-valued function $\eta(k)$ $(k = 1, 2, \ldots)$ such that $\eta(k) > k$ and

$$\int_{D_k - D_{\eta(k)}} |f[s, y_k(s)]|^{1/\beta} ds > \gamma \quad (k = 1, 2, \ldots).$$

Put

$$k_1 = 1, \; k_2 = \eta(k_1), \ldots, k_n = \eta(k_{n-1}), \ldots$$

The measures of the sets

$$\Omega_n = D_{k_n} - D_{k_{n+1}} \quad (n = 1, 2, \ldots)$$

converge to zero as $n \to \infty$; obviously these sets are mutually disjoint.

It is easy to see that the functions

$$x_n(s) = y_{k_n}(s) \quad (n = 1, 2, \ldots)$$

satisfy the inequality

$$\int_{\Omega_n} |f[s, x_n(s)]|^{1/\beta} ds > \gamma \quad (n = 1, 2, \ldots). \tag{17.5}$$

The functions x_n have equi-absolutely continuous norms. Hence it can be assumed without loss of generality that

$$\sum_{n=1}^{\infty} \int_{\Omega_n} |x_n(s)|^{1/\alpha} ds < \infty.$$

Define the function $u(s) \in L_\alpha$ by the relation

$$u(s) = \begin{cases} x_n(s), & \text{if } s \in \Omega_n \\ 0, & \text{if } s \notin \bigcup_n \Omega_n. \end{cases}$$

The property of partial additivity of the superposition operator implies that[1]

$$\mathfrak{f}u(s) = \sum_{n=1}^{\infty} \boldsymbol{P}_{\Omega_n} \mathfrak{f}x_n(s) + \boldsymbol{P}_{\Omega_0} \mathfrak{f}\theta, \tag{17.6}$$

where $\Omega_0 = \Omega - \bigcup_{n=1}^{\infty} \Omega_n$. From (17.6) and (17.5) it follows that

$$\int_{\Omega} |f[s, u(s)]|^{1/\beta} ds \geqq \sum_{n=1}^{\infty} \int_{\Omega_n} |f[s, x_n(s)]|^{1/\beta} ds > \sum_{n=1}^{\infty} \gamma = \infty.$$

Thus we have constructed a function $u(s) \in L_\alpha$ for which $\mathfrak{f}u(s) \notin L_\beta$. This contradicts the hypothesis of the lemma. The lemma has been proved.

LEMMA 17.4: *Let the operator $\mathfrak{f}$ act from L_0 to L_β, with $\beta > 0$. Then the operator $\mathfrak{f}$ transforms each bounded set $\mathfrak{M} \subset L_0$ to a set of functions with equi-absolutely continuous norms in L_β.*

This assertion follows from Lemma 17.2.

17.3 *Fundamental theorems*

In this and following section it is assumed that $f(s, u)$ satisfies the Caratheodory condition.

LEMMA 17.5:[2] *The superposition operator $\mathfrak{f}$ transforms a sequence of functions converging in measure into a sequence of functions converging in measure.*

PROOF: Let $x_n(s)$ converge to $x^*(s)$ in measure. Then it is possible to choose from each sequence $x_{n_k}(s)$ a subsequence $x_{n_{k'}}(s)$, converging to $x^*(s)$ almost everywhere. The continuity of $f(s, u)$ in the variable u implies that the sequence $\mathfrak{f}x_{n_{k'}}(s)$ also converges almost everywhere to the function $\mathfrak{f}x^*(s)$. This means that it is possible to choose from each sequence $\mathfrak{f}x_{n_k}(s)$ a sub-

---

[1] As usual, $\boldsymbol{P}_D$ denotes the multiplication operator by the characteristic function of the set $\boldsymbol{D}$.

[2] C. Caratheodory [1] (see also V. V. Nemyckii [1]).

sequence converging almost everywhere to $\mathfrak{f}x^*(s)$. Hence the sequence $\mathfrak{f}x_n(s)$ converges to $\mathfrak{f}x^*(s)$ in measure. The lemma has been proved.

THEOREM 17.1:[1] *If the superposition operator $\mathfrak{f}$ acts from L_α to L_β, with $\beta > 0$, then it is continuous.*

PROOF: Let the sequence $x_n(s)$ converge to a function $x^*(s)$ in L_α. Then it converges to $x^*(s)$ in measure. By Lemma 17.5 the sequence $\mathfrak{f}x_n(s)$ converges to $\mathfrak{f}x^*(s)$ in measure.

To complete the proof, it remains only to point out that the terms of the sequence $\mathfrak{f}x_n(s)$ have equi-absolutely continuous norms in L_β—for $\alpha > 0$, this follows from Lemma 17.3, while for $\alpha = 0$ it follows from Lemma 17.4. The theorem has been proved.

This theorem immediately implies

THEOREM 17.2: *If the superposition operator $\mathfrak{f}$ acts from L_α to L_β, with $\beta > 0$, then the norms of its values are bounded on each ball of L_α.*

Let us now analyze superposition operators $\mathfrak{f}$ acting from L_α to L_0. Obviously (for $\alpha > 0$) such operators are defined by bounded functions $f(s, u)$. If $f(s, u)$ actually depends on u then the operator does not possess the property of continuity. Note further that a superposition operator $\mathfrak{f}$ acts from L_0 to L_0 and is continuous if and only if the function $f(s, u)$ is equi-uniformly continuous with respect to u for almost all $s \in \Omega$, whenever u is restricted to a finite interval.

17.4 *Examples*

Consider as our first example the non-linear superposition operator $\mathfrak{f}$ defined by the function

$$f(s, u) = a(s)|u|^\nu, \tag{17.7}$$

[1] Theorems 17.1 and 17.2 were proved by M. A. Krasnoselskii [3, 7]. They were generalized to Orlicz spaces by Y. B. Rutickii [1] and to spaces of abstract functions by M. A. Krasnoselskii, Y. B. Rutickii and R. H. Sultanov [1].

where $\nu > 0$ and $a(s)$ is a measurable function. If $a(s) \in L_\gamma$, then Hölder's inequality implies that $L(\mathfrak{f}; \text{cont.})$ contains all points $\{\alpha, \beta\}$ for which

$$\beta \geqq \gamma + \nu\alpha$$

(Fig. 17.1). This observation permits us to draw the entire L-characteristic $L(\mathfrak{f}; \text{cont.})$; if $a(s) \in L_{\gamma_0}$ but $a(s) \notin L_\gamma$ for $\gamma < \gamma_0$, then $L(\mathfrak{f}; \text{cont.})$ coincides with the set of points $\{\alpha, \beta\}$ for which $\beta \geqq \gamma_0 + \nu\alpha$; if $a(s) \notin L_{\gamma_0}$ but belongs to all L_γ for $\gamma > \gamma_0$, then $L(\mathfrak{f}; \text{cont.})$ coincides with the set of points $\{\alpha, \beta\}$ for which $\beta > \gamma_0 + \nu\alpha$.

Now assume that the operator $\mathfrak{f}$ is defined by a function of the form

$$f(s, u) = a_1(s)|u|^{\nu_1} + \ldots + a_n(s)|u|^{\nu_n} \quad (\nu_1 < \nu_2 < \ldots < \nu_n). \tag{17.8}$$

In this case the L-characteristic $L(\mathfrak{f}; \text{cont.})$ coincides with the intersection of the L-characteristics $L(\mathfrak{f}_i; \text{cont.})$ $(i = 1, 2, \ldots, n)$, where

$$\mathfrak{f}_i x(s) = a_i(s)|x(s)|^{\nu_i} \quad (i = 1, 2, \ldots, n).$$

Consequently, the L-characteristic $L(\mathfrak{f}; \text{cont.})$ of the superposition operator generated by the function (17.8) is a polygon (Fig. 17.2). It is interesting to note that the set of points in the L-characteristic $L(\mathfrak{f}; \text{cont.})$ lying on any ray from the origin is always an infinite interval (if it is not empty). In other words, each ray from the origin cuts the broken line $\beta = \xi(\alpha)$ which is the

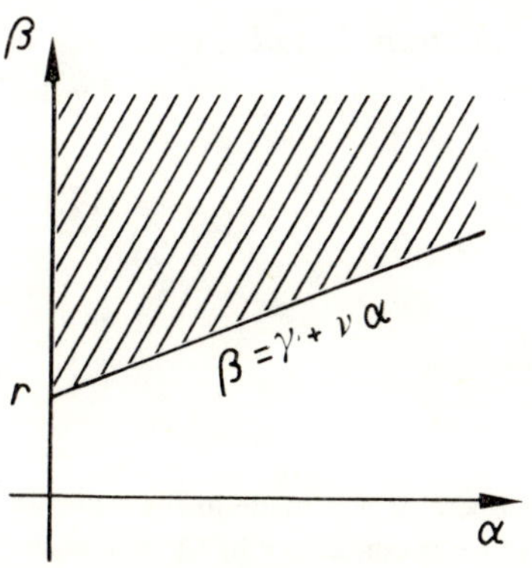

Figure 17.1

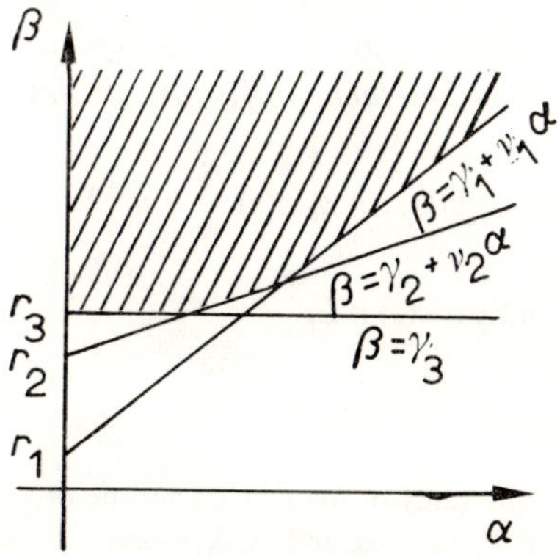

Figure 17.2

lower boundary of the L-characteristic $L(\mathfrak{f}$; cont.) at no more than one point. The only exception is the case in which the broken line $\beta = \xi(\alpha)$ contains an infinite segment on some line $\beta = \nu^*\alpha$ (this means that the coefficient $a_n(s)$ corresponding to the largest exponent $\nu^* = \nu_n$ on the right side of formula (17.8) is bounded). Note also that the L-characteristic of the superposition operator defined by the function (17.8) is convex.

These arguments immediately suggest a method for construction of the L-characteristic $L(\mathfrak{f}$; cont.) in the case in which the function $f(s, u)$ has the form

$$f(s, u) = \sum_{i=1}^{n} a_i(s)\varphi_i(u) \tag{17.9}$$

where the $a_i(s)$ are measurable and the $\varphi_i(u)$ are continuous functions.

As our next example consider the superposition operator $\mathfrak{f}$, defined by the function

$$f(s, u) = \min\{a(s), u^2\} \tag{17.10}$$

where $a(s)$ is a positive function in $L_{\gamma_0} - \bigcup_{\gamma<\gamma_0} L_\gamma$. It is easy to see that in this example the L-characteristic does not possess the property of convexity—it coincides with the set hatched in Fig. 17.3.

17.5 *General form of L-characteristics of superposition operators*[1]

LEMMA 17.6: *Let the superposition operator $\mathfrak{f}$ act from L_α to L_β, where $\alpha > 0$. Then the function $f(s, u)$ satisfies the inequality*

$$|f(s, u)| \leqq a(s) + b|u|^{\beta/\alpha} \tag{17.11}$$

where $a(s) \in L_\beta$.

---

[1] A study of the L-characteristic of a superposition operator was made in the report of P. P. Zabreiko and M. A. Krasnoselskii [1]. Lemma 17.6 was established by M. A. Krasnoselskii (see Y. B. Rutickii [1]); the present proof was indicated by Y. B. Rutickii.

PROOF: By Theorem 17.2 there is a positive number $\tilde{b}$ such that the inequality $\|x\|_\alpha \leqq 1$ implies the inequality

$$\|\mathfrak{f}x\|_\beta \leqq \tilde{b}.$$

Define the function

$$\varphi(s, u) = \begin{cases} |f(s, u)|^{1/\beta} - \tilde{b}^{1/\beta}|u|^{1/\alpha}, & \text{if } |f(s, u)| \geqq \tilde{b}\,|u|^{\beta/\alpha} \\ 0, & \text{if } |f(s, u)| < \tilde{b}\,|u|^{\beta/\alpha}. \end{cases}$$

Let $x(s) \in L_\alpha$. Denote by $\hat{\Omega}$ the set of those $s \in \Omega$ for which $\varphi[s, x(s)] > 0$. Let

$$\int_{\hat{\Omega}} |x(s)|^{1/\alpha} ds = n + \varepsilon$$

where n is an integer and $0 \leqq \varepsilon < 1$. Partition the set $\hat{\Omega}$ to $n + 1$ parts $\Omega_0, \ldots, \Omega_n$ such that the following inequalities hold:

$$\int_{\Omega_i} |x(s)|^{1/\alpha} ds < 1 \qquad (i = 0, 1, \ldots, n).$$

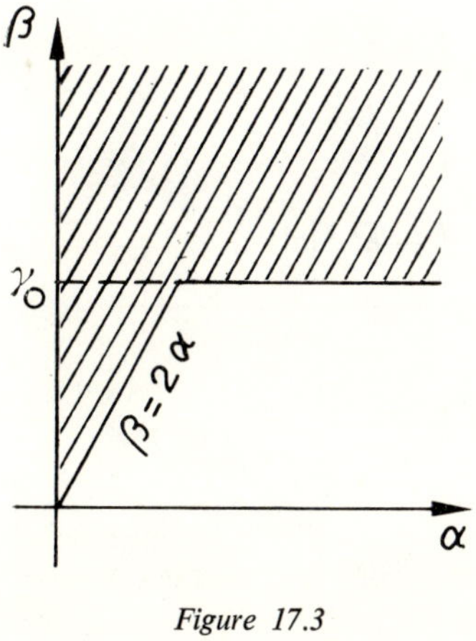

Figure 17.3

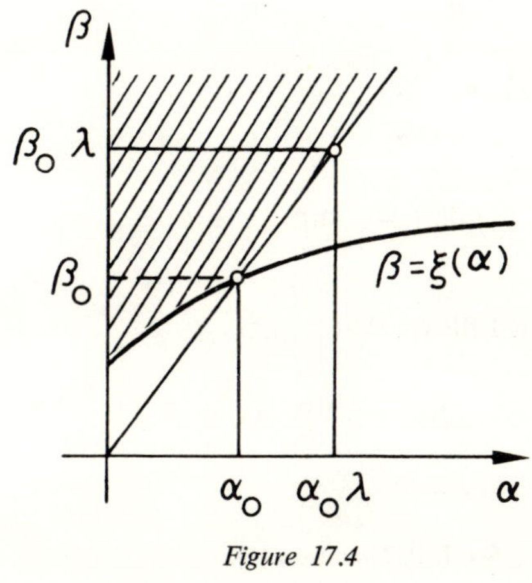

Figure 17.4

Then

$$\int_{\hat{\Omega}} |f[s, x(s)]|^{1/\beta} ds = \sum_{i=0}^{n} \int_{\Omega_i} |f[s, x(s)]|^{1/\beta} ds \leqq (n + 1)\tilde{b}^{1/\beta}$$

and

$$\int_{\Omega} \varphi[s, x(s)] ds \leqq \int_{\hat{\Omega}} |f[s, x(s)]|^{1/\beta} ds - \tilde{b}^{1/\beta} \int_{\hat{\Omega}} |x(s)|^{1/\alpha} ds$$

$$\leqq (n + 1)\tilde{b}^{1/\beta} - (n + \varepsilon)\tilde{b}^{1/\beta} \leqq \tilde{b}^{1/\beta}. \tag{17.12}$$

From Lemma 17.2 it follows that there exists a sequence of functions $u_k(s)$, $|u_k(s)| \leqq k$ such that

$$\varphi[s, u_k(s)] = \max_{|u| \leqq k} \varphi(s, u).$$

Put

$$\tilde{a}(s) = \sup_{-\infty < u < \infty} \varphi(s, u) = \lim_{k \to \infty} \varphi[s, u_k(s)].$$

By (17.12) and Fatou's lemma [1]

$$\int_{\Omega} \tilde{a}(s) ds \leqq \sup_{n} \int_{\Omega} \varphi[s, u_n(s)] ds \leqq \tilde{b}^{1/\beta}.$$

Hence $\tilde{a}(s) \in L_1$.

Since

$$\tilde{a}(s) = \sup_{-\infty < u < \infty} \varphi(s, u) \geqq \sup_{-\infty < u < \infty} \{|f(s, u)|^{1/\beta} - \tilde{b}^{1/\beta} |u|^{1/\alpha}\},$$

it follows that

$$|f(s, u)|^{1/\beta} \leqq \tilde{a}(s) + \tilde{b}^{1/\beta} |u|^{1/\alpha};$$

---

[1] See I. P. Natanson [1].

that is,

$$|f(s, u)| \leqq a(s) + b|u|^{\beta/\alpha}$$

where $a(s) = 2^{\beta}\tilde{a}(s)^{\beta}$ and $b = 2^{\beta}\tilde{b}$. The lemma has been proved.

The assertion of this lemma does not hold in case $\alpha = 0$.

LEMMA 17.7: *Let* $\{\alpha_0, \beta_0\} \in L(\mathfrak{f}; \text{cont.})$. *Then all points on the ray* $\alpha = \alpha_0\lambda$, $\beta = \beta_0\lambda$ *for* $\lambda \geqq 1$ *also belong to* $L(\mathfrak{f}; \text{cont.})$.

PROOF: In the case $\alpha_0 = 0$ the assertion of the lemma is obvious. Let $\alpha_0 > 0$. By Lemma 17.6 the inequality (17.11) is valid. Hence the operator $\mathfrak{f}$ acts from L_{α} to L_{β}, where $\beta = (\beta_0/\alpha_0)\alpha$ for $\alpha \geqq \alpha_0$. It remains only to use Theorem 17.1. The lemma has been proved.

A set G situated in the quadrant $\alpha, \beta \geqq 0$ is said to possess the *zero-concavity property* if it contains together with each non-zero point $\{\alpha_0, \beta_0\}$ all the points $\{\alpha_0\lambda, \beta_0\lambda\}$ for $\lambda \geqq 1$. Obviously the set T of all points $\{\alpha, \beta\}$ where $\beta \geqq \xi(\alpha)$ (or $\beta > \xi(\alpha)$) and $\xi(\alpha)$ is a non-negative concave function (Fig. 17.4) is zero-concave. Convex sets can possess the zero-concavity property; in the preceding section it was noted that the L-characteristic of the superposition operator $\mathfrak{f}$ defined by the function (17.8) is convex and zero-concave.

Lemma 17.7 means that the following is valid.

THEOREM 17.3: *The L-characteristic* $L(\mathfrak{f}; \text{cont.})$ *of each superposition operator possesses the zero-concavity property.*

The L-characteristic $L(\mathfrak{f}; \text{cont.})$ obviously possesses the extrapolation property introduced in 1.6°. Hence this L-characteristic can be described with the aid of the function

$$\xi(\alpha) = \inf_{\{\alpha, \beta\} \in L(\mathfrak{f}; \text{cont.})} \beta. \tag{17.13}$$

This function is non-decreasing. From Theorem 17.3 follows that the function

$$\eta(\alpha) = \frac{1}{\alpha}\,\xi(\alpha) \tag{17.14}$$

is non-increasing. This implies that $\xi(\alpha)$ is continuous (from the monotonicity of $\xi(\alpha)$ follows that it has only discontinuities of the first kind while from the monotonicity of $\eta(\alpha)$ it follows that such discontinuities do not exist). In the sequel a continuous, non-decreasing and non-negative function $\xi(\alpha)$ will be called *zero-concave* if the function (17.14) is non-increasing.

It turns out that *for each zero-concave function* $\xi_0(\alpha)$ *it is possible to construct a function* $f_1(s, u)$ *such that the L-characteristic* $L(\mathfrak{f}_1; \text{cont.})$ *of the corresponding operator* $\mathfrak{f}_1$ *coincides with the set* $G_1 = \{\{\alpha, \beta\} : \beta \geqq \xi_0(\alpha)\}$. *Similarly, it is possible to construct a function* $f_2(s, u)$ *such that* $L(\mathfrak{f}_2; \text{cont.})$ *coincides with the set* $G_2 = \{\{\alpha, \beta\} : \beta > \xi_0(\alpha)\}$.

We give here the construction of the function $f_1(s, u)$.

Let $a_0(s)$ be a function, satisfying the conditions

$$a_0(s) > 1, \quad a_0(s) \in L_1 - \bigcup_{\alpha < 1} L_\alpha. \tag{17.15}$$

Put

$$f_1(s, u) = \inf_{0 < \alpha < \infty} \{a_0(s)^{\xi_0(\alpha)} + (2 + |u|)^{\xi_0(\alpha)/\alpha}\}. \tag{17.16}$$

The function $f_1(s, u)$ satisfies the Caratheodory condition. We shall show that for all s, $f_1(s, u)$ is continuous in u in the case

$$\lim_{\alpha \to \infty} \xi_0(\alpha) = \lim_{\alpha \to 0} \frac{\xi_0(\alpha)}{\alpha} = \infty \tag{17.17}$$

(the remaining cases are treated similarly).

Equations (17.16) and (17.17) imply that

$$f_1(s, u) = a_0(s)^{\xi_0(\bar{\alpha})} + (2 + |u|)^{\xi_0(\bar{\alpha})/\bar{\alpha}} \tag{17.18}$$

where $\bar{\alpha}$ is some number in the interval $(0, \infty)$ which depends on s and u. Fix s. Since the functions $\xi_0(\alpha)$ and $\xi_0(\alpha)/\alpha$ are continuous it is possible for each u_0 and $\varepsilon > 0$ to take $\delta > 0$ such that $|u - u_0| < \delta$ implies

$$f_1(s, u) < f_1(s, u_0) + \varepsilon.$$

This means that for each s the function $f_1(s, u)$ is upper semi-continuous.

Obviously the function $f_1(s, u)$ is monotone in u. Hence the upper semicontinuity implies

$$\lim_{u\to u_0+0} f_1(s, u) = f_1(s, u).$$

Now let $u_n \to u_0$, $u_n < u_0$ and suppose

$$f_1(s, u_n) \leqq m < f_1(s, u_0).$$

Denote by $\bar{\alpha}_n$ the sequence of numbers, for which

$$f_1(s, u_n) = a_0(s)^{\xi_0(\bar{\alpha}_n)} + (2 + |u_n|)^{\xi_0(\bar{\alpha}_n)/\bar{\alpha}_n}. \tag{17.19}$$

Equation (17.17) implies that the numbers $\bar{\alpha}_n$ satisfy the inequalities $0 < a \leqq \bar{\alpha}_n \leqq A < \infty$. Hence it can be assumed without loss of generality that $\bar{\alpha}_n \to \alpha^* \neq 0$. Passing to limit as $n \to \infty$ in the relation (17.19) we obtain

$$a_0(s)^{\xi_0(\alpha^*)} + (2 + |u_0|)^{\xi_0(\alpha^*)/\alpha^*} < f_1(s, u_0)$$

which contradicts (17.16). We have proved that

$$\lim_{u\to u_0-0} f_1(s, u) = f_1(s, u_0).$$

The continuity of the function $f_1(s, u)$ in u has been proved.

The L-characteristic of the superposition operator $\mathfrak{f}_1$ defined by the function (17.16) contains all points $\{\alpha, \beta\}$ for which $\beta \geqq \xi_0(\alpha)$. Furthermore for each α_0

$$\mathfrak{f}_1[s, a_0(s)^{\alpha_0}] = \inf_{\alpha} \{a_0(s)^{\xi_0(\alpha)} + [2 + a_0(s)^{\alpha_0}]^{\xi_0(\alpha)/\alpha}\}$$

$$\geqq \inf_{\alpha} \max \{a_0(s)^{\xi_0(\alpha)},\ a_0(s)^{\alpha_0\xi_0(\alpha)/\alpha}\} = a_0(s)^{\xi_0(\alpha_0)}.$$

This inequality and (17.15) imply that $\mathfrak{f}_1$ does not act from L_{α_0} to L_β for any $\beta < \xi_0(\alpha_0)$.

We have shown that the L-characteristic $L(\mathfrak{f}_1; \text{cont.})$ coincides with the set of points $\{\alpha, \beta\}$ for which $\beta \geqq \xi_0(\alpha)$.

Now consider the function

$$f_2(s, u) = \ln [e + c(s)] \cdot f_1(s, u),$$

where $c(s)$ is a non-negative summable function, such that $a_0(s) \ln \ln [e + c(s)]$ is not summable. It is easy to see that the L-characteristic $L(\mathfrak{f}_2; \text{cont.})$ consists of the points $\{\alpha, \beta\}$ for which $\beta > \xi_0(\alpha)$.

17.6 *Uniform continuity of the superposition operator*

Simple examples show that a superposition operator $\mathfrak{f}$ is not, in general, uniformly continuous on bounded sets of a space L_α.

For instance, put

$$f(u) = u \sin u,$$

and consider the corresponding operator $\mathfrak{f}$. For any $\alpha \geqq 0$ this operator obviously acts from L_α to L_α and is continuous. Let us show that on any ball $\|x\|_\alpha \leqq r$ it does not possess the property of uniform continuity.

Let Ω_n $(n = 1, 2, \ldots)$ be an arbitrary sequence of sets for which

$$\operatorname{mes} \Omega_n = \left(\frac{r}{4\pi n}\right)^{1/\alpha}.$$

Define the functions $x_n(s)$ and $y_n(s)$ by the relations

$$x_n(s) = \begin{cases} (4n + 1)\pi/2, & \text{if } s \in \Omega_n \\ 0, & \text{if } s \notin \Omega_n; \end{cases}$$

$$y_n(s) = \begin{cases} (4n - 1)\pi/2, & \text{if } s \in \Omega_n \\ 0, & \text{if } s \notin \Omega_n. \end{cases}$$

It is clear that $x_n(s), y_n(s) \in L_\alpha$

$$\|x_n(s)\|_\alpha, \ \|y_n(s)\|_\alpha \leqq r$$

and

$$\|x_n(s) - y_n(s)\|_\alpha \to 0.$$

However

$$\mathfrak{f}x_n(s) - \mathfrak{f}y_n(s) = \begin{cases} 4\pi n, & \text{if } s \in \Omega_n \\ 0, & \text{if } s \notin \Omega_n, \end{cases}$$

whence it follows that $\|\mathfrak{f}x_n - \mathfrak{f}y_n\|_\alpha = r$.

Denote by $L(A$; unif. cont.) the set of points $\{\alpha, \beta\}$ such that A acts from L_α to L_β and is uniformly continuous on each ball $\|x\|_\alpha \leqq R$.

THEOREM 17.4: *Let* $\{\alpha_0, \beta_0\} \in L(\mathfrak{f}; \text{cont.})$. *Then all points* $\{\alpha, \beta\}$ *where* $\alpha \leqq \alpha_0$, $\beta \geqq \beta_0$ *and* $\{\alpha, \beta\} \neq \{\alpha_0, \beta_0\}$ *belong to* $L(\mathfrak{f}$; unif. cont.).

PROOF: Let us show that the operator $\mathfrak{f}$ is uniformly continuous in measure on each ball of any space L_α. This means that for any two sequences $u_n(s)$, $v_n(s)$ in the ball $\|x\|_\alpha \leqq R$ for which the sequence $w_n(s) = u_n(s) - v_n(s)$ converges to zero in measure, the sequence of functions $\mathfrak{f}u_n(s) - \mathfrak{f}v_n(s)$ also converges to zero in measure.

Let positive numbers h and ε be given. By Lemma 17.1 it is possible to take a closed set $\Omega_0 \subset \Omega$ such that $\operatorname{mes}(\Omega - \Omega_0) < \varepsilon/4$ and the function $f(s, u)$ is jointly continuous in the variables $s \in \Omega_0$, $-\infty < u < \infty$. Let

$$M = \left(\frac{8}{\varepsilon}\right)^\alpha R.$$

The function $f(s, u)$ is jointly uniformly continuous in the variables $s \in \Omega_0$, $|u| \leqq M$. Hence there is a $\delta > 0$ such that $|u - v| < \delta$ and $|u|, |v| \leqq M$ imply the inequality

$$|f(s, u) - f(s, v)| \leqq h.$$

Now consider sequences $u_n(s)$ and $v_n(s)$ in the ball $\|x(s)\|_\alpha \leqq R$ such that $w_n(s) = u_n(s) - v_n(s)$ forms a sequence convergent to zero in measure. Denote by Ω_n the set of points $s \in \Omega_0$ for which $|u_n(s)|, |v_n(s)| \leqq M$.

Obviously $\operatorname{mes}(\Omega_0 - \Omega_n) < \varepsilon/4$ and $\operatorname{mes}(\Omega - \Omega_n) < \varepsilon/2$ for n sufficiently large. The convergence of the sequence $w_n(s)$ to zero in measure implies the existence of an N such that for $n \geqq N$ the inequality $\operatorname{mes}(G_n) < \varepsilon/2$ is valid, where

$$G_n = \{s; |u_n(s) - v_n(s)| \geqq \delta\}.$$

Then for $s \in \Omega_n - G_n$ the inequality $|f[s, u_n(s)] - f[s, v_n(s)]| \leqq h$ holds. It remains only to note that

$$\operatorname{mes}[\Omega - (\Omega_n - G_n)] \leqq \operatorname{mes}(\Omega - \Omega_n) + \operatorname{mes} G_n < \varepsilon.$$

Let us now prove the assertion of the theorem. We have to show that for any sequence of functions $u_n(s)$, $v_n(s)$ in the ball $\|x\|_\alpha \leqq R$ satisfying $\|u_n(s) - v_n(s)\|_\alpha \to 0$ the following relation is valid:

$$\lim_{n\to\infty} \|\mathfrak{f}u_n(s) - \mathfrak{f}v_n(s)\|_\beta = 0. \tag{17.20}$$

Since the difference $\mathfrak{f}u_n(s) - \mathfrak{f}v_n(s)$ converges to zero in measure, it suffices to show that these differences have equi-absolutely continuous norms. This last assertion is obvious if $\beta > \beta_0$; it follows from Lemma 17.3, if $\alpha < \alpha_0$. The theorem has been proved.

Next, suppose that a function $f(s, u)$ satisfies the inequality

$$|f(s, u_1) - f(s, u_2)| \leqq g(s, R)|u_1 - u_2|^\delta \quad (|u_1|, |u_2| \leqq R), \tag{17.21}$$

where $g(s, u)$ is a function satisfying the Caratheodory condition and $\delta > 0$. Suppose further that the superposition operator $\mathfrak{f}$ defined by the function $f(s, u)$ acts from L_α to L_β, where $\beta \geqq \alpha\delta$, and that the superposition operator $\mathfrak{g}$ defined by $g(s, u)$ acts from L_α to $L_{\beta-\alpha\delta}$. Then the Hölder inequality implies that

$$\|\mathfrak{f}x_1 - \mathfrak{f}x_2\|_\beta \leqq \|g[s, \max\{|x_1(s)|, |x_2(s)|\}]|x_1(s) - x_2(s)|^\delta\|_\beta$$

$$\leqq \|g[s, \max\{|x_1(s)|, |x_2(s)|]\|_{\beta-\alpha\delta}\|x_1 - x_2\|_\alpha^\delta,$$

i.e.

$$\|\mathfrak{f}x_1 - \mathfrak{f}x_2\|_\beta \leqq k(R)\cdot\|x_1 - x_2\|_\alpha^\delta \quad (\|x_1\|_\alpha, \|x_2\|_\alpha \leqq R)$$

where

$$k(R) = \sup_{\|x\| \leqq 2R} \|\mathfrak{g}x\|_{\beta-\alpha\delta}.$$

It thus follows from (17.21) that the indicated requirement on $g(t, u)$ implies the uniform continuity of $\mathfrak{f}$ as an operator from L_α to L_β, on each ball of the space L_α.

Suppose that we know the L-characteristics of the operator $\mathfrak{f}$ and of the operator $\mathfrak{g}$ defined by the function $g(s, u)$ on the right side of inequality (17.21). The L-characteristic of the operator $\mathfrak{f}$ is indicated in Fig. 17.5 a) while the curve $\beta = \xi_1(\alpha)$ in Fig. 17.5 b) denotes the boundary of the L-characteristic of the operator $\mathfrak{g}$. Then the estimate (17.21) implies that the set hatched in Fig. 17.5 c) belongs to $L(\mathfrak{f}$; unif. cont.). This hatched set consists of those points in $L(\mathfrak{f}$; cont.) not lying below the curve $\beta = \xi_1(\alpha) + \delta\alpha$. We note that inequality (17.21) thus permits us to show that some points on the boundary of $L(\mathfrak{f}$; cont.) belong to $L(\mathfrak{f}$; unif. cont.) (which cannot be accomplished by use of Theorem 17.4).

Condition (17.21) can be replaced by a more general inequality of the type

$$|f(s, u_1) - f(s, u_2)| \leqq \sum_{i=1}^{n} g_i(s, R)|u_1 - u_2|^{\delta_i} \quad (|u_1|, |u_2| < R), \tag{17.22}$$

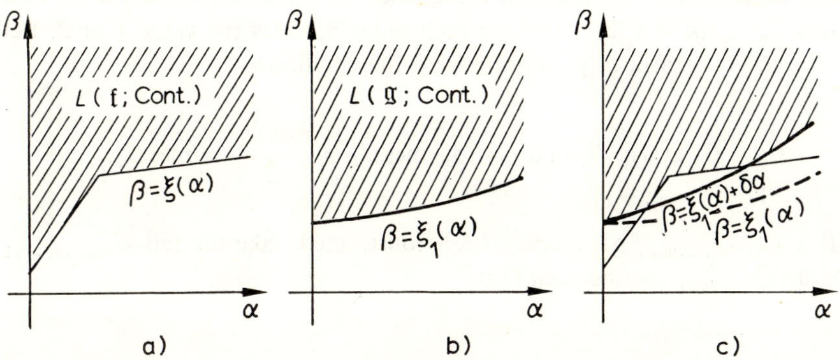

Figure 17.5

where the functions $g_i(s, u)$ satisfy the Caratheodory condition and the exponents δ_i are positive. If $\mathfrak{f}$ acts from L_α to L_β, $\beta \geqq \delta_i\alpha$, and each operator $\mathfrak{g}_i$ defined by the function $g_i(s, u)$, acts from L_α to $L_{\beta-\delta_i\alpha}$, then the following estimate is valid:

$$\|\mathfrak{f}x_1 - \mathfrak{f}x_2\|_\beta \leqq \sum_{i=1}^{n} k_i(R)\|x_1 - x_2\|_\alpha^{\delta_i} \quad (\|x_1\|_\alpha, \|x_2\|_\alpha \leqq R).$$

It is interesting to note that in condition (17.21) the number δ has to satisfy the inequality $\delta \leqq 1$—in the contrary case the function $f(s, u)$ does not depend on u. In condition (17.22) some of numbers δ_i can be greater than 1.

17.7 *Improvement of superposition operators*[1]

In applications, superposition operators possessing a compactness property would be particularly convenient. If $f(s, u)$ actually depends on u then it turns out that the operator $\mathfrak{f}$ acting from L_α to L_β transforms each ball of the space L_α into a set which is not compact in L_β. Furthermore, on each ball of the space L_α the range of such an operator $\mathfrak{f}$ forms a set of functions not even possessing the property of compactness in measure.

In fact, let $\mathfrak{f}u_1(s) \neq \mathfrak{f}u_2(s)$ for some fixed functions $u_1(s)$ and $u_2(s)$ in L_α. This means that there exists a number $\gamma > 0$ and a set $\Omega_0 \subset \Omega$ of non-zero measure such that

$$|f[s, u_1(s)] - f[s, u_2(s)]| \geqq \gamma \quad (s \in \Omega_0).$$

Let us construct by induction a sequence of partitions of the set Ω_0 into sets $G_{\alpha_1\ldots\alpha_k}$ $(k = 1, 2, \ldots)$ where each index α_i takes the value 1 or 2. Let G_1 and G_2 be arbitrary sets satisfying the conditions

$$\Omega_0 = G_1 \cup G_2, \quad G_1 \cap G_2 = \varnothing, \quad \operatorname{mes} G_1 = \frac{\operatorname{mes} \Omega_0}{2} = \operatorname{mes} G_2.$$

If a set $G_{\alpha_1,\ldots,\alpha_{k-1}}$ has already been constructed, take the sets $G_{\alpha_1,\ldots,\alpha_{k-1},1}$ and $G_{\alpha_1,\ldots,\alpha_{k-1},2}$ to be such that

[1] P. P. Zabreiko and E. P. Pustylnik [1].

$$G_{\alpha_1,\dots,\alpha_{k-1},1} \cup G_{\alpha_1,\dots,\alpha_{k-1},2} = G_{\alpha_1,\dots,\alpha_k},$$

$$G_{\alpha_1,\dots,\alpha_{k-1},1} \cap G_{\alpha_1,\dots,\alpha_{k-1},2} = \varnothing,$$

$$\operatorname{mes} G_{\alpha_1,\dots,\alpha_{k-1},1} = \operatorname{mes} G_{\alpha_1,\dots,\alpha_{k-1},2} = \frac{\operatorname{mes} \Omega_0}{2^k}.$$

Now define the function $x_k(s)$ by the relation

$$x_k(s) = \begin{cases} u_1(s), & \text{if } s \in \bigcup\limits_{\alpha_1,\dots,\alpha_{k-1}} G_{\alpha_1,\dots,\alpha_{k-1},1} \\ u_2(s), & \text{if } s \in \bigcup\limits_{\alpha_1,\dots,\alpha_{k-1}} G_{\alpha_1,\dots,\alpha_{k-1},2} \\ 0, & \text{if } s \notin \Omega_0. \end{cases}$$

It is easy to see that the set of functions $x_k(s)$ lies in some ball of the space L_α. At the same time

$$|\mathfrak{f}x_k(s) - \mathfrak{f}x_l(s)|$$

$$= \begin{cases} |f[s, u_1(s)] - f[s, u_2(s)]|, & \text{if } s \in F_{k,l} \\ 0, & \text{if } s \notin F_{k,l} \end{cases}$$

where

$$F_{k,l} = \bigcup_{\alpha_k \neq \alpha_l} G_{\alpha_1,\dots,\alpha_k,\dots,\alpha_l}.$$

Since $\operatorname{mes}(F_{k,l}) = \operatorname{mes}(\Omega_0)/2$, it follows that the sequence $\mathfrak{f}x_k(s)$ is not compact in measure.

An operator $\mathfrak{f}$ will be said to act from L_α to L_β ($\beta > 0$) as an *improving* operator if it transforms each ball of the space L_α to a set of functions with equi-absolutely continuous norms in L_β. The notion of the L-characteristic $L(\mathfrak{f}$; imp.) is defined in the obvious way. From the Hölder inequality and Lemma 17.3 it follows that all interior points of the L-characteristic $L(\mathfrak{f}$; cont.) belong to the L-characteristic $L(\mathfrak{f}$; imp.). It is also not difficult to see that $L(\mathfrak{f}$; imp.) $\subset L(\mathfrak{f}$; unif. cont.).

Theorem 17.1 and Lemma 17.6 imply that the operator $\mathfrak{f}$ acts from L_α to L_β if and only if the following inequality holds:

$$|f(s, u)| \leqq a_0(s) + b|u|^{\beta/\alpha} \qquad (s \in \Omega, -\infty < u < \infty), \tag{17.23}$$

where $a_0(s) \in L_\beta$. It is easily proved that $\mathfrak{f}$ acts from L_α to L_β as an improving operator if

$$|f(s, u)| \leqq a_0(s) + b(s)|u|^{(\beta-\varepsilon)/\alpha} \qquad (s \in \Omega, -\infty < u < \infty), \tag{17.24}$$

where $a_0(s) \in L_\beta$, $\varepsilon \in (0, \beta)$ and $b(s) \in L_\varepsilon$. Clearly inequality (17.24) is a more severe requirement than inequality (17.23).

THEOREM 17.5: *The operator* $\mathfrak{f}$ *acts from* L_α *to* L_β $(\alpha, \beta > 0)$ *as an improving operator if and only if there exists a continuous even function* $M(u)$ *satisfying the condition*

$$\lim_{u \to \infty} \frac{M(u)}{u} = \infty, \tag{17.25}$$

such that

$$M[f(s, u)] \leqq a_0(s) + b|u|^{\beta/\alpha} \qquad (s \in \Omega, -\infty < u < \infty) \tag{17.26}$$

where $a_0(s) \in L_\beta$.

PROOF: Let $\mathfrak{f}$ act from L_α to L_β as an improving operator. Denote by $\mathfrak{M}$ its range on the ball $\|x\|_\alpha \leqq 1$. The set $\mathfrak{M}$ has equi-absolutely continuous norms in L_β. By the Vallée-Poussin criterion (see 1.2°) there exists a continuous even function $M(u)$ satisfying condition (17.25), such that $M[y(s)] \in L_\beta$ for $y(s) \in \mathfrak{M}$. This means that the superposition operator

$$\mathfrak{f}_1 x(s) = M\{f[s, x(s)]\}$$

acts from the unit ball of the space L_α to L_β. Hence it acts from the whole of L_α to L_β. Lemma 17.5 now implies the estimate (17.26).

Conversely, if condition (17.26) is fulfilled, then the equi-absolute continuity in (L_β-)norm of the range of $\mathfrak{f}$ on each ball of the space L_α is also a consequence of the Vallée-Poussin criterion. The theorem has been proved.

17.8 *Supplementary remarks*

In this section we consider functions $f(s, u)$ not satisfying the Caratheodory condition.

A function $f(s, u)$ will be called *superpositionally measurable* if the superposition operator $\mathfrak{f}$ transforms each measurable function to a measurable one, i.e. if the superposition $f[s, x(s)]$ is measurable for each measurable function $x(s)$. It is not difficult to construct functions $f(s, u)$ which are jointly measurable in the variables $s \in \Omega$, $-\infty < u < \infty$ and possess for each fixed s no more than one discontinuity point in u, but do not possess the property of superpositional measurability. The simplest example is a function $f(s, u)$ which is almost everywhere equal to zero and is the characteristic function of the graph of the function $u = \kappa_D(s)$, where κ_D is the characteristic function of a non-measurable set in Ω.

However, functions satisfying the Caratheodory condition are not the only superpositionally measurable functions. It is not difficult to see, for instance, that a function $f(s, u) = f_0(u)$ is superpositionally measurable if $f_0(u)$ is Borel measurable (B-measurable, see I. P. Natanson [1]).

It follows directly from the definition that a function $f(s, u)$ is superpositionally measurable if for almost all $s \in \Omega$ and for all u it is of the form

$$f(s, u) = \lim_{n\to\infty} f_n(s, u) \tag{17.27}$$

where the $f_n(s, u)$ are superpositionally measurable. This remark permits us to define certain classes $B(\alpha)$ of superpositionally measurable functions, where α is a transfinite number with countable power. The class $B(0)$ consists of those functions satisfying the Caratheodory condition. The class $B(\alpha_0)$ consists of those functions which can be obtained via the relation (17.27) with $f_n(s, u) \in B(\alpha)$, $\alpha < \alpha_0$.

Note further that for the superpositional measurability of a function $f(s, u)$ it is sufficient (by Lusin's theorem) that the operator $\mathfrak{f}$ transform each continuous function to a measurable one. In particular, if $\mathfrak{f}$ acts from L_α to L_β then the function $f(s, u)$ is superpositionally measurable.

LEMMA 17.8: *Let the function $f(s, u)$ be superpositionally measurable and let the function $x_0(s)$ be measurable. Then the operator $\mathfrak{f}$ is continuous in measure at the point $x_0(s)$ if and only if, for each $h > 0$*

$$\lim_{u \to 0} \text{mes}\,\{s : |f[s, x_0(s) + u] - f[s, x_0(s)]| \geqq h\} = 0. \tag{17.28}$$

PROOF: Let a sequence $x_n(s)$ converge in measure to the function $x_0(s)$ and let h be a fixed number. For each $\varepsilon > 0$, select the number $\delta > 0$ such that for $|u| \leqq \delta$ the following inequality holds:

$$\text{mes}\,\{s : |f[s, x_0(s) + u] - f[s, x_0(s)]| \geqq h\} < \varepsilon/2.$$

Further, let n_0 be a number such that for $n \geqq n_0$ the inequality

$$\text{mes}\,\{s : |x_n(s) - x_0(s)| \geqq \delta\} < \varepsilon/2$$

holds. Then for each $n \geqq n_0$ the inequality

$$|f[t, x_n(s)] - f[t, x_0(s)]| \geqq h$$

holds only for points in a set, whose measure is less than ε. The sufficiency has been proved. The necessity is trivial. The lemma has been proved.

Lemmas 17.3 and 17.8 imply:

THEOREM 17.6: *Let the operator $\mathfrak{f}$ act from L_α to L_β, $\beta > 0$. Then the operator $\mathfrak{f}$ is continuous at a point $x_0(s) \in L_\alpha$ if and only if the following condition holds for each $h > 0$:*

$$\lim_{u \to 0} \text{mes}\,\{s : |f[s, x_0(s) + u] - f[s, x_0(s)]| \geqq h\} = 0.$$

The most important case in applications is that in which $f(s, u)$ is jointly measurable in the two variables and is continuous in u for almost all $s \in \Omega$, except at a finite number of points $u_1(s), u_2(s), \ldots, u_n(s)$, where the $u_i(s)$ $(i = 1, \ldots, n)$ are measurable functions. It is clear that such a function is superpositionally measurable. The corresponding operator $\mathfrak{f}$ is continuous in measure at each point $x(s)$ for which

$$\text{mes}\,\{s : x(s) = u_i(s)\} = 0 \quad (i = 1, 2, \ldots, n).$$

The arguments of this section can be applied without difficulty to superposition operators $\mathfrak{f}$ corresponding to functions $f(s, u)$, which are not defined for all u. Such operators cannot be considered on any set in L_α ($\alpha > 0$) containing interior points. Let us consider the most important special case.

Let $v_1(s)$ and $v_2(s)$ be measurable functions (possibly, taking the value $+\infty$ or $-\infty$ on sets of positive measure) with $v_1(s) \leqq v_2(s)$. Assume that a function $f(s, u)$ is defined for $s \in \Omega$, $v_1(s) \leqq u \leqq v_2(s)$. In this case the superposition $f[s, x(s)]$ is defined for those functions $x(s)$ satisfying for almost all $s \in \Omega$ the inequality $v_1(s) \leqq x(s) \leqq v_2(s)$. Thus the operator $\mathfrak{f}$ can be applied only to functions $x(s)$ in L_α satisfying the supplementary conditions $v_1(s) \leqq x(s) \leqq v_2(s)$. It is convenient to denote this set of functions by $L_\alpha\langle v_1, v_2\rangle$. Let us formulate the fundamental result.

If a function $f(s, u)$ is measurable in s (for all u) and continuous in u for $v_1(s) < u < v_2(s)$ and for almost all s, then $\mathfrak{f}$ is continuous at all points of the set $L_\alpha\langle v_1, v_2\rangle$ as an operator from L_α to L_β, if $\mathfrak{f}$ acts from $L_\alpha\langle v_1, v_2\rangle$ to L_β.

§ 18 Conditions for continuity of integral operators[1]

18.1 *Definitions and simple properties*

Let Ω and Ω^* be two sets of finite Lebesgue measure in a finite dimensional space. Let a function $K(t, s, u)$ be defined for $t \in \Omega^*$, $s \in \Omega$, $-\infty < u < \infty$. The non-linear integral operator

$$Ax(t) = \int_\Omega K[t, s, x(s)]\,ds \tag{18.1}$$

---

[1] There are papers by many authors in connection with the study of various classes of non-linear integral operators (A. M. Lyapunov, E. Schmidt, P. S. Uryson, L. Lichtenstein, A. Hammerstein, N. Golomb, V. V. Nemyckii, A. I. Guseinov, M. M. Vainberg, M. A. Krasnoselskii, L. A. Ladyzenskii, Y. B. Rutickii, E. N. Pustylnik, P. P. Zabreiko, Wang Sheng-Wang and others.)

A special analysis of continuity conditions for general integral operators of the form (18.1) was given by P. P. Zabreiko [1, 4]. His results form the fundamental part of this paragraph. Also many important constructions have been taken from the papers of M. A. Krasnoselskii [1, 7] and M. A. Krasnoselskii and L. A. Ladyzenskii [1].

is called an *Uryson integral operator* and the function $K(t, s, u)$ is called its kernel.

An Uryson operator with a kernel $K(t, s, u)$ is defined on functions $x(s)$ for which the function $K[t, s, x(s)]$ is measurable. In the sequel it will be assumed that the function $K(t, s, u)$ satisfies the Caratheodory condition: for all u it is jointly measurable in the variables $(t, s) \in \Omega^* \times \Omega$ and for almost all $(t, s) \in \Omega^* \times \Omega$ it is continuous in u. This assumption (see Lemma 17.1) guarantees measurability of the function $K[t, s, x(s)]$ if the function $x(s)$ is measurable.

The operator (18.1) is non-linear. However, (as a superposition operator; see 17.2°) it possesses a partial additivity property: if the functions $x_1(s), \ldots, x_n(s)$ have disjoint supports, then

$$A(x_1 + \ldots + x_n) = Ax_1 + \ldots + Ax_n - (n - 1)A\theta. \tag{18.2}$$

The partial additivity property implies a number of simple assertions on Uryson operators.

First, each operator A, defined on some ball $\|x - x_0\|_\alpha \leqq \gamma$ and taking its values in a space L_β, will be defined on the entire space L_α, provided $\alpha > 0$. Second, each Uryson operator A acting from L_α $(\alpha > 0)$ to L_β and continuous at points of some ball of the space L_α, will be continuous at all point of the space L_α. Finally an Uryson operator acting from L_α $(\alpha > 0)$ to L_β and bounded on some ball of the space L_α, will be bounded on each ball of the space L_α.

Of course, these assertions do not hold for operators acting from L_0 to L_β.

The operator (18.1) can be described as the superposition $A = J\mathfrak{K}^0$, where

$$\mathfrak{K}^0 x(s) = K[t, s, x(s)] \tag{18.3}$$

and

$$Ju(t, s) = \int_\Omega u(t, s)\,\mathrm{d}s. \tag{18.4}$$

The operator $\mathfrak{K}^0$ acts from spaces of functions of a single variable to spaces of functions of two variables, while the operator J acts from spaces of functions of two variables to spaces of functions of a single variable. If the operator $\mathfrak{K}^0$ acts from some space $L_\alpha(\Omega)$ to a space $L_\beta(\Omega^* \times \Omega)$ $(0 \leqq \beta \leqq 1)$, then the operator $A = J\mathfrak{K}^0$ acts from $L_\alpha(\Omega)$ to $L_\beta(\Omega^*)$, since J is a continuous linear operator acting from $L_\beta(\Omega^* \times \Omega)$ to $L_\beta(\Omega^*)$. Similarly, if $\mathfrak{K}^0$ is continuous and bounded on each ball of the space $L_\alpha(\Omega)$, then the operator A is also continuous and bounded on each ball of the space $L_\alpha(\Omega)$.

Consider along with the operator (18.3), the operator

$$\mathfrak{K}u(t, s) = K[t, s, u(t, s)], \tag{18.5}$$

which acts in spaces of functions of two variables. If this operator acts from $L_\alpha(\Omega^* \times \Omega)$ to $L_\beta(\Omega^* \times \Omega)$, then by Theorem 17.1 it is continuous. In this case the operator $\mathfrak{K}^0$ will be continuous as an operator from $L_\alpha(\Omega)$ to $L_\beta(\Omega^* \times \Omega)$. Hence conditions, under which the operator (18.5) acts from $L_\alpha(\Omega^* \times \Omega)$ to $L_\beta(\Omega^* \times \Omega)$ yield conditions for continuity of the Uryson operator as an operator from $L_\alpha(\Omega)$ to $L_\beta(\Omega^*)$. This means that all points $\{\alpha, \beta\}$, $0 < \beta \leqq 1$, of the L-characteristic $L(\mathfrak{K}$; def.) belong to the L-characteristic $L(A$; cont.). In the sections which follow it will be shown that the above considerations yield only a part of the L-characteristic $L(A$; cont.) of the Uryson operator A.

18.2 *Conditions for continuity of Uryson operators*

We often encounter operators (18.1) with kernels having the special form

$$K(t, s, u) = K_0(t, s) \cdot f(s, u). \tag{18.6}$$

Integral operators

$$Ax(t) = \int_\Omega K_0(t, s) \cdot f[s, x(s)] \, \mathrm{d}s \tag{18.7}$$

possessing such kernels are called *Hammerstein operators*. Each Hammerstein integral operator admits a representation of the form

$$A = K_0 \mathfrak{f}, \tag{18.8}$$

where K_0 is a linear integral operator with the kernel $K_0(t, s)$:

$$\boldsymbol{K}_0 x(t) = \int_{\Omega} K_0(t, s) x(s) \mathrm{d}s, \tag{18.9}$$

and $\mathfrak{f}$ is the non-linear superposition operator

$$\mathfrak{f} x(s) = f[s, x(s)]. \tag{18.10}$$

Hence the study of a Hammerstein operator can be reduced to the study of the linear operator in (18.9) and the non-linear operator $\mathfrak{f}$ in (18.10). In particular, the following result is valid:

THEOREM 18.1: *Let $\mathfrak{f}$ be an operator acting from L_α to L_γ ($\gamma > 0$), and let $\boldsymbol{K}_0$ be a continuous operator acting from L_γ to L_β. Then the Hammerstein operator $\boldsymbol{A} = \boldsymbol{K}_0 \mathfrak{f}$ acts from L_α to L_β and is continuous.*

THEOREM 18.2: *Let $\mathfrak{f}$ act from L_α to L_0 and let $\boldsymbol{K}_0$ be a regular operator, acting from L_0 to L_β ($\beta > 0$). Then the Hammerstein operator $\boldsymbol{A} = \boldsymbol{K}_0 \mathfrak{f}$ acts from L_α to L_β and is continuous.*

Only Theorem 18.2 needs proof. Lemma 17.5 implies that under the conditions of the theorem the operator $\mathfrak{f}$ transforms any convergent sequence in L_α to a bounded sequence, converging in measure; the regular integral operator $\boldsymbol{K}_0$ transforms each such sequence to a sequence converging in the norm of L_β (it suffices to observe that $\|\boldsymbol{K}_0 \boldsymbol{P}_D\|_{0 \to \beta} \to 0$ as mes $D \to 0$).

By combining conditions for the continuity of the operator $\mathfrak{f}$ (see § 17) with conditions for the continuity of linear integral operators (see § 4–8) it is possible to formulate various criteria for the continuity of a Hammerstein operator.

Note that Theorems 18.1 and 18.2 permit us to construct a part of the L-characteristic $L(\boldsymbol{A}$; cont.) of a Hammerstein operator $\boldsymbol{A}$ if the L-characteristics $L(\boldsymbol{K}_0$; cont.) and $L(\mathfrak{f}$; cont.) are known. For this purpose, it is convenient to construct first the functions

$$\xi(\alpha; \boldsymbol{K}_0; \text{cont.}) = \inf_{\{\alpha, \beta\} \in L(\boldsymbol{K}_0; \text{cont.})} \beta$$

$$\xi(\alpha; \mathfrak{f}; \text{cont.}) = \inf_{\{\alpha, \beta\} \in L(\mathfrak{f}; \text{cont.})} \beta$$

and then to construct the function

$$\xi(\alpha) = \xi[\xi(\alpha; \mathfrak{f}; \text{cont.}); \boldsymbol{K}_0; \text{cont.}].$$

The entire set of points $\{\alpha, \beta\}$ for which $\beta > \xi(\alpha)$ will be contained in the L-characteristic $L(\boldsymbol{A}; \text{cont.})$ (see 1.6°). By means of Theorem 18.1 it is possible to show that some of the points $\{\alpha, \xi(\alpha)\}$ also belong to $L(\boldsymbol{A}; \text{cont.})$. Naturally, that part of the L-characteristic $L(\boldsymbol{A}; \text{cont.})$ of the Hammerstein operator $\boldsymbol{A}$ which is constructed by means of Theorems 18.1 and 18.2 may turn out to be only a 'proper' subset of the set $L(\boldsymbol{A}; \text{cont.})$.

In investigating the Hammerstein operator (18.7) it is sometimes convenient to represent it as the superposition

$$\boldsymbol{A} = \boldsymbol{K}_1 \mathfrak{f}_1$$

of operators

$$\boldsymbol{K}_1 x(t) = \int_\Omega K(t, s) v_0(s) x(s) \, \mathrm{d}s$$

and

$$\mathfrak{f}_1 x(s) = \frac{f[s, x(s)]}{v_0(s)},$$

where $v_0(s)$ is a function not taking on the value zero. A successful choice of $v_0(s)$ permit one to find additional points of the L-characteristic $L(\boldsymbol{A}; \text{cont.})$.

Suppose that

$$f(s, u) = f_1(s, u) + \dots + f_n(s, u). \tag{18.11}$$

Then the operator (18.7) can be written in the form of a sum of operators

$$\boldsymbol{A}_i x(t) = \int_\Omega K(t, s) f_i[s, x(s)] \, \mathrm{d}s \qquad (i = 1, 2, \dots, n).$$

The L-characteristic $L(\boldsymbol{A}; \text{cont.})$ will contain the intersection of the L-characteristics of operators $\boldsymbol{A}_i$. That part of the L-characteristic $L(\boldsymbol{A}; \text{cont.})$

constructed in this way will, generally speaking, be larger than the part which can be obtained by means of Theorem 18.1 without splitting $\mathfrak{f}$ into a sum of operators.

Suppose that the function $f(s, u)$ admits the estimate

$$|f(s, u)| \leqq g_1(s, u) + \dots + g_n(s, u) \tag{18.12}$$

where all the functions $g_i(s, u)$ are non-negative and satisfy the Caratheodory condition. Then $f(s, u)$ admits the representation (18.11) in which each function $f_i(s, u)$ satisfies the inequality

$$|f_i(s, u)| \leqq g_i(s, u). \tag{18.13}$$

Such functions $f_i(s, u)$ can be defined, for instance, by the relations

$$f_1(s, u) = \min\{|f(s, u)|, g_1(s, u)\} \operatorname{sign} f(s, u)$$

$$f_2(s, u) = \min\{|f(s, u) - f_1(s, u)|, g_2(s, u)\} \operatorname{sign}[f(s, u) - f_1(s, u)]$$

$$\dots$$

$$f_n(s, u) = f(s, u) - f_1(s, u) - \dots - f_{n-1}(s, u).$$

A general theorem on the structure of the L-characteristics $L(A; \text{cont.})$ of Hammerstein operators (similar to theorems on the convexity of the L-characteristics of linear operators or on the zero-concavity of the L-characteristics of superposition operators) is not known to us.

It is possible to give examples of Hammerstein operators with convex L-characteristics by examining operators with the simplest non-linearity

$$f(s, u) = |u|^k \operatorname{sign} u.$$

To conclude this section, let us give an example of a Hammerstein operator whose L-characteristic $L(A; \text{cont.})$ does not possess the property of convexity.

Consider the operator

$$A_0 x(t) = \int_0^1 |t - s|^{-\delta} \min\{a_0(s), x^2(s)\}\, ds, \tag{18.14}$$

where $\delta \in (\frac{1}{2}, 1)$, and the function $a_0(s)$ satisfies the conditions

$$a_0(s) > 1, a_0(s) \in L_{\frac{1}{2}} - \bigcup_{\alpha < \frac{1}{2}} L_\alpha$$

$$\int_0^1 |t - s|^{-\delta} a_0(s) \mathrm{d}s \in L_{\delta - \frac{1}{2}} - \bigcup_{\beta < \delta - \frac{1}{2}} L_\beta$$

(a proof of the existence of such a function is left to readers). The L-characteristics $L(\boldsymbol{K}_0; \text{cont.})$ and $L(\mathfrak{f}; \text{cont.})$ of the operators

$$\boldsymbol{K}_0 x(t) = \int_0^1 |t - s|^{-\delta} x(s) \mathrm{d}s,$$

$$\mathfrak{f} x(s) = \min\{a_0(s), x^2(s)\}$$

are shown in Fig. 18.1. Theorem 18.1 implies that the L-characteristic $L(\boldsymbol{A}; \text{cont.})$ contains the set of points, hatched in Fig. 18.2.

Consider now the value of the operator (18.14) on the function $x_0(s) = \sqrt{a_0(s)} \in L_{\frac{1}{4}}$. It is clear that

$$\boldsymbol{A}_0 x_0(t) = \boldsymbol{K}_0 a_0(t) \notin L_\beta \quad (\beta < \delta - \tfrac{1}{2}).$$

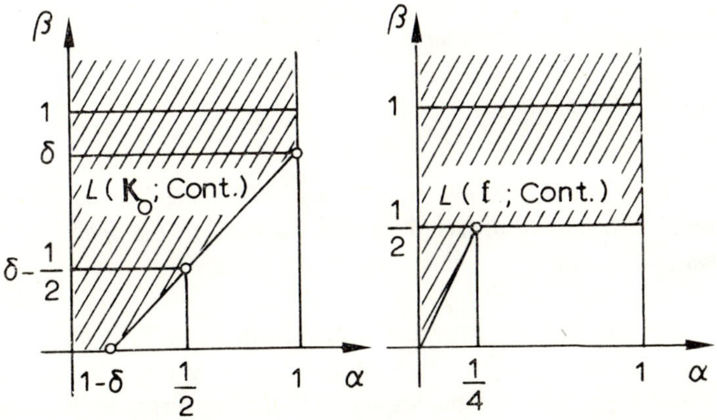

Figure 18.1

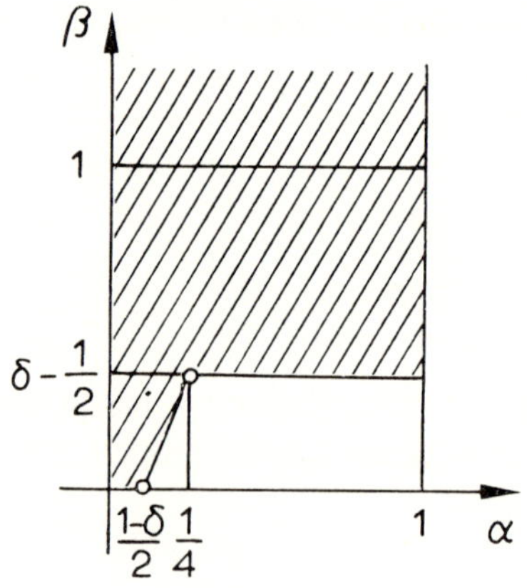

Figure 18.2

This means that points $\{\frac{1}{4}, \beta\}$, where $\beta < \delta - \frac{1}{2}$, do not belong to the L-characteristic $L(A_0; \text{cont.})$. Hence it follows that $L(A_0; \text{cont.})$ is not a convex set.

18.3 *General theorem on continuity of Uryson operators*

THEOREM 18.3: *Let the functions $K(t, s, u)$ and $R(t, s, u)$ satisfy the Caratheodory condition. Let*

$$|K(t, s, u)| \leqq R(t, s, u) \quad (t \in \Omega^*, s \in \Omega, -\infty < u < \infty) \tag{18.15}$$

and suppose the integral operator

$$\boldsymbol{B}x(t) = \int_{\Omega} R[t, s, x(s)]\mathrm{d}s \tag{18.16}$$

acts from L_α to L_β, $\beta > 0$, and is continuous. Then the integral operator $\boldsymbol{A}$ with kernel $K(t, s, u)$ also acts from the space L_α to the space L_β and is continuous.

PROOF: The fact that the operator $\boldsymbol{A}$ acts from L_α to L_β is clear.

Let a sequence of functions $x_n(s)$ converge to a function $x_0(s)$ in the norm of the space L_α. Then the sequence $\boldsymbol{B}x_n(t)$ converges to $\boldsymbol{B}x_0(t)$ in the norm of the space L_β and consequently converges to $\boldsymbol{B}x_0(t)$ in measure.

Hence it is possible to choose a subsequence $x_{n_k}(s)$ such that the functions

$$\boldsymbol{B}x_{n_k}(t) = \int_{\Omega} R[t, s, x_{n_k}(s)]\,ds \tag{18.17}$$

converge to $\boldsymbol{B}x_0(t)$ for all t in a subset $\Omega_0^* \subset \Omega^*$ of full measure. The functions $R[t, s, x_n(s)]$ converge to the function $R[t, s, x_0(s)]$ in measure (on $\Omega^* \times \Omega$). It can be assumed that the $R[t, s, x_{n_k}(s)]$ converge to $R[t, s, x_0(s)]$ almost everywhere on $\Omega^* \times \Omega$ and hence, for each $t \in \Omega_0^*$, converge to $R[t, s, x_0(s)]$ almost everywhere on Ω. Vitali's theorem[1] implies that for each $t \in \Omega_0^*$ the integrals

$$\boldsymbol{B}x_{n_k}(t) = \int_{\Omega} R[t, s, x_{n_k}(s)]\,ds \qquad (k = 1, 2, \ldots).$$

are equi-absolutely continuous. The inequality (18.15) then implies that for each $t \in \Omega_0^*$ the integrals

$$\boldsymbol{A}x_{n_k}(t) = \int_{\Omega} K[t, s, x_{n_k}(s)]\,ds \qquad (k = 1, 2, \ldots) \tag{18.18}$$

are also equi-absolutely continuous.

The sequence $K[t, s, x_{n_k}(s)]$ converges to $K[t, s, x_0(s)]$ in measure. It may be further assumed that for each $t \in \Omega_0^*$ it converges almost everywhere. The equi-absolute continuity of the integrals (18.18) proved above implies that the functions $\boldsymbol{A}x_{n_k}(t)$ converge to $\boldsymbol{A}x_0(t)$ for almost all $t \in \Omega^*$.

The norms of the sequence of functions $\boldsymbol{B}x_{n_k}(t)$ are equi-absolutely continuous in L_β. The inequality (18.15) implies that the norms of the functions $\boldsymbol{A}x_{n_k}(t)$ are also equi-absolutely continuous. Hence the sequence $\boldsymbol{A}x_{n_k}(t)$ converges to $\boldsymbol{A}x_0(t)$ in the norm of L_β.

We have proved that for each sequence $x_n(s)$ converging to $x_0(s)$ in L_α it is possible to take a subsequence such that functions $\boldsymbol{A}x_{n_k}(t)$ converge to the function $\boldsymbol{A}x_0(t)$ in the norm of L_β. This means that $\boldsymbol{A}$ is continuous as an operator from L_α to L_β. The theorem has been proved.

Theorem 18.3 means that $L(\boldsymbol{A}; \text{cont.})$ contains $L(\boldsymbol{B}; \text{cont.})$, if inequality (18.15) is fulfilled. As $R(t, s, u)$ we usually take a function defining

---

[1] I. P. Natanson [1].

a Hammerstein operator, or a sum of such functions, and then apply Theorems 18.1 and 18.2 to study the resulting operator $\boldsymbol{B}$.

18.4 *On a property of Uryson operators*

Let $\beta > 0$. Denote by $\mathcal{M}_\beta$[1] the set of all measurable functions $u(t, s)$ of two variables t, s for which

$$\int_{\Omega^*} [\int_{\Omega} |u(t, s)| \, ds]^{1/\beta} dt < \infty.$$

Obviously the set $\mathcal{M}_\beta$ forms a linear manifold. For $\beta \in (0, 1]$ the set $\mathcal{M}_\beta$ becomes a complete normed space if the norm is defined by the relation

$$\|u(t, s)\|_{\mathcal{M}_\beta} = \|\int_{\Omega} |u(t, s)| \, ds\|_{L_\beta}. \tag{18.19}$$

For $\beta > 1$ this space is not normed; it is, however, a complete metric space with the metric

$$\rho(u, v) = \int_{\Omega^*} [\int_{\Omega} |u(t, s) - v(t, s)| \, ds]^{1/\beta} dt.$$

The proof of these assertions can be accomplished by the usual methods.

LEMMA 18.1: *If* $u(t, s) \in \mathcal{M}_\beta$ $(\beta > 0)$, *then*

$$\lim_{\text{mes}\, D \to 0} \|\int_D |u(t, s)| \, ds\|_\beta = 0. \tag{18.20}$$

In fact, for almost all $t \in \Omega^*$ the function $u(t, s)$ is summable in s; hence the integrals $\int_{\Omega} |u(t, s)| \, ds$ are absolutely continuous for almost all t. This means that for almost all $t \in \Omega^*$, the functions $\varphi(D; t) = \int_D |u(t, s)| \, ds$ converge to zero as mes $D \to 0$. On the other hand the inequality

$$\varphi(D; t) \leqq \int_{\Omega} |u(t, s)| \, ds,$$

[1] A. Benedek, R. Panzone [1].

holds from which it follows that the functions $\varphi(D; t)$ in the space L_β have equi-absolutely continuous norms. Hence as mes $D \to 0$ the functions $\varphi(D; t)$ converge to zero in L_β-norm, i.e. relation (18.20) is valid. The lemma has been proved.

THEOREM 18.4: *Let the function* $K(t, s, u)$ $(t \in \Omega^*, s \in \Omega, -\infty < u < \infty)$ *satisfy the Caratheodory condition. Suppose that for each function* $x(s) \in L_\alpha$ $(\alpha > 0)$ *the function* $K[t, s, x(s)]$ *belongs to the space* $\mathcal{M}_\beta$ $(0 < \beta \leqq 1)$:

$$\int_{\Omega^*} [\int_{\Omega} |K[t, s, x(s)]| \, ds]^{1/\beta} \, dt < \infty.$$

Then the Uryson operator $\boldsymbol{A}$ *with the kernel* $K(t, s, u)$ *acts from* L_α *to* L_β *and possesses the following property: for each set* $\mathfrak{M}$ *of functions in* L_α *with equi-absolutely continuous norms and for each* $\varepsilon > 0$ *it is possible to find a number* $\delta > 0$ *such that the inequality* mes $D < \delta$ *implies the inequality*

$$\| \int_D |K[t, s, x(s)]| \, ds \|_\beta < \varepsilon \qquad (x \in \mathfrak{M}). \tag{18.21}$$

PROOF: Suppose the contrary. Then there are an $\varepsilon_0 > 0$, a sequence of functions $y_k(t) \in L_\alpha$ with equi-absolutely norms, and a sequence of sets $F_k \subset \Omega$ with mes $F_k \to 0$ as $k \to \infty$, such that

$$\| \int_{F_k} |K[t, s, y_k(s)]| \, ds \|_\beta > \varepsilon_0 \qquad (k = 1, 2, \ldots).$$

It can be assumed without loss of generality that

$$\sum_{k=1}^{\infty} \text{mes } F_k < \infty \tag{18.22}$$

and

$$\sum_{k=1}^{\infty} \|\boldsymbol{P}_{F_k} y_k\|_\alpha < \infty. \tag{18.23}$$

Let us proceed further as in the proof of Lemma 17.3. Introduce the notation

$$D_k = \bigcup_{i=k}^{\infty} F_i \qquad (k = 1, 2, \ldots).$$

The relation (18.22) implies that

$$\lim_{k\to\infty} \operatorname{mes} D_k = 0.$$

Hence Lemma 18.1 implies the existence of an integer-valued function $\eta(k)$ such that $\eta(k) > k$ and

$$\Big\| \int_{F_k - D_{\eta(k)}} |K[t, s, y_k(s)]| \, \mathrm{d}s \Big\|_\beta > \varepsilon_0 \qquad (k = 1, 2, \ldots).$$

Put

$$k_1 = 1,\ k_2 = \eta(k_1), \ldots,\ k_n = \eta(k_{n-1}), \ldots$$

The measures of the sets

$$\Omega_n = F_{k_n} - D_{k_{n+1}} \qquad (n = 1, 2, \ldots)$$

converge to zero as $n \to \infty$; these sets are obviously mutually disjoint.

Put

$$x_n(s) = y_{k_n}(s) \qquad (n = 1, 2, \ldots).$$

These functions satisfy the condition

$$\sum_{n=1}^{\infty} \|\boldsymbol{P}_{\Omega_n} x_n\|_\alpha < \infty \tag{18.24}$$

and the inequality

$$\Big\| \int_{\Omega_n} |K[t, s, x_n(s)]| \, \mathrm{d}s \Big\|_\beta > \varepsilon_0 \qquad (n = 1, 2, \ldots). \tag{18.25}$$

Now consider the functions

$$u(s) = \begin{cases} x_n(s), & \text{if } s \in \Omega_n \\ 0, & \text{if } s \notin \bigcup_{n=1}^{\infty} \Omega_n. \end{cases}$$

Clearly (18.24) implies that $u(s) \in L_\alpha$ and hence

$$K[t, s, u(s)] \in \mathscr{M}_\beta.$$

On the other hand, (18.25) and the inequality

$$\sum_{i=1}^{\infty} a_i^p \leqq \left(\sum_{i=1}^{\infty} a_i\right)^p \quad (a_i \geqq 0,\ p \geqq 1) \tag{18.26}$$

imply that

$$\int_{\Omega^*} \left[\int_{\Omega} |K[t, s, u(s)]|\, ds\right]^{1/\beta} dt$$

$$\geqq \sum_{n=1}^{\infty} \int_{\Omega^*} \left[\int_{\Omega_n} |K[t, s, x_n(s)]|\, ds\right]^{1/\beta} dt = \infty.$$

We have arrived at a contradiction. The theorem has been proved.

Non-essential charges in these arguments show that the assertion of Theorem 18.4 remains true even in cases in which one or both of the numbers α, β is greater than 1.

18.5 *Regular Uryson operators*

We here study another class of spaces of functions of two variables.

Let $\alpha > 0$. Denote by $\mathscr{N}_\alpha$ the set of all jointly measurable functions $u(t, s)$ for which

$$\int_{\Omega} \operatorname*{ess.\,sup}_{t \in \Omega^*} |u(t, s)|^{1/\alpha} ds < \infty.$$

Obviously $\mathscr{N}_\alpha$ is a linear manifold. For $\alpha \in (0, 1]$ the set $\mathscr{N}_\alpha$ form sa complete normed space if the norm is defined by the relation

$$\|u(t, s)\|_{\mathscr{N}_\alpha} = \|\operatorname*{ess.\,sup}_{t \in \Omega^*} |u(t, s)|\,\|_{L_\alpha}. \tag{18.27}$$

In the case $\alpha > 1$ the set $\mathscr{N}_\alpha$ is a complete metric space under the metric

$$\rho(u, v) = \int_{\Omega} \operatorname*{ess.\,sup}_{t \in \Omega^*} |u(t, s) - v(t, s)|^{1/\alpha} ds.$$

THEOREM 18.5: *Let a function* $K(t, s, u)$ $(t \in \Omega^*, s \in \Omega, -\infty < u < \infty)$ *satisfy the Caratheodory condition. Let the operator*

$$\mathfrak{K}u(t, s) = K[t, s, u(t, s)] \tag{18.28}$$

act from a space $\mathcal{N}_\alpha$ *to a space* $\mathcal{M}_\beta$ *with* $\alpha \geqq 0$, $0 < \beta \leqq 1$. *Then the Uryson integral operator* A *with kernel* $K(t, s, u)$ *acts from the space* L_α *to the space* L_β *and is continuous.*

PROOF: The fact that the operator A acts from L_α to L_β is obvious. Let us show the continuity of the operator A.

Let a sequence of functions $x_n(s)$ converge to a function $x_0(s)$ in norm of L_α, and suppose

$$|x_n(s)| \leqq h \tag{18.29}$$

where h is some fixed number. Let us show that the sequence of functions $Ax_n(t)$ converges to the function $Ax_0(t)$ in L_β-norm.

The hypothesis of the theorem implies that the operator

$$\mathfrak{K}u(t, s) = K[t, s, u(t, s)]$$

acts from $L_0(\Omega^* \times \Omega)$ to $L_1(\Omega^* \times \Omega)$. Hence (see Lemma 17.4) the following relation holds:

$$\lim_{n\to\infty} \int_{\Omega^*} \int_{\Omega} |K[t, s, x_n(s)] - K[t, s, x_0(s)]| \, ds \, dt = 0.$$

This implies that

$$\lim_{n\to\infty} \int_{\Omega^*} |Ax_n(t) - Ax_0(t)| \, dt = 0$$

and further that the sequence of functions $Ax_n(t)$ converges to the function $Ax_0(t)$ in measure.

Now let us show that the functions $Ax_n(t)$ have equi-absolutely continuous norms in L_β. It will then follow that

$$\lim_{n\to\infty} \|Ax_n - Ax_0\|_\beta = 0. \tag{18.30}$$

Suppose that the norms of the functions $Ax_n(t)$ $(n = 1, 2, \ldots)$ do not possess the property of equi-absolute continuity. Then it can be assumed without loss of generality that there exist sets $F_n \subset \Omega^*$, whose measures satisfy the inequalities

$$\operatorname{mes} F_n < 1/2^n \quad (n = 1, 2, \ldots)$$

such that

$$\|\boldsymbol{P}_{F_n} A x_n\|_\beta > \varepsilon_0 > 0 \quad (n = 1, 2, \ldots).$$

Put

$$D_n = \bigcup_{i=n}^{\infty} F_i \quad (n = 1, 2, \ldots).$$

It is clear that $\operatorname{mes} D_n \to 0$ and

$$\|\boldsymbol{P}_{D_n} A x_n\|_\beta > \varepsilon_0 \quad (n = 1, 2, \ldots).$$

As in the proof of Lemma 17.3 and Theorem 18.3, let us construct an integer-valued function $\eta(n)$ such that $\eta(n) > n$ and

$$\|\boldsymbol{P}_{D_n - D_{\eta(n)}} A x_n\|_\beta > \varepsilon_0 \quad (n = 1, 2, \ldots).$$

Put

$$n_1 = 1,\ n_2 = \eta(n_1),\ \ldots,\ n_k = \eta(n_{k-1}),\ \ldots$$

and

$$\Omega_k^* = D_{n_k} - D_{n_{k+1}} \quad (k = 1, 2, \ldots).$$

Obviously the sets Ω_k^* are mutually disjoint and

$$\|\boldsymbol{P}_{\Omega^*_k} A x_{n_k}\|_\beta > \varepsilon_0 \quad (k = 1, 2, \ldots).$$

Now let

$$u(t, s) = \begin{cases} x_{n_k}(s), & \text{if } t \in \Omega_k^* \\ 0, & \text{if } t \notin \bigcup_k \Omega_k^*. \end{cases}$$

Then (18.29) implies that the function $u(t, s)$ is bounded, hence that $\mathfrak{K}u(t, s) \in \mathcal{M}_\beta$. On the other hand, $\mathfrak{K}u(t, s) \notin \mathcal{M}_\beta$ because

$$\int_{\Omega^*} \Big[\int_{\Omega} |K[t, s, u(t, s)]| \, ds\Big]^{1/\beta} dt$$

$$\geqq \sum_{k=1}^{\infty} \int_{\Omega^*} \Big[\int_{\Omega} |K[t, s, x_{n_k}(s)]| \, ds\Big]^{1/\beta} dt = \infty.$$

We have arrived at a contradiction. Consequently (18.30) follows from $\|x_n(s) - x_0(s)\|_\alpha \to 0$ under the supplementary condition (18.29).

Now let $x_n(s)$ be an arbitrary sequence of functions in L_α converging to $x_0(s)$ in L_α-norm. Let $\|x_n(s)\|_\alpha \leqq a$.

For each $h > a$ the sequence of functions

$$T_h x_n(s) = \begin{cases} x_n(s), & \text{if } |x_n(s)| \leqq h \\ h \cdot \operatorname{sign} x_n(s), & \text{if } |x_n(s)| > h, \end{cases}$$

converges to the function $T_h x_0(s)$ in L_α-norm as $n \to \infty$; in addition, $|T_h x_n(s)| \leqq h$. Hence

$$\lim_{n \to \infty} \|A T_h x_n - A T_h x_0\|_\beta = 0. \tag{18.31}$$

Let $\varepsilon > 0$ be given. Theorem 18.4 implies that there exists $\delta > 0$ such that for mes $D < \delta$ the following inequality holds:

$$\Big\| \int_D |K[t, s, T_h x_n(s)]| \, ds \Big\|_\beta \leqq \varepsilon/5 \quad (0 < h < \infty; n = 1, 2, \ldots). \tag{18.32}$$

Further let $h_0 = a\delta^{-\alpha}$. Then

$$\operatorname{mes}\{s: |x_n(s)| \geqq h_0\} < \delta \qquad (n = 0, 1, 2, \ldots).$$

Now let us estimate $\|Ax_n(s) - Ax_0(s)\|_\beta$. It is clear that

$$\|Ax_n(t) - Ax_0(t)\|_\beta \leqq \|AT_{h_0}x_n(t) - AT_{h_0}x_0(t)\|_\beta$$

$$+ \|\int_{|x_n(s)| \geqq h_0} [|K[t, s, x_n(s)]| + |K[t, s, T_{h_0}x_n(s)]|]\, ds\|_\beta$$

$$+ \|\int_{|x_0(s)| \geqq h_0} [|K[t, s, x_0(s)]| + |K[t, s, T_{h_0}x_0(s)]|]\, ds\|_\beta.$$

However (18.31) and (18.32) imply that for large n the inequality

$$\|Ax_n(t) - Ax_0(t)\|_\beta < \varepsilon$$

is satisfied. The theorem has been completely proved.

This theorem is also valid for the case $\beta > 1$.

Note that the proof of Theorem 18.5 does not use in its entirety the assumption that the operator (18.28) acts from $\mathcal{N}_\alpha$ to $\mathcal{M}_\beta$. We need only to restrict the behavior of this operator on bounded functions $u(t, s)$ of two variables and on functions $u(t, s) = x(s)$, where $x(s) \in L_\alpha$.

An Uryson operator acting from L_α to L_β is called *regular*, if the operator (18.28) acts from $\mathcal{N}_\alpha$ to $\mathcal{M}_\beta$. Thus the notion of regularity is connected with the indices α and β.

It can be shown without difficulty that for linear integral operators the notion of regularity of an Uryson operator coincides with the notion of regularity of linear operators introduced in 2.2°. Theorem 18.5 can thus be considered as an immediate generalization to non-linear integral operators of the fact that regular linear operators are continuous.

The simplest criterion for the regularity of an Uryson operator with kernel $K(t, s, u)$ is the assumption that the operator

$$\mathfrak{K}u(t, s) = K[t, s, u(t, s)]$$

acts from $L_\alpha(\Omega^* \times \Omega)$ to $L_\beta(\Omega^* \times \Omega)$.

Sufficient conditions for regularity can also be formulated in the form of the conditions (18.15) of Theorem 18.3. In other words, if

$$|K(t, s, u)| \leqq R(t, s, u) \tag{18.15}$$

and the Uryson operator with kernel $R(t, s, u)$ is regular, then the Uryson operator with kernel $K(t, s, u)$ is also regular. In particular, the Uryson operator with kernel $K(t, s, u)$ is regular as an operator from L_α to L_β if inequality (18.15) is fulfilled with a function $R(t, s, u)$ which is even and monotone in u ($u > 0$) such that the Uryson operator with kernel $R(t, s, u)$ acts from L_α to L_β.

The regularity of a Hammerstein operator

$$Ax(t) = \int_\Omega K(t, s) f[s, x(s)]\,ds$$

as an operator from L_α to L_β clearly follows if the superposition operator $\mathfrak{f}x(s) = f[s, x(s)]$ acts from L_α to L_γ and the linear integral operator with kernel $K(t, s)$ acts from L_γ to L_β and is regular.

18.6 *Special examples*

In § 4 it was shown that for a linear integral operator continuity automatically results from the fact that the operator acts from L_α to L_β. In the preceding paragraph similar assertions were proved for superposition operators. Unfortunately an Uryson operator (even one with a non-negative kernel) which acts from L_α to L_β may not possess the property of continuity. Let us give an example corresponding to this.

Define the non-negative function $K_1(t, s, u)$, even in u, ($t, s \in [0, 1]$, $-\infty < u < \infty$) by the relation:

$$K_1(t, s, u) = \begin{cases} (2 - 2^n|u|)k_n(t) + (2^n|u| - 1)k_{n-1}(t) \\ \qquad\qquad \text{if } 1/2^n < |u| \leqq 1/2^{n-1} \\ 0, \qquad\quad \text{if } |u| \leqq 0 \text{ or } |u| > 1, \end{cases} \tag{18.33}$$

where

$$k_0(t) \equiv 0$$

$$k_n(t) = \begin{cases} 2^n, \text{ if } 1/2^n \leqq t < 1/2^{n-1}, & (n = 1, 2, \ldots) \\ 0, \text{ if } t < 1/2^n \text{ or } t \geqq 1/2^{n-1}. \end{cases} \tag{18.34}$$

The function $K_1(t, s, u)$ obviously satisfies the Caratheodory condition. Let us show that the operator

$$A_1x(t) = \int_0^1 K_1[t, s, x(s)]\,ds \tag{18.35}$$

transforms each measurable function into a summable one.

Define the sets D_n by

$$D_n = \{s;\ 1/2^n < |x(s)| \leqq 1/2^{n-1}\} \qquad (n = 1, 2, \ldots) \tag{18.36}$$

$$D_0 = \{s;\ x(s) = 0\}.$$

Then

$$A_1x(t) = \sum_{n=0}^{\infty} \int_{D_n} K_1[t, s, x(s)]\,ds$$

$$= \sum_{n=0}^{\infty} \int_{D_n} [(2 - 2^n|x(s)|)k_n(t) + (2^n|x(s)| - 1)k_{n-1}(t)]\,ds. \tag{18.37}$$

We have thus obtained an expression for $A_1x(t)$ in the form of a series with non-negative summands. Hence

$$\int_0^1 A_1x(t)\,dt$$

$$= \sum_{n=1}^{\infty} \int_0^1 \{\int_{D_n} [(2 - 2^n|x(s)|)k_n(t) + (2^n|x(s)| - 1)k_{n-1}(t)]\,ds\}\,dt$$

$$= \sum_{n=1}^{\infty} \int_{D_n} \{\int_0^1 [(2 - 2^n|x(s)|)k_n(t) + (2^n|x(s)| - 1)k_{n-1}(t)]\,dt\}\,ds$$

$$= \sum_{n=1}^{\infty} \int_{D_n} \{2 - 2^n|x(s)| + 2^n|x(s)| - 1\}\,ds = \sum_{n=1}^{\infty} \operatorname{mes} D_n.$$

Thus the function $A_1x(t)$ is summable.

In particular, A_1 can be considered as an operator acting from any L_α to L_1. This operator transforms the zero function to the zero function, while it transforms each function $x(s)$ satisfying the inequalities $0 < x(s) < 1$ to a function $A_1 x(t)$ such that $\|A_1 x\|_1 = 1$. Hence the operator A_1 is discontinuous at the origin of the space L_α.

Next let us give an example of a non-linear Uryson operator with non-negative kernel which does not possesses the property of regularity.

Define the kernel $K_2(t, s, u)$ $(t, s \in [0, 1], -\infty < u < \infty)$ by the relation:

$$K_2(t, s, u) = \begin{cases} (2 - 2^n|u|)k_n(t)/n + (2^n|u| - 1)k_{n-1}(t)/(n-1) \\ \qquad \text{if } 1/2^n < |u| \leqq 1/2^{n-1} \\ 0, \qquad \text{if } u = 0 \text{ or } |u| \geqq 1, \end{cases} \tag{18.38}$$

where the $k_n(t)$ are the functions defined in (18.34). This kernel satisfies the inequalities

$$0 \leqq K_2(t, s, u) \leqq K_1(t, s, u), \tag{18.39}$$

where $K_1(t, s, u)$ is the kernel given in (18.33). Hence the operator

$$A_2 x(t) = \int_0^1 K_2[t, s, x(s)]\,\mathrm{d}s \tag{18.40}$$

acts from each L_α to L_1.

Let $x(s)$ be an arbitrary function in L_α and let the D_n be the sets defined in (18.36). Denote by F_n the set of points $\{t, s\}$ in the square $0 \leqq t, s \leqq 1$, for which the second coordinate s belongs to D_n. For each set Q in the square $0 \leqq t, s \leqq 1$ the following inequality holds:

$$\iint_{Q \cap F_n} [(2 - 2^n|x(s)|)k_n(t)/n + (2^n|x(s)| - 1)k_{n-1}(t)/(n-1)]\,\mathrm{d}s\,\mathrm{d}t$$

$$\leqq \frac{1}{n-1} \iint_{Q \cap F_n} [k_n(t) + k_{n-1}(t)]\,\mathrm{d}s\,\mathrm{d}t \leqq \frac{2 \operatorname{mes} D_n}{n-1},$$

consequently for each $x(s)$ in L_α,

$$\iint_Q K_2[t, s, x(s)]\,ds\,dt \leqq \sum_{n=1}^{\infty} \iint_{Q\cap F_n} K_2[t, s, x(s)]\,ds\,dt$$

$$\leqq \sum_{n=1}^{l} \iint_{Q\cap F_n} K_2[t, s, x(s)]\,ds\,dt + \frac{2}{l} \sum_{n=l+1}^{\infty} \operatorname{mes} D_n.$$

Let $\varepsilon > 0$ be given. Then for $l \geqq 4/\varepsilon$

$$\iint_Q K_2[t, s, x(s)]\,ds\,dt \leqq \sum_{n=1}^{l} \iint_Q [k_n(t) + k_{n-1}(t)]\,ds\,dt + \varepsilon/2.$$

Hence there exists a $\delta > 0$ such that mes $Q < \delta$ implies the inequality

$$\iint_Q K_2[t, s, x(s)]\,ds\,dt < \varepsilon. \tag{18.41}$$

We emphasize that δ depends on ε but is independent of the functions $x(s)$. In other words, we have shown that the operator

$$\mathfrak{K}_2^0[x(s)] = K_2[t, s, x(s)]$$

transforms the entire space L_α into a set of functions of two variables having equi-absolutely continuous norms in L_1.

Let a sequence $x_n(s) \in L_\alpha$ converge to $x_0(s)$. The sequence $\mathfrak{K}_2^0[x_n(s)]$ converges to $\mathfrak{K}_2^0[x_0(s)]$ in measure. Hence

$$\lim_{n\to\infty} \int_0^1 \int_0^1 |K_2[t, s, x_n(s)] - K_2[t, s, x_0(s)]|\,ds\,dt = 0.$$

This relation implies that

$$\lim_{n\to\infty} \|A_2 x_n(t) - A_2 x_0(t)\|_1 = 0.$$

Thus, A_2 is continuous as an operator from L_α to L_1. But A_2 does not possess the regularity property. This is obvious, because

$$\int_0^1 \int_0^1 K[t, s, u_0(t, s)]\,ds\,dt = \infty$$

where

$$u_0(t, s) = \begin{cases} 1/2^n, & \text{if } 1/2^n \leqq t < 1/2^{n-1} \\ 0, & \text{if } t = 0. \end{cases}$$

18.7 *Uryson operators with values in the space of bounded functions*

THEOREM 18.6: *Let a function* $K(t, s, u)$ $(t \in \Omega^*, s \in \Omega, -\infty < u < \infty)$ *satisfy the Caratheodory condition. Let the following conditions be satisfied:*

a) *there exists a set* $\Omega_0^* \subset \Omega^*$ *of full measure such that for each set* $\mathfrak{M}$ *of functions with equi-absolutely continuous norms in* L_α *the integrals*

$$Ax(t) = \int_\Omega K[t, s, x(s)]\,ds \qquad (t \in \Omega_0^*) \tag{18.42}$$

are equi-absolutely continuous (as integrals over Ω*)*

b) *for each* $\varepsilon > 0$ *and for each* R *there exists an* η *such that*

$$\int_\Omega \sup_{|u_1|, |u_2| \leqq R, |u_1 - u_2| \leqq \eta} |K(t, s, u_1) - K(t, s, u_2)|\,ds < \varepsilon \qquad (t \in \Omega_0^*). \tag{18.43}$$

Then the non-linear integral operator A *with kernel* $K(t, s, u)$ *acts from* L_α *to* L_0 *and is continuous.*

Before proceeding with the proof of this theorem note that for linear integral operators acting from L_α to L_0 the conditions of the theorem are not only sufficient but also necessary. In fact, an integral operator with a kernel $K(t, s)$ acts from L_α to L_0 if and only if (see Theorem 6.2)

$$\varphi(t) = \|K(t, s)\|_{1-\alpha} \in L_0.$$

Hence for each set $D \subset \Omega$ and for each function $x(s) \in L_\alpha$ the inequality

$$\left| \int_D K(t, s)x(s)\,ds \right| \leqq \varphi(t) \cdot \|\boldsymbol{P}_D x\|_\alpha$$

holds from which condition a) follows. Condition b) is obvious.

PROOF of Theorem 18.6: Condition a) implies that the operator A is defined on each function $x(s) \in L_\alpha$ and that its value belongs to the space L_0.

Let a sequence of functions $x_n(s) \in L_\alpha$ converge to a function $x_0(s)$ in norm and let $\varepsilon > 0$ be given. Select a number $\delta > 0$ such that $\operatorname{mes}(D) < \delta$ implies the inequality

$$\|\int_D K[t, s, x_n(s)]\,ds\|_0 < \varepsilon \qquad (n = 0, 1, 2, \ldots).$$

Next select a number R such that the measure of each set $\{s: |x_n(s)| \geqq R\}$ does not exceed δ.

Let η be a number such that $|u_1 - u_2| \leqq \eta$, $|u_1|, |u_2| \leqq R$ implies the inequality (18.43). Finally, take a number n_0 such that $n \geqq n_0$ implies the inequality

$$\operatorname{mes}\{s: |x_n(s) - x_0(s)| \geqq \eta\} < \delta.$$

Then for $n \geqq n_0$ and $t \in \Omega_0^*$

$$|Ax_n(t) - Ax_0(t)| = |\int_\Omega \{K[t, s, x_n(s)] - K[t, s, x_0(s)]\}\,ds|$$

$$\leqq |\int_{\substack{|x_n(s) - x_0(s)| \leqq \eta \\ |x_n(s)|,\, |x_0(s)| \leqq R}} \{K[t, s, x_n(s)] - K[t, s, x_0(s)]\}\,ds| +$$

$$+ |\int_{\substack{|x_n(s) - x_0(s)| > \eta \\ |x_n(s)|,\, |x_0(s)| < R}} \{K[t, s, x_n(s)] - K[t, s, x_0(s)]\}\,ds|$$

$$+ |\int_{|x_n(s)| \geqq R} K[t, s, x_n(s)]\,ds| + |\int_{|x_0(s)| \geqq R} K[t, s, x_0(s)]\,ds|$$

$$\leqq \varepsilon + 2\varepsilon + \varepsilon + \varepsilon = 5\varepsilon.$$

From the arbitrariness of ε there follows:

$$\lim_{n \to \infty} \|Ax_n(t) - Ax_0(t)\|_0 = 0.$$

The theorem has been proved.

Condition b) of Theorem 18.6 is an 'integral continuity' requirement (in u) on the kernel $K(t, s, u)$. Sufficient criteria for condition a) can be formulated, for instance, in the form of an inequality

$$|K(t, s, u)| \leqq K_0(t, s) + \sum_{i=1}^{n} K_i(t, s) \cdot |u|^{\delta_i} + b|u|^{1/\alpha},$$

where $K_0(t, s)$ is a function such that

$$\lim_{\text{mes}\, D \to 0} \| \int_D K_0(t, s) \, ds \|_0 = 0,$$

and the exponents δ_i belong to the interval $(0, 1/\alpha)$ while the functions K_i satisfy the conditions

$$\psi_i(t) = \int_\Omega |K_i(t, s)|^{1/(1-\delta\alpha_i)} ds \in L_0 \qquad (i = 1, \ldots, n).$$

It is clear that an Uryson operator acts from L_0 to L_0 and is continuous if the kernel $K(t, s, u)$ is jointly continuous in all its variables and the sets Ω and Ω^* are closed and bounded. The reader can formulate without difficulty a variety of more general assertions (it is easy, for instance, to formulate an analogue of Theorem 18.6). Note further that in the study of an Uryson operator on a ball $\|x\|_0 \leqq a$ only the values of the function $K(t, s, u)$ for $|u| \leqq a$ are relevant.

18.8 *On uniform continuity of Uryson operators*

Recall again that continuous Uryson operators do not in general possess the property of uniform continuity on bounded sets. The reader can construct examples of this type using, for instance, the methods in 17.6°.

Let the Uryson operator A with kernel $K(t, s, u)$ act from L_α to L_β, where $0 \leqq \alpha, \beta < \infty$. Suppose that $K(t, s, u)$ satisfies the Hölder condition

$$|K(t, s, u_1) - K(t, s, u_2)| \leqq R(t, s, u)|u_1 - u_2|^\delta \quad (|u_1|, |u_2| \leqq u), \tag{18.44}$$

where $0 < \delta \leqq 1$, $\alpha\delta < 1$, and $R(t, s, u)$ not only satisfies the Caratheodory condition but is such that the operator

$$\boldsymbol{R}^{[1-\alpha\delta]}x(t) = \int_{\Omega} |R[t, s, x(s)]|^{1/(1-\alpha\delta)} \mathrm{d}s \tag{18.45}$$

acts from L_α to $L_{\beta/(1-\alpha\delta)}$ and is bounded on each ball $\|x\|_\alpha \leqq \rho$. Then Hölder's inequality implies that

$$|Ax_1(t) - Ax_2(t)| \leqq \int_{\Omega} |K[t, s, x_1(s)] - K[t, s, x_2(s)]| \, \mathrm{d}s$$

$$\leqq \int_{\Omega} |R\{t, s, \max[|x_1(s)|, |x_2(s)|]\} \, |x_1(s) - x_2(s)|^{\delta} \mathrm{d}s$$

$$\leqq \{\int_{\Omega} |R\{t, s, \max[|x_1(s)|, |x_2(s)|]\}|^{1/(1-\alpha\delta)} \mathrm{d}s\}^{1-\alpha\delta} \|x_1 - x_2\|_\alpha^\delta;$$

hence

$$\|Ax_1 - Ax_2\|_\beta \leqq k(\rho) \|x_1 - x_2\|_\alpha^\delta, \tag{18.46}$$

where

$$k(\rho) = \sup_{\|x\|_\alpha \leqq 2\rho} \|\boldsymbol{R}^{[1-\alpha\delta]}x\|_{\beta/(1-\alpha\delta)}.$$

Consequently (18.44) implies that the Uryson operator A is not only uniformly continuous on each ball, but also satisfies a Hölder condition.

The Hölder condition (18.44) can sometimes be replaced by the inequality

$$|K(t, s, u_1) - K(t, s, u_2)| \leqq v(s) |u_1 - u_2|^{1/\alpha} \tag{18.47}$$

where $v(t) \in L_\beta$, $\alpha \geqq 1$.

In conclusion let us present a general criterion for uniform continuity of an Uryson operator whose proof can be obtained similarly to the proof of Theorem 18.5.

THEOREM 18.7: *Let A be a regular Uryson operator acting from L_α to L_β ($\alpha > 0$, $0 < \beta \leqq 1$). Let*

$$\lim_{\text{mes}\, D\to 0} \sup_{\|x\|_\alpha \leqq 1} \|\int_D K[t, s, x(s)]\,\mathrm{d}s\|_\beta = 0. \tag{18.48}$$

Then the operator A is uniformly continuous on each ball of the space L_α.

Denote by $L(A$; unif. cont.) the set of points $\{\alpha, \beta\}$ such that A acts from L_α to L_β and is uniformly continuous on each ball of the space L_α. Theorems 18.7 and 18.3 imply that $\{\alpha, \beta\} \in L(A$; unif. cont.) if $\beta \in (0, 1]$ and if there is $\alpha_1 > \alpha$ such that $\{\alpha_1, \beta\} \in L(A$; reg.). In particular, all interior points of the L-characteristic $L(A$; reg.) are points in the L-characteristic $L(A$; unif. cont.).

§ 19 Conditions for completely continuity of an Uryson operator [1]

19.1 *Problem setting*

Recall that a non-linear operator A is completely continuous if it is continuous and if it transforms each bounded set to a compact set. In this paragraph we will establish criteria for the complete continuity of an Uryson operator

$$Ax(t) = \int_\Omega K[t, s, x(s)]\,\mathrm{d}s \tag{19.1}$$

acting from L_α to L_β. Assume that $K(t, s, u)$ satisfies the Caratheodory condition. In the fundamental theorems of this paragraph, the continuity of A follows from assertions proved previously. Hence our attention is mainly focused on criteria for compactness, i.e. criteria ensuring that A transforms each bounded set to a compact set.

LEMMA 19.1: *Let an Uryson operator A with kernel $K(t, s, u)$ transform some ball $\|x - x_0\|_\alpha \leqq r$ of a space L_α, where $\alpha > 0$, to a set of functions, compact*

---

[1] This paragraph is based on the papers of M. A. Krasnoselskii and L. A. Ladyzenskii [1], M. A. Krasnoselskii and E. I. Pustylnik [2–4], E. I. Pustylnik [3], P. P. Zabreiko [4], P. P. Zabreiko and E. I. Pustylnik [1]. See also M. A. Krasnoselskii [1, 7, 9], M. M. Vainberg [6], M. A. Krasnoselskii and Y. B. Rutickii [2, 4–6] and Y. B. Rutickii [2].

in L_β, $0 \leqq \beta \leqq 1$. Then the operator A transforms each ball of the space L_α to a set of functions compact in L_β.

PROOF: Consider the auxiliary operator

$$A_1 x(t) = \int_\Omega K_1[t, s, x(s)]\,\mathrm{d}s,$$

where

$$K_1(t, s, u) = K[t, s, x_0(s) + u] - K[t, s, x_0(s)].$$

The range of the operator A_1 on the ball $\|x\|_\alpha \leqq r$ then also forms a compact set in L_β.

Consider the ball $T = \{x\colon \|x\|_\alpha \leqq R\}$, where R is an arbitrary number. Denote by n an integer such that $rn^\alpha \geqq R$. Then for each function $x(s) \in T$ it is possible to take a partition of Ω to disjoint sets $\Omega_1, \Omega_2, \ldots, \Omega_n$, such that

$$\|P_{\Omega_i} x\|_\alpha \leqq r \qquad (i = 1, 2, \ldots, n).$$

The relation $K_1(t, s, 0) = 0$ implies that

$$A_1 x(t) = \sum_{i=1}^{n} A_1 P_{\Omega_i} x.$$

This means that all functions $(1/n) A_1 x(t)$ $(x \in T)$ belong to the convex hull of the range of the operator A_1 on the ball $\|x\|_\alpha \leqq r$, which is also compact. The lemma has been proved.

This lemma permits one to study the complete continuity of the Uryson operator (19.1) by studying its behavior only on the unit ball of the space L_α.

In most of the results which follow, the compactness of the range of an Uryson operator on a ball is established with the aid of Lemma 1.1. In other words, the proof of compactness will be reduced to establishing two independent facts: first, equi-absolutely continuity in norm for the range of the operator on the unit ball and second, the compactness in measure of the operator. Here by compactness in measure of the operator A we mean compactness (in measure) of its range on each bounded set.

19.2 *Hammerstein operators*

Consider the operator

$$Ax(t) = \int_{\Omega} K(t, s) f[s, x(s)] \, ds. \tag{19.2}$$

As in the preceding paragraph, denote by $\boldsymbol{K}$ and $\mathfrak{f}$ the operators defined by the relations:

$$\boldsymbol{K}x(t) = \int_{\Omega} K(t, s) x(s) \, ds \tag{19.3}$$

$$\mathfrak{f}x(s) = f[s, x(s)]. \tag{19.4}$$

The following assertion on complete continuity of a Hammerstein operator follows from the representation of this operator in the form of a superposition

$$\boldsymbol{A} = \boldsymbol{K}\mathfrak{f}. \tag{19.5}$$

THEOREM 19.1: *Let the superposition operator* $\mathfrak{f}$ *act from* L_α $(0 \leqq \alpha < \infty)$ *to* L_γ $(0 \leqq \gamma \leqq 1)$, *and let the integral operator* $\boldsymbol{K}$ *act from* L_γ *to* L_β $(0 \leqq \beta < \infty)$. *Let one of the following conditions be satisfied*:

a) $\gamma > 0$; *the operator* $\boldsymbol{K}$ *is compact.*

b) $\gamma > 0$, $0 < \beta < \infty$; *the operator* $\boldsymbol{K}$ *is regular and the operator* $\mathfrak{f}$ *is 'improving'.*

c) $\gamma = 0$, $\beta > 0$; *the operator* $\boldsymbol{K}$ *is regular.*

Then the Hammerstein operator $\boldsymbol{A} = \boldsymbol{K}\mathfrak{f}$ *is a completely continuous operator acting from* L_α *to* L_β.

PROOF: In case a) the assertion of the theorem is obvious.

In case b) the assertion of the theorem with $\gamma = 1$ follows from Lemmas 5.2 and 1.3. If $\gamma < 1$, the following argument can be used. From Theorem 2.8 it follows that the values of the operator $\boldsymbol{A}$ on the unit ball of the space L_α are equi-absolutely continuous in norm. Hence it is only necessary to prove that $\boldsymbol{A}$ is compact in measure—this compactness follows from Lemma 5.1 (for $0 < \beta \leqq 1$) or from Theorem 5.8 (for $\beta > 1$).

In case c) the continuity of the operator A was proved in 18.2°. Compactness follows from the compactness of a regular integral operator, acting from L_0 to L_β ($\beta > 0$) (see 5.2°). The theorem has been proved.

On applying Theorem 19.1 we can use the criteria established in Chapter 2 for regularity, for compactness in measure, or for compactness of a linear integral operator $\boldsymbol{K}$ as well as the theorems established in § 17 for the superposition operator $\mathfrak{f}$. Note also that in investigating the compactness of a concrete Hammerstein operator we can use the method discussed at the end of 18.2° for analyzing the continuity of such operators.

19.3 *Complete continuity of regular Uryson operators acting from L_0 to L_β, $\beta \in (0, 1]$*

We begin our study of the complete continuity of Uryson operators with an investigation of such operators acting from L_0 to L_β. This is justified by the fact that an investigation of complete continuity for Uryson operators acting from other spaces L_α to L_β can be reduced to this special case.

THEOREM 19.2: *Let a function $K(t, s, u)$ ($t \in \Omega^*$, $s \in \Omega$, $|u| \leqq a$) satisfy the Caratheodory condition. Let the Uryson operator A with kernel $K(t, s, u)$ act from the ball $\|x\|_0 \leqq a$ to L_β, where $\beta \in (0, 1]$ and be a regular operator. Then the operator A is completely continuous.*

PROOF: Continuity follows from Theorem 18.5.

To prove complete continuity let us show that the functions Ax ($\|x\|_0 \leqq a$) form a compact set in L_β.

The condition of the theorem implies that the operator

$$\mathfrak{K}u(t, s) = K[t, s, u(t, s)]$$

acts from $L_0(\Omega^* \times \Omega)$ to $M_\beta(\Omega^* \times \Omega)$. Hence (see Lemma 17.2) there exists a jointly measurable function $R(t, s)$, such that

$$|K(t, s, u)| \leqq R(t, s) \quad (|u| \leqq a), \tag{19.6}$$

and

$$\int_{\Omega^*} \left[\int_{\Omega} R(t, s)\,\mathrm{d}s\right]^{1/\beta} \mathrm{d}t < \infty, \tag{19.7}$$

i.e. $R(t, s) \in \mathscr{M}_\beta(\Omega^* \times \Omega)$.

Let $\varepsilon > 0$ be given. From the absolute continuity of the integral $\int\int_{\Omega^* \times \Omega} R(t, s)\,\mathrm{d}s\,\mathrm{d}t$ it follows that there exists a $\delta > 0$ such that for sets $E \subset \Omega^* \times \Omega$, mes $E < \delta$ implies the inequality

$$\int\int_E |K[t, s, x(s)]|\,\mathrm{d}s\,\mathrm{d}t < \varepsilon/2 \qquad (\|x\|_0 \leqq a). \tag{19.8}$$

By Lemma 17.1 it is possible to take a closed set $Q_0 \subset \Omega^* \times \Omega$, whose measure satisfies the inequality

$$\operatorname{mes}(\Omega^* \times \Omega - Q_0) < \delta \tag{19.9}$$

and such that the function $K(t, s, u)$ is jointly continuous in the variables t, s, u for $\{t, s\} \in Q_0$, $|u| \leqq a$.

Let

$$b = \max_{|u| \leqq a,\ \{t, s\} \in Q_0} |K(t, s, u)|.$$

Let Q_1 be an open subset of $\Omega^* \times \Omega$, containing Q_0 and such that

$$\operatorname{mes}(Q_1 - Q_0) < \varepsilon/2b. \tag{19.10}$$

Define the jointly continuous function $K_1(t, s, u)$ for $\{t, s\} \in \Omega^* \times \Omega$, $|u| \leqq a$, by putting it equal to $K(t, s, u)$ for $\{t, s\} \in Q_0$ and equal to zero for $\{t, s\} \notin Q_1$ and extending it continuously so that the following relation holds:

$$\max_{\{t, s\} \in \Omega^* \times \Omega,\ |u| \leqq a} |K_1(t, s, u)| = b. \tag{19.11}$$

The Uryson operator A_1 with kernel $K_1(t, s, u)$

$$A_1 x(t) = \int_{\Omega} K_1[t, s, x(s)]\,\mathrm{d}s,$$

is defined on the ball $\|x\|_0 \leqq a$ of the space L_0 and acts into C, the space of continuous functions on Ω^*. Since the function $K_1(t, s, u)$ is uniformly continuous in its variables for $\{t, s\} \in \Omega^* \times \Omega$, $|u| \leqq a$, the set of functions $A_1x(t)$ ($\|x\|_0 \leqq a$) is bounded in the norm of the space C and its functions are equi-continuous. By Arzela's theorem this set is compact in the space C, hence in the space L_1.

Let us show that

$$\|Ax - A_1x\|_1 < \varepsilon \quad (\|x\|_0 \leqq a).$$

It will then follow from the arbitrariness of ε that the set of functions $Ax(t)$ ($\|x\|_0 \leqq a$) is compact in L_1.

It is clear that

$$\|Ax - A_1x\|_1 \leqq \int\int_{\Omega^* \times \Omega} |K[t, s, x(s)] - K_1[t, s, x(s)]| \, ds \, dt$$

$$\leqq \int\int_{Q_0} |K[t, s, x(s)] - K_1[t, s, x(s)]| \, ds \, dt$$

$$+ \int\int_{\Omega^* \times \Omega - Q_0} |K[t, s, x(s)]| \, ds \, dt + \int\int_{\Omega^* \times \Omega - Q_0} |K_1[t, s, x(s)]| \, ds \, dt.$$

The first term is equal to zero. By (19.8) and (19.9) the second term satisfies the inequality

$$\int\int_{\Omega^* \times \Omega - Q_0} |K[t, s, x(s)]| \, ds \, dt < \varepsilon/2.$$

The third term is estimated with the aid of (19.10) and (19.11)

$$\int\int_{\Omega^* \times \Omega - Q_0} |K_1[t, s, x(s)]| \, ds \, dt$$

$$= \int\int_{Q_1 - Q_0} |K_1[t, s, x(s)]| \, ds \, dt \leqq \frac{b\varepsilon}{2b} = \varepsilon/2.$$

The compactness in L_1 of the set of functions $Ax(t)$ ($\|x\|_0 \leqq a$) implies its compactness in measure. From (19.6) it follows that

$$|Ax(t)| \leqq \varphi(t) = \int_{\Omega} R(t, s) \, ds \quad (t \in \Omega^*).$$

Since $\varphi(t) \in L_\beta$, the functions Ax are equi-absolutely continuous in norm. Hence the set of functions $Ax(t)$ ($\|x\|_0 \leqq a$) is compact in L_β. The theorem has been proved.

To apply Theorem 19.2 it is important to realize that the regularity assumption on the Uryson operator A acting from L_0 to L_β is actually equivalent to the estimate (19.6), where $R(t, s) \in M_\beta$.

19.4 *Complete continuity of regular Uryson operators acting from L_α to L_β, $\alpha > 0$, $0 < \beta \leqq 1$*

THEOREM 19.3: *Let the function $K(t, s, u)$ ($t \in \Omega^*$, $s \in \Omega$, $-\infty < u < \infty$) satisfy the Caratheodory condition and let the Uryson operator A with kernel $K(t, s, u)$ act from L_α to L_β, where $\alpha > 0$, $0 < \beta \leqq 1$ and be a regular operator. Suppose that*

$$\lim_{\operatorname{mes} D \to 0} \sup_{\|x\|_\alpha \leqq 1} \left\| \int_D K[t, s, x(s)]\, ds \right\|_\beta = 0. \tag{19.12}$$

Then the Uryson operator A is completely continuous.

PROOF: The continuity of the operator A results from Theorem 18.5. Let us show that the values Ax assumed by the operator A on the ball $\|x\|_\alpha \leqq 1$ lie in a compact set of the space L_β.

Let ε be an arbitrary positive number. By the hypothesis of the theorem we can select a number $\delta > 0$ such that if $\operatorname{mes} D < \delta$ then:

$$\sup_{\|x\|_\alpha \leqq 1} \left\| \int_D K[t, s, x(s)]\, ds \right\|_\beta < \varepsilon/2. \tag{19.13}$$

Theorem 19.2 implies that the set of functions $Ax(t)$ ($\|x\|_0 \leqq \delta^{-\alpha}$) is compact in L_β. Let us now show that for each function x in the ball $\|x\|_\alpha \leqq 1$ we can find a function $Tx(t)$ in the ball $\|z\|_0 \leqq \delta^{-\alpha}$ such that

$$\|Ax - ATx\|_\beta \leqq \varepsilon. \tag{19.14}$$

The compactness of the range of the operator A on the ball $\|x\|_\alpha \leqq 1$ will follow.

Let $\|x\|_\alpha \leqq 1$. Define the function $Tx(t)$ by the relation

$$Tx(s) = \min\{|x(s)|, \delta^{-\alpha}\} \operatorname{sgn} x(s).$$

Obviously $\|Tx(s)\|_\alpha \leqq 1$ and

$$\text{mes}\{s\colon x(s) \neq Tx(s)\} \leqq \delta.$$

Hence (by (19.13))

$$\|Ax - ATx\|_\beta$$

$$= \| \int_{x(s)\neq Tx(s)} \{K[t, s, x(s)] - K[t, s, Tx(s)]\} \mathrm{d}s\|_\beta$$

$$\leqq \| \int_{x(s)\neq Tx(s)} K[t, s, x(s)]\mathrm{d}s\|_\beta + \| \int_{x(s)\neq Tx(s)} K[t, s, Tx(s)]\mathrm{d}s\|_\beta.$$

The theorem has been proved.

It is seen without difficulty that condition (19.12) can be replaced by the relation

$$\lim_{\text{mes}\, D\to 0} \sup_{\|x\|_\alpha \leqq a} \| \int_D K[t, s, x(s)]\mathrm{d}s\|_\beta = 0, \tag{19.15}$$

where a is any positive number.

Suppose that the kernel $K(t, s, u)$ satisfies the additional condition:

$$K(t, s, 0) \equiv 0 \quad (t \in \Omega^*, s \in \Omega). \tag{19.16}$$

Then condition (19.12) can be written in the form

$$\lim_{\text{mes}\, D\to 0} \sup_{\|x\|_\alpha \leqq 1} \|AP_D x\|_\beta = 0. \tag{19.17}$$

Recall that (19.17) is always satisfied by compact linear operators acting from L_α, $0 < \alpha < 1$, to L_β, $0 < \beta \leqq 1$ (see Theorem 3.3). In other words, the sufficient conditions for complete continuity given in Theorem 19.3 are also necessary in the case of linear integral operators (for $\alpha < 1$).

Let us prove that under the conditions of Theorem 19.3 the Uryson operator A possesses an important additional property—*it transforms each bounded sequence $x_n \in L_\alpha$ converging to $x_0 \in L_\alpha$ in measure into a sequence Ax_n converging to Ax_0 in L_β.*

It can be assumed without loss of generality that $\|x_n\|_\alpha \leqq 1$ ($n = 0, 1, 2, \ldots$) and that $K(t, s, 0) = 0$. Let $\varepsilon > 0$ be given. Denote by δ a positive number such that mes $D < \delta$ implies the inequality:

$$\sup_{\|x\|_\alpha \leqq 1} \|\int_D K[t, s, x(s)] \mathrm{d}s\|_\beta < \varepsilon/5.$$

Put $h = \delta^{-\alpha}$. Then mes $\Omega_n \leqq \delta$, where

$$\Omega_n = \{s : |x_n(s)| \geqq h\} \qquad (n = 0, 1, 2, \ldots).$$

The sequence $y_n(s) = \min\{|x_n(s)|, h\} \operatorname{sgn} x_n(s)$ converges to $y_0(s) = \min\{|x_0(s)|, h\} \operatorname{sgn} x_0(s)$ in L_α-norm. Since by Theorem 19.3 the operator A is continuous, there exists an n_0 such that for $n \geqq n_0$

$$\|Ay_n - Ay_0\|_\beta < \varepsilon/5.$$

For such n we have

$$\|Ax_n - Ax_0\|_\beta \leqq \|\int_\Omega \{K[t, s, y_n(s)] - K[t, s, y_0(s)]\} \mathrm{d}s\|_\beta$$

$$+ \|\int_{\Omega_n} K[t, s, x_n(s)] \mathrm{d}s\|_\beta + \|\int_{\Omega_0} K[t, s, x_0(s)] \mathrm{d}s\|_\beta$$

$$+ \|\int_{\Omega_n} K[t, s, y_n(s)] \mathrm{d}s\|_\beta + \|\int_{\Omega_0} K[t, s, y_0(s)] \mathrm{d}s\|_\beta$$

$$\leqq \|Ay_n - Ay_0\|_\beta + 4\varepsilon/5 \leqq \varepsilon,$$

consequently $\|Ax_n - Ax_0\|_\beta \to 0$ as $n \to \infty$.

This assertion admits a converse: *if an Uryson operator transforms each bounded sequence x_n in L_α converging to $x_0 \in L_\alpha$ in measure into a sequence Ax_n converging to Ax_0 in L_β-norm, then condition* (19.12) *is satisfied.*

The simplest proof is attained by contradiction. Suppose that there exists a sequence of functions $x_n(s) \in L_\alpha$ and a sequence of sets $D_n \subset \Omega$ such that $\|x_n(s)\|_\alpha \leqq 1$, mes $D_n \to 0$ and

$$\|\int_{D_n} K[t, s, x_n(s)] \mathrm{d}s\|_\beta \geqq \varepsilon_0 > 0. \tag{19.18}$$

The sequence of functions $P_{D_n}x_n$ converges in measure to the function θ, which is identically equal to zero. Hence

$$\lim_{n\to\infty} \|AP_{D_n}x_n - A\theta\|_\beta = 0.$$

Lemma 18.1 implies that

$$\lim_{n\to\infty} \|AP_{D_n}\theta\|_\beta = 0.$$

Hence from the relation

$$\int_{D_n} K[t, s, x_n(s)]\,\mathrm{d}s = AP_{D_n}x_n - A\theta + AP_{D_n}\theta$$

it follows that

$$\lim_{n\to\infty} \|\int_{D_n} K[t, s, x_n(s)]\,\mathrm{d}s\|_\beta = 0,$$

but this contradicts (19.18).

19.5 *Special criteria for complete continuity*

The general theorem 19.3 is not very convenient in the study of concrete Uryson operators. The greatest difficulty arises in testing the regularity and the validity of condition (19.12).

Usually a check for these conditions reduces to establishing an estimate of the form

$$|K(t, s, u)| \leqq R(t, s, u) \qquad (t \in \Omega^*,\ s \in \Omega,\ -\infty < u < \infty), \tag{19.19}$$

where $R(t, s, u)$ defines an Uryson operator

$$Bx(t) = \int_\Omega R[t, s, x(s)]\,\mathrm{d}s \tag{19.20}$$

which can be investigated 'more simply'. As we have already noted (see 18.5°), the regularity of the operator (19.20) implies the regularity of the

Uryson operator A with kernel $K(t, s, u)$. It is clear that condition (19.12) is fulfilled for the operator A if it is fulfilled for the operator $\boldsymbol{B}$. Thus we have proved:

THEOREM 19.4: *Let the inequality* (19.19) *be satisfied and let the operator* (19.20) *act from* L_α *to* L_β $(0 < \alpha < \infty)$ *and be a regular operator satisfying condition* (19.12). *Then the Uryson operator* A *with kernel* $K(t, s, u)$ *acts from* L_α *to* L_β *and is completely continuous.*

This theorem implies a more specialized, but also more convenient criterion for complete continuity.

THEOREM 19.5: *Let a function* $K(t, s, u)$ $(t \in \Omega^*,\ s \in \Omega,\ -\infty < u < \infty)$ *satisfy the Caratheodory condition and let*

$$|K(t, s, u)| \leqq \sum_{i=1}^{n} R_i(t, s) \cdot f_i(s, u) \quad (t \in \Omega^*, s \in \Omega, -\infty < u < \infty), \tag{19.21}$$

where the functions $f_i(s, u)$ *define superposition operators* $\mathfrak{f}_i$ *acting from* $L_\alpha(\alpha > 0)$ *to* L_{γ_i} *and the non-negative kernels* $R_i(t, s)$ *define linear integral operators acting from* L_{γ_i} *to* L_β $(0 < \beta \leqq 1)$. *Suppose that for each* i $(i = 1, 2, \ldots, n)$ *one of the following conditions holds:*

a) $0 < \gamma_i \leqq 1$; *the operator* $\mathfrak{f}_i$ *is 'improving';*

b) $0 < \gamma_i < 1$; *the operator* $\boldsymbol{R}_i$ *is compact.*

Then the Uryson operator A *with the kernel* $K(t, s, u)$ *acts from* L_α *to* L_β *and is completely continuous.*

For the proof it suffices to study the Hammerstein operators defined by each term on the right side of inequality (19.21). In other words, it suffices to show that each Hammerstein operator

$$\boldsymbol{B}_i x(t) = \int_\Omega R_i(t, s) f_i[s, x(s)] \mathrm{d}s \tag{19.22}$$

is regular and satisfies condition (19.12).

The regularity of the operators (19.22) under the conditions of Theorem 19.5 is obvious.

To prove relation (19.12) note first the trivial estimate:

$$\|\int_D R_i(t, s)\cdot f_i[s, x(s)]\,\mathrm{d}s\|_\beta \leqq \|\boldsymbol{R}_i\boldsymbol{P}_D\|_{\gamma_i\to\beta}\|\boldsymbol{P}_D\mathfrak{f}_i x\|_{\gamma_i}, \tag{19.23}$$

where

$$\boldsymbol{R}_i x(t) = \int_\Omega R_i(t, s)x(s)\,\mathrm{d}s,$$

$$\mathfrak{f}_i x(s) = f_i[s, x(s)].$$

If the operator $\mathfrak{f}_i$ is 'improving' then (see 17.7°)

$$\lim_{\mathrm{mes}\,D\to 0}\ \sup_{\|x\|_\alpha\leqq 1}\|\boldsymbol{P}_D\mathfrak{f}_i x\|_{\gamma_i} = 0,$$

and (19.12) follows from (19.23). If $\gamma_i < 1$ but the operator $\boldsymbol{R}_i$ is compact, then by Theorem 3.3

$$\lim_{\mathrm{mes}\,D\to 0}\|\boldsymbol{R}_i\boldsymbol{P}_D\|_{\gamma_i\to\beta} = 0$$

and again (19.12) follows from (19.23). The theorem has been proved.

In concluding this section let us present one more criterion of a special character for complete continuity.

Introduce the notation

$$h(s, u) = \int_{\Omega^*}|K(t, s, u)|\,\mathrm{d}t \tag{19.24}$$

and denote by $\mathfrak{h}$ the superposition operator defined by this function:

$$\mathfrak{h}x(s) = h[s, x(s)]. \tag{19.25}$$

THEOREM 19.6: *Let an Uryson operator A with kernel $K(t, s, u)$ act from L_α to L_β ($0 < \alpha < \infty$, $0 < \beta \leqq 1$) and be a regular operator. Let*

$$\lim_{\mathrm{mes}\,D\to 0}\ \sup_{\|x\|_\alpha\leqq 1}\|\boldsymbol{P}_D A x\|_\beta = 0. \tag{19.26}$$

Finally, let the superposition operator $\mathfrak{h}$ *defined by the function* (19.24) *which acts from* L_α *to* L_1 *be an 'improving' operator. Then the Uryson operator* A *is completely continuous.*

Proof: Lemma 1.1 implies that it suffices to establish complete continuity as an operator from L_α to L_1. For this one can use the trivial estimate:

$$\int_{\Omega^*} \left| \int_D K[t, s, x(s)]\,ds \right| dt \leqq \int_D \left[\int_{\Omega^*} |K[t, s, x(s)]|\,dt \right] ds,$$

which implies, since the operator $\mathfrak{h}$ is 'improving', that

$$\lim_{\operatorname{mes} D \to 0} \sup_{\|x\|_\alpha \leqq 1} \left\| \int_D K[t, s, x(s)]\,ds \right\|_1 = 0.$$

It remains to use Theorem 19.3. The theorem has been proved.

Note that condition (19.26) is automatically fulfilled if A is bounded on the unit ball of the space L_α as an operator with values in L_{β_1}, where $\beta_1 < \beta$.

19.6 *On L-characteristics of Uryson operators*

The authors do not know of any general assertion on the structure of the L-characteristics $L(A$; comp. cont.) of Uryson operators. Here the situation is just as it was in the case of the L-characteristics $L(A$; cont.).

Theorem 19.7: *Let an Uryson operator* A *act from* L_α *to* L_β, *where* $0 < \alpha < \infty$, $0 < \beta \leqq 1$, *and be a regular operator. Then* A *is completely continuous as an operator from each* L_{α_1}, *where* $\alpha_1 < \alpha$, *to* L_β.

Proof: The functions in the unit ball of a space L_{α_1} have equi-absolutely continuous norms in the space L_α. Hence Theorem 18.4 implies that

$$\lim_{\operatorname{mes} D \to 0} \sup_{\|x\|_\alpha \leqq 1} \left\| \int_D K[t, s, x(s)]\,ds \right\|_\beta = 0.$$

It remains to use Theorem 19.3. The theorem has been proved.

In particular, it follows from this theorem that the interior points of the L-characteristic $L(A$; reg.) of an Uryson operator A belong to the L-

characteristic $L(A;\text{comp. cont.})$. Furthermore, Theorem 19.7 implies that a point $\{\alpha_0, \beta_0\}$ $(0 < \alpha_0 < \infty, 0 < \beta_0 \leqq 1)$ belongs to the L-characteristic $L(A;\text{comp. cont.})$, if the L-characteristic $L(A;\text{reg.})$ contains any point of the form $\{\alpha, \beta_0\}$, where $\alpha_0 < \alpha$.

It is natural to raise the question of whether $\{\alpha_0, \beta_0\} \in L(A;\text{reg.})$ $(0 < \alpha_0 < \infty,\ 0 < \beta_0 \leqq 1)$ implies that each point $\{\alpha_0, \beta\}$ for $\beta > \beta_0$ belongs to $L(A;\text{comp. cont.})$. In the general case the answer to this problem is negative (we leave it to the reader to find an example). The answer is positive if A is compact in measure in L_{α_0}; note, for instance, that by Theorem 19.6 the operator A is compact in measure if the operator (19.25) acts from L_{α_0} to L_1 and is an improving operator.

19.7 *Weakening of singularities*

Let us continue with the study of an Uryson operator

$$Ax(t) = \int_{\Omega} K[t, s, x(s)]\,ds, \tag{19.27}$$

acting from L_α to L_β $(0 \leqq \alpha, \beta < \infty)$. The following problem is of interest: how do the properties of the operator (19.27) change if the singularity of its kernel is slightly 'weakened' or 'strengthened'?

Let a function $\Phi(t, s, u)$ be defined for $t \in \Omega^*$, $s \in \Omega$, $-\infty < u < \infty$. Denote by $\boldsymbol{B}$ the integral operator with kernel $K(t, s, u)\cdot\Phi(t, s, u)$:

$$\boldsymbol{B}x(t) = \int_{\Omega} K[t, s, x(s)]\cdot\Phi[t, s, x(s)]\,ds. \tag{19.28}$$

Lemma 19.2: *Let an Uryson integral operator with a non-negative kernel $K(t, s, u)$ act from L_α to L_β $(0 \leqq \alpha, \beta \leqq 1)$ as a compact operator. Let the function $\Phi(t, s, u)$ $(t \in \Omega^*, s \in \Omega, -\infty < u < \infty)$ be bounded and uniformly continuous in the variables $t \in \Omega^*$, $s \in \Omega$, uniformly with respect to u. Then the operator* (19.28) *also acts from L_α to L_β as a compact operator.*

Proof: Let $\Phi_0(t, s, u)$ be the characteristic function of a set $\Omega_0^* \times \Omega_0 \times R$, where $\Omega_0 \subset \Omega$, $\Omega_0^* \subset \Omega^*$, $R \subset (-\infty, \infty)$. Let us show that the operator

$$\boldsymbol{B}_0x(t) = \int_{\Omega} K[t, s, x(s)]\cdot\Phi_0[t, s, x(s)]\,ds$$

is compact. It can be assumed without loss of generality that $K(t, s, 0) = 0$. Consider the operators

$$\boldsymbol{Q}v(t) = \begin{cases} v(t) & \text{for } t \in \Omega_0^* \\ 0 & \text{for } t \notin \Omega_0^* \end{cases}$$

$$\boldsymbol{P}u(s) = \begin{cases} u(s) & \text{for } s \in \Omega_0,\ u(s) \in R \\ 0 & \text{for other } s, \end{cases}$$

then $\boldsymbol{B}_0 = \boldsymbol{QAP}$. The operator $\boldsymbol{P}$ is bounded in the space L_α, the operator $\boldsymbol{A}$ is compact as an operator from L_α to L_β, and the linear operator $\boldsymbol{Q}$ is continuous in the space L_β. Consequently, the operator $\boldsymbol{B}_0$ is also compact.

Let the function $\Phi(t, s, u)$ satisfy the conditions of the lemma and let ε be a given positive number. Partition the sets Ω and Ω^*

$$\Omega = \Omega_1 \cup \ldots \cup \Omega_n, \quad \Omega^* = \Omega_1^* \cup \ldots \cup \Omega_n^*$$

so that for any fixed i, j

$$|\Phi(t', s', u) - \Phi(t'', s'', u)| < \varepsilon \qquad (t', t'' \in \Omega_i^*;\ s', s'' \in \Omega_j).$$

Select points t_i, s_j in each Ω_i^* and in each Ω_j, respectively.

Let $|\Phi(t, s, u)| \leqq b$ and $k_0 = [2b\varepsilon^{-1}] + 1$. Put

$$\varphi_k = -b + k\varepsilon \qquad (k = 0, 1, \ldots, k_0).$$

Consider the sets

$$R_{ijk} = \{u \colon \varphi_{k-1} \leqq \Phi(t_i, s_j, u) < \varphi_k\} \qquad (k = 1, 2, \ldots, k_0).$$

It is clear that for fixed i, j, k

$$|\Phi(t', s', u') - \Phi(t'', s'', u'')| < 2\varepsilon$$

$$(t', t'' \in \Omega_i^*;\ s', s'' \in \Omega_j;\ u', u'' \in R_{ijk}). \tag{19.29}$$

Put

$$\Phi_\varepsilon(t, s, u) = \inf_{\substack{\tau\in\Omega_i^*, \sigma\in\Omega_j \\ v\in R_{ijk}}} \Phi(\tau, \sigma, v) \quad (t \in \Omega_i^*,\ s \in \Omega_j,\ u \in R_{ijk}). \tag{19.30}$$

The discussion given at the beginning of this proof shows that the operator

$$\boldsymbol{B}_\varepsilon x(t) = \int_\Omega K[t, s, x(s)]\,\Phi_\varepsilon[t, s, x(s)]\,\mathrm{d}s$$

is compact. But by (19.29)

$$|\boldsymbol{B}_\varepsilon x(t) - \boldsymbol{B}x(t)| \leqq 3\varepsilon|\boldsymbol{A}x(t)| \quad (t \in \Omega^*),$$

which implies the compactness of the operator $\boldsymbol{B}$. The lemma has been proved.

A bounded function $\Phi(t, s, u)$ which is uniformly continuous in joint variables $t \in \Omega^*$, $s \in \Omega$, uniformly with respect to u, is called a *regularizer*[1] of the Uryson operator $\boldsymbol{A}$ with kernel $K(t, s, u)$ if, for each $\varepsilon > 0$, the operator

$$\boldsymbol{A}_\varepsilon x(t) = K[t, s, x(s)]\cdot\kappa_\varepsilon[t, s, x(s)]\,\mathrm{d}s \tag{19.31}$$

acts from L_α to L_β and is compact; here $\kappa_\varepsilon(t, s, u)$ denotes the characteristic function of the set of points $\{t, s, u\} \in \Omega^* \times \Omega \times (-\infty, \infty)$ for which $|\Phi(t, s, u)| \geqq \varepsilon$.

THEOREM 19.8: *Let an Uryson operator* $\boldsymbol{A}$ *with a non-negative kernel* $K(t, s, u)$ *act from* L_α *to* L_β $(0 \leqq \alpha, \beta \leqq 1)$ *and be bounded on each ball. Then for any regularizer* $\Phi(t, s, u)$ *the Uryson operator* $\boldsymbol{B}$ *with kernel* $K(t, s, u)\cdot\Phi(t, s, u)$ *acts from* L_α *to* L_β *as a compact operator.*

[1] Regularizers which are independent of u are often convenient.

PROOF: Let an arbitrary positive number ε be given. Put

$$B_1x(t) = \int_\Omega K[t, s, x(s)] \cdot \Phi[t, s, x(s)] \cdot \kappa_\varepsilon[t, s, x(s)] \, ds,$$

$$B_2x(t) = \int_\Omega K[t, s, x(s)] \cdot \Phi[t, s, x(s)] \{1 - \kappa_\varepsilon[t, s, x(s)]\} \, ds.$$

Obviously $B = B_1 + B_2$.

By Lemma 19.2 the operator B_1 is compact. Furthermore,

$$|B_2x(t)| \leqq \varepsilon \int_\Omega K[t, s, x(s)] \, ds \leqq \varepsilon \cdot Ax(t),$$

whence it follows that

$$\|B_2x(t)\|_\beta \leqq \varepsilon \|Ax(t)\|_\beta.$$

Hence the assertion of the theorem follows from the arbitrariness of ε. The theorem has been proved.

Theorem 19.8 means that a regularizer is a function which, under multiplication, 'weakens' the singularity of the kernel at those points which may 'disturb' the compactness of the operator (19.27). This definition of a regularizer generalizes the notion of a regularizer of a linear operator (see 5.8°).

19.8 *On two criteria for compactness (in measure) of operators*

A kernel $K(t, s, u)$ $(t \in \Omega^*, s \in \Omega, -\infty < u < \infty)$ is called *quasi-monotone* in t if it is possible to select a family of sets $\Omega^*(c) \subset \Omega^*$ $(0 \leqq c \leqq \text{mes}\, \Omega^*)$ such that

$$\Omega^*(c_1) \subset \Omega^*(c_2) \text{ for } c_1 < c_2, \quad \text{mes}\, \Omega^*(c) = c \tag{19.32}$$

and for fixed s, u

$$K(t', s, u) \leqq K(t'', s, u) \qquad (t' \in \Omega^*(c),\ t'' \notin \Omega^*(c)).$$

THEOREM 19.9:[1] *Let an Uryson operator A with kernel $K(t, s, u)$ act from L_α to L_β ($0 \leqq \alpha, \beta < \infty$) and be bounded on each ball of L_α. Let the kernel $K(t, s, u)$ be quasi-monotone in t. Then the operator A is compact in measure on L_α.*

This theorem follows immediately from the compactness in measure of each bounded family $\mathfrak{M}$ of functions in L_β for which it is possible to select a family of sets $\Omega^*(c)$ satisfying the conditions (19.32), such that

$$y(t') \leqq y(t'') \qquad (t' \in \Omega^*(c),\ t'' \notin \Omega^*(c))$$

for all functions $y(t) \in \mathfrak{M}$. The proof of compactness of such families of functions is left to the reader.

Introduce the notation

$$F_h(s, u) = \int_{\Omega^*} |K(t + h, s, u) - K(t, s, u)|\, dt, \tag{19.33}$$

where

$$K(t + h, s, u) = \begin{cases} K(t + h, s, u), & \text{if } t + h \in \Omega^* \\ 0, & \text{if } t + h \notin \Omega^*. \end{cases}$$

Denote by $\mathfrak{F}_h$ the superposition operators defined by the functions $F_h(s, u)$.

THEOREM 19.10: *Let an Uryson operator A with kernel $K(t, s, u)$ act from L_α to L_β ($0 \leqq \alpha < \infty, 0 \leqq \beta \leqq 1$) and be bounded on each ball of L_α. Suppose that*

$$\lim_{h \to 0} \sup_{\|x\|_\alpha \leqq 1} \|\mathfrak{F}_h x\|_1 = 0. \tag{19.34}$$

Then the operator A is compact in measure on L_α.

This theorem follows from the well-known Riesz criterion for compactness of sets in the space L_1 (see N. Dunford and J. T. Schwartz [1]).

[1] A closely related theorem was established by T. Nurekenov [1].

19.9 *Complete continuity of Uryson operators with values in* L_0

Let us begin by studying operators acting from L_0 to L_0.

THEOREM 19.11:[1] *Let the kernel* $K(t, s, u)$ $(t \in \Omega^*,\ s \in \Omega,\ |u| \leqq a)$ *satisfy the conditions*:

a) *for almost all* $t \in \Omega^*$

$$\int_{\Omega} \sup_{|u| \leqq a} |K(t, s, u)| \, ds < \infty; \tag{19.35}$$

b) *for each* $\varepsilon > 0$ *it is possible to form a partition of* Ω^* *into sets* $\Omega_1^*, \ldots, \Omega_m^*$ *such that*

$$\sup_{t', t'' \in \Omega_i^*} \int_{D} \sup_{|u| \leqq a} |K(t', s, u) - K(t'', s, u)| \, ds < \varepsilon \quad (i = 1, 2, \ldots, n). \tag{19.36}$$

Then the Uryson operator A *with kernel* $K(t, s, u)$ *which is defined on the ball* $\|x\|_0 \leqq a$ *of the space* L_0, *acts in* L_0 *and is completely continuous.*

PROOF: Conditions a) and b) imply (details are left to the reader) that

$$\int_{\Omega} \sup_{|u| \leqq a} |K(t, s, u)| \, ds \leqq b < \infty \quad (t \in \Omega^*).$$

Hence the values of the operator on the ball $\|x\|_0 \leqq a$ satisfy the inequality

$$|Ax(t)| \leqq b \qquad (t \in \Omega^*). \tag{19.37}$$

Thus condition b) implies that the hypothesis of Lemma 1.2 holds for the set of functions $\{Ax(t)\colon \|x\|_0 \leqq a\}$. Hence this set is compact in L_0, i.e. the operator A is compact.

Let a sequence $x_n(s)$ $(\|x_n(s)\|_0 \leqq a)$ converge to a function $x_0(s)$ in L_0. Lemma 17.5 implies that for almost all $t \in \Omega^*$ the functions $K[t, s, x_n(s)]$ converge in measure to $K[t, s, x_0(s)]$. By condition a) it can be assumed

[1] L. A. Ladyzenskii [1].

that for such t the functions $|K[t, s, x_n(s)]|$ are bounded by the summable function $\sup_{|u| \leq a} |K(t, s, u)|$. Hence for such t

$$\int_\Omega K[t, s, x_0(s)]\,ds = \lim_{n \to \infty} \int_\Omega K[t, s, x_n(s)]\,ds.$$

This means that the operator A transforms convergent sequences in L_0 into almost everywhere convergent sequences. These latter sequences converge in L_0 norm because the operator A is compact. The theorem has been proved.

Let us now examine operators acting from L_α ($\alpha > 0$) to L_0. We give a criterion for the complete continuity of such Uryson operators which can be proved in almost the same way as Theorem 19.3.

THEOREM 19.12: *Let a kernel* $K(t, s, u)$ $(t \in \Omega^*,\ s \in \Omega,\ -\infty < u < \infty)$ *satisfy the conditions of Theorem 19.11 for each* $a > 0$. *Moreover, let*

$$\lim_{\text{mes}\, D \to 0} \sup_{\|x\|_\alpha \leq 1} \left\| \int_D K[t, s, x(s)]\,ds \right\|_0 = 0. \tag{19.38}$$

Then the Uryson operator A *with kernel* $K(t, s, u)$ *acts from* L_α *to* L_0 *and is completely continuous.*

§ 20 Differentiation of non-linear operators

20.1 *Derivative of a non-linear operator*

Let a non-linear operator A act from a Banach space E_1 to a Banach space E_2. The operator A is said to be *differentiable*[1] at a point $x_0 \in E_1$, if it is

---

[1] The fundamental facts on differential calculus in Banach spaces are presented in L. A. Ljusternik and V. I. Sobolev [1], L. V. Kantorovic and G. P. Akilov [1], E. Hille and R. Phillips [1], M. M. Vainberg [2, 6]. Sufficient conditions for the differentiability of superposition operators and of non-linear integral operators have been established and applied by many authors. In this paragraph we present in essence (20.4, 20.5, 20.7–20.10) certain new results established by P. P. Zabreiko and M. A. Krasnoselskii (see P. P. Zabreiko [5]).

Throughout this paragraph Fréchet derivatives are used.

defined on some ball $\|x - x_0\|_{E_1} \leqq r$ and if there exists a continuous linear operator $\boldsymbol{B}$ acting from E_1 to E_2 such that

$$\lim_{\|h\|_{E_1} \to 0} \frac{\|\boldsymbol{A}(x_0 + h) - \boldsymbol{A}x_0 - \boldsymbol{B}h\|_{E_2}}{\|h\|_{E_1}} = 0. \tag{20.1}$$

The above relation is equivalent to requiring that the increment

$$\boldsymbol{A}(x_0 + h) - \boldsymbol{A}x$$

can be written in the form

$$\boldsymbol{A}(x_0 + h) - \boldsymbol{A}x_0 = \boldsymbol{B}h + \omega(x_0, h) \quad (\|h\| \leqq r), \tag{20.2}$$

where $\omega(x_0, h) = o(\|h\|)$, i.e.

$$\lim_{\|h\|_{E_1} \to 0} \frac{\|\omega(x_0, h)\|_{E_2}}{\|h\|_{E_1}} = 0. \tag{20.3}$$

The operator $\boldsymbol{B}$ is called the *derivative* of the operator $\boldsymbol{A}$ at the point x_0 and is denoted by $\boldsymbol{A}'(x_0)$.

Assume that an operator $\boldsymbol{A}$ is differentiable on a set $G \subset E_1$ (this means that the derivative $\boldsymbol{A}'(x_0)$ exists for all $x_0 \in G$). The derivative can then be considered as an operator defined on G with values in the space $B(E_1 \to E_2)$ of continuous linear operators acting from E_1 to E_2. The operator $\boldsymbol{A}$ is called *continuously differentiable* if $\boldsymbol{A}'(x)$ is continuous as an operator from E_1 to $B(E_1 \to E_2)$.

Consider, as a simple example, the superposition operator

$$\mathfrak{f}x(s) = f[s, x(s)]. \tag{20.4}$$

Assume that the function $f(s, u)$ has a continuous derivative in u,

$$g(s, u) = f'_u(s, u). \tag{20.5}$$

The function $g(s, u)$, as is easily checked, then satisfies the Caratheodory condition; it defines the superposition operator

$$\mathfrak{g}x(s) = f'_u[s, x(s)]. \tag{20.6}$$

If the functions $f(s, u)$ and $g(s, u)$ are jointly continuous in $\{s, u\}$ and if the operator $\mathfrak{f}$ is examined in the space C of continuous functions, then it is obviously continuously differentiable and its derivative is defined by the relation

$$\mathfrak{f}'(x_0)h(s) = \mathfrak{g}(x_0)h(s). \tag{20.7}$$

Going to the spaces L_α makes the situation more complicated. In the sections to follow there will be given examples of continuously differentiable functions $f(s, u)$ defining superposition operators $\mathfrak{f}$ which act in some space L_α but which are differentiable only at certain points of this space.

The notion of derivative was introduced above for the case of operators, acting in Banach spaces. It extends without change to the case of an operator acting from a space L_α to a space L_β for arbitrary non-negative α, β.

20.2 *General form of the derivative of a superposition operator*

THEOREM 20.1: *Let a superposition operator $\mathfrak{f}$ defined by a function $f(s, u)$ act from L_α to L_β ($\alpha > 0$) and have the derivative $\mathfrak{f}'(x_0)$ at a point $x_0 \in L_\alpha$. Then this derivative has the form*

$$\mathfrak{f}'(x_0)h(s) = g(s)h(s) \quad (h(s) \in L_\alpha), \tag{20.8}$$

where $g(s)$ is defined by the relation[1]

$$g(s) = \underset{u \to 0}{\text{limmeas}} \frac{f[s, x_0(s) + u] - f[s, x_0(s)]}{u}. \tag{20.9}$$

Here: if $\alpha > \beta$, then $g(s) \equiv 0$ and

$$f(s, u) = b(s) \in L_\beta; \tag{20.10}$$

if $\alpha = \beta$, then $g(s) \in L_0$ and

$$f(s, u) = b(s) + g(s)u, \tag{20.11}$$

where $b(s) \in L_\alpha$; if $\alpha < \beta$, then $g(s) \in L_{\beta-\alpha}$.

[1] The symbol limmeas denotes limit in measure.

PROOF: Put $h = u\kappa_0$ in (20.2), where $\kappa_0(t) = 1$. Then (20.2) implies that $g(s) = \mathfrak{f}'(x_0)\kappa_0$ is represented by the relation (20.9).

It is easy to see that $\mathfrak{f}'(x_0)\kappa_D = g(s)\kappa_D$ for characteristic functions κ_D of sets $D \subset \Omega$. This implies the validity of relation (20.8) for step functions $h(s)$.

Put $\boldsymbol{B}h(s) = g(s)h(s)$. Obviously the operator $\boldsymbol{B}$ is closed. It coincides with the continuous operator $\mathfrak{f}'(x_0)$ on the set of step functions, which is dense in L_α. Hence it coincides with $\mathfrak{f}'(x_0)$ on the whole of L_α. Thus the relation (20.8) has been proved.

Consider the function

$$\omega(s, u) = f[s, x_0(s) + u] - f[s, x_0(s)] - g(s)u \tag{20.12}$$

and suppose that it is not identically equal to zero. Then there exist a $u_0 \neq 0$ and a set $\Omega_0 \subset \Omega$ with positive measure such that

$$|\omega(s, u_0)| \geqq c > 0 \quad (s \in \Omega_0). \tag{20.13}$$

Denote by Ω_n a sequence of subsets of Ω_0 such that $\text{mes}\,\Omega_n \to 0$. Then (20.13) implies that

$$\frac{c}{|u_0|}(\text{mes}\,\Omega_n)^{\beta-\alpha} \leqq \frac{\|\omega[s, u_0\kappa_{\Omega_n}(s)]\|_\beta}{\|u_0\kappa_{\Omega_n}(s)\|_\alpha},$$

hence by (20.3), $\beta > \alpha$.

Hence $\omega(s, u) \equiv 0$, if $\alpha \geqq \beta$. This implies the representations (20.10) and (20.11).

If $\alpha < \beta$, then $g(s) \in L_{\beta-\alpha}$, because otherwise the operator (20.8) does not act from L_α to L_β. The theorem has been proved. [1]

20.3 *Conditions for the differentiability of a superposition operator on the whole space*

THEOREM 20.2: *Let a superposition operator $\mathfrak{f}$ defined by a function $f(s, u)$ act from L_α to L_β, where $0 < \alpha < \beta$. Let the function $f(s, u)$ be differentiable*

[1] S. W. Wang [1] (see also M. M. Vainberg [2, 6], M. A. Krasnoselskii and Y. B. Rutickii [2, 5, 6]).

in u and suppose that the function $f'_u(s, u)$ satisfies the Caratheodory condition. It is necessary and sufficient for the differentiability of the operator $\mathfrak{f}$ at each point of the space L_α that the superposition operator $\mathfrak{g}$ defined by the function $f'_u(s, u)$ act from L_α to $L_{\beta-\alpha}$.

PROOF: Necessity follows from Theorem 20.1. Let us prove the sufficiency. Let $x_0(s)$ be a fixed function in L_α. Put

$$\omega(s, u) = f[s, x_0(s) + u] - f[s, x_0(s)] - f'_u[s, x_0(s)]u.$$

It is clear that for each function $h(s) \in L_\alpha$

$$\omega[s, h(s)] = \{f'_u[s, x_0(s) + \theta_h(s)h(s)] - f'[s, x_0(s)]\}h(s), \tag{20.14}$$

here $\theta_h(s)$ satisfies $0 \leqq \theta_h(s) \leqq 1$ and can be assumed to be measurable.[1] The relation (20.14) implies that

$$\frac{\|\omega[s, h(s)]\|_\beta}{\|h(s)\|_\alpha} \leqq \|\mathfrak{g}(x_0 + \theta_h h) - \mathfrak{g}(x_0)\|_{\beta-\alpha}.$$

The operator $\mathfrak{g}$ acts from L_α to $L_{\beta-\alpha}$ and is continuous by Theorem 17.1. Hence

$$\lim_{\|h\|_\alpha \to 0} \frac{\|\omega[s, h(s)]\|_\beta}{\|h\|_\alpha} = \lim_{\|h\|_\alpha \to 0} \|\mathfrak{g}(x_0 + \theta_h h) - \mathfrak{g}(x_0)\|_{\beta-\alpha} = 0.$$

The theorem has been proved.

Since in our hypothesis the function $f'_u(s, u)$ satisfies the Caratheodory condition, the superposition operator $\mathfrak{g}$ defined by it is continuous as an operator from L_α to $L_{\beta-\alpha}$ if it acts from L_α to $L_{\beta-\alpha}$ (see Theorem 17.1).

---

[1] In order that $\theta_h(s)$ is measurable, it is sufficient to define its value at s as the smallest θ for which:

$$h(s) \cdot f_u'[s, x_0(s) + \theta h(s)] = f[s, x_0(s) + h(s)] - f[s, x_0(s)]. \tag{20.15}$$

Then $\theta_h(s)$ is lower semi-continuous.

The continuity of the operator $\mathfrak{g}$ implies that the operator $\mathfrak{f}$ is continuously differentiable.

Denote by $L(\mathfrak{f}; \text{diff.})$ the set of points $\{\alpha, \beta\}$ such that $\mathfrak{f}$ as an operator from L_α to L_β is differentiable on the whole of L_α. Theorem 20.2 means that $L(\mathfrak{f}; \text{diff.})$ coincides with the intersection of the L-characteristic $L(\mathfrak{f}; \text{def.})$ with the set L^* of those points $\{\alpha, \beta\}$ $(0 < \alpha < \beta)$ for which $\{\alpha, \beta - \alpha\} \in L(\mathfrak{g}; \text{def.})$. In Fig. 20.1 the bold line shows the lower boundary of the L-characteristic $L(\mathfrak{f}; \text{def.})$; while the dotted line shows the lower-boundary of the L-characteristic $L(\mathfrak{g}; \text{def.})$; the L-characteristic $L(\mathfrak{f}; \text{diff.})$ is hatched.

Consider as our first example a superposition operator $\mathfrak{f}$ defined by a polynomial

$$f(s, u) = a_0(s) + a_1(s)u + \dots + a_n(s)u^n,$$

where $a_i(s) \in L_{\alpha_i}$. The function $f'_u(s, u)$ is also a polynomial.

$$f'_u(s, u) = a_1(s) + 2a_2(s)u + \dots + na_n(s)u^{n-1}.$$

If the operator $\mathfrak{f}$ acts from L_α to L_β $(0 < \alpha < \beta)$ then the operator $\mathfrak{g}$ acts from L_α to $L_{\beta-\alpha}$. Hence for a superposition operator defined by a polynomial, the L-characteristics $L(\mathfrak{f}; \text{cont.})$ and $L(\mathfrak{f}; \text{diff.})$ coincide.

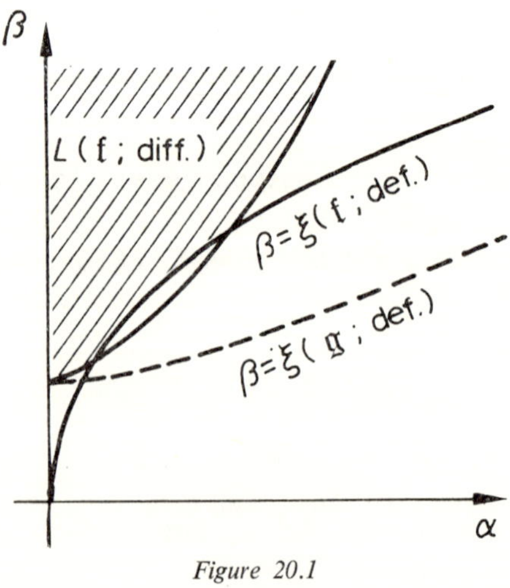

Figure 20.1

Consider as a second example the operator

$$\mathfrak{f}x(s) = \sin x^n(s).$$

Here $f(s, u) = \sin u^n$, $f'(s, u) = nu^{n-1} \cos u^n$. For this operator the L-characteristic $L(\mathfrak{f};\text{ def.})$ contains all points $\{\alpha, \beta\}$ with $\alpha, \beta \geqq 0$, while the L-characteristic $L(\mathfrak{g};\text{ def.})$ consists of points $\{\alpha, \beta\}$, for which $\beta \geqq (n-1)\alpha$. Theorem 20.2 implies that $L(\mathfrak{f};\text{ diff.})$ consists of those points $\{\alpha, \beta\}$ for which $\beta \geqq n\alpha$. Thus in this example the L-characteristic $L(\mathfrak{f};\text{ diff.})$ is a proper subset of the L-characteristic $L(\mathfrak{f};\text{ cont.})$.

An operator $\mathfrak{f}$ can be differentiable at some points of the space L_α $(\alpha < \beta)$ even in a case where $\{\alpha, \beta\} \notin L(\mathfrak{f};\text{ diff.})$. For instance, the operator $\mathfrak{f}x(s) = \sin x^n(s)$ considered above has a derivative (equal to zero) at the point $x_0(s) \equiv 0$, as an operator acting from L_α to any L_β, $\beta > \alpha$. As will be shown in the sections to follow, such operators are differentiable at the points of a dense subset of L_α.

20.4 *Sufficient criteria for the differentiability of a superposition operator*

Consider the question of obtaining sufficient conditions for the differentiability of a superposition operator $\mathfrak{f}x(s) = f[s, x(s)]$ at a given point[1] $x_0(s) \in L_\alpha$ $(\alpha > 0)$. Assume that the operator $\mathfrak{f}$ acts from L_α to L_β, where $\beta > \alpha$; this case is the only one of interest by Theorem 20.1. Moreover, assume that the function

$$g(s) = \operatorname*{limmeas}_{u \to 0} \frac{f[s, x_0(s) + u] - f[s, x_0(s)]}{u} \tag{20.16}$$

is defined and that it belongs to the space $L_{\beta-\alpha}$. These conditions are necessary (by Theorem 20.1) for the existence of the derivative $\mathfrak{f}'(x_0)$.

Consider the function

---

[1] Criteria for the differentiability of non-linear operators at distinguished points of function spaces were obtained, perhaps for the first time, by M. A. Krasnoselskii and Y. B. Rutickii [2, 5]. Theorem 20.3 was proved by S. W. Wang [1].

$$a(s,u) = \begin{cases} \dfrac{f[s, x_0(s)+u] - f[s, x_0(s)]}{u} - g(s), & \text{if } u \neq 0 \\ 0, & \text{if } u = 0. \end{cases} \tag{20.17}$$

This function is continuous in u for $u \neq 0$. Hence it is superpositionally measurable, i.e. transforms measurable functions $h(s)$ into measurable functions $a[s, h(s)]$.

THEOREM 20.3: *Let the operator* $\mathfrak{a}h(s) = a[s, h(s)]$ *act from* L_α *to* $L_{\beta-\alpha}$. *Then the operator* $\mathfrak{f}$ *is differentiable at the point* x_0 *and*

$$\mathfrak{f}'(x_0)h(s) = g(s)h(s) \qquad (h(s) \in L_\alpha).$$

PROOF: We have to prove

$$\lim_{\|h\|_\alpha \to 0} \frac{\|\omega[s, h(s)]\|_\beta}{\|h(s)\|_\alpha} = 0, \tag{20.18}$$

where the function $\omega(s, u)$ is defined by the relation (20.12).

Since

$$\omega[s, h(s)] = a[s, h(s)]h(s),$$

it follows by the Hölder inequality that

$$\frac{\|\omega[s, h(s)]\|_\beta}{\|h(s)\|_\alpha} \leqq \|\mathfrak{a}h(s)\|_{\beta-\alpha}.$$

The right side of this inequality converges to zero as $\|h\|_\alpha \to 0$ because the operator $\mathfrak{a}$ is continuous at the point zero by Theorem 17.6. The theorem has been proved.

Theorem 20.3 gives only a sufficient condition for the differentiability of an operator $\mathfrak{f}$ at specified points. Consider, for instance, the superposition operator $\mathfrak{f}$, defined by the function

$$f(s,u) = \begin{cases} 0, & \text{if } |u| \leqq 1,\ 0 \leqq s \leqq 1 \\ \left|s \ln \dfrac{s}{2}\right|^{-\frac{1}{2}} (|u| - 1), & \text{if } 1 < |u| \leqq 2,\ 0 < s \leqq 1 \\ \left|s \ln \dfrac{s}{2}\right|^{-\frac{1}{2}}, & \text{if } |u| > 2,\ 0 < s \leqq 1. \end{cases}$$

This operator acts from $L_{\frac{1}{2}}$ to L_1. It is differentiable at the point θ and its derivative is equal to zero. In fact,

$$\mathfrak{f}h(s) = \begin{cases} 0, & \text{if } |h(s)| \leqq 1 \\ \left|s \ln \dfrac{s}{2}\right|^{-\frac{1}{2}} (|h(s)| - 1), & \text{if } 1 < |h(s)| \leqq 2 \\ \left|s \ln \dfrac{s}{2}\right|^{-\frac{1}{2}}, & \text{if } |h(s)| > 2, \end{cases}$$

whence it follows that

$$\|\mathfrak{f}h(s)\|_1 \leqq \int\limits_{\Omega(h)} \left|s \ln \frac{s}{2}\right|^{-\frac{1}{2}} ds,$$

where $\Omega(h)$ is the set of points s, for which $|h(s)| \geqq 1$. Now $\|h\|_{\frac{1}{2}} \leqq \delta$ implies that mes $\Omega(h) \leqq \delta^2$. Hence

$$\|\mathfrak{f}h(s)\|_1 \leqq \int\limits_0^{\delta^2} \frac{ds}{\sqrt{\left|s \ln \dfrac{s}{2}\right|}},$$

so that

$$\|\mathfrak{f}h(s)\|_1 \leqq \delta \int\limits_0^1 \frac{du}{\sqrt{\left|u \ln \dfrac{\delta^2 u}{2}\right|}} \leqq \frac{\delta}{\sqrt{2\left|\ln \dfrac{\delta}{2}\right|}} \int\limits_0^1 \frac{du}{\sqrt{u}} = o(\delta).$$

On the other hand, the hypotheses of Theorem 20.3 are not satisfied since the function $h_0(s) \equiv 0$, which belongs to $L_{\frac{1}{2}}$, is transformed by the operator $\mathfrak{a}$ to the function $\mathfrak{a}h_0(s) = \frac{1}{2}|s \ln s/2|^{-\frac{1}{2}}$, which does not belong to $L_{\frac{1}{2}}$.

The hypothesis of Theorem 20.3 that the operator $\mathfrak{a}$ acts from L_α to $L_{\beta-\alpha}$ is by Lemma 17.6[1] equivalent to the inequality

$$|\omega(s, u)| \leqq a(s)|u| + b|u|^{\beta/\alpha}, \tag{20.19}$$

where $a(s) \in L_{\beta-\alpha}$. Unfortunately, a direct verification of the inequality (20.19) is often difficult. It is considerably simpler to obtain separate estimates of $\omega(s, u)$ for large and for small values of u. There then arises the problem of how to deduce the inequality (20.19) from these independent estimates.

LEMMA 20.1: *Let a function $\omega(s, u)$ satisfy, for $|u| \leqq u_0$, the inequality*

$$|\omega(s, u)| \leqq c(s)|u|^k \quad (s \in \Omega), \tag{20.20}$$

where $k > 1$, $c(s) \in L_\gamma$, while it satisfies for all $u \in (-\infty, \infty)$ the inequality

$$|\omega(s, u)| \leqq c_0(s) \cdot |u|^{\gamma_0} + b|u|^{\beta/\alpha} \quad (s \in \Omega), \tag{20.21}$$

where $0 \leqq \gamma_0 \leqq 1$, $c_0(s) \in L_q$. Let

$$\beta - \alpha \geqq q, \ \beta - \alpha \geqq \frac{k-1}{k-\gamma_0} q + \frac{1-\gamma_0}{k-\gamma_0} \gamma. \tag{20.22}$$

Then the function $\omega(s, u)$ satisfies the inequality (20.19).

PROOF: Consider the functions

$$c_1(s) = [c_0(s) + b|u_0|^{\beta/\alpha-\gamma_0}]^{(k-1)/(k-\gamma_0)} c(s)^{(1-\gamma_0)/(k-\gamma_0)}$$

$$c_2(s) = c_0(s)|u_0|^{\gamma_0 - 1}.$$

[1] Actually, the function a does not necessarily satisfy the Caratheodory condition assumed in that lemma unless we change (20.16) to the requirement of convergence a.e. (Ed.).

Both of these functions belong to the space $L_{\beta-\alpha}$. For the first, this follows from the fact that $c_1(s) \in L_{((k-1)/(k-\gamma_0))q+((1-\gamma_0)/(k-\gamma_0))\gamma}$ together with the second inequality of (20.22), while for the second, it follows from the first inequality of (20.22). Denote by $a(s)$ the sum

$$a(s) = c_1(s) + c_2(s). \tag{20.23}$$

This function also belongs to the space $L_{\beta-\alpha}$.

Let $|u| \leqq u_0$. Then from inequalities (20.20) and (20.21) it follows that

$$|\omega(s, u)| \leqq \min\{c(s)|u|^k, c_0(s)|u|^{\gamma_0} + b|u|^{\beta/\alpha}\}$$

$$\leqq [c_0(s)|u|^{\gamma_0} + b|u|^{\beta/\alpha}]^{(k-1)/(k-\gamma_0)}[c(s)|u|^k]^{(1-\gamma_0)/(k-\gamma_0)}$$

$$\leqq [c_0(s) + b|u_0|^{\beta/\alpha-\gamma_0}]^{(k-1)/(k-\gamma_0)}\cdot c(s)^{(1-\gamma_0)/(k-\gamma_0)}|u| = c_1(s)|u|.$$

Thus inequality (20.19) is fulfilled for $|u| \leqq u_0$.

Let $|u| \geqq u_0$. Then by inequality (20.21)

$$|\omega(s, u)| \leqq c_0(s)|u|^{\gamma_0} + b|u|^{\beta/\alpha} \leqq c_2(s)|u| + b|u|^{\beta/\alpha},$$

consequently the inequality (20.19) also holds for $|u| \geqq u_0$. The lemma has been proved.

This lemma permits us to use estimates of a fairly general type in proving the differentiability of a superposition operator.

THEOREM 20.4: *Let a superposition operator* $\mathfrak{f}$ *defined by a function* $f(s, u)$ *act from* L_α *to* L_β*, where* $0 < \alpha < \beta$. *Let the function* $\omega(s, u)$ *satisfy the two conditions*:

$$|\omega(s, u)| \leqq c(s)|u|^k \quad (s \in \Omega,\ |u| \leqq u_0), \tag{20.24}$$

and

$$|\omega(s, u)| \leqq \sum_{i=1}^{k} c_i(s)|u|^{\gamma_i} + b|u|^{\beta/\alpha} \quad (s \in \Omega,\ -\infty < u < \infty), \tag{20.25}$$

where $c(s) \in L_r$, $c_i(s) \in L_{q_i}$, $k > 1$, $0 \leqq \gamma_i \leqq 1$ *and*

$$\beta - \alpha \geqq q_i, \; \beta - \alpha \geqq \frac{k-1}{k-\gamma_i} q_i + \frac{1-\gamma_i}{k-\gamma_i} r \quad (i = 1, \ldots, k). \tag{20.26}$$

Then the operator $\mathfrak{f}$ *is differentiable at the point* x_0.

For the proof it suffices to represent $\omega(s, u)$ in the form of a sum of functions each of which is estimated by one of the terms on the right side of inequality (20.25) (a similar construction was applied in 18.2°). Each of these functions admits an estimate of the form (20.19) by Lemma 20.1. Hence the function $\omega(s, u)$ itself admits the estimate (20.19). It remains only to apply Theorem 20.3.

20.5 *Differentiability of superposition operators on dense sets*

Let a superposition operator $\mathfrak{f}$ act from L_α to L_β $(0 < \alpha < \beta)$. There arises the natural question of how to describe the set of points $x_0 \in L_\alpha$ at which the operator $\mathfrak{f}$ is differentiable. This problem has not been solved.

In this section we indicate certain sufficient conditions under which $\mathfrak{f}$, as an operator from L_α to L_β, is differentiable at all points x_0 in some space L_γ, where $0 \leqq \gamma < \alpha$.

Assume that the function $f'_u(s, u)$ exists and satisfies the Caratheodory condition. If the operator $\mathfrak{f}$ is differentiable at all points $x \in L_\gamma$, then by Theorem 20.1 its derivative has the form $\mathfrak{f}'(x_0)h = f'_u[s, x_0(s)]h(s)$, and $f'_u[s, x_0(s)] \in L_{\beta-\alpha}$. This means that the superposition operator

$$\mathfrak{g}x(s) = f'_u[s, x(s)]$$

acts from L_γ to $L_{\beta-\alpha}$. We have thus obtained a necessary condition for the differentiability of the superposition operator $\mathfrak{f}$ at all points of L_γ. We do not know whether this condition is sufficient.

Assume that this necessary condition is fulfilled. Then it is natural to use Theorem 20.3 to obtain a sufficient condition. In other words, one must find a condition under which the function

$$\Delta f(s, u) = f[s, x_0(s) + u] - f[s, x_0(s)],$$

constructed for any fixed function $x_0(s) \in L_\gamma$, satisfies the estimate:

$$|\Delta f(s, u)| \leqq a(s)|u| + b|u|^{\beta/\alpha}, \tag{20.27}$$

where $a(s) \in L_{\beta-\alpha}$.

Since the operator $\mathfrak{g}$ acts from L_γ to $L_{\beta-\alpha}$ it follows by Lemma 17.6 that

$$|f'_u(s, u)| \leqq a_0(s) + b|u|^{(\beta-\alpha)/\gamma},$$

where $a_0(s) \in L_{\beta-\alpha}$. Hence

$$|\Delta f(s, u)| \leqq \{a_0(s) + b[|x_0(s)| + |u|]^{(\beta-\alpha)/\gamma}\}|u|.$$

Consequently for each fixed $u_0 > 0$

$$|\Delta f(s, u)| \leqq a(s, u_0)|u| \quad (|u| \leqq u_0)$$

where

$$a(s, u_0) = a_0(s) + b[|x_0(s)| + |u_0|]^{(\beta-\alpha)/\gamma};$$

that is an inequality of type (20.27) is satisfied for $|u| \leqq u_0$.

Thus any conditions under which an inequality of form (20.27) is also satisfied for large u are sufficient conditions for the differentiability of $\mathfrak{f}$ (as an operator from L_α to L_β) at all points $x_0(s) \in L_\gamma$.

For instance, suppose

$$|f(s, u)| \leqq a_0(s) + b|u|^{\beta/\alpha-\varepsilon} \tag{20.28}$$

where $a_0(s) \in L_{\beta-\alpha}$ and $\varepsilon \in [0, \beta/\alpha]$. Then

$$|\Delta f(s, u)| \leqq 2a_0(s) + c_1|x_0(s)|^{\beta/\alpha-\varepsilon} + c_2|u|^{\beta/\alpha-\varepsilon}. \tag{20.29}$$

If

$$\gamma \leqq \alpha\frac{\beta-\alpha}{\beta-\varepsilon\alpha}, \tag{20.30}$$

then (20.29) implies that an inequality of the form (20.27) holds for $|u| \geqq 1$. Hence the inequality (20.28) is a sufficient condition for the differentiability of $\mathfrak{f}$ as an operator from L_α to L_β at all points in L_γ, where γ satisfies the inequality (20.30) (provided that the operator $\mathfrak{g}$ acts from L_γ to $L_{\beta-\alpha}$).

A second sufficient condition can be given in the form of an inequality:

$$|f(s, u + v) - f(s, u)| \leqq \varphi(s, u)|v| + b|v|^{\beta/\alpha}$$

$$(-\infty < u < \infty, |v| \geqq v_0),$$

where the function $\varphi(s, u)$ defines a superposition operator acting from L_γ to $L_{\beta-\alpha}$ and v_0 is a positive number. The verification is left to the reader.

20.6 *Derivatives of Hammerstein operators*

Consider a Hammerstein operator

$$Ax(t) = \int_\Omega K(t, s)f[s, x(s)]ds. \tag{20.31}$$

Assume that it acts from L_α to L_β and write it in the form of a superposition

$$A = K \cdot \mathfrak{f} \tag{20.32}$$

where the nonlinear operator

$$\mathfrak{f}x(s) = f[s, x(s)] \tag{20.33}$$

acts from L_α to some space L_γ $(\gamma > \alpha)$ and the continuous linear integral operator

$$Kx(t) = \int_\Omega K(t, s)x(s)ds, \tag{20.34}$$

acts from L_γ to L_β.

Suppose that $\mathfrak{f}$ is differentiable at a point $x_0 \in L_\alpha$ as an operator acting from L_α to L_γ. Its derivative is then a linear operator acting from L_α to L_γ:

$$\mathfrak{f}'(x_0)h(s) = g(s)h(s).$$

Hence

$$\boldsymbol{B}h(t) = \int_{\Omega} K(t, s)g(s)h(s)\,\mathrm{d}s,$$

defines a continuous linear operator which acts from L_α to L_β. It is clear that

$$\lim_{\|h\|_\alpha \to 0} \|\boldsymbol{A}(x_0 + h) - \boldsymbol{A}x_0 - \boldsymbol{B}h\|_\beta$$

$$\leqq \|\boldsymbol{K}\|_{\gamma\to\beta} \cdot \lim_{\|h\|_\gamma \to 0} \|\mathfrak{f}(x_0 + h) - \mathfrak{f}x_0 - gh\|_\gamma = 0.$$

This means that the operator (20.31) is differentiable at the point x_0 and that its derivative coincides with the operator $\boldsymbol{B}$.

Thus we can obtain from each of the criteria for differentiability of the operator $\mathfrak{f}$ obtained in the preceding sections a sufficient criterion for the differentiability of the Hammerstein operator.

For instance, let the derivative $f'_u(s, u)$ exist and define a superposition operator acting from L_α to $L_{\gamma-\alpha}$. Then the operator (20.31) is continuously differentiable on the entire space L_α and

$$\boldsymbol{A}'(x_0)h = \int_{\Omega} K(t, s)\cdot f'_u[s, x_0(s)]h(s)\,\mathrm{d}s. \tag{20.35}$$

It is similarly possible to obtain conditions for the existence of derivatives at individual points or on dense subsets of L_α.

20.7 *Derivatives of Uryson operators*

Now consider the problem of differentiability for an Uryson operator

$$\boldsymbol{A}x(t) = \int_{\Omega} K[t, s, x(s)]\,\mathrm{d}s. \tag{20.36}$$

In what follows it will be assumed that the operator (20.36) acts from $L_\alpha = L_\alpha(\Omega)$ to $L_\beta = L_\beta(\Omega^*)$ $(0 \leqq \alpha \leqq 1;\ 0 \leqq \beta < \infty)$ and is continuous (see § 18).

Suppose that the derivative $\boldsymbol{A}'(x_0)$ of the operator (20.36) exists at a point $x_0 \in L_\alpha$. We do not know whether this derivative is an integral operator. In fact, we do not know whether this derivative is an integral operator even in those cases where the derivative $K'_u(t, s, u)$ exists and the linear operator

$$\boldsymbol{B}h(t) = \int_{\Omega} K'_u[t, s, x_0(s)]h(s)\,\mathrm{d}s \tag{20.37}$$

acts from L_α to L_β.

In what follows we will establish sufficient conditions for the differentiability of Uryson operators. Note that under these conditions the derivative $A'(x_0)h$ will always coincide with the operator (20.37).

Consider the Uryson operator (20.36) as a superposition

$$\boldsymbol{A} = \boldsymbol{J}\mathfrak{K}^0, \tag{20.38}$$

where

$$\boldsymbol{J}[u(t, s)] = \int_{\Omega} u(t, s)\,\mathrm{d}s \tag{20.39}$$

and

$$\mathfrak{K}^0 x(s) = K[t, s, x(s)]. \tag{20.40}$$

Assume that the operator $\mathfrak{K}^0$ acts from a space L_α to a space[1] $\mathcal{M}_\beta$ of functions of two variables $t \in \Omega^*$, $s \in \Omega$.

It is not difficult to see that the differentiability of $\mathfrak{K}^0$ as an operator from L_α to $\mathcal{M}_\beta$ implies the differentiability of $\boldsymbol{A}$ as an operator from L_α to L_β and in addition that

$$A'(x_0)h = \boldsymbol{J}[\mathfrak{K}^0]'(x_0)h = \int_{\Omega} [\mathfrak{K}^0]'(x_0)[h(s)]\,\mathrm{d}s. \tag{20.41}$$

---

[1] Recall (see 18.4°) that $\mathcal{M}_\beta$ $(0 < \beta \leqq 1)$ is the Banach space of functions $u(t, s)$ of two variables with norm

$$\|u(t, s)\|_{\mathcal{M}_\beta} = \|\int_{\Omega} |u(t, s)|\,\mathrm{d}s\|_\beta.$$

It can be shown that the derivative of the operator $\mathfrak{K}^0$, if it exists, always has the form

$$[\mathfrak{K}^0]'(x_0)[h(s)] = R(t, s)h(s);$$

here the function $R(t, s)$ is defined by the relation

$$R(t, s) = \operatorname*{limmeas}_{u\to 0} \frac{K[t, s, x_0(s) + u] - K[t, s, x_0(s)]}{u}.$$

In particular, if the derivative $K_u'(t, s, u)$ exists then the function $R(t, s)$ coincides with the function $K_u'[t, s, x_0(s)]$. This implies that the derivative $A'(x_0)$ is defined by formula (20.37) if the operator $\mathfrak{K}^0$ is differentiable at the point $x_0(s) \in L_\alpha$.

Denote by $\mathfrak{B}(\alpha, \beta)$ the set of measurable functions $k(t, s)$ $(t \in \Omega^*, s \in \Omega)$, satisfying for all $x \in L_\alpha$

$$\| \int_\Omega |k(t, s)| x(s) \mathrm{d}s \|_\beta < \infty.$$

For $0 \leqq \beta \leqq 1$ the set $\mathfrak{B}(\alpha, \beta)$ is a Banach space[1] with the usual operation of addition and scalar multiplication and with the norm

$$\|k(t, s)\|_{\mathfrak{B}(\alpha,\beta)} = \sup_{\|x\|_\alpha \leqq 1} \| \int_\Omega |k(t, s)| x(s) \mathrm{d}s \|_\beta.$$

Each function $k(t, s) \in \mathfrak{B}(\alpha, \beta)$ defines a regular integral operator

$$\boldsymbol{K}x(t) = \int_\Omega k(t, s) x(s) \mathrm{d}s$$

acting from L_α to L_β, and

$$\|k(t, s)\|_{\mathfrak{B}(\alpha,\beta)} = \| |\boldsymbol{K}| \|_{\alpha\to\beta},$$

where $|\boldsymbol{K}|$ is the linear integral operator with the kernel $|k(t, s)|$.

---

[1] A. Zaanen [3].

Consider next the superposition operator:

$$\mathfrak{H}u(t, s) = K'_u[t, s, u(t, s)]. \tag{20.42}$$

THEOREM 20.5: *Let the operator* $\mathfrak{H}$ *act continuously from the space*[1] $\mathcal{N}_\alpha$ *to the space* $\mathfrak{B}(\alpha, \beta)$. *Then the Uryson operator* (20.36) *is differentiable at all points of the space* L_α *and*

$$A'(x_0)h = \int_\Omega K'_u[t \cdot s, x_0(s)]h(s)\,\mathrm{d}s.$$

PROOF: It suffices to prove that for any fixed $x_0(s) \in L_\alpha$

$$\lim_{\|h\|_\alpha \to 0} \frac{1}{\|h\|_\alpha} \left\| \int_\Omega \{K[t, s, x_0(s) + h(s)] - K[t, s, x_0(s)] \right.$$

$$\left. - K'_u[t, s, x_0(s)]h(s)\}\,\mathrm{d}s \right\|_\beta = 0. \tag{20.43}$$

The difference

$$\Delta = K[t, s, x_0(s) + h(s)] - K[t, s, x_0(s)] - K'_u[t, s, x_0(s)]h(s)$$

can be written in the form

$$\Delta = \{K'_u[t, s, x_0(s) + \theta_h(t, s)h(s)] - K'_u[t, s, x_0(s)]\}\,h(s),$$

where the function $\theta_h(t, s)$ is measurable[2] and satisfies the inequality $0 \leqq \theta_h(t, s) \leqq 1$. This implies that

---

[1] Recall (see 18.5°) that $\mathcal{N}_\alpha$ $(0 \leqq \alpha \leqq 1)$ is a Banach space of functions $u(t, s)$ of two variables $t \in \Omega^*$, $s \in \Omega$ with norm

$$\|u(t, s)\|_{\mathcal{N}_\alpha} = \|\operatorname*{ess\,sup}_{t \in \Omega^*} |u(t, s)|\,\|_\alpha.$$

[2] See the footnote on p. 421.

$$\frac{1}{\|h\|_\alpha}\left\|\int_\Omega \{K[t, s, x_0(s) + h(s)]\right.$$

$$\left. - K[t, s, x_0(s)] - K'_u[t, s, x_0(s)]h(s)\}\,\mathrm{d}s\right\|_\beta$$

$$= \frac{1}{\|h\|_\alpha}\left\|\int_\Omega \{K'_u[t, s, x_0(s) + \theta_h(t, s)h(s)]\right.$$

$$\left. - K'_u[t, s, x_0(s)]\}h(s)\,\mathrm{d}s\right\|_\beta$$

$$\leqq \|\mathfrak{H}[x_0(s) + \theta_h(t, s)h(s)] - \mathfrak{H}[x_0(s)]\|_{\mathfrak{B}(\alpha, \beta)}. \tag{20.44}$$

Let $\|h_n\|_\alpha \to 0$. Then $\|\theta_{h_n}h_n\|_\alpha \to 0$ and (20.43) follows from (20.44) and the continuity of $\mathfrak{H}$. The theorem has been proved.

Let us now present some simple sufficient conditions for continuity of the operator $\mathfrak{H}$. Assume that the function $K'_u(t, s, u)$ satisfies the inequality:

$$|K'_u(t, s, u_1) - K'_u(t, s, u_2)|$$

$$\leqq R(t, s, u)|u_1 - u_2|^\delta \quad (|u_1|, |u_2| \leqq u), \tag{20.45}$$

where $0 < \delta \leqq 1$ and suppose that the function $R(t, s, u)$ satisfies the Caratheodory condition while the superposition operator

$$\mathfrak{R}u(t, s) = R[t, s, u(t, s)]$$

acts from $\mathcal{N}_\alpha$ to $\mathfrak{B}(\alpha + \delta\alpha, \beta)$. Then

$$|K'_u[t, s, u_1(t, s)] - K'_u[t, s, u_2(t, s)]|$$

$$\leqq R[t, s, u(t, s)]|u_1(t, s) - u_2(t, s)|^\delta,$$

where $u(t, s) = \max\{|u_1(t, s)|, |u_2(t, s)|\}$. Hence

$$\|\mathfrak{H}u_1 - \mathfrak{H}u_2\|_{\mathfrak{B}(\alpha, \beta)} \leqq \|\mathfrak{R}u|u_1 - u_2|^\delta\|_{\mathfrak{B}(\alpha, \beta)}$$

$$\leqq \|\mathfrak{R}u\|_{\mathfrak{B}(\alpha + \alpha\delta, \beta)}\|u_1 - u_2\|^\delta_{\mathcal{N}_\alpha}.$$

This inequality implies that under the condition (20.45) $\mathfrak{H}$ is continuous as an operator from $\mathcal{N}_\alpha$ to $\mathfrak{B}(\alpha, \beta)$.

Let us give another example. Let $0 < \alpha < 1$, $\beta > 0$. Assume that the operator $\mathfrak{H}$ acts from $L_\alpha(\Omega^* \times \Omega)$ to the space $L_\delta(\Omega^* \times \Omega)$ where $\delta = \min\{1 - \alpha, \beta\}$. In other words, assume that the inequality

$$|K_u'(t, s, u)| \leqq a(t, s) + b|u|^{\delta/\alpha} \tag{20.46}$$

holds where $a(t, s) \in L_\delta(\Omega^* \times \Omega)$. Theorem 17.1 implies that $\mathfrak{H}$ is continuous as an operator from $L_\alpha(\Omega^* \times \Omega)$ to $L_\delta(\Omega^* \times \Omega)$. Hence it is a continuous operator acting from $\mathcal{N}_\alpha$ to $L_\delta(\Omega^* \times \Omega)$. Theorem 6.1 implies that $L_\delta(\Omega^* \times \Omega) \subset \mathfrak{B}(\alpha, \beta)$ and that

$$\|k(t, s)\|_{\mathfrak{B}(\alpha, \beta)} \leqq c\|k(t, s)\|_{L_\delta(\Omega^* \times \Omega)}.$$

Hence $\mathfrak{H}$ is a continuous operator acting from $\mathcal{N}_\alpha$ to $\mathfrak{B}(\alpha, \beta)$.

Thus *inequality* (20.46) *implies the differentiability of the operator* (20.36) *at each point of the space* L_α. This assertion can be substantially improved.

20.8 *A general theorem*

THEOREM 20.6: *Let a function* $Q(t, s, u)$ $(t \in \Omega^*,\ s \in \Omega,\ -\infty < u < \infty)$ *satisfy the Caratheodory condition. Suppose that for any function* $u(t, s) \in \mathcal{N}_\alpha$ $(0 \leqq \alpha < 1)$ *the kernel* $|Q[t, s, u(t, s)]|$ *defines a compact linear operator acting from* L_α *to* L_β $(0 < \beta < \infty)$. *Then the superposition operator*

$$\mathfrak{Q}u(t, s) = Q[t, s, u(t, s)], \tag{20.47}$$

acts continuously from $\mathcal{N}_\alpha$ *to* $\mathfrak{B}(\alpha, \beta)$.

PROOF BY CONTRADICTION: Suppose that the function $Q(t, s, u)$ satisfies the conditions of Theorem 20.6, but that the superposition operator (20.47) acting from $\mathcal{N}_\alpha$ to $\mathfrak{B}(\alpha, \beta)$ does not possess the property of continuity. Then there is a function $u_0(t, s) \in \mathcal{N}_\alpha$ and a sequence of functions $u_n(t, s) \in \mathcal{N}_\alpha$ converging to $u_0(t, s)$ in norm of the space $\mathcal{N}_\alpha$, such that

$$\|\mathfrak{Q}u_n(t, s) - \mathfrak{Q}u_0(t, s)\|_{\mathfrak{B}(\alpha, \beta)} > \varepsilon_0 > 0 \quad (n = 1, 2, \ldots). \tag{20.48}$$

It can be assumed without loss of generality that $Q(t, s, 0) = 0$, that $u_0(t, s) = 0$ and that

$$\sum_{n=1}^{\infty} \|u_n(t, s)\|_{\mathcal{N}_\alpha} < \infty. \tag{20.49}$$

In this case inequality (20.48) takes the form

$$\|\mathfrak{Q}u_n(t, s)\|_{\mathfrak{B}(\alpha, \beta)} > \varepsilon_0 > 0 \quad (n = 1, 2, \ldots). \tag{20.50}$$

The function

$$u^*(t, s) = \sum_{n=1}^{\infty} |u_n(t, s)| \tag{20.51}$$

belongs to the space $\mathcal{N}_\alpha$ by (20.49). By Lemma 17.2 the function

$$R_0(t, s) = \sup_{|u| \leqq u^*(t, s)} |Q(t, s, u)| \tag{20.52}$$

belongs to the space $\mathfrak{B}(\alpha, \beta)$, and the linear integral operator

$$\boldsymbol{R}_0 x(s) = \int_\Omega R_0(t, s) x(s) \, \mathrm{d}s$$

acts from L_α to L_β and is compact. Theorems 3.3 and 5.7 imply that

$$\lim_{\operatorname{mes} D \to 0} \|\boldsymbol{R}_0 \boldsymbol{P}_D\|_{\alpha \to \beta} = 0. \tag{20.53}$$

From Lemma 17.5 it follows that the functions $Q[t, s, u_n(t, s)]$ converge to zero in measure on $\Omega^* \times \Omega$ as $n \to \infty$. Here (by taking a subsequence, if necessary) it can be assumed without loss of generality that for almost all $t \in \Omega^*$ the functions $Q[t, s, u_n(t, s)]$ converge in measure to zero.

Equation (20.52) implies that

$$|Q[t, s, u_n(t, s)]| \leqq R_0(t, s). \tag{20.54}$$

For almost all $t \in \Omega^*$ the function $R_0(t, s)$ is summable in s on Ω. Hence for almost all $t \in \Omega^*$ functions

$$\psi_n(t) = \int_\Omega |Q[t, s, u_n(t, s)]| \, ds \qquad (n = 1, 2, \dots)$$

converge to zero. Furthermore, from the relation

$$\varphi(t) = \int_\Omega R_0(t, s) \, ds \in L_\beta$$

and from (20.54) it follows that the functions $\psi_n(t)$ converge to zero in norm of the space L_β.

Hereafter $\Omega(x, h)$, where $x = x(s)$ is a function in the unit ball of the space L_α, will denote the set of points $s \in \Omega$ for which $|x(s)| \geqq h$. It is clear that

$$\operatorname{mes} \Omega(x; h) \leqq h^{-1/\alpha} \qquad (0 < h < \infty). \tag{20.55}$$

The inequality

$$\int_\Omega |Q[t, s, u_n(t, s)] x(s)| \, ds$$

$$= \int_{\Omega - \Omega(x; h)} |Q[t, s, u_n(t, s)] x(s)| \, ds$$

$$+ \int_{\Omega(x; h)} |Q[t, s, u_n(t, s)] x(s)| \, ds$$

$$\leqq h\psi_n(t) + \int_{\Omega(x; h)} R_0(t, s) |x(s)| \, ds \qquad (\|x\|_\alpha \leqq 1)$$

implies that

$$\| \int_\Omega |Q[t, s, u_n(t, s)] x(s)| \, ds \|_{L_\beta}$$

$$\leqq 2^\beta h \| \psi_n(t) \|_{L_\beta} + 2^\beta \| \boldsymbol{R}_0 \cdot \boldsymbol{P}_{\Omega(x; h)} x \|_{L_\beta}$$

or by (20.55)

$$\|\int_{\Omega} |Q[t, s, u_n(t, s)]x(s)| \, ds\|_{L_\beta}$$

$$\leqq 2^\beta h \|\psi_n(t)\|_{L_\beta} + \sup_{\text{mes } D \leqq h^{-1/\alpha}} 2^\beta \|\boldsymbol{R}_0 \boldsymbol{P}_D\|_{\alpha \to \beta}. \tag{20.56}$$

Let $\varepsilon > 0$ be given. Denote by h_0 a number such that

$$\sup_{\text{mes } D \leqq h_0^{-1/\alpha}} \|\boldsymbol{R}_0 \boldsymbol{P}_D\|_{\alpha \to \beta} < \varepsilon / 2^{1+\beta}.$$

Then (20.56) implies that

$$\|\int_{\Omega} |Q[t, s, u_n(t, s)]x(s)| \, ds\|_{L_\beta} \leqq 2^\beta h_0 \|\psi_n(t)\|_{L_\beta} + \varepsilon/2.$$

Select a number n_0 such that for $n \geqq n_0$ the inequality

$$\|\psi_n(t)\|_{L_\beta} \leqq \varepsilon / 2^{1+\beta} h_0$$

holds. Then for $n \geqq n_0$ the following inequality will be satisfied:

$$\|\int_{\Omega} |Q[t, s, u_n(t, s)]x(s)| \, ds\|_{L_\beta} < \varepsilon. \tag{20.57}$$

The number n_0 was chosen independently of the functions $x(s)$, $\|x(s)\|_\alpha \leqq 1$. Hence (20.57) means that

$$\|Q[t, s, u_n(t, s)]\|_{\mathfrak{B}(\alpha, \beta)} \leqq \varepsilon.$$

For small ε this inequality contradicts the inequality (20.50). The theorem has been proved.

Let the operator (20.47) act from $\mathcal{N}_\alpha$ to $\mathfrak{B}(\alpha, \beta)$, where $\alpha < \beta$. Then the compactness of the operators with kernels $Q[t, s, u(t, s)]$ $(u(t, s) \in \mathcal{N}_\alpha)$ follows from Theorems 5.5 and 5.8. Hence Theorem 20.6 implies that the operator (20.47) is continuous (as an operator from $\mathcal{N}_\alpha$ to $\mathfrak{B}(\alpha, \beta)$). The analogous assertion does not hold if $\alpha \geqq \beta$, as is shown by the operator

$$\mathfrak{Q}_0 u(t, s) = Q_0[t, s, u(t, s)] \tag{20.58}$$

defined as follows. Let α, β be fixed numbers, $0 < \beta \leqq \alpha < 1$. Put

$$Q_0(t, s, u) = \begin{cases} (2 - 2^n|u|)Q_n(t, s) + (2^n|u| - 1)Q_{n-1}(t, s) \\ \qquad\qquad \text{if } 1/2^n \leqq |u| < 1/2^{n-1} \\ 0 \qquad\qquad \text{for other } u, \end{cases}$$

where

$$Q_0(t, s) = 0,$$

$$Q_n(t, s) = \begin{cases} 2^{n(1-\alpha+\beta)}, & \text{if } 1/2^n \leqq t, s < 1/2^{n-1} \\ 0 & \text{for other } t, s \end{cases}$$

$(n = 1, 2, \ldots)$.

It is not difficult to see that the function $Q_0(t, s, u)$ satisfies the Caratheodory condition. Moreover, the following inequality holds:

$$|Q_0(t, s, u)| \leqq K_0(t, s), \tag{20.59}$$

where

$$K_0(t, s) = \begin{cases} 2^{n(1-\alpha+\beta)}, & \text{if } 1/2^n \leqq t, s < 1/2^{n-1} \\ 0 & \text{for other } t, s. \end{cases}$$

The kernel $K_0(t, s)$ is the kernel (7.19) with $\theta = \beta$, $r(1 - v) = 1 - \alpha$. In 7.2° it was shown that the kernel $K_0(t, s)$ defines a linear integral operator acting from L_α to L_β. Hence (20.59) implies that the operator (20.58) acts from $\mathcal{N}_\alpha$ to $\mathfrak{B}(\alpha, \beta)$. Let us now show that it does not possess continuity property.

Put

$$u_n(t, s) = 1/2^n \qquad (t, s \in [0, 1];\ n = 1, 2, \ldots).$$

The functions $u_n(t, s)$ converge to the function $\theta(t, s) \equiv 0$ in $\mathcal{N}_\alpha$ norm.

But the functions

$$\mathfrak{D}_0 u_n(t, s) = Q_n(t, s) \qquad (n = 1, 2, \ldots)$$

do not converge in norm of the space $\mathfrak{B}(\alpha, \beta)$ to the function

$$\mathfrak{D}_0 Q(t, s) = 0.$$

To see this, consider the functions

$$x_n(s) = \begin{cases} 2^{n\alpha}, & \text{if } 1/2^n \leqq s < 1/2^{n-1} \\ 0 & \text{for other } s. \end{cases}$$

Obviously $\|x_n\|_\alpha = 1$ and

$$\int_0^1 Q_n(t, s) x_n(s) \, ds = \begin{cases} 2^{n\beta}, & \text{if } 1/2^n \leqq t < 1/2^{n-1} \\ 0 & \text{for other } t. \end{cases}$$

Hence

$$\|Q_n(t, s)\|_{\mathfrak{B}(\alpha, \beta)} \geqq \|\int_0^1 Q_n(t, s) x_n(s) \, ds\|_\beta = 1 \qquad (n = 1, 2, \ldots).$$

Theorems 20.5 and 20.6 immediately imply:

THEOREM 20.7: *Let the functions $K(t, s, u)$ and $K'_u(t, s, u)$ $(t \in \Omega^*,\ s \in \Omega,$ $-\infty < u < \infty)$ satisfy the Caratheodory condition. Let the non-linear Uryson operator A with kernel $K(t, s, u)$ act from L_α to L_β. Suppose that for any function $u(t, s) \in \mathcal{N}_\alpha$, the kernel $K'_u[t, s, u(t, s)]$ defines a compact linear operator acting from L_α to L_β, $0 \leqq \alpha < 1$, $0 < \beta < \infty$. Then the operator A is continuously differentiable on the entire space L_α, and*

$$A'(x_0)h = \int_\Omega K'_u[\cdot, s, x_0(s)] h(s) \, ds \qquad (x_0(s), h(s) \in L_\alpha).$$

20.9 *Partial criteria for differentiability of Uryson operators*

Applications of Theorem 20.7 require proof of the compactness of the linear integral operators

$$\boldsymbol{B}(u)h = \int_{\Omega} |K'_u[\cdot, s, u(t, s)]|\, h(s)\,\mathrm{d}s \qquad (u(t, s) \in \mathcal{N}_\alpha). \tag{20.60}$$

Here of course one can apply various criteria for the compactness of linear integral operators which were established in § 4–8. Let us present an example.

THEOREM 20.8: *Let functions* $K(t, s, u)$ *and* $K'_u(t, s, u)$ $(t \in \Omega^*,\ s \in \Omega,\ -\infty < u < \infty)$ *satisfy the Caratheodory condition. Let the non-linear Uryson operator* $\boldsymbol{A}$ *with kernel* $K(t, s, u)$ *act from* L_α *to* L_β, *where* $0 \leqq \alpha < 1$, $0 < \beta < \infty$. *Suppose that the function* $K'_u(t, s, u)$ *satisfies the inequality*:

$$|K'_u(t, s, u)| \leqq \sum_{i=1}^{n} R_i(t, s)\cdot f_i(s, u), \tag{20.61}$$

where each function $f_i(s, u)$ *is non-negative, satisfies the Caratheodory condition and defines a superposition operator acting from* L_α *to* L_{γ_i}, $0 \leqq \gamma_i \leqq 1-\alpha$, *and each non-negative kernel* $R_i(t, s)$ *defines a linear integral operator acting from* $L_{\alpha+\gamma_i}$ *to* L_β, *where this linear operator is to be compact if* $\gamma_i = 0$. *Then the operator* $\boldsymbol{A}$ *is continuously differentiable on the entire space* L_α, *and*

$$\boldsymbol{A}'(x)h = \int_{\Omega} K'_u[\cdot, s, x(s)]\, h(s)\,\mathrm{d}s \qquad (x(s), h(s) \in L_\alpha).$$

PROOF: By Theorem 20.7 it suffices to show that each of the operators (20.60) is a compact operator acting from L_α to L_β.

By Lemma 17.6 each of the functions $f_i(s, u)$ $(i = 1, \ldots, n)$ satisfies the inequality

$$|f_i(s, u)| \leqq a_i(s) + b\,|u|^{\gamma_i/\alpha},$$

where $a_i(s) \in L_{\gamma_i}$. Hence for any fixed function $u(t, s) \in \mathcal{N}_\alpha$ the inequality

$$|f_i[s, u(t, s)]| \leqq a_i(s) + b\,|v(s)|^{\gamma_i/\alpha} = g_i(s), \tag{20.62}$$

holds, where $g_i(s) \in L_{\gamma_i}$ and

$$v(s) = \operatorname*{ess\,sup}_{t \in \Omega^*} |u(t, s)| \in L_\alpha.$$

Inequalities (20.61) and (20.62) imply the inequality

$$|K'_u[t, s, u(t, s)]| \leqq \sum_{i=1}^{n} R_i(t, s) \cdot g_i(s).$$

Since $0 \leqq \alpha < 1$, $0 < \beta < \infty$, we see by Theorem 5.10 that to prove compactness of the linear integral operator with kernel $K'_u[t, s, u(t, s)]$ it suffices to remark that each of the operators

$$\boldsymbol{B}_i h(t) = \int_\Omega R_i(t, s) g_i(s) h(s) \mathrm{d}s$$

is compact (for instance, by Theorem 19.1). The theorem has been proved.

20.10 *Differentiability of Uryson operators at distinguished points*

The question of differentiability of an Uryson operator $\boldsymbol{A}$ acting from L_α to L_β at distinguished points of the space L_α or on dense sets in L_α, has not been investigated satisfactorily. Let us present here a specialized result.

THEOREM 20.9: *Let a non-linear Uryson operator $\boldsymbol{A}$ with kernel $K(t, s, u)$ act from a space L_α, $0 \leqq \alpha < 1$, to a space L_β. Let the function $K'_u(t, s, u)$ satisfy the inequality*

$$|K'_u(t, s, u + h) - K'_u(t, s, u)| \leqq \sum_{i=1}^{n} R_i(t, s, u) |h|^{\delta_i}, \tag{20.63}$$

where $0 < \delta_1, \ldots, \delta_n < (1 - \alpha)/\alpha$ and the non-negative functions $R_i(t, s, u)$ are such that for any fixed $x(s) \in L_\gamma$, $\gamma \leqq \alpha$, the kernel $R_i[t, s, x(s)]$ defines a linear integral operator acting from $L_{\alpha(1+\delta_i)}$ to L_β. Finally, let the kernel $|K'_u(t, s, 0)|$ define a linear integral operator acting from L_α to L_β. Then the operator $\boldsymbol{A}$ is differentiable at each point of the space L_γ and

$$\boldsymbol{A}'(x_0)h = \int_\Omega K'_u[\cdot, s, x_0(s)] h(s) \mathrm{d}s \qquad (x_0(s) \in L_\gamma,\ h(s) \in L_\alpha).$$

PROOF: The inequality (20.63) implies that

$$|K_u'[t, s, x_0(s)]| \leqq |K_u'(t, s, 0)| + \sum_{i=1}^{n} R_i(t, s, 0)|x_0(s)|^{\delta_i}.$$

It is not difficult to see that under the conditions of the theorem the linear integral operators with the kernels $|K_u'(t, s, 0)|$, $R_1(t, s, 0)|x_0(s)|^{\delta_1}, \ldots,$ $R_n[t, s, x_0(s)]|x_0(s)|^{\delta_n}$ $(x_0(s) \in L_\gamma)$ act from L_α to L_β. Hence each integral operator

$$\boldsymbol{B}[x_0]h = \int_\Omega |K_u'[\cdot, s, x_0(s)]|\, h(s)\, ds$$

acts from L_α to L_β and is continuous.

Let $x_0(s)$ be a fixed function in L_γ, and let $h(s) \in L_\alpha$. The mean value formula together with inequality (20.63) implies that

$$|\boldsymbol{A}(x_0 + h) - \boldsymbol{A}x_0 - \boldsymbol{B}[x_0]h| \leqq \sum_{i=1}^{n} \int_\Omega R_i[\cdot, s, x_0(s)]|h(s)|^{1+\delta_i} ds. \quad (20.64)$$

The functions $|h(s)|^{1+\delta_i}$ belong to the spaces $L_{\alpha(1+\delta_i)}$, respectively. Hence inequality (20.64) implies that

$$\|\boldsymbol{A}(x_0 + h) - \boldsymbol{A}x_0 - \boldsymbol{B}[x_0]h\|_{L_\beta} \leqq \sum_{i=1}^{n} \|\boldsymbol{R}_i(x_0)\|_{\alpha(1+\delta_i)\to\beta} \|h\|_{L_\alpha}^{1+\delta_i},$$

where $\boldsymbol{R}_i(x_0)$ is the linear integral operator with kernel $R_i[t, s, x_0(s)]$. In particular,

$$\|\boldsymbol{A}(x_0 + h) - \boldsymbol{A}x - \boldsymbol{B}[x_0]h\|_{L_\beta} = o(\|h\|_{L_\alpha}).$$

The theorem has been proved.

Note that under the conditions of this theorem the derivative $\boldsymbol{A}'(x_0)$ satisfies the Hölder type condition:

$$\|\boldsymbol{A}'(x_0 + h) - \boldsymbol{A}'(x_0)\|_{\mathfrak{B}(\alpha,\beta)} \leqq \sum_{i=1}^{n} k_i(r)\|h(s)\|_{L_\alpha}^{\delta_i} \quad (\|x_0\|_{L_\gamma} \leqq r). \quad (20.65)$$

20.11 *Asymptotic derivatives of non-linear operators*

A non-linear operator $\boldsymbol{A}$ acting from a space E_1 to another space E_2 is called *asymptotically linear* if it is defined on all elements lying outside some ball and if there exists a linear operator $\boldsymbol{B}$ such that

$$\lim_{\|x\|_{E_1}\to\infty} \frac{\|\boldsymbol{A}x - \boldsymbol{B}x\|_{E_2}}{\|x\|_{E_1}} = 0. \tag{20.66}$$

The operator $\boldsymbol{B}$ is called the *asymptotic derivative* (or *derivative at infinity*) of the operator $\boldsymbol{A}$; it is often denoted by $\boldsymbol{A}'(\infty)$.

Consider first a superposition operator

$$\mathfrak{f}x(s) = f[s, x(s)] \tag{20.67}$$

which acts from L_α to L_β. Suppose that $\beta < \alpha$; then Lemma 17.6 implies that $\mathfrak{f}$ is asymptotically linear and that its asymptotic derivative is equal to zero. If $\beta \geqq \alpha$ then $\mathfrak{f}$ is asymptotically linear, provided, for instance,

$$|f(s, u) - g(s)u| \leqq a(s) + b(s)|u|^\gamma,$$

where $g(s) \in L_{\beta-\alpha}$, $a(s) \in L_\beta$, $\gamma \in (0, 1)$ and $b(s) \in L_{\beta-\gamma\alpha}$, in this case

$$\mathfrak{f}'(\infty)h(s) = g(s)h(s).$$

To investigate the asymptotic linearity of a superposition operator, it is convenient to proceed along the following lines. First, it is necessary to test whether the limit of the quotient $f(s, u)/u$ exists as $u \to \infty$. If this limit $g(s)$ exists it is necessary to form the difference $\omega(s, u) = f(s, u) - g(s)u$ and to try to prove that

$$\lim_{\|h\|_\alpha\to\infty} \frac{\|\omega[s, h(s)]\|_\beta}{\|h(s)\|_\alpha} = 0.$$

Let us now examine Hammerstein operators:

$$\boldsymbol{A}x(t) = \int_\Omega K(t, s) f[s, x(s)]\,\mathrm{d}s. \tag{20.68}$$

Assume that the superposition operator

$$\mathfrak{f}x(s) = f[s, x(s)]$$

acts from L_α to L_γ and has the asymptotic derivative

$$\mathfrak{f}'(\infty)h(s) = g(s)h(s),$$

while the linear operator

$$\boldsymbol{K}x(t) = \int_\Omega K(t, s)h(s)\,\mathrm{d}s$$

acts from L_γ to L_β and is continuous. Then the operator A is asymptotically linear as an operator from L_α to L_β and

$$A'(\infty)h(t) = \int_\Omega K(t, s)g(s)h(s)\,\mathrm{d}s. \tag{20.69}$$

This assertion follows from the relation:

$$\lim_{\|h\|_\alpha \to \infty} \frac{\|Ah(t) - \boldsymbol{K}[g(s)h(s)]\|_\beta}{\|h(s)\|_\alpha}$$

$$\leqq \|\boldsymbol{K}\|_{\gamma\to\beta} \lim_{\|h\|_\alpha \to \infty} \frac{\|\mathfrak{f}h(s) - g(s)h(s)\|_\gamma}{\|h(s)\|_\alpha} = 0.$$

The problem of finding conditions for the existence of an asymptotic derivative for an Uryson operator

$$Ax(t) = \int_\Omega K[t, s, x(s)]\,\mathrm{d}s \tag{20.70}$$

is more complicated. Let us confine ourselves to a simple assertion.

Assume that the limit

$$K_0(t, s) = \lim_{u\to\infty} \frac{K(t, s, u)}{u} \tag{20.71}$$

exists. Represent the function $K(t, s, u)$ in the form

$$K(t, s, u) = K_0(t, s)u + \omega(t, s, u). \tag{20.72}$$

Assume that the operator $\boldsymbol{A}$ and the linear operator

$$\boldsymbol{B}h(t) = \int_{\Omega} K_0(t, s)h(s)\mathrm{d}s \tag{20.73}$$

act from L_α to L_β and are continuous. Assume further that the function $\omega(t, s, u)$ satisfies the inequality

$$|\omega(t, s, u)| \leqq R_0(t, s)\cdot f(s, u) \quad (t \in \Omega^*, s \in \Omega, -\infty < u < \infty), \tag{20.74}$$

where $R_0(t, s)$ is the kernel of a continuous linear operator acting from L_γ $(0 \leqq \gamma \leqq 1)$ to L_β, and $f(s, u)$ defines a superposition operator $\boldsymbol{\mathfrak{f}}$ acting from L_α to L_γ and having the asymptotic derivative zero. A simple verification shows that under these conditions the operator $\boldsymbol{A}$ is asymptotically linear and its asymptotic derivative coincides with the operator (20.73).

20.12 *On higher order derivatives*

We define higher order derivatives of non-linear operators by the aid of Taylor's formula.

Suppose that in some neighborhood $\|x - x_0\|_{E_1} \leqq r$ of a point $x_0 \in E_1$ a non-linear operator $\boldsymbol{A}$ can be written in the form

$$\boldsymbol{A}(x_0 + h) = \boldsymbol{A}x_0 + \boldsymbol{B}_1 h + \frac{1}{2!}\boldsymbol{B}_2 h + \ldots + \frac{1}{n!}\boldsymbol{B}_n h + \omega_n(h), \tag{20.75}$$

where each operator $\boldsymbol{B}_k$ is continuous and possesses k^{th} order homogeneity:

$$\boldsymbol{B}_k(\lambda h) = \lambda^k \boldsymbol{B}_k h \quad (k = 1, 2, \ldots, n),$$

while the remainder $\omega_n(h)$ satisfies the condition

$$\lim_{\|h\|_{E_1}\to 0} \frac{\|\omega_n(h)\|_{E_2}}{\|h\|_{E_1}^n} = 0. \tag{20.76}$$

Then A is said to have a derivative of order n at the point x_0. Here the derivative $A^{(k)}(x)$ of order k is defined by the relation:

$$A^{(k)}(x_0)h = \boldsymbol{B}_k h. \tag{20.77}$$

Consider a superposition operator

$$\mathfrak{f}x(s) = f[s, x(s)]. \tag{20.78}$$

It can be proved without difficulty that the derivatives $\mathfrak{f}^{(k)}(x_0)$ (if they exist) have the form

$$\mathfrak{f}^{(k)}(x_0)h = a_k(s)[h(s)]^k, \tag{20.79}$$

where

$$\frac{1}{k!}\, a_k(s) =$$

$$= \operatorname*{limmeas}_{u\to 0} \frac{f[s, x_0(s)+u] - f[s, x_0(s)] - a_1(s)u - \ldots - \dfrac{a_{k-1}(s)}{(k-1)!}\, u^{k-1}}{u^k}. \tag{20.80}$$

Suppose that an operator $\mathfrak{f}$ acts from L_α to L_β. Suppose also that $\mathfrak{f}$ has derivatives up to order[1] k_0 and that the limit (20.80) exists for $k = k_0 + 1$. Then we can seek the derivative $\mathfrak{f}^{(k_0+1)}(x_0)h$ in the form

$$\mathfrak{f}^{(k_0+1)}(x_0)h = a_{k_0+1}(s)[h(s)]^{k_0+1}. \tag{20.81}$$

To see whether formula (20.81) is valid it is necessary (in view of the definition of higher order derivatives) to form the function

$$\omega_{k_0+1}(s, u) = f[s, x_0(s)+u] - f[s, x_0(s)] - a_1(s)u - \ldots - \frac{1}{k_0!}\, a_{k_0}(s)u^{k_0}$$

---

[1] If $\beta \leqq k_0\alpha$ ($\beta < k_0\alpha$), $f(s, u)$ is then a polynomial (in u) of degree not higher than k_0 (less than k_0).

and to see whether the following relation is satisfied:

$$\lim_{\|h\|_\alpha \to 0} \frac{\|\omega_{k_0+1}[s, h(s)]\|_\beta}{\|h(s)\|_\alpha^{k_0+1}} = 0. \tag{20.82}$$

Consider the auxiliary function

$$\omega^0_{k_0+1}(s, u) = \omega_{k_0+1}(s, |u|^{1/(k_0+1)} \operatorname{sgn} u).$$

It is easy to see that relation (20.82) is equivalent to the relation

$$\lim_{\|h\|_{\alpha(k_0+1)} \to 0} \frac{\|\omega^0_{k_0+1}[s, h(s)]\|_\beta}{\|h(s)\|_{\alpha(k_0+1)}} = 0. \tag{20.83}$$

Conditions under which relations of the type (20.83) hold have been considered in detail in 20.1°–20.5°.

The reader can formulate and prove without difficulty analogue for higher order derivatives of all the assertions in 20.1°–20.5°.

Similarly one can study the problem of the existence of higher order derivatives of non-linear operators.

*   *

*

The theorems presented in Chapter 5 show that the scale of spaces L_α is, generally speaking, a convenient one for studying integral operators with polynomial non-linearities. In the case of non-linearities of exponential type, Orlicz spaces can be utilized (M. A. Krasnoselskii and Y. B. Rutickii [5, 6]).

It is natural to give an axiomatic description of function spaces in which a rich theory of non-linear integral operators can be constructed. It has been shown (P. P. Zabreiko) that such a theory can be constructed for spaces in which the cone of non-negative functions is normal in the sense of M. G. Krein as well as reproducing and strongly minihedral. Analogues of the theorems in Chapter 5 can be obtained, provided the cone of non-negative functions in the range space is regular. Such a theory includes operators acting in Orlicz, Lorentz, Marcinkiewicz and other symmetric spaces, as well as in Honda spaces etc.

6

Some applications

§ 21 Equations with completely continuous operators

21.1 *Linear equations*

Consider a linear integral operator

$$Ax(t) = \int_{\Omega} K(t, s)x(s)\mathrm{d}s. \tag{21.1}$$

Let us suppose that its compactness in some normed space E has already been established (for this the results of Chapter 2 can be used). Then the general theory of compact linear operators implies (see, for instance, L. V. Kantorovic and G. P. Akilov [1]) that the integral operator (21.1) has either a finite or countable number of non-zero eigenvalues $\lambda_1, \lambda_2, \ldots, \lambda_n, \ldots$, to which there are associated eigenfunctions in E. If the set of eigenvalues is countable, then $\lambda_n \to 0$ as $n \to \infty$. Each of the non-zero eigenvalues has finite order and finite multiplicity.

Recall that the *order* of an eigenvalue of an operator A is the dimension of the subspace of solutions of the equation $Ax = \lambda x$ or, equivalently, the maximum number of linearly independent eigenfunctions (vectors) of the operator A. The *multiplicity* of an eigenvalue λ is the dimension of the subspace of all solutions of any of the equations $(\lambda \boldsymbol{I} - A)^n x = 0$ $(n = 1, 2, \ldots)$. For a compact operator an integer $n_0 = n_0(\lambda)$ (λ being a non-zero eigenvalue) can be found such that all solutions of any equation $(\lambda \boldsymbol{I} - A)^n x = 0$ for $n > n_0$ are also solutions of the equation $(\lambda \boldsymbol{I} - A)^{n_0} x = 0$.

Suppose that the L-characteristic $L(A; \text{cont.})$ of the operator (21.1) contains some neighborhood of the segment $\beta = \alpha$ $(0 \leqq \alpha \leqq 1)$. Assume, for definiteness, that it contains the segment $\beta = \alpha - \alpha_0$ $(\alpha_0 \leqq \alpha \leqq 1)$, where $\alpha_0 > 0$. It is then easy to see that $L(A^2; \text{cont.})$ contains the entire lower-triangle of the unit square if $\alpha_0 \geqq \frac{1}{2}$, while it contains the segment $\beta = \alpha - 2\alpha_0$ $(2\alpha_0 \leqq \alpha \leqq 1)$ if $\alpha_0 < \frac{1}{2}$. Similarly, $L(A^3; \text{cont.})$ contains the segment $\beta = \alpha - 3\alpha_0$ $(3\alpha_0 \leqq \alpha \leqq 1)$. Thus for $n > 1/\alpha_0$ the operator A^n transforms each function in any L_α $(0 \leqq \alpha \leqq 1)$ to a bounded function. Hence it follows, in particular, that all eigenfunctions of the integral operator (21.1) (considered in any space L_α) are bounded if they correspond to non-zero eigenvalues. A similar argument shows that all the associated functions[1] of this integral operator are bounded.

If it is further known that the operator A (or some operator A^n) acts from L_0 to a space W of smoother functions, then the eigenfunctions and associated functions belong, of course, to W. Such situation occurs (see § 16) if A is an operator inverse to an elliptic differential operator.

As is well known, not every linear operator has eigenfunctions. In proving the existence of eigenfunctions for the operator (21.1), it is often helpful to utilize, the compactness of this operator. For instance, let the kernel $K(t, s)$ satisfy an inequality of the form

$$K(t, s) \geqq a_0 > 0 \qquad (t, s \in \Omega) \tag{21.2}$$

and let the operator (21.1) be compact in some space L_α. Then this operator A takes the cone $\mathfrak{K}$ of non-negative functions in L_α into itself and satisfies the inequality

$$Ax_0(t) \geqq a_0 \operatorname{mes}(\Omega) \cdot x_0(t) \qquad (t \in \Omega),$$

where $x_0(t) \equiv 1$. Hence the general theory of positive linear operators (see, for instance, M. G. Krein and M. A. Rutman [1], M. A. Krasnoselskii [9]) shows that the operator (21.1) has a positive eigenfunction. This eigenfunction corresponds to a positive eigenvalue λ_0 which coincides with the spectral radius of the operator (21.1).

Let us now pass to an inhomogeneous linear equation of the form

$$Ax(t) = \int_\Omega K(t, s)x(s)\,ds + f(t). \tag{21.3}$$

---

[1] Translator's note. A vector x is an associated vector if $(\lambda I - A)^n x = 0$ for some n.

Let the operator (21.1) be compact in a space E and let $f \in E$. Then from the general theory of compact operators (see, for instance, L. V. Kantorovic and G. P. Akilov [1]) it follows that the Fredholm alternative is valid for equation (21.3).

The compactness of the linear integral operator (21.1) is also important in many other problems.

21.2 *On approximate solutions of equations*

Consider the problem of approximating the solution of an integral equation

$$x(t) = \mu \int_{\Omega} K(t, s)x(s)\mathrm{d}s + f(t). \tag{21.4}$$

Assume that the operator (21.1) is defined and continuous in a space E and that $f \in E$. Then for sufficiently small μ the spectral radius of the operator μA is less than 1, whence equation (21.4) has a unique solution x^* in E. The successive approximations

$$x_{n+1}(t) = \mu \int_{\Omega} K(t, s)x_n(s)\mathrm{d}s + f(t) \quad (n = 0, 1, 2, \ldots) \tag{21.5}$$

converge to this solution in the norm of the space E. In many cases it is desirable to establish the convergence of the successive approximations in a norm $\|x\|_{E_1}$ stronger than the norm $\|x\|_E$ of the space E. For instance, if $E = L_\alpha$ it is often desirable to establish that the successive approximations converge uniformly. Such a theorem can be obtained easily when some power A^{k_0} of the operator (21.1) acts from the space E into the space E_1.

In fact, a simple verification shows that

$$x^* - x_n = \mu^{k_0} A^{k_0}(x^* - x_{n-k_0}) \quad (n = k_0, k_0 + 1, \ldots).$$

Hence

$$\|x^* - x_n\|_{E_1} \leqq |\mu|^{k_0} \|A^{k_0}\|_{E \to E_1} \|x^* - x_{n-k_0}\|_E \quad (n = k_0, k_0 + 1, \ldots)$$

and it follows from

$$\lim_{n \to \infty} \|x^* - x_n\|_E = 0$$

that

$$\lim_{n\to\infty} \|x^* - x_n\|_{E_1} = 0.$$

For instance, if $K(t, s)$ is a kernel of potential type (see § 8) then the successive approximations (21.5) converge to the solution of equation (21.4) uniformly (notwithstanding the fact that the solution itself may be unbounded if the function $f(t)$ is unbounded!).

This observation suggests a method for 'regularization' of the iteration process (21.5). To solve the equation

$$x = \mu A x + f, \tag{21.6}$$

it is appropriate, first, to find the function

$$x_{k_0} = f + \mu A f + \ldots + \mu^{k_0 - 1} A^{k_0 - 1} f, \tag{21.7}$$

and then, to substitute

$$y = x - x_{k_0}. \tag{21.8}$$

By the definition of y we obtain the equation

$$y = \mu A y + f_1 \tag{21.9}$$

with a 'well behaved' forcing term

$$f_1 = \mu^{k_0} A^{k_0} f \tag{21.10}$$

which belongs to the space E_1. The successive approximations

$$y_{n+1} = \mu A y_n + f \quad (n = 0, 1, 2, \ldots)$$

will then belong to E_1 and converge in the norm of the space E_1.

Note that it is sometimes possible to use as E_1 various spaces of differentiable functions. In such cases, the successive approximations converge together with their derivatives up to a definite order.

To obtain successive approximations of solutions of linear integral equations the so-called projection methods (Galerkin method, Petrov method, Ritz method, method of least square etc.) are also widely applied. Let us describe the general scheme of such projection methods (S. G. Mikhlin [1, 2], L. V. Kantorovic and G. P. Akilov [1], M. A. Krasnoselskii [7]).

Let there be given a sequence of linear projection operators in E

$$\boldsymbol{P}_1, \boldsymbol{P}_2, \dots, \boldsymbol{P}_n, \dots, \tag{21.11}$$

each of which projects E onto a corresponding finite-dimensional subspace $E_n = \boldsymbol{P}_n E$. Suppose that the norms of operators $\boldsymbol{P}_n$ are uniformly bounded and that for each $x \in E$

$$\lim_{n \to \infty} \|x - \boldsymbol{P}_n x\| = 0. \tag{21.12}$$

For instance, let $e_1, e_2, \dots, e_n, \dots$ be a basis of the space E. Then each element $x \in E$ is represented uniquely by a series

$$x = \sum_{i=1}^{\infty} \xi_i(x) e_i, \tag{21.13}$$

where the $\xi_i(x)$ are linear functionals. The operators

$$\boldsymbol{P}_n x = \sum_{i=1}^{n} \xi_i(x) e_i \quad (n = 1, 2, \dots) \tag{21.14}$$

satisfy the conditions indicated above.

Consider an equation of the form

$$x = \mu \boldsymbol{A} x + f \tag{21.15}$$

with $\boldsymbol{A}$ a compact operator. Galerkin's approximate equations are defined by the relations

$$x = \mu \boldsymbol{P}_n \boldsymbol{A} x + \boldsymbol{P}_n f \quad (n = 1, 2, \dots). \tag{21.16}$$

Exact solutions x_n of these equations are approximate solutions of the equations (21.15). It is clear that the solutions x_n are to be sought in the finite dimensional subspaces E_n. In other words, seeking the solution x_n is equivalent to solving a finite system of linear algebraic equations. For instance, if the operators $\boldsymbol{P}_n$ are defined by the formulas (21.14) then the solution x_n has the form

$$x_n = \alpha_1 e_1 + \alpha_2 e_2 + \ldots + \alpha_n e_n, \tag{21.17}$$

where the coefficients $\alpha_1, \alpha_2, \ldots, \alpha_n$ satisfy the system of equations

$$\left.\begin{array}{l} \alpha_1 = \mu\xi_1(Ae_1)\alpha_1 + \mu\xi_1(Ae_2)\alpha_2 + \ldots + \mu\xi_1(Ae_n)\alpha_n + \xi_1(f) \\ \cdots\cdots\cdots\cdots\cdots\cdots\cdots\cdots \\ \alpha_n = \mu\xi_n(Ae_1)\alpha_1 + \mu\xi_n(Ae_2)\alpha_2 + \ldots + \mu\xi_n(Ae_n)\alpha_n + \xi_n(f) \end{array}\right\} \tag{21.18}$$

Let equation (21.15) have the unique solution $x^* \in E$. Then the following assertion is valid (see, for instance, L. V. Kantorovic and G. P. Akilov [1], M. A. Krasnoselskii [7]).

Each Galerkin equation (21.16) *has, for sufficiently large n, the unique solution* x_n, *and*

$$\lim_{n\to\infty} \|x_n - x^*\|_E = 0. \tag{21.19}$$

The rate of convergence of the Galerkin method is characterized by the inequality

$$\|x_n - x^*\|_E \leqq (1 + \varepsilon_n)\|\boldsymbol{P}_n x^* - x^*\| \quad (n = 1, 2, \ldots), \tag{21.20}$$

where $\varepsilon_n \to 0$.

In cases where the projection operators are defined by the formulas (21.14), the difference $\boldsymbol{P}_n x^* - x^*$ coincides with the remainder term $\sum_{i=n+1}^{\infty} \xi_i(x^*)e_i$ in the Fourier expansion (21.13) of the solution x^* of the equation (21.15). Hence formula (21.20) implies that the Galerkin method converges at the same rate as the converging Fourier expansion of the solution.

Of course, the solution of equation (21.15) is not known to us. However in many cases, one can find the rate of convergence of the Fourier expansion of the solution by an apriori estimate of the smoothness of the solution. To estimate the smoothness of the solution it is convenient to apply the methods discussed in the preceding section, which we there applied to prove the boundedness of eigenfunctions of integral operators. In practical applications of projection methods it is appropriate, where possible, to go over to equivalent equations whose solutions are expanded in 'rapidly converging' Fourier series. To accomplish this, it is often convenient to transform equation (21.6) to equation (21.9) with the aid of the substitution (21.8). By this replacement we obtain an equation whose solution y^* is equal to $\mu_0^{k_0} A^{k_0} x^*$. If the operator A^{k_0} 'essentially improves' functions then the Fourier series of the function y^* converges more rapidly[1] than the Fourier series of the solution x^* of equation (21.6). We emphasize that in applying Galerkin's method to obtain a solution of equation (21.6), the transformation to equation (21.9) is not connected with an assumption of smallness for the parameter μ.

Projection methods (in particular, Galerkin's method) are not iterative, because the Galerkin approximation x_n cannot be used directly to construct the Galerkin approximation x_{n+1}. It is natural to think of combining the idea of Galerkin's method with the previous idea of constructing convergent successive approximations.

Consider again a linear equation

$$x = Ax + f, \tag{21.21}$$

where A is a compact linear operator acting in a space E. Let $\boldsymbol{P}$ be a projection operator onto a finite dimensional subspace $E_0 \subset E$, and put $\boldsymbol{Q} = \boldsymbol{I} - \boldsymbol{P}$. Each element $x \in E$ can be written in the form

$$x = u + v \quad (u = \boldsymbol{P}x,\ v = \boldsymbol{Q}x). \tag{21.22}$$

Equation (21.21) is then equivalent to the system of two equations

$$\begin{aligned} u &= \boldsymbol{P}Au + \boldsymbol{P}Av + \boldsymbol{P}f, \\ v &= \boldsymbol{Q}Au + \boldsymbol{Q}Av + \boldsymbol{Q}f. \end{aligned} \tag{21.23}$$

---

[1] The question of the rate of convergence of Fourier series relative to eigenfunctions of selfadjoint operators will be discussed in § 22.

Since the space E_0 is finite-dimensional, it is, in general fairly simple to solve the first of the equations in (21.23) for u. This solution has the form

$$u = (I - PA)^{-1}PAv + f_1, \tag{21.24}$$

where

$$f_1 = (I - PA)^{-1}Pf. \tag{21.25}$$

Substituting this value of u into the second of the equations in (21.23), we obtain

$$v = QA(I - PA)^{-1}v + f_2, \tag{21.26}$$

where

$$f_2 = QAf_1 + Qf. \tag{21.27}$$

By the compactness of the operator A it follows that with an appropriate choice of the finite-dimensional operator P, the operator $QA(I - PA)^{-1}$ will have small norm. Hence the solution of equation (21.6) can be obtained by successive approximations

$$v_{n+1} = QA(I - PA)^{-1}v_n + f_2 \quad (n = 0, 1, \ldots). \tag{21.28}$$

The method given by (21.28) is equivalent to the construction of successive approximations v_n, z_n by the recursion formulas

$$\begin{aligned} z_n &= PAz_n + v_n \\ v_{n+1} &= QAz_n + f_2 \end{aligned} \qquad (n = 0, 1, 2, \ldots).$$

The method in (21.28) can be considered as the usual iteration process applied to the solution of the following equation which is equivalent to (21.21),

$$x = (I - S)^{-1}(A - S)x + (I - S)^{-1}f, \tag{21.29}$$

where S is a linear operator, such that the difference $A - S$ is small, and such that $(I - S)^{-1}$ exists and can be 'simply' calculated.

21.3 *Existence of solutions of non-linear integral equations*

Many problems require solving or studying non-linear operator equations. In particular, non-linear integral equations often arise. Many non-linear boundary value problems for ordinary differential equations and for partial differential equations can be reduced to such integral equations.

We confine ourselves here to some of the simplest assertions concerning non-linear integral equations of the form

$$x(t) = \mu \int_{\Omega} K[t, s, x(s)]\mathrm{d}s + f(t). \tag{21.30}$$

Suppose that it is possible to establish the complete continuity of the operator

$$Ax(t) = \int_{\Omega} K[t, s, x(s)]\mathrm{d}s \tag{21.31}$$

in a function space E (for this the results in § 19 can be used). Then *for sufficiently small* μ *equation* (21.30) *has at least one solution in* E. For the proof it suffices to consider the operator $\mu Ax + f$ on an arbitrary ball $\|x - f\|_E \leqq \rho_0$ and to remark that it transforms this ball into itself. Following this, it only remains to use the Schauder fixed point principle.

Suppose that the operator (21.31) has an asymptotic derivative $A'(\infty)$ (see § 20). This derivative is also a compact operator, whose eigenvalues will be denoted by

$$\lambda_1, \lambda_2, \ldots, \lambda_n, \ldots . \tag{21.32}$$

If

$$\mu \neq 1/\lambda_i \quad (i = 1, 2, \ldots) \tag{21.33}$$

then equation (21.30) *has at least one solution in* E. For the proof it is only necessary to replace equation (21.30) by the equivalent equation:

$$x = \mu[I - \mu A'(\infty)]^{-1}[Ax - A'(\infty)x] + [I - \mu A'(\infty)]^{-1}f \tag{21.34}$$

and to remark that the operator on the right side is completely continuous and transforms each ball $\|x\| \leqq \rho$ into itself for sufficiently large radius ρ. Following this, it remains to use Schauder's fixed point principle.

To prove theorems on the existence of solutions of the equation (21.30) with a completely continuous operator various fixed point principles besides Schauder's principle can be applied (see M. A. Krasnoselskii [7]); alternatively, the theory of positive operators can be applied (M. A. Krasnoselskii [9]). These methods can be used to study the dependence of solutions on a parameter, to prove the existence of solutions different from known trivial solutions, to estimate the number of solutions, etc.

Let us present one of these simple assertions. Suppose that one solution $x^*(t)$ of equation (21.30) is known to us. Let the operator A have the derivative $A'(\infty)$ and be differentiable at the point $x^*(t)$ (see § 20), and let $0 < 1/\mu$ not be an eigenvalue of either the operator $A'(\infty)$ or the operator $A'(x^*)$. Then *for the existence in E of at least one solution of equation* (21.30) *different from* x^*, *it is sufficient that the sum of the multiplicities of those real eigenvalues of the operators* $A'(\infty)$ *and* $A'(x^*)$ *which are greater than* $1/\mu$, *be odd.* For the proof, it is necessary to use the theory of completely continuous vector fields (see M. A. Krasnoselskii [7]) and to remark that on the boundary of the domain obtained from the ball $\|x\| \leqq \rho_0$ (with large radius ρ_0) by excising any small neighborhood of the point x^*, the vector field $x - \mu Ax - f$ has a non-zero mapping degree (it is equal to 2 or -2).

Note that in many cases it is convenient to use the following known principle on fixed points in order to prove existence theorems.

An operator $A(x; \lambda)$ ($x \in E$, $0 \leqq \lambda \leqq 1$) is called *completely continuous* if it is jointly continuous in its variables and if each set $\{y: y = A(x; \lambda), \|x\| \leqq \rho_0, 0 \leqq \lambda \leqq 1\}$ is compact in E.

Suppose that for all solutions of every equation of the form

$$x = A(x; \lambda) \quad (0 \leqq \lambda \leqq 1), \tag{21.35}$$

with $A(x; \lambda)$ a completely continuous operator, the following general apriori estimate can be established:

$$\|x(\lambda)\| \leqq R_0 \quad (0 \leqq \lambda \leqq 1). \tag{21.36}$$

Let the operator $A(x; 1)$ satisfy the condition

$$\|A(x; 1)\| < R_0 \quad (\|x\| = R_0). \tag{21.37}$$

Then the equation

$$x = A(x; 0) \tag{21.38}$$

has at least one solution.

For the proof, consider the auxiliary operator

$$Bx = \begin{cases} A(x; 0), & \text{if } \|x\| \leqq R_0 \\ A\left(x; \dfrac{\|x\|}{R_0} - 1\right), & \text{if } R_0 \leqq \|x\| \leqq 2R_0 \\ \dfrac{\|x\|}{4R_0 - \|x\|} A\left(\dfrac{4R_0 - \|x\|}{\|x\|} x; 1\right), & \text{if } 2R_0 \leqq \|x\| \leqq 3R_0 \\ 3A\left(\dfrac{R_0 x}{\|x\|}; 1\right), & \text{if } 3R_0 \leqq \|x\| < \infty. \end{cases} \tag{21.39}$$

It is easy to see that the operator $\boldsymbol{B}$ is completely continuous and maps the whole space E into a compact set. Schauder's principle shows that this operator has at least one fixed point x^*. As is shown by a simple check, the estimates (21.36) and (21.37) imply the inequality $\|x^*\| \leqq R_0$. Hence x^* is a solution of equation (21.38).

In conclusion we note that in order to approximate solutions of equations of the type (21.30) (or more general equations with completely continuous operators) one can apply projection methods, which were discussed in the last section for linear equations. These projection methods (M. A. Krasnoselskii [7]) converge to the solution x^*, provided the Fréchet derivative of the corresponding operator at the point x^* exists and does not possess the eigenvalue 1. In these cases, the rate of convergence is characterized, as in the case of linear equations, by inequalities of the form (21.20). Thus, in order to investigate the convergence of projection methods it is useful to know that the integral operator (21.31) is differentiable at all points of some set Π to which the solution x^* belongs. Theorems on the differentiability of integral operators on some sets which are dense in L_α were established in § 20.

21.4 *Eigenfunctions of non-linear integral operators*

Consider a non-linear equation of the form

$$x = A(x; \mu), \tag{21.40}$$

where μ is a scalar parameter. There arises the problem of how to find those values μ for which the equation (21.40) has non-zero solutions. The operator $A(x; \mu)$ often satisfies the condition

$$A(0; \mu) \equiv 0; \tag{21.41}$$

in other words, for all values of the parameter equation (21.40) has the zero solution.

Equations of the form (21.40) appear in many problems of non-linear mechanics; in the search for critical load and the form of loss of stability of elastic systems, in the investigation of vibrating processes, in the investigation of process of wave growth in moving liquids etc. In such problems the role of the parameter μ can be played by the load, the frequency of vibration, the speed of motion of liquid, etc.

We confine ourselves to the case where the equation (21.40) has the special form

$$x = \mu A x. \tag{21.42}$$

Non-zero solutions of this equation are called *eigenfunctions* because of the formal analogy to linear problems, and the corresponding parameter values μ are called *characteristic* values. Of course, in the theory of non-linear operators characteristic values and eigenfunctions do not play the same role as that with which we are familiar in linear analysis.

Various methods have been developed for the proof of existence of non-zero solutions of the equation (21.40) or (21.42) and for the investigation of their dependence on the parameter μ. In particular, a number of general theorems concerning equations with completely continuous operators have been proved (see M. A. Krasnoselskii [7] and [9]). The results of Chapter 5 permit us to apply all these theorems to the investigation of integral equations of the form

$$x(t) = \int_{\Omega} K[t, s, x(s), \mu]\,\mathrm{d}s \tag{21.43}$$

and

$$x(t) = \mu \int_{\Omega} K[t, s, x(s)]\,\mathrm{d}s. \tag{21.44}$$

As illustrations, we present only a few of the simplest assertions.

Assume that the operator

$$Ax(t) = \int_{\Omega} K[t, s, x(s)]\,\mathrm{d}s \tag{21.45}$$

is completely continuous in some space L_α (see § 19). *Suppose that the kernel $K(t, s, u)$ satisfies the inequality*

$$K(t, s, u) \geqq k_0 u \qquad (t, s \in \Omega, 0 < u < \infty), \tag{21.46}$$

where $k_0 > 0$. Then equation (21.44) *has a continuum of positive solutions belonging to certain values of the parameter μ.* Among these solutions there are functions of arbitrarily large and of arbitrarily small norm. This assertion follows immediately from general theorems on monotone minorants. If it is further assumed that $K(t, s, u)$ is non-decreasing and concave in the parameter u, then positive solutions of equation (21.44) exist for all values of μ in some interval (μ_1, μ_2), and in fact to each $\mu \in (\mu_1, \mu_2)$ there is a unique positive solution $x(\mu)$. These assertions follow immediately from general theorems on equations with concave operators; they are an extension of the famous results of P. S. Uryson [1].

A number μ_0 is called a *bifurcation point* of equation (21.44) if to each $\varepsilon > 0$ there corresponds at least one point $\mu \in (\mu_0 - \varepsilon, \mu_0 + \varepsilon)$ for which equation (21.44) has a non-zero solution in the ball $\|x\| < \varepsilon$. Suppose the operator defined by the right side of the equation (21.44) transforms the zero-function to the zero-function, is completely continuous in a space L_α, and is differentiable (see § 20) at the zero point of this space. Let the linearized version of equation (21.44) at zero have the form

$$x(t) = \mu \int_{\Omega} K_0(t, s)x(s)\,\mathrm{d}s. \tag{21.47}$$

General theorems on bifurcation points (M. A. Krasnoselskii [7, 9]) imply that each odd-multiple characteristic value μ_0 of the linear equation (21.47) is a bifurcation point for the non-linear equation (21.44). This simple assertion can be supplemented by criteria which provide estimates for the number of small solutions of the non-linear equation (21.44) for values of μ close to μ_0. Note further that in the case of even-multiple characteristic values μ_0 it is necessary to calculate the index γ (M. A. Krasnoselskii [7], V. B. Melamed [1], M. A. Krasnoselskii, A. I. Perov, A. J. Povolockii, P. P. Zabreiko [1], P. P. Zabreiko and M. A. Krasnoselskii [1]) of the zeros, or singular points of the vector field $\Phi x = x - \mu_0 Ax$; if $|\gamma| \neq 1$, then μ_0 is a bifurcation point of the non-linear equation (21.44).

Similar assertions apply to equation (21.43).

§ 22 Convergence of Fourier's method [1]

22.1 *General theorems on convergence of Fourier's method*

Let A be a strictly positive definite selfadjoint operator in the Hilbert space H (see § 11) which possesses a compact inverse A^{-1}. Then the operators A and A^{-1} admit representations of the form

$$Au = \sum_{i=1}^{\infty} \lambda_i (u, e_i) e_i \qquad (u \in D(A)), \tag{22.1}$$

$$A^{-1}u = \sum_{i=1}^{\infty} \frac{1}{\lambda_i} (u, e_i) e_i \qquad (u \in H). \tag{22.2}$$

Here $\{e_i\}$ denotes the complete orthonormal system of eigenfunctions and $\{\lambda_i\}$ denotes the sequence of eigenvalues of the operator A. In what follows it will be convenient for us to assume that the eigenvalues are in non-decreasing order ($\lambda_i \leqq \lambda_{i+1}$). The compactness of the operator A^{-1} is equivalent to the requirement that $\lambda_i \to \infty$.

---

[1] The main considerations of this paragraph were indicated in the review by M. A. Krasnoselskii of the doctoral thesis of V. A. Ilin. They were then detailed in the paper of M. A. Krasnoselskii and E. I. Pustylnik [1].

Each element $u \in H$ can be expanded in a Fourier series

$$u = \sum_{i=1}^{\infty} (u, e_i) e_i. \tag{22.3}$$

This series converges in the norm of the space H. In particular, if $H = L_{\frac{1}{2}}$, then the series (22.3) converges in mean-square.

There are various well known theorems (see, for instance, N. K. Bari [1], O. A. Ladyzenskaya [1], V. A. Ilin [1, 2]) giving conditions which guarantee the convergence of a Fourier series not only in mean-square, but in some stronger sense. For instance, we speak of the uniform convergence of a Fourier series, of its absolute convergence, of the convergence of a Fourier series after formal differentiation, etc. It turns out that a number of important results on Fourier series can be obtained quite simply from certain general theorems on fractional powers of an operator which were established in Chapters 3 and 4.

In this section we present two general theorems concerning abstract operators A of the form (22.1).

THEOREM 22.1: *Let an operator $A^{-\sigma}$ ($\sigma \geqq 0$) act continuously from H to some Banach space E:*

$$\|A^{-\sigma} u\|_E \leqq a \|u\|_H \quad (u \in H). \tag{22.4}$$

Let $u \in D(A^{\sigma+\tau})$, where $\tau > 0$. Then $u \in E$ and the series (22.3) *converges to u in the norm of the space E. Here the rate of convergence is characterized by the estimate*

$$\|u - \sum_{i=1}^{n} (u, e_i) e_i\|_E = o(\lambda_n^{-\tau}). \tag{22.5}$$

PROOF: First, we remark that (22.4) implies that all eigenfunctions e_i belong to the space E. Hence the partial sums of the Fourier series (22.3) of an element $u \in H$ always belong to the space E.

Since $u \in D(A^{\sigma+\tau})$, there exists an element $v \in H$ such that

$$u = A^{-\sigma-\tau} v. \tag{22.6}$$

Thus (22.4) implies that $u \in E$.

Expand the element $A^{-\tau}v$ into a Fourier series:

$$A^{-\tau}v = \sum_{i=1}^{\infty} (A^{-\tau}v, e_i)e_i. \tag{22.7}$$

This expansion converges in the norm of the space H. Furthermore

$$\|A^{-\tau}v - \sum_{i=1}^{n} (A^{-\tau}v, e_i)e_i\|_H = \| \sum_{i=n+1}^{\infty} (A^{-\tau}v, e_i)e_i\|_H$$

$$= \| \sum_{i=n+1}^{\infty} (v, A^{-\tau}e_i)e_i\|_H = \| \sum_{i=n+1}^{\infty} \lambda_i^{-\tau}(v, e_i)e_i\|_H,$$

whence

$$\|A^{-\tau}v - \sum_{i=1}^{n} (A^{-\tau}v, e_i)e_i\|_H \leqq \lambda_{n+1}^{-\tau} \| \sum_{i=n+1}^{\infty} (v, e_i)e_i\|_H$$

i.e.

$$\|A^{-\tau}v - \sum_{i=1}^{n} (A^{-\tau}v, e_i)e_i\|_H = o(\lambda_n^{-\tau}). \tag{22.8}$$

Equations (22.6) and (22.4) imply the inequality

$$\|u - \sum_{i=1}^{n} (u, e_i)e_i\|_E = \|A^{-\sigma}[A^{-\tau}v - \sum_{i=1}^{n} (A^{-\tau}v, e_i)e_i]\|_E$$

$$\leqq a\|A^{-\tau}v - \sum_{i=1}^{n} (A^{-\tau}v, e_i)e_i\|_H,$$

which, when combined with (22.8), leads to the relation (22.5). The theorem has been proved.

The series (22.7) converges in the space H under any permutation of its terms, because these terms are mutually orthogonal. This implies that under the conditions of Theorem 22.1 the series (22.3) also converges in the space E under any permutation of its terms. For instance, if E is the space C of continuous functions, then the possibility of permuting terms of the Fourier series implies that the series (22.3) converges absolutely.

THEOREM 22.2: *Let a linear operator* $\boldsymbol{B}$ *be defined on the domain* $D(A^\sigma)$ *of the operator* A^σ, *where* $\sigma \geqq 0$, *and satisfy the inequality*

$$\|\boldsymbol{B}A^{-\sigma}u\|_E \leqq b\,\|u\|_H \qquad (u \in H). \tag{22.9}$$

Let $u \in D(A^{\sigma+\tau})$, *where* $\tau > 0$. *Then the operator* $\boldsymbol{B}$ *can operate term by term on the series*

$$u = \sum_{i=1}^{\infty} (u, e_i)e_i, \tag{22.10}$$

and the resulting series

$$\boldsymbol{B}u = \sum_{i=1}^{\infty} (u, e_i)\boldsymbol{B}e_i \tag{22.11}$$

will converge in the norm of the space E. Here the rate of convergence is characterized by the estimate

$$\|\boldsymbol{B}u - \sum_{i=1}^{n} (u, e_i)\boldsymbol{B}e_i\|_E = o(\lambda_n^{-\tau}). \tag{22.12}$$

The proof is similar to the proof of Theorem 22.1. Note that the terms of the series (22.11) can be permutated in an arbitrary manner.

In concluding this section let us point out that Theorems 22.1 and 22.2 can be carried over to cases where the operator $\boldsymbol{A}$ acts not in a Hilbert space but in a Banach space E_0. In such cases it is necessary to assume that the eigenfunctions e_i of the operator $\boldsymbol{A}$ form a basis in E_0. It may happen that this basis turns out to be a conditional basis (i.e., terms in expansions with respect to this basis cannot always be permuted); then the expansions (22.3) and (22.11) will also converge conditionally in the space E.

22.2 *Convergence of Fourier series with respect to eigenfunctions of elliptic operators*

In this section $\boldsymbol{A}$ denotes a selfadjoint, strictly positive definite elliptic operator (see § 16) of order $2k$ which is defined on functions of N variables. We will assume that the coefficients of the differential expression (16.5) and the boundary conditions (16.9) are sufficiently smooth. The boundary F of the corresponding domain Ω will also be assumed to be sufficiently smooth (see 16.1°).

$\boldsymbol{D}^r$ will denote, as usual, a differentiation operator of order $|r|$.

LEMMA 22.1: *When $\sigma > \sigma(|r|)$, where*

$$\sigma(|r|) = |r|/2k + N/4k \qquad (|r| = 0, 1, \ldots, 2k-1), \tag{22.13}$$

then the operator $\boldsymbol{D}^r\boldsymbol{A}^{-\sigma}$ acts from $L_{\frac{1}{2}}$ to C and is compact.

PROOF: For each $i = 0, 1, \ldots$ the following inequalities hold for the positive-type elliptic operator $\boldsymbol{A}$:

$$\|\boldsymbol{A}u\|_{W_{\frac{1}{2}}^{i}} \geqq a_i \|u\|_{W_{\frac{1}{2}}^{i+2k}} \qquad (u \in D\boldsymbol{A}) \cap W_{\frac{1}{2}}^{i+2k}) \tag{22.14}$$

(see, for instance, S. Agmon, A. Douglis, L. Nirenberg [1]). Consequently, for any $m = 1, 2, \ldots$ the following inequalities are valid:

$$\|\boldsymbol{A}^m u\|_{L_{\frac{1}{2}}} \geqq c_m \|u\|_{W_{\frac{1}{2}}^{2km}} \qquad (u \in D(\boldsymbol{A}^m)). \tag{22.15}$$

The inequalities (22.15) imply, in particular, that the operators $\boldsymbol{A}^{-m}$ act from $L_{\frac{1}{2}}$ to $W_{\frac{1}{2}}^{2km}$ and are continuous. Hence the operators $\boldsymbol{D}^r\boldsymbol{A}^{-m}$ ($|r| = 0, 1, \ldots, 2k-1$) act from $L_{\frac{1}{2}}$ to $W_{\frac{1}{2}}^{2km-|r|}$ and are continuous.

The imbedding theorem of S. L. Sobolev then implies that for

$$m > \frac{1}{2k}\left(\frac{N}{2} + |r|\right),$$

the operators $\boldsymbol{D}^r\boldsymbol{A}^{-m}$ act from $L_{\frac{1}{2}}$ in C and are compact.

Set

$$m_0 = \left[\frac{N}{4k} + \frac{|r|}{2k}\right] + 1.$$

It follows from the multiplicative inequality (16.35) with $l = |r|$, $l_0 = 2km_0$, that

$$\|u\|_{C^{|r|}} \leqq M[\|u\|_{W_{\frac{1}{2}}^{2km_0}}]^{\tau_0}[\|u\|_{L_{\frac{1}{2}}}]^{1-\tau_0}, \tag{22.16}$$

where

$$\tau_0 = \frac{|r|}{2km_0} + \frac{N}{4km_0}.$$

Together (22.16) and (22.15) imply the inequality

$$\|\boldsymbol{D}^r u\|_C \leqq M_1[\|\boldsymbol{A}^{m_0}u\|_{L_{\frac{1}{2}}}]^{\tau_0}[\|u\|_{\frac{1}{2}}]^{1-\tau_0}. \tag{22.17}$$

Now apply Theorem 16.3 to the operators $\boldsymbol{A}^{m_0}$ and $\boldsymbol{D}^r$. By this theorem the operators $\boldsymbol{D}^r(\boldsymbol{A}^{m_0})^{-\tau}$, with $\tau > \tau_0$ act from $L_{\frac{1}{2}}$ to C and are compact. Since

$$(\boldsymbol{A}^{m_0})^{-\tau} = \boldsymbol{A}^{-m_0\tau},$$

the above assertion implies that the operators $\boldsymbol{D}^r\boldsymbol{A}^{-\sigma}$, $\sigma > m_0\tau_0 = \sigma(|r|)$, act from $L_{\frac{1}{2}}$ to C and are compact. The lemma has been proved.

Combining this lemma with Theorem 22.2, we arrive at the following assertion.

THEOREM 22.3: *Let $u \in D(\boldsymbol{A}^\omega)$, where $\omega > N/4k$, and let $0 \leqq |r| < 2k\omega - N/2$. Then the Fourier series* (22.2) *can be differentiated term by term $|r|$ times, and the resulting series*

$$\boldsymbol{D}^r u = \sum_{i=1}^{\infty} (u, e_i)\boldsymbol{D}^r e_i \tag{22.18}$$

will converge absolutely and uniformly. The rate of convergence is characterized by the estimate

$$\|\boldsymbol{D}^r u - \sum_{i=1}^{n} (u, e_i)\boldsymbol{D}^r e_i\|_C = o(\lambda_n^{-\omega+N/4k+|r|/2k+\varepsilon}), \tag{22.19}$$

where ε is an arbitrary positive number.

If a function u belongs to the domain of an operator $\boldsymbol{A}^\tau$, where τ is so small that Theorem 22.3 cannot be applied to study the Fourier series of the function, then it can still be deduced from Theorem 22.1 that the Fourier series (22.18) converges not only in mean square but in the norm of space L_α, where $\alpha < \frac{1}{2}$. We remark that in many cases Theorem 16.6 permits us to indicate the precise value of the smallest such index α. This calculation is left to the reader.

From Lemma 22.1 it is possible to obtain[1] estimates for the eigenfunctions of elliptic operators A of order $2k$, as well as estimates for the derivatives of these eigenfunctions. In fact, for each $\varepsilon > 0$ the inequality:

$$\max|e_i(x)| = \lambda_i^{N/4k+\varepsilon}\|A^{-N/4k-\varepsilon}e_i\|_C \leqq a(\varepsilon)\cdot\lambda_i^{N/4k+\varepsilon}\|e_i\|_H$$

is valid, i.e.

$$|e_i(x)| \leqq a(\varepsilon)\lambda_i^{N/4k+\varepsilon} \qquad (i = 1, 2, \ldots). \tag{22.20}$$

The same lemma implies that

$$|D^r e_i(x)| \leqq b(\varepsilon, r)\lambda^{N/4k+|r|/2k+\varepsilon}$$
$$(i = 1, 2, \ldots; |r| = 1, 2, \ldots, 2k-1). \tag{22.21}$$

More precise estimates for these eigenfunctions and their derivatives have been obtained by Ju. M. Berezanskii [1–3] by the use of other methods.

In connection with the above estimates for the rate of convergence of a Fourier series, there arises the natural question of the asymptotic behavior of the eigenvalues λ_i. There is a great deal of literature concerning this problem. We present here only a rather rough estimate.

Lemma 22.1 and Theorem 6.2 imply that for $\varepsilon > 0$, $A^{-(N/4k+\varepsilon)}$ is an integral operator

$$A^{-(N/4k+\varepsilon)}u(x) = \int_\Omega K(x, y)u(y)\,dy \tag{22.22}$$

whose kernel is square summable. The Hilbert-Schmidt theorem then implies that the series consisting of the squares of the eigenvalues of the integral operator (22.22) converges. Consequently

$$\sum_{i=1}^{\infty}\lambda^{-N/2k-2\varepsilon} < \infty. \tag{22.23}$$

---

[1] Similar considerations (based on the use of multiplicative inequalities) have been applied by S. G. Krein to obtain estimates of eigenfunctions.

22.3 *Fourier's method for hyperbolic equations*

The fractional powers A^τ of a strictly positive definite selfadjoint operator A can also be used in formulating Fourier's method for solution of hyperbolic and parabolic equations.

Consider first a hyperbolic equation

$$\frac{d^2u(t)}{dt^2} + Au(t) = f(t), \tag{22.24}$$

where A is as before a strictly positive definite selfadjoint operator

$$Au = \sum_{i=1}^{\infty} \lambda_i(u, e_i)e_i, \tag{22.25}$$

possessing compact inverse on the Hilbert space H, and where t varies on an interval $[0, T]$.

A solution of equation (22.24) satisfying the initial conditions

$$u(0) = u_0, \quad u_t'(0) = v_0 \tag{22.26}$$

can be written in the form

$$u(t) = \cos(A^{\frac{1}{2}}t)u_0 + A^{-\frac{1}{2}} \sin(A^{\frac{1}{2}}t)v_0$$

$$+ A^{-\frac{1}{2}} \int_0^t \sin(A^{\frac{1}{2}}t - A^{\frac{1}{2}}s)f(s)\,ds, \tag{22.27}$$

where

$$\cos(A^{\frac{1}{2}}t)u = \sum_{i=1}^{\infty} \cos(t\sqrt{\lambda_i})(u, e_i)e_i$$

$$\sin(A^{\frac{1}{2}}t)u = \sum_{i=1}^{\infty} \sin(t\sqrt{\lambda_i})(u, e_i)e_i.$$

It is easy to see that the operators $\cos(A^{\frac{1}{2}}t)$ and $\sin(A^{\frac{1}{2}}t)$ are bounded in H for any t and that their norms do not exceed one.

Fourier's method is often applied in solving equation (22.24). In this method the solution (22.27) is replaced by the successive approximations:

$$u_n(t) = \boldsymbol{P}_n\{\cos(A^{\frac{1}{2}}t)u_0 + A^{-\frac{1}{2}}\sin(A^{\frac{1}{2}}t)v_0$$

$$+ A^{-\frac{1}{2}}\int_0^t \sin(A^{\frac{1}{2}}t - A^{\frac{1}{2}}s)f(s)\,\mathrm{d}s\}, \tag{22.28}$$

where

$$\boldsymbol{P}_n u = \sum_{i=1}^{n} (u, e_i)e_i \qquad (u \in H). \tag{22.29}$$

THEOREM 22.4: *Let an operator $A^{-\sigma}$ ($\sigma > 0$) act continuously from H to a space E_0. Let $u_0 \in D(A^{\sigma+\tau})$, $v_0 \in D(A^{\sigma+\tau-\frac{1}{2}})$ and $f(t) \in D(A^{\sigma+\tau-\frac{1}{2}})$ for all $t \in [0, T]$, and let the function*

$$\varphi(t) = A^{\sigma+\tau-\frac{1}{2}}f(t) \tag{22.30}$$

be continuous on $[0, T]$ in the norm of the space H. Then the Fourier approximations (22.8) converge to the solution of problem (22.24), (22.26) in the norm of the space E, uniformly with respect to $t \in [0, T]$. The rate of convergence is characterized by the inequality

$$\|u(t) - u_n(t)\|_E = o(\lambda_n^{-\tau}). \tag{22.31}$$

PROOF: Introduce the notation

$$u_1 = A^{\sigma+\tau}u_0, \quad v_1 = A^{\sigma+\tau-\frac{1}{2}}v_0.$$

Then the vector-functions (22.27) and (22.28) can be written in the form

$$u(t) = A^{-\sigma}u^{(1)}(t), \quad u_n(t) = A^{-\sigma}u_n^{(1)}(t),$$

where

$$u^{(1)}(t) = \cos(A^{\frac{1}{2}}t)A^{-\tau}u_1 + \sin(A^{\frac{1}{2}}t)A^{-\tau}v_1$$

$$+ \int_0^t \sin(A^{\frac{1}{2}}t - A^{\frac{1}{2}}s)A^{-\tau}\varphi(s)\,\mathrm{d}s,$$

$$u_n^{(1)}(t) = \cos(A^{\frac{1}{2}}t)P_nA^{-\tau}u_1 + \sin(A^{\frac{1}{2}}t)P_nA^{-\tau}v_1$$

$$+ \int_0^t \sin(A^{\frac{1}{2}}t - A^{\frac{1}{2}}s)P_nA^{-\tau}\varphi(s)\,\mathrm{d}s.$$

The continuity of the function $\varphi(s)$ implies the following estimate, uniform for $s \in [0, T]$,

$$\|A^{-\tau}\varphi(s) - P_nA^{-\tau}\varphi(s)\|_H \leqq \lambda_{n+1}^{-\tau}\|\varphi(s) - P_n\varphi(s)\|_H \leqq \lambda_{n+1}^{-\tau}\cdot\alpha(n)$$

where $\alpha(n) \to 0$ as $n \to \infty$. It is clear that

$$\|A^{-\tau}v_1 - P_nA^{-\tau}v_1\|_H = o(\lambda_n^{-\tau}),\ \ \|A^{-\tau}u_1 - P_nA^{-\tau}u_1\|_H = o(\lambda_n^{-\tau}).$$

Hence

$$\|u^{(1)}(t) - u_n^{(1)}(t)\|_H \leqq \|\cos(A^{\frac{1}{2}}t)(A^{-\tau}u_1 - P_nA^{-\tau}u_1)\|_H$$

$$+ \|\sin(A^{\frac{1}{2}}t)(A^{-\tau}v_1 - P_nA^{-\tau}v_1)\|_H$$

$$+ \|\int_0^t \sin(A^{\frac{1}{2}}t - A^{\frac{1}{2}}s)[A^{-\tau}\varphi(s) - P_nA^{-\tau}\varphi(s)]\,\mathrm{d}s\|_H$$

$$\leqq \|A^{-\tau}u_1 - P_nA^{-\tau}u_1\|_H + \|A^{-\tau}v_1 - P_nA^{-\tau}v_1\|_H$$

$$+ T\sup_{0\leqq s\leqq T}\|A^{-\tau}\varphi(s) - P_nA^{-\tau}\varphi(s)\|_H = o(\lambda_n^{-\tau}).$$

Consequently

$$\|u(t) - u_n(t)\|_E \leqq \|A^{-\sigma}\|_{H\to E}\|u^{(1)}(t) - u_n^{(1)}(t)\|_H = o(\lambda_n^{-\tau}).$$

The theorem has been proved.

If A is an elliptic operator of order $2k$ then the above theorem implies that the approximations (22.28) converge absolutely and uniformly to the exact solution (22.27) if

$$\sigma > N/4k.$$

The following assertion can be proved in the same manner as Theorem 22.4.

THEOREM 22.5: *Let a linear operator* $\boldsymbol{B}$ *be defined on* $D(A^{\sigma})$, *where* $\sigma > 0$, *and satisfy the inequality* (22.9). *Let* $u_0 \in D(A^{\sigma+\tau})$, $v_0 \in D(A^{\sigma+\tau-\frac{1}{2}})$ *and* $f(t) \in D(A^{\sigma+\tau-\frac{1}{2}})$ *for all* $t \in [0, T]$, *and let the function* (22.30) *be continuous in the norm of the space E. Then the operator* $\boldsymbol{B}$ *can be applied to the Fourier approximations* (22.28), *and the vector-function* $\boldsymbol{B}u_n(t)$ *converges in the norm of the space E to the vector-function* $\boldsymbol{B}u(t)$, *where* $u(t)$ *is the exact solution* (22.27), *the convergence being uniform with respect to* $t \in [0, T]$. *The rate of convergence is characterized by the estimate*

$$\|\boldsymbol{B}u(t) - \boldsymbol{B}u_n(t)\|_E = o(\lambda_n^{-\tau}). \tag{22.32}$$

If in the hypothesis of this theorem the differentiation operators $\boldsymbol{D}^r$ are taken as $\boldsymbol{B}$, and the space C of continuous function is taken as E, then we obtain theorems on the uniform and absolute convergence of the Fourier approximations together with their derivatives up to definite orders.

22.4 *Fourier's method for parabolic equations*

The arguments of the preceding section apply, almost without change, to equations of parabolic type

$$\frac{\mathrm{d}u(t)}{\mathrm{d}t} + Au(t) = f(t). \tag{22.33}$$

Here, as before, A is a strictly positive definite selfadjoint operator with eigenfunctions e_i and eigenvalues λ_i. Equation (22.33) is to be solved with an initial condition

$$u(0) = u_0, \tag{22.34}$$

where u_0 is an arbitrary element of the space H.

A solution of problem (22.33)–(22.34) is a continuous vector-function on $(0, T]$ which satisfies equation (22.33) on $(0, T]$ and satisfies condition (22.34). Such a solution is produced by the formula

$$u(t) = e^{-At}u_0 + \int_0^t e^{-A(t-s)}f(s)\,\mathrm{d}s \tag{22.35}$$

(the representation (22.35) will be considered in greater detail in the next paragraph).

Fourier's method for a solution of the problem (22.33)–(22.34) consists in constructing successive approximations

$$u_n(t) = \boldsymbol{P}_n\{e^{-At}u_0 + \int_0^t e^{-A(t-s)}f(s)\mathrm{d}s\}, \tag{22.36}$$

where $\boldsymbol{P}_n$ is the operator (22.29). The Fourier approximations (22.36) converge to the exact solution (22.35) in the norm of the space H. As in the case of hyperbolic equations, it is easy to formulate conditions under which Fourier's method for parabolic equations converges in some stronger norm.

THEOREM 22.6: *Let an operator $A^{-\sigma}$ ($\sigma > 0$) act continuously from H to a space E. Let $u_0 \in D(A^{\sigma+\tau})$ and $f(t) \in D(A^{\sigma+\tau_1})$, $\tau_1 > \tau - 1$, for all $t \in [0, T]$, and let the function*

$$\varphi(t) = A^{\sigma+\tau_1}f(t) \tag{22.37}$$

be continuous in the norm of the space H on $[0, T]$. Then the Fourier approximations (22.36) converge to the solution (22.35) of problem (22.33)–(22.34) in the norm of the space E, uniformly with respect to $t \in [0, T]$. The rate of convergence is characterized by the inequality

$$\|u(t) - u_n(t)\|_E = o(\lambda_n^{-\tau}). \tag{22.38}$$

It is easy to formulate assertions which bear the same relation to Theorem 22.6, as Theorems 22.2 and 22.5 bear to Theorems 22.1 and 22.4. It is also easy to apply general theorems on the convergence of Fourier's method for abstract parabolic equations to the case of equations with elliptic differential operators A of order $2k$.

The operator e^{-At} possesses an 'improving' property: for $t > 0$ and all $u_0 \in H$ the elements $e^{-At}u_0$ belong (see § 11) to the domains of all the positive powers of the operator A. Even, if the assumption $u_0 \in D(A^{\sigma+\tau})$ is omitted from the hypotheses of Theorem 22.6, the above remark still permits us to assert that Fourier's method converges in the norm of the space E on the interval $(0, T]$ and that this convergence is uniform (with estimates similar to (22.38)) on each interval $[t_1, T] \subset (0, T]$.

Let us make one final remark. In Theorems 22.1, 22.4 and 22.6 it is possible to take for E various spaces of differentiable functions. By this approach these theorems can be used to obtain results on the convergence of differentiated Fourier series which do not rely on assertions of the type appearing in Theorems 22.2 and 22.5.

§ 23 Translation operators along trajectories of differential equations[1]

23.1 *Linear equations*

Consider the problem

$$\frac{dv}{dt} + Av = f(t) \qquad (0 < t \leqq t_0) \tag{23.1}$$

$$v(0) = v_0 \tag{23.2}$$

in a Banach space E. Here $v(t)$ is unknown, while $f(t)$ is a given function with values in E defined on the interval $[0, t_0]$; dv/dt is the derivative, defined as the limit in E norm of the corresponding difference quotients; and A is a linear operator acting in E.

---

[1] In recent years the theory of linear equations with unbounded operators in Banach spaces has been developed with great completeness. We will only refer to the classical monographs: E. Hille and R. S. Phillips [1], N. Dunford and J. T. Schwartz [1]. In this paragraph we will basically use results obtained in the following works: M. A. Krasnoselskii, S. G. Krein and P. E. Sobolevskii [1, 2], M. Z. Solomjak [1, 2], P. E. Sobolevskii [7, 8].

Non-linear equations with unbounded operators in Banach spaces have been investigated mainly by mathematicians in Voronezh (M. A. Krasnoselskii and S. G. Krein [2], M. A. Krasnoselskii, S. G. Krein and P. E. Sobolevskii [1, 2], M. A. Krasnoselskii [10], P. E. Sobolevskii [1, 2, 4, 6, 9–13]).

Translation and quasi-translation operators along trajectories of differential equations with unbounded operators were constructed and investigated by M. A. Krasnoselskii and P. E. Sobolevskii [3]. Note also the work of U. S. Kolesov [1].

A function $v(t)$ is called a *solution of the problem* (23.1)–(23.2) if

1°) $v(t)$ is continuous on $[0, t_0]$;
2°) $dv(t)/dt$ and $Av(t)$ are continuous on $(0, t_0]$;
3°) $v(t)$ satisfies equation (23.1) on $(0, t_0]$;
4°) $v(t)$ satisfies the initial condition (23.2).

If the operator A is bounded and the function $f(t)$ is continuous on $[0, t_0]$ then the solution of problem (23.1)–(23.2) exists and is determined uniquely by the formula

$$v(t) = T(t)v_0 + \int_0^t T(t-s)f(s)\,ds. \tag{23.3}$$

Here $T(t) = e^{-At}$ (see formula (13.20)).

It turns out that formula (23.3) for the solutions of problem (23.1)–(23.2) is also valid in the case of unbounded operators A.

LEMMA 23.1: *Let $v(t)$ be a solution of problem* (23.1)–(23.2) *and let $-A$ be the generator of a strongly continuous semi-group. Then formula* (23.3) *is valid.*

PROOF: Consider the operators

$$A_n = nA(nI + A)^{-1}.$$

As was shown in 13.6°, the operators A_n are defined and bounded for sufficiently large positive n. These operators are uniformly subordinate to the operator A, i.e.

$$\|A_n u\| \leqq c\|Au\| \quad (u \in D(A)),$$

where c is independent of n. The operators A_n approximate (see (13.41)) the operator A on its domain:

$$\lim_{n\to\infty} \|A_n u - Au\| = 0 \quad (u \in D(A)).$$

Finally, the semi-groups $T_n(t) = e^{-A_n t}$ are bounded uniformly in n and converge strongly to the semi-group $T(t)$ (see 13.6°).

Let $v(t)$ be a solution of problem (23.1)–(23.2). Then for $t > 0$

$$\frac{\mathrm{d}v}{\mathrm{d}t} + A_n v = f(t) + (A_n - A)v(t).$$

Since the operators A_n are bounded, we have for any $0 < \tau < t \leqq t_0$

$$v(t) = T_n(t - \tau)v(\tau) + \int_\tau^t T_n(t - s)[f(s) + (A_n - A)v(s)]\mathrm{d}s.$$

Now let $n \to \infty$. The first summand on the right side converges to $T(t - \tau)v(\tau)$. In the second summand the integrated expressions converge to $T(t - s)f(s)$ and are bounded uniformly in n. Hence it is possible to proceed to the limit under the integration sign. Consequently

$$v(t) = T(t - \tau)v(\tau) + \int_\tau^t T(t - s)f(s)\mathrm{d}s.$$

Finally, letting τ tend to zero we arrive at formula (23.3). The lemma has been proved.

This lemma implies, in particular, that a solution of problem (23.1)–(23.2) is unique (of course, under the hypothesis that $-A$ is the generator of a strongly continuous semi-group).

THEOREM 23.1: *Let the operator* A *be of strongly positive type and let the function* $f(t)$ *satisfy the Hölder condition*

$$\|f(t) - f(\tau)\| \leqq c(\varepsilon)|t - \tau|^\delta \quad (t, \tau \in [\varepsilon, t_0];\ \varepsilon, \delta > 0). \tag{23.4}$$

Then problem (23.1)–(23.2) *has a solution.*

PROOF: As is easily seen (see § 13), the function $v = T(t)v_0$ is a solution of the homogeneous problem

$$\frac{\mathrm{d}v}{\mathrm{d}t} + Av = 0, \quad v(0) = v_0.$$

The function

$$Qf(t) = \int_0^t T(t-s)f(s)\,\mathrm{d}s$$

is continuous on $[0, t_0]$ and is equal to zero at $t = 0$.

Let $t \geqq t_1 > 0$. Consider the function

$$Q_h f(t) = \int_0^{t-h} T(t-s)f(s)\,\mathrm{d}s \quad (0 < h \leqq t_1).$$

Since a semi-group generated by a strongly positive-type operator A is continuously differentiable for $t > 0$, the function $Q_h f(t)$ is continuously differentiable and

$$\frac{\mathrm{d}Q_h f(t)}{\mathrm{d}t} = T(h)f(t-h) - \int_0^{t-h} AT(t-s)f(s)\,\mathrm{d}s$$

$$= T(h)f(t-h) - AQ_h f(t). \tag{23.5}$$

Now let h tend to zero. Then it is clear that

$$Q_h f(t) \to Qf(t),$$

and the convergence is uniform with respect to $t \in [t_1, t_0]$. Since

$$\int_0^{t-h} AT(t-s)f(s)\,\mathrm{d}s = \int_0^{t-h} AT(t-s)f(t)\,\mathrm{d}s$$

$$+ \int_0^{t-h} AT(t-s)[f(s) - f(t)]\,\mathrm{d}s$$

$$= [T(h) - T(t)]f(t) + \int_0^{t-h} AT(t-s)[f(s) - f(t)]\,\mathrm{d}s,$$

inequality (23.4) and the estimate (see (13.64))

$$\|AT(t)\| \leqq \frac{c}{t}$$

imply that as $h \to 0$

$$\int_0^{t-h} AT(t-s)f(s)\,\mathrm{d}s \to$$

$$\to [I - T(t)]f(t) + \int_0^t AT(t-s)[f(s) - f(t)]\,\mathrm{d}s,$$

and the convergence is uniform with respect to $t \in [t_1, t_0]$. It follows from this and from (23.5) that the function $Qf(t)$ is continuously differentiable for $t > 0$ and

$$\frac{\mathrm{d}}{\mathrm{d}t} Qf(t) = T(t)f(t) - \int_0^t AT(t-s)[f(s) - f(t)]\,\mathrm{d}s.$$

Further, the closedness of the operator A implies that the values of the function $Qf(t)$ belong to $D(A)$ for $t > 0$ and

$$AQf(t) = [I - T(t)]f(t) + \int_0^t AT(t-s)[f(s) - f(t)]\,\mathrm{d}s.$$

It follows from these two relations that

$$\frac{\mathrm{d}}{\mathrm{d}t} Qf(t) + AQf(t) = f(t).$$

Thus, $T(t)v_0 + Qf(t)$ is a solution of the problem (23.1)–(23.2). The theorem has been proved.

23.2 *The Cauchy operator*

To investigate non-linear differential equations we have to study the properties of the *Cauchy operator*, introduced in the preceding section:

$$\boldsymbol{Q}f(t) = \int_0^t \boldsymbol{T}(t-s)f(s)\,\mathrm{d}s. \tag{23.6}$$

LEMMA 23.2: *Let an operator* $\boldsymbol{A}$ *be of strongly positive-type and let* $f(t)$ *be continuous on* $[0, t_0]$. *Then the following inequalities are valid*:

$$\|\boldsymbol{Q}f(t) - \boldsymbol{Q}f(\tau)\| \leqq c|t-\tau|(1 + |\ln|t-\tau||) \max_{0 \leqq s \leqq \max\{t,\tau\}} \|f(s)\|, \tag{23.7}$$

$$\|\boldsymbol{A}^{\alpha}\boldsymbol{Q}f(t) - \boldsymbol{A}^{\alpha}\boldsymbol{Q}f(\tau)\|$$

$$\leqq c(\alpha)|t-\tau|^{1-\alpha} \max_{0 \leqq s \leqq \max\{t,\tau\}} \|f(s)\| \quad (0 < \alpha < 1). \tag{23.8}$$

PROOF: Assume for definiteness that $\tau \leqq t$. First let $\tau \leqq t/2$. Then

$$\|\boldsymbol{Q}f(t) - \boldsymbol{Q}f(\tau)\| \leqq \|\boldsymbol{Q}f(t)\| + \|\boldsymbol{Q}f(\tau)\|$$

$$\leqq \int_0^t \|\boldsymbol{T}(t-s)\|\,\|f(s)\|\,\mathrm{d}s + \int_0^\tau \|\boldsymbol{T}(\tau-s)\|\cdot\|f(s)\|\,\mathrm{d}s.$$

Since the semi-group $\boldsymbol{T}(t)$ is bounded and $t + \tau \leqq 3(t-\tau)$,

$$\|\boldsymbol{Q}f(t) - \boldsymbol{Q}f(\tau)\| \leqq ct \max_{0 \leqq s \leqq t} \|f(s)\| + c\tau \max_{0 \leqq s \leqq \tau} \|f(s)\|$$

$$\leqq 3c(t-\tau) \max_{0 \leqq s \leqq t} \|f(s)\| \tag{23.9}$$

Now let $\tau \geqq t/2$. Then

$$\boldsymbol{Q}f(t) - \boldsymbol{Q}f(\tau) = \int_{2\tau-t}^{t} \boldsymbol{T}(t-s)f(s)\,\mathrm{d}s - \int_{2\tau-t}^{\tau} \boldsymbol{T}(\tau-s)f(s)\,\mathrm{d}s$$

$$+ \int_0^{2\tau-t} [\boldsymbol{T}(t-s) - \boldsymbol{T}(\tau-s)]f(s)\,\mathrm{d}s. \tag{23.10}$$

Using the identity

$$T(t-s) - T(\tau - s) = \int_{\tau-s}^{t-s} AT(\xi)\,d\xi \tag{23.11}$$

and the estimate

$$\|AT(\xi)\| \leqq c_1/\xi \quad (\xi > 0), \tag{23.12}$$

which is valid for strongly positive-type operators (see 13.8°), we obtain

$$\|T(t-s) - T(\tau - s)\| \leqq \frac{c_1|t-\tau|}{\tau - s}. \tag{23.13}$$

From this and (23.10) there follows the inequality

$$\begin{aligned}
&\|Qf(t) - Qf(\tau)\| \\
&\leqq 2c(t-\tau) \max_{2\tau-t \leqq s \leqq t} \|f(s)\| + c(t-\tau) \max_{2\tau-t \leqq s \leqq t} \|f(s)\| \\
&+ c_1(t-\tau)|\ln \tau - \ln(t-\tau)| \max_{0 \leqq s \leqq 2\tau - t} \|f(s)\| \\
&\leqq c_3|t-\tau|(1 + |\ln|t-\tau||) \max_{0 \leqq s \leqq t} \|f(s)\|.
\end{aligned} \tag{23.14}$$

Clearly (23.9) and (23.14) imply the first assertion of the lemma.

Now let us prove the inequality (23.8).

First let $\tau \leqq t/2$. Then, using the estimate

$$\|A^\alpha T(\xi)\| \leqq \frac{c(\alpha)}{\xi^\alpha} \quad (\xi > 0), \tag{23.15}$$

which is valid for strongly positive-type operators (see Theorem 14.11), we obtain

$$\|A^\alpha Qf(t) - A^\alpha Qf(\tau)\| \leqq \|A^\alpha Qf(t)\| + \|A^\alpha Qf(\tau)\|$$

$$\leqq \int_0^t \frac{c(\alpha)}{(t-s)^\alpha}\,ds \cdot \max_{0\leqq s\leqq t} \|f(s)\| + \int_0^\tau \frac{c(\alpha)}{(\tau-s)^\alpha}\,ds \cdot \max_{0\leqq s\leqq \tau} \|f(s)\|$$

$$\leqq \frac{c(\alpha)}{1-\alpha}(t^{1-\alpha} + \tau^{1-\alpha}) \cdot \max_{0\leqq s\leqq t} \|f(s)\|$$

$$\leqq \frac{6^\alpha c(\alpha)}{1-\alpha}(t-\tau)^{1-\alpha} \cdot \max_{0\leqq s\leqq t} \|f(s)\|. \tag{23.16}$$

Now let $\tau \geqq t/2$. Using the identity (23.11) and the estimate (23.15), we obtain

$$\|A^\alpha T(t-s) - A^\alpha T(\tau - s)\| \leqq \frac{c(1+\alpha)}{(\tau-s)^{1+\alpha}}|t-\tau|. \tag{23.17}$$

From this and the identity (23.10) there follows the inequality

$$\|A^\alpha Qf(t) - A^\alpha Qf(\tau)\| \leqq \int_{2\tau-t}^t \frac{c(\alpha)}{(t-s)^\alpha}\,ds \cdot \max_{2\tau-t\leqq s\leqq t} \|f(s)\|$$

$$+ \int_{2\tau-t}^\tau \frac{c(\alpha)}{(\tau-s)^\alpha}\,ds \max_{2\tau-t\leqq s\leqq t} \|f(s)\|$$

$$+ \int_0^{2\tau-t} \frac{c(1+\alpha)|t-\tau|}{(\tau-s)^{1+\alpha}}\,ds \cdot \max_{0\leqq s\leqq 2\tau-t} \|f(s)\|$$

$$\leqq \left\{\frac{2^{1-\alpha}c(\alpha)}{1-\alpha}|t-\tau|^{1-\alpha} + \frac{c(\alpha)}{1-\alpha}|t-\tau|^{1-\alpha}\right.$$

$$\left. + \frac{c(1+\alpha)|t-\tau|}{\alpha}\left[\frac{1}{|t-\tau|^\alpha} - \frac{1}{\tau^\alpha}\right]\right\} \max_{0\leqq s\leqq t} \|f(s)\|$$

$$\leqq c_1(\alpha)|t-\tau|^{1-\alpha} \cdot \max_{0\leqq s\leqq t} \|f(s)\|. \tag{23.18}$$

Now (23.8) follows from (23.16) and (23.18). The lemma has been proved.

Consider a set $\mathfrak{M}$ of functions $f(t)$ which are continuous on an interval $[0, t_1] \subset [0, t_0]$. Let this set be uniformly bounded, i.e. there exists a number $M > 0$ such that for any function $f(t) \in \mathfrak{M}$ the inequality:

$$\|f(t)\| \leqq M \quad (0 \leqq t \leqq t_1)$$

is valid. Then it follows from Lemma 23.2 that for any $\alpha \in [0, 1)$ the set of functions $\{A^\alpha Q f(t)\}$ is uniformly bounded and equi-continuous (i.e. for any $\varepsilon > 0$ there can be selected a $\delta > 0$, depending only on ε, such that

$$\|A^\alpha Q f(t) - A^\alpha Q f(\tau)\| < \varepsilon \quad (t, \tau \in [0, t_1]),$$

if $|t - \tau| < \delta$).

Let the operator A^{-1} be compact in E. Then (see Theorem 14.12) all the operators $A^{-\varepsilon}$ $(\varepsilon > 0)$ are compact in E. Hence for each fixed t the set of elements $A^\alpha Q f(t)$ $(f \in \mathfrak{M})$ is compact in E. In fact,

$$A^\alpha Q f(t) = A^{-\varepsilon} A^{\alpha+\varepsilon} Q f(t)$$

and for $\varepsilon \in (0, 1 - \alpha)$ the set $A^{\alpha+\varepsilon} Q f(t)$ is bounded in E.

Thus the set of functions $A^\alpha Q f(t)$ is equi-continuous, and for each t the set $A^\alpha Q f(t)$ is compact in E. From the general theorem of Arzela there follows:

LEMMA 23.3: *Let A be a strongly positive-type operator and let A^{-1} be compact. Then for any $\alpha \in [0, 1)$ the operator $A^\alpha Q$ is a compact operator in the space $C\{[0, t_1], E\}$ of continuous E-valued functions on* $[0, t_1]$.

23.3 *Non-linear equations*

Consider the non-linear problem

$$\frac{dv}{dt} + Av = f(t, v) \quad (0 < t \leqq t_0) \tag{23.19}$$

$$v(0) = v_0. \tag{23.20}$$

Here for each $t \in [0, t_0]$ $f(t, v)$ is a non-linear operator acting in E.

A function $v(t)$ is called a *solution of problem* (23.19)–(23.20) if the function $f[t, v(t)]$ is continuous on $[0, t_0]$ and if the function $v(t)$ is a solution of the linear problem (23.1)–(23.2) with right side $f(t)$ equal to $f[t, v(t)]$.

If the operator $-\boldsymbol{A}$ generates a strongly continuous semi-group $\boldsymbol{T}(t)$, then Lemma 23.1 implies that any solution $v(t)$ of problem (23.19)–(23.20) satisfies the integral equation

$$v(t) = \boldsymbol{T}(t)v_0 + \int_0^t \boldsymbol{T}(t-s)f[s, v(s)]\,\mathrm{d}s. \tag{23.21}$$

Hence it is natural to call any solution $v(t)$ of the integral equation (23.21) possessing the property that the function $f[t, v(t)]$ is continuous on $[0, t_0]$, a *generalized solution* of the problem (23.19)–(23.20). By Theorem 23.1 a generalized solution of this problem will be a true solution if the function $f[t, v(t)]$ satisfies the Hölder condition (23.4).

THEOREM 23.2: *Let $\boldsymbol{T}(t)$ be a strongly continuous semi-group. Let the function $f(t, v)$ be continuous in t on $[0, t_0]$ for each $v \in E$, and let*

$$\|f(t, v_1) - f(t, v_2)\| \leqq C(R)\|v_1 - v_2\| \qquad (\|v_1\|, \|v_2\| \leqq R). \tag{23.22}$$

Then there exists a unique continuous solution $v(t)$ of the equation (23.21), *defined on some interval $[0, t^*] \subset [0, t_0]$. This function $v(t)$ will be the unique generalized solution of the problem* (23.19)–(23.20).

The proof of existence and uniqueness of a continuous solution $v(t)$ of equation (23.21) can be achieved by the method of successive approximation. Since the operator $f(t, v)$ is jointly continuous in the variables $t \in [0, t_0]$ and $v \in E$, the function $f[t, v(t)]$ is continuous on $[0, t^*]$. Consequently, the solution of equation (23.21) which is obtained will be a generalized solution of problem (23.19)–(23.20). Finally, since every generalized solution of this problem is a continuous solution of equation (23.21), problem (23.19)–(23.20) has a unique generalized solution.

THEOREM 23.3: *Let the operator $\boldsymbol{A}$ be of strongly positive-type and let the operator $\boldsymbol{A}^{-1}$ be compact. Let the operator $f(t, v)$ be jointly continuous in the variables $t \in [0, t_0]$ and $v \in E$, and be bounded on each bounded set in E, i.e.*

$$\|f(t, v)\| \leqq M(R) < \infty \qquad (0 \leqq t \leqq t_0, \|v\| \leqq R). \tag{23.23}$$

Then there exists at least one continuous solution $v(t)$ *of equation* (23.21) *defined on some interval* $[0, t^*] \subset [0, t]$. *This function* $v(t)$ *will be a generalized solution of the problem* (23.19)–(23.20) *on this interval.*

PROOF: Consider the operator

$$\boldsymbol{B}v(t) = \boldsymbol{T}(t)v_0 + \int_0^t \boldsymbol{T}(t-s)f[s, v(s)]\,\mathrm{d}s. \tag{23.24}$$

It follows from Lemma 23.3 that this operator is completely continuous in the space $C\{[0, t^*], E\}$.

Denote by $Ш(R)$ the ball with radius R and center at the origin in the space $C\{[0, t^*], E\}$, i.e. the family of functions $v(t)$ continuous on $[0, t^*]$ and satisfying the condition $\|v(t)\| \leqq R$. It follows from (23.23) that the operator $\boldsymbol{B}$ maps this ball into itself provided that

$$t^* \leqq \frac{R - c\|v_0\|}{cM(R)}, \quad R > c\|v_0\|, \quad c = \sup_{0 \leqq t \leqq t_0} \|T(t)\|. \tag{23.25}$$

Thus the equation

$$v(t) = \boldsymbol{B}v(t)$$

satisfies the hypothesis of Schauder's fixed point principle on the Banach space $C\{[0, t^*], E\}$. Hence there exists at least one solution $v(t)$.

Finally it follows from the continuity of $f(t, v)$ that $v(t)$ is a generalized solution of the problem (23.19)–(23.20). The theorem has been proved.

As was shown earlier, every solution of problem (23.19)–(23.20) will also be a generalized solution. Let us investigate under what circumstances every generalized solution of problem (23.19)–(23.20) will also be an ordinary solution.

THEOREM 23.4: *Let the operator* $\boldsymbol{A}$ *be of strongly positive-type. Let*

$$\|f(t_1, v_1) - f(t_2, v_2)\| \leqq C(R, \varepsilon)\{|t_1 - t_2|^\delta + \|v_1 - v_2\|^\rho\}$$

$$(0 < \varepsilon \leqq t_1, t_2 \leqq t_0;\ \|v_1\|, \|v_2\| \leqq R;\ \delta, \rho \in (0, 1]). \tag{23.26}$$

Then every generalized solution of problem (23.19)–(23.20) *will also be an ordinary solution.*

PROOF: Let $v(t)$ be a generalized solution of problem (23.19)–(23.20). From Lemma 23.2 it follows that for $t > 0$ the function $v(t)$ satisfies the Hölder condition with any exponent less than 1. This fact together with (23.26) implies that for $t > 0$ the function $f[t, v(t)]$ will satisfy the Hölder condition with any exponent less than $\min\{\rho, \delta\}$. From the remark at the beginning of this section it follows that $v(t)$ is a solution of problem (23.19)–(23.20). The theorem has been proved.

Theorems 23.2 and 23.3 are of local character. One can raise the question: when does problem (23.19)–(23.20) have a generalized solution defined on the entire interval $[0, t_0]$.

Using the method of successive approximations (Theorem 23.2), together with Schauder's fixed point principle (Theorem 23.3), it is possible to guarantee the existence of a solution on an interval $[0, t^*]$ with length

$$t^* = \frac{R - c\|v_0\|}{cM(R)} \tag{23.27}$$

(see (23.25)).

Now (23.27) implies that for any $v_0 \in E$ there will exist a generalized solution of the problem (23.19)–(23.20) defined on the entire interval $[0, t_0]$ if

$$\overline{\lim_{R\to\infty}} \frac{R}{M(R)} = \infty. \tag{23.28}$$

If only small $v_0 \in E$ are considered, it suffices to assume that

$$\sup_R \frac{R}{M(R)} > t_0 c. \tag{23.29}$$

Let $v(t)$ be a continuous solution of equation (23.21) for $t \geqq \tau \geqq 0$. Then $v(t)$ satisfies the relation

$$v(t) = \boldsymbol{T}(t - \tau)v(\tau) + \int_\tau^t \boldsymbol{T}(t - s)f[s, v(s)]\,\mathrm{d}s, \tag{23.30}$$

which results from the chain of relations

$$v(t) = T(t)v_0 + \int_0^t T(t-s)f[s, v(s)]\,ds = T(t-\tau)T(\tau)v_0$$

$$+ \int_0^\tau T(t-\tau)T(\tau-s)f[s, v(s)]\,ds + \int_\tau^t T(t-s)f[s, v(s)]\,ds$$

$$= T(t-\tau)\{T(\tau)v_0 + \int_0^\tau T(\tau-s)f[s, v(s)]\,ds\}$$

$$+ \int_\tau^t T(t-s)f[s, v(s)]\,ds$$

$$= T(t-\tau)v(\tau) + \int_\tau^t T(t-s)f[s, v(s)]\,ds.$$

The identity (23.30) permits to solve equation (23.21) step by step by proceeding from each interval to the next. The following result is proved by this method.

THEOREM 23.5: *Let the conditions of Theorem 23.2 or 23.3 be satisfied. Suppose any possible solution $v(t)$ of the equation* (23.21) *admits the a priori estimate*

$$\|v(t)\| \leqq c(v_0). \tag{23.31}$$

Then a solution exists on the entire interval $[0, t_0]$.

The a priori estimate (23.31) is usually established with the aid of differential or integral inequalities.

23.4 *Equations with unbounded non-linearities*

In the preceding section the non-linear problem (23.19)–(23.20) was investigated under the assumption that the operator $f(t, v)$ is continuous in v. For many applications this assumption is too restrictive. The theory of fractional powers of operators sometimes permits us to investigate problem (23.19)–(23.20) for unbounded operators $f(t, v)$.

Throughout this entire section, α is a fixed number in the interval $[0, 1)$.

THEOREM 23.6: *Let the operator A be of strongly positive type. Let the function $f(t, A^{-\alpha}w)$ be continuous in t on $[0, t_0]$ for each fixed $w \in E$, and let*

$$\|f(t, A^{-\alpha}w_1) - f(t, A^{-\alpha}w_2)\|$$

$$\leqq c(R)\|w_1 - w_2\| \qquad (\|w_1\|, \|w_2\| \leqq R). \tag{23.32}$$

Finally, let $v_0 \in D(A^\alpha)$. Then there exists a unique continuous solution $w(t)$ of the integral equation

$$w(t) = T(t)A^\alpha v_0 + \int_0^t A^\alpha T(t-s) f[s, A^{-\alpha}w(s)]\,\mathrm{d}s, \tag{23.33}$$

defined on some interval $[0, t^*] \subset [0, t_0]$. The function

$$v(t) = A^{-\alpha}w(t) \tag{23.34}$$

is then a generalized solution of the problem (23.19)–(23.20).

The proof of existence and uniqueness of a continuous solution of equation (23.33) defined on some interval can be achieved by the method of successive approximations. This can be carried out since by the estimate (23.15) the kernel of the integral equation has weak singularity ($\alpha < 1$), while the non-linear operator f satisfies the Lipschitz condition (23.32).

It is easy to see that the function $v(t)$ defined by relation (23.34) is a continuous solution of equation (23.21). Hence, since the function $f[t, A^{-\alpha}w(t)]$ is continuous on $[0, t^*]$, $v(t)$ is a generalized solution of the problem (23.19)–(23.20).

Finally, if $v(t)$ is a generalized solution of problem (23.19)–(23.20), $v(t) \in D(A^\alpha)$ for each $t \in [0, t^*]$ and the function

$$w(t) = A^\alpha v(t)$$

is a continuous solution of the equation (23.33). Consequently, the generalized solution of problem (23.19)–(23.20) is unique. The theorem has been proved.

The following result is an analogue to Theorem 23.3.

THEOREM 23.7: *Let the operator A be of strongly positive type. Let the operator $f(t, A^{-\alpha}w)$ be jointly continuous in the variables $t \in [0, t_0]$ and $w \in E$ and be bounded on each bounded set in E, i.e.*

$$\|f(t, A^{-\alpha}w)\| \leqq M_\alpha(R) \qquad (t \in [0, t_0],\ \|w\| \leqq R). \tag{23.35}$$

Let A^{-1} be compact. Finally, let $v_0 \in D(A^\alpha)$. Then there exists at least one continuous solution $w(t)$ of equation (23.33) *defined on some interval* $[0, t^*] \subset [0, t_0]$. *Formula* (23.34) *provides a generalized solution $v(t)$ of the problem* (23.19)–(23.20).

Theorems 23.6 and 23.7 are of local character. The length t^* of the interval $[0, t^*]$ on which there exists a solution satisfies the inequality

$$t^* \geqq \left[\frac{R - d_\alpha \|A^\alpha v_0\|}{d_\alpha M_\alpha(R)}\right]^{1/(1-\alpha)}, \tag{23.36}$$

where

$$d_\alpha = \frac{1}{1-\alpha} \sup_{0<t\leqq t_0} t^\alpha \|A^\alpha T(t)\|.$$

Formula (23.36) implies the following assertion. If

$$\overline{\lim_{R\to\infty}} \frac{R}{M_\alpha(R)} = \infty, \tag{23.37}$$

then a solution $v(t)$ of equation (23.33) exists on the entire interval $[0, t_0]$ for any $v_0 \in D(A^\alpha)$. If

$$\sup_R \frac{R}{M_\alpha(R)} > d_\alpha t_0^{1-\alpha}, \tag{23.38}$$

then a solution $v(t)$ of the equation (23.33) exists on the entire interval $[0, t_0]$ for $\|A^\alpha v_0\|$ sufficiently small.

The following result is proved with the aid of the identity (23.30).

THEOREM 23.8: *Let the hypotheses of Theorem* 23.6 *or* 23.7 *be fulfilled. Let each possible solution $v(t)$ of the equation* (23.33) *admit the a priori estimate*

$$\|A^\alpha v(t)\| \leqq c_\alpha(A^\alpha v_0). \tag{23.39}$$

Then the solution is defined on the whole interval $[0, t_0]$.

Finally, let us present an analogue to Theorem 23.4.

THEOREM 23.9: *Let the operator A be of strongly positive type. Let*

$$\|f(t_1, A^{-\alpha}w_1) - f(t_2, A^{-\alpha}w_2)\|$$

$$\leqq c(R, \varepsilon)\{|t_1 - t_2|^\delta + \|w_1 - w_2\|^\rho\} \tag{23.40}$$

$$(0 < \varepsilon \leqq t_1, t_2 \leqq t_0;\ \|w_1\|, \|w_2\| \leqq R;\ \delta, \rho \in (0, 1]).$$

Then every generalized solution of problem (23.19)–(23.20) *will also be an ordinary solution.*

PROOF: Let $v(t)$ be a generalized solution of the problem (23.19)–(23.20). Then it follows from Lemma 23.2 that for $t > 0$ the function $A^\alpha v(t)$ will satisfy the Hölder condition with exponent $1 - \alpha$. Then (23.40) implies that for $t > 0$ the function $f[t, v(t)]$ will satisfy the Hölder condition with exponent $\min\{\delta, (1-\alpha)\rho\}$. Hence it follows that $v(t)$ is a solution of problem (23.19)–(23.20). The theorem has been proved.

23.5 *The translation operator*

Consider the problem (23.19)–(23.20). Assume that each v in some given set determines a unique solution (or generalized solution) $v(t, v_0)$ of this problem on the interval $[0, t_0]$. Then for each $t \in [0, t_0]$ define the operator

$$V(t)v_0 = v(t, v_0), \tag{23.41}$$

which assign to each v_0 the value at time t of the corresponding solution (or generalized solution) for the problem (23.19)–(23.20). This operator is called the *translation operator along the trajectory of the differential equation* (23.19).

THEOREM 23.10: *Let the hypotheses of Theorem 23.2 be fulfilled and let the operator $f(t, v)$ be written in the form*

$$f(t, v) = \boldsymbol{B}(t)v + \omega(t, v), \tag{23.42}$$

where $\boldsymbol{B}(t)$ is a bounded linear operator in E depending strongly continuously on $t \in [0, t_0]$, and where the non-linear operator $\omega(t, v)$ satisfies the condition

$$\lim_{R\to 0} \sup_{0 \leqq t \leqq t_0,\ \|v\| \leqq R} \frac{\|\omega(t, v)\|}{\|v\|} = 0. \tag{23.43}$$

Then the translation operator is defined on some ball $\|v\| \leqq r$.

PROOF: Consider the equation

$$z(t) = \int_0^t \boldsymbol{T}(t - s)\boldsymbol{B}(s)z(s)\,\mathrm{d}s + F(t), \tag{23.44}$$

where $F(t)$ is a continuous function on $[0, t_0]$. There exists a unique continuous solution of this equation; it can be found by the method of successive approximations and is written in the form

$$z(t) = F(t) + \int_0^t \boldsymbol{K}(t, s)F(s)\,\mathrm{d}s. \tag{23.45}$$

Here $\boldsymbol{K}(t, s)$ is some linear operator jointly continuous in the variables t, s. Using the representation (23.42) and the relations (23.44) and (23.45) we find that every generalized solution of the problem (23.19)–(23.20) satisfies the identity

$$v(t) = \boldsymbol{T}(t)v_0 + \int_0^t \boldsymbol{K}(t, s)\boldsymbol{T}(s)v_0\,\mathrm{d}s$$

$$+ \int_0^t \Big[\boldsymbol{T}(t - \tau) + \int_\tau^t \boldsymbol{K}(t, s)\boldsymbol{T}(s - \tau)\,\mathrm{d}s\Big]\omega[\tau, v(\tau)]\,\mathrm{d}\tau. \tag{23.46}$$

Conversely, each continuous solution of equation (23.46) will be a generalized solution of the problem (23.19)–(23.20).

By (23.42) and the conditions of Theorem 23.2 the non-linear operator $\omega(t, v)$ satisfies the Lipschitz condition. Hence for equation (23.46) a local theorem on existence and uniqueness of a continuous solution is valid.

The relation (23.43) implies that a condition of the form (23.29) is valid for equation (23.46). It follows that for v_0 sufficiently small in norm a non-local theorem on existence is valid for equation (23.46). The theorem has been proved.

It is seen from the proof of Theorem 23.10 that the translation operator $v(t, v_0)$ is not only defined on some ball $\|v_0\| \leqq r$, but is bounded on this ball, i.e.

$$\max_{0 \leqq t \leqq t_0} \|v(t, v_0)\| \leqq R(r) < \infty \qquad (\|v_0\| \leqq r). \tag{23.47}$$

THEOREM 23.11: *Let the hypotheses of Theorem* 23.10 *be fulfilled. Then the translation operator is continuous on the ball* $\|v_0\| \leqq r$.

PROOF: Let $\|v_{01}\|, \|v_{02}\| \leqq r$. Using (23.21) and (23.22) we obtain the inequality

$$\|v(t, v_{01}) - v(t, v_{02})\| \leqq \|\boldsymbol{T}(t)\| \cdot \|v_{01} - v_{02}\|$$

$$+ c(r) \int_0^t \|\boldsymbol{T}(t-s)\| \cdot \|v(s, v_{01}) - v(s, v_{02})\| \, ds.$$

Hence it follows that

$$\|v(t, v_{01}) - v(t, v_{02})\| \leqq k \|v_{01} - v_{02}\|,$$

i.e. the translation operator satisfies the Lipschitz condition. The theorem has been proved.

THEOREM 23.12: *Let the hypotheses of Theorem* 23.10 *be fulfilled. Let the operator A be of strongly positive type, and suppose the operator A^{-1} is compact in* E. *Then the translation operator for* $t > 0$ *is completely continous on the ball* $\|v_0\| \leqq r$ *as an operator in the space* E.

PROOF: The estimates (23.15) and (23.47) permit us to show that for any $0 \leqq \alpha < 1$ and $t > 0$

$$\|A^{\alpha}v(t, v_0)\| \leqq R(r, t) \quad (\|v_0\| \leqq r).$$

From this and the compactness of the operator $A^{-\alpha}$ there follows the compactness of the translation operator for any $t > 0$. Its continuity was proved previously. The theorem has been proved.

23.6 *Differentiability of the translation operator*

It was proved above that the translation operator satisfies the Lipschitz condition on some ball $\|v_0\| \leqq r$.

THEOREM 23.13: *Let the hypotheses of Theorem* 23.10 *be fulfilled. Then the following representation is valid*:

$$v(t, v_0) = \boldsymbol{B}_1(t)v_0 + \omega_1(t, v_0), \tag{23.48}$$

where $\boldsymbol{B}_1(t)$ is a bounded linear operator in the space E depending strongly continuously on $t \in [0, t_0]$, and $\omega_1(t, v_0)$ is a non-linear operator, satisfying the condition

$$\lim_{\rho \to 0} \sup_{0 \leqq t \leqq t_0, \|v_0\| \leqq \rho} \frac{\|\omega_1(t, v_0)\|}{\|v_0\|} = 0. \tag{23.49}$$

PROOF: Consider the equation

$$z(t) = \boldsymbol{T}(t)v_0 + \int_0^t \boldsymbol{T}(t-s)\boldsymbol{B}(s)z(s)\,ds. \tag{23.50}$$

Its solution can be written (see (23.45)) in the form

$$z(t, v_0) = \boldsymbol{T}(t)v_0 + \int_0^t \boldsymbol{K}(t, s)\boldsymbol{T}(s)v_0\,ds = \boldsymbol{B}_1(t)v_0.$$

It is easy to see that for each fixed $t \in [0, t_0]$ the operator $\boldsymbol{B}_1(t)$ is bounded in the space E and depends strongly continuously on $t \in [0, t_0]$. Using the representation (23.42), we obtain the identity

$$v(t, v_0) - z(t, v_0) = \int_0^t \boldsymbol{T}(t-s)\boldsymbol{B}(s)[v(s, v_0) - z(s, v_0)]\,ds$$

$$+ \int_0^t \boldsymbol{T}(t-s)\omega[s, v(s, v_0)]\,ds.$$

It follows from this and (23.44), (23.45) that

$$\omega_1(t, v_0) = v(t, v_0) - z(t, v_0) = \int_0^t \boldsymbol{T}(t-s)\omega[s, v(s, v_0)]\,ds$$

$$+ \int_0^t \boldsymbol{K}(t, \tau)\{\int_0^\tau \boldsymbol{T}(\tau - s)\omega[s, v(s, v_0)]\,ds\}\,d\tau. \tag{23.51}$$

Since $v(t, 0) = 0$ and $v(t, v_0)$ satisfies the Lipschitz condition, $\|v(t, v_0)\| \leqq \leqq k\|v_0\|$. Hence

$$\|\omega[s, v(s, v_0)]\| \leqq \sigma(\rho) \quad (\|v_0\| \leqq \rho \leqq r),$$

where $\lim_{\rho \to 0} \sigma(\rho)/\rho = 0$. This and (23.51) imply that

$$\lim_{\rho \to 0} \sup_{0 \leqq t \leqq t_0,\ \|v_0\| \leqq \rho} \frac{\|v(t, v_0) - z(t, v_0)\|}{\|v_0\|} = 0.$$

The theorem has been proved.

Supplementary information about the smoothness of the function $f(t, v)$ at $v = 0$ permit us to estimate the smoothness of the translation operator at zero more precisely. For instance, the following result is valid.

THEOREM 23.14: *Let the hypotheses of Theorem 23.2 hold and let the operator $f(t, v)$ be written in the form*

$$f(t, v) = \boldsymbol{B}(t)v + \boldsymbol{C}(t)[v, v] + \boldsymbol{D}(t, v), \tag{23.52}$$

where $\boldsymbol{B}(t)$ is a bounded linear operator in E depending strongly continuously *on* $t \in [0, t_0]$, $\boldsymbol{C}(t)[v_1, v_2]$ *is a bounded bilinear operator*[1] *depending strongly continuously on* $t \in [0, t_0]$, *and the non-linear operator* $\boldsymbol{D}(t, v)$ *satisfies the condition*

$$\lim_{\rho \to 0} \sup_{0 \leqq t \leqq t_0,\ \|v\| \leqq \rho} \frac{\|\boldsymbol{D}(t, v)\|}{\|v\|^2} = 0. \tag{23.53}$$

Then the translation operator can be written in the form

$$v(t, v_0) = \boldsymbol{B}_1(t)v_0 + \boldsymbol{C}_1(t)[v_0, v_0] + \boldsymbol{D}_1(t, v_0), \tag{23.54}$$

where $\boldsymbol{B}_1(t)$ *is a bounded linear operator depending strongly continuously on* $t \in [0, t_0]$, $\boldsymbol{C}_1(t)[v_1, v_2]$ *is a bounded bilinear operator and depending strongly continuously on* $t \in [0, t_0]$, *and the non-linear operator* $\boldsymbol{D}_1(t, v_0)$ *satisfies the condition*

$$\lim_{\rho \to 0} \sup_{0 \leqq t \leqq t_0,\ \|v_0\| \leqq \rho} \frac{\|\boldsymbol{D}_1(t, v_0)\|}{\|v_0\|^2} = 0. \tag{23.55}$$

PROOF: A solution of the equation

$$v(t, v_0, w_0) = \int_0^t \boldsymbol{T}(t - s)\boldsymbol{C}(s)[z(s, v_0), w_0]\,\mathrm{d}s$$

$$+ \int_0^t \boldsymbol{T}(t - s)\boldsymbol{B}_1(s)v(s, v_0, w_0)\,\mathrm{d}s, \tag{23.56}$$

where $z(t, v_0)$ is the solution of equation (23.50), has the form

$$v(t, v_0, w_0) = \int_0^t \boldsymbol{T}(t - s)\boldsymbol{C}(s)[z(s, v_0), w_0]\,\mathrm{d}s$$

$$+ \int_0^t \boldsymbol{K}(t, \tau)\Big\{\int_0^\tau \boldsymbol{T}(\tau - s)\boldsymbol{C}(s)[z(s, v_0), w_0]\,\mathrm{d}s\Big\}\,\mathrm{d}\tau \equiv \tilde{\boldsymbol{C}}(t)[v_0, w_0]. \tag{23.57}$$

---

[1] An operator $\boldsymbol{C}[v_1, v_2]$ is *bilinear*, if it is linear in each argument v_1, v_2. The operator $\boldsymbol{C}(t)[v_1, v_2]$ depends strongly continuously on t if for each fixed $v_1, v_2 \in E$ and $t^* \in [0, t_0]$

$$\lim_{t \to t^*} \|\boldsymbol{C}(t)[v_1, v_2] - \boldsymbol{C}(t^*)[v_1, v_2]\| = 0.$$

It is easy to see that $\tilde{C}(t)[v_0, w_0]$ is a bilinear operator which is strongly continuous in $t \in [0, t_0]$.

Put $C_1(t)[v_0, w_0] = \tilde{C}(t)[v_0, B_1(t)w_0]$. This is also a bilinear operator which is strongly continuous in $t \in [0, t_0]$. Using the representation (23.52) and the relations (23.50) and (23.56), we obtain

$$v(t, v_0) - B_1(t)v_0 - C_1(t)[v_0, v_0]$$

$$= \int_0^t T(t-s)B(s)\{v(s, v_0) - B_1(s)v_0 - C_1(s)[v_0, v_0]\}\,ds$$

$$+ \int_0^t T(t-s)\{C(s)[v(s, v_0), v(s, v_0)]$$

$$- C(s)[B_1(s)v_0, B_1(s)v_0] + D[s, v(s, v_0)]\}\,ds. \tag{23.58}$$

Hence

$$D_1(t, v_0) = v(t, v_0) - B_1(t)v_0 - C_1(t)[v_0, v_0]$$

$$= \chi(t) + \int_0^t K(t, \tau)\chi(\tau)\,d\tau. \tag{23.59}$$

where $\chi(t)$ is equal to the second summand on the right side of (23.58). Using (23.48), we obtain

$$\chi(t) = \int_0^t T(t-s)\{C(s)[B(s)v_0, \omega_1(s, v_0)]$$

$$+ C(s)[\omega_1(s, v_0), B(s)v_0] + C(s)[\omega_1(s, v_0), \omega_1(s, v_0)]$$

$$+ D[s, v(s, v_0)]\}\,ds.$$

It follows from this, (23.59) and the estimates (23.49) and (23.53) that the function $D_1(t, v_0) = v(t, v_0) - B_1(t)v_0 - C_1(t)[v_0, v_0]$ satisfies the relation (23.55). The theorem has been proved.

By the same method it is possible to obtain an expansion of the translation operator up to terms of arbitrary order in a neighborhood of zero, provided such an expansion holds for the non-linear operator $f(t, v)$.

23.7 The quasi-translation operator

In cases where the operator $f(t, v)$ is unbounded, the translation operator is generally speaking defined only on a dense set (see 23.4°). Hence if the operator $\varphi(t, w) = f(t, A^{-\alpha}w)$ is continuous for some $\alpha < 1$, we consider instead of the equation with an unbounded non-linear operator, the equation

$$w(t) = \boldsymbol{T}(t)w_0 + \int_0^t \boldsymbol{A}^\alpha \boldsymbol{T}(t-s)\varphi[s, w(s)]\,ds. \tag{23.60}$$

Here, of course, it is assumed that the operator $\boldsymbol{A}$ is of strongly positive type.

If it is possible to assign to each w_0 in some set a unique continuous solution $\boldsymbol{W}(t)w_0 = w(t, w_0)$ of the equation (23.60) defined on $[0, t_0]$, it is said that on this set the *quasi-translation operator* $\boldsymbol{W}(t)$ of the problem (23.19)–(23.20) is defined.

Let the hypotheses of Theorem 23.6 be fulfilled. Let the operator $\varphi(t, w)$ admit the representation

$$\varphi(t, w) = \boldsymbol{B}(t)w + \omega(t, w), \tag{23.61}$$

where $\boldsymbol{B}(t)$ is a bounded linear operator depending strongly continuously on $t \in [0, t_0]$, and the operator $\omega(t, w)$ satisfies the condition

$$\lim_{\rho\to 0} \sup_{0 \leqq t \leqq t_0,\, \|w\| \leqq \rho} \frac{\|\omega(t, w)\|}{\|w\|} = 0. \tag{23.62}$$

Then the quasi-translation operator $w(t, w_0)$ is defined on some ball $\|w_0\| \leqq r$. On this ball $w(t, w_0)$ satisfies a Lipschitz condition (uniformly in t) and admits a representation

$$w(t, w_0) = \boldsymbol{B}_1(t)w_0 + \omega_1(t, w_0), \tag{23.63}$$

where $\boldsymbol{B}_1(t)$ is a bounded linear operator depending strongly continuously on $t \in [0, t_0]$, and the operator $\omega_1(t, w_0)$ satisfies the condition

$$\lim_{\rho\to 0} \sup_{0 \leqq t \leqq t_0,\, \|w_0\| \leqq \rho} \frac{\|\omega_1(t, w_0)\|}{\|w_0\|} = 0. \tag{23.64}$$

If, further, the operator A^{-1} is compact in E then the quasi-translation operator $w(t, w_0)$ is completely continuous for each $t \in [0, t_0]$.

All these assertions are proved in the same manner as Theorems 23.10–23.13.

Finally, let

$$\varphi(t, w) = \boldsymbol{B}(t)w + \boldsymbol{C}(t)[w, w] + \boldsymbol{D}(t, w), \tag{23.65}$$

where $\boldsymbol{B}(t)$ is a bounded linear operator depending strongly continuously on $t \in [0, t_0]$, $\boldsymbol{C}(t)[w_1, w_2]$ is a bilinear operator depending strongly continuously on $t \in [0, t_0]$, and the non-linear operator $\boldsymbol{D}(t, w)$ satisfies the condition

$$\lim_{\rho \to 0} \sup_{0 \leqq t \leqq t_0,\, \|w\| \leqq \rho} \frac{\|\boldsymbol{D}(t, w)\|}{\|w\|^2} = 0. \tag{23.66}$$

Then the quasi-translation operator admits a representation

$$w(t, w_0) = \boldsymbol{B}_1(t)w + \boldsymbol{C}_1(t)[w_0, w_0] + \boldsymbol{D}_1(t, w_0), \tag{23.67}$$

where $\boldsymbol{B}_1(t)w$, $\boldsymbol{C}_1(t)[w_1, w_2]$, $\boldsymbol{D}_1(t, w)$ possess the same properties on the ball $\|w\| \leqq r$, as $\boldsymbol{B}(t)w$, $\boldsymbol{C}(t)[w_1, w_2]$, $\boldsymbol{D}(t, w)$ respectively.

The proof of this assertion is similar to the proof of Theorem 23.14.

23.8 *Equations with variable operators*[1]

Now consider the equation

$$\frac{dv}{dt} + \boldsymbol{A}(t)v = f(t, v) \qquad (0 < t \leqq t_0) \tag{23.68}$$

with a variable unbounded operator $\boldsymbol{A}(t)$. An investigation of this equation can be made by the method presented above if it is possible to construct

[1] The theory of such equations was initiated by T. Kato [1, 2]. Further results are due to P. E. Sobolevskii [2, 6] and T. Kato [6]. A complete exposition of the theory of equations with variable operators with a constant domain is found in P. E. Sobolevskii [7].

the solution

$$v(t) = U(t, s)v_s \qquad (0 \leqq s \leqq t \leqq t_0)$$

of the linear equations

$$\frac{dv}{dt} + A(t)v = 0, \quad v(s) = v_s,$$

and if the operator $U(t, s)$ possess sufficiently good properties.

Once the operator $U(t, s)$ has been constructed, a generalized solution of equation (23.68) satisfying the initial condition $v(0) = v_0$, is determined as a continuous solution of the following equation which is analogous to (23.21):

$$v(t) = U(t, 0)v + \int_0^t U(t, s)f[s, v(s)]ds. \tag{23.69}$$

If in addition the operator $f(t, v)$ is unbounded, it is natural to seek a solution $v(t)$ in the form (23.34), where $w(t)$ is determined from the following equation, which is analogous to (23.33)

$$w(t) = U(t, 0)A^{\alpha}(0)v_0$$

$$+ \int_0^t A^{\alpha}(0)U(t, s)f[s, A^{-\alpha}(0)w(s)]ds. \tag{23.70}$$

The principal difficulties with this approach arise in the construction and investigation of the operator $U(t, s)$. Such a construction and investigation can be successfully achieved under fairly general assumptions: for instance it suffices, if for each fixed t, $-A(t)$ is the generator of a strongly continuous semi-group $T_t(\tau)$ satisfying the inequality $\|T_t(\tau)\| \leqq e^{-\delta\tau}$; if the domains of the operators $A(t)$ do not depend on t; and if the operator-function $A(t)A(0)^{-1}$ is strongly continuously differentiable in t. Transforming to the equation (23.70) can only be justified under more severe restrictions on $A(t)$; for instance, it is sufficient that the operators $-A(t)$ are of strongly positive type and that the operator-function $A(t)A(0)^{-1}$ satisfies the Hölder condition.

If a solution of equation (23.68) is defined, the translation operator along the trajectory of this equation can be defined. The properties of this operator are similar to the properties of the translation operator along the trajectory of an equation with a constant operator, which was studied in 23.5°–23.7°.

23.9 *The translation operator and periodic solutions of parabolic equations*[1]

Let us return to the study of the non-linear equation

$$\frac{dv}{dt} + Av = f(t, v) \tag{23.71}$$

considered in 23.3°–23.6°.

Assume that the right side of equation (23.71) is periodic in t:

$$f(t + \omega, v) = f(t, v). \tag{23.72}$$

It is easy to see that the problem of existence of ω-periodic solutions of the equation (23.71) is equivalent to the problem of existence of fixed points for the translation operator $V(t)$ of amount ω along the trajectory of equation (23.71).

Suppose that the displacement operator is completely continuous (see Theorem 23.12). In addition, let the following condition be fulfilled

$$\lim_{R\to\infty} \sup_{0\leqq t\leqq\omega,\ \|v\|\leqq R} \frac{\|f(t, v)\|}{R} = 0. \tag{23.73}$$

Then, as is easily verified, the translation operator satisfies the hypotheses for Schauder's fixed point principle. Hence the translation operator has at least one fixed point, which guarantees the existence of a periodic solution of the equation (23.71). Thus condition (23.73) is sufficient for the existence of at least one periodic solution.

More refined methods of non-linear functional analysis can also be applied to the study of the translation operator: the theory of completely

---

[1] M. A. Krasnoselskii [10], M. A. Krasnoselskii and P. E. Sobolevskii [4].

continuous vector fields, the methods of the theory of cones, the theory of concave operators, etc. These methods permit one to estimate the number of periodic solutions from above and below, to study the dependence of these solutions on a parameter, to investigate bifurcation problems, etc. A more detailed analysis of the translation operator also permits one to establish the stability or instability of periodic solutions and other properties.

A similar situation occurs in cases where only the quasi-translation operator $W(t)$ can be constructed, for the trajectories of equation (23.71) with the unbounded operator $f(t, v)$.

As was already mentioned in 23.7°, the translation operator $V(t)$ in this case is defined, generally speaking, only on a dense set in E. This circumstance makes it difficult to investigate the operator $V(t)$. But the quasi-translation operator $W(t)$ is defined, under reasonably general conditions, on the entire space. It is easy to see that the operators $V(t)$ and $W(t)$ are connected by the relation

$$W(t)v_0 = A^{\alpha}V(t)A^{-\alpha}v_0 \quad (v_0 \in E). \tag{23.74}$$

This relation implies that each fixed point of the quasi-translation operator determines the initial value

$$v(0) = A^{-\alpha}v_0$$

of a ω-periodic solution of the equation (23.71). Thus, the analysis of various problems, associated with ω-periodic solutions of the equation (23.71) can be achieved by investigating the quasi-translation operator.

It is clear that all these considerations also apply to equations with a variable operator $A(t)$.

The equations in Banach spaces considered in this paragraph contain as is known, various boundary problems for partial differental equations. For instance, they contain mixed boundary value problems for quasi-linear parabolic equations. Hence the methods presented can be applied to the proof of existence of periodic solutions of quasi-linear parabolic equations. Unfortunately, various technical difficulties arise in the course of such proofs, and a large amount of calculation is necessary—such matters are beyond the scope of this book.

Bibliography

AGMON, S.

[1] On the eigenfunctions and on the eigenvalues of general elliptic boundary value problems, *Comm. Pure Appl. Math.* 15 (1962) 119–147. MR 26 # 5288.

AGMON, S., A. DOUGLIS and L. NIRENBERG

[1] Estimates near the boundary for solutions of elliptic partial differential equations satisfying boundary conditions I–II, *Comm. Pure Appl. Math.* 12 (1959) 623–727; 17 (1964) 35–92. MR 23 # A2610, 28 # 5252.

AKHIEZER, N. I. and I. M. GLAZMAN

[1] *Theory of linear operators in Hilbert space* (English translation), Ungar, New York, 1962.

ANDO, T.

[1] On compactness of integral operators, *Indag. Math.* 24 (1962) 235–239. MR 25 # 464.

BABIC, V. M.

[1] A problem on imbedding theorems in case of limiting exponents, *Vestnik Leningrad. Univ.* 11 (1956) 186–188. MR 18–877.

BAHTIN, I. A., M. A. KRASNOSELSKII and V. Y. STECENKO

[1] On the continuity of positive linear operators, *Sibirsk. Mat. Z.* 3 (1962) 156–160. MR 25 # 2451.

BALAKRISHNAN, A. V.

[1] An operational calculus for infinitesimal generators of semi-groups, *Trans. Amer. Math. Soc.* 91 (1959) 330–353. MR 21 # 5904.

[2] Fractional powers of closed operators and semi-groups generated by them, *Pacific J. Math.* 10 (1960) 419–437. MR 22 # 5899.

BANACH, S.

[1] *Théorie des opérations linéaires* (Reprint), Chelsea, New York, 1955.

BARI, N. K.

[1] *Trigonometric series* (English translation), Pergamon, New York, 1965.

BENEDEK, A. and R. PANZONE

[1] The spaces L^p with mixed norm, *Duke Math. J.* 28 (1961) 301–324. MR 23 # A3451.

BEREZANSKII, JU. M.

[1] Eigenfunction expansions of selfadjoint operators, *Mat. Sb.* (*N.S.*) 43 (85) (1957) 75–126. MR 21 # 3646.

[2] On eigenfunction expansions of selfadjoint operators, *Ukrain. Mat. Z.* 11 (1959) 16–24. MR 23 # A518.

[3] *Expansions in eigenfunctions of selfadjoint operators* (English translation), Amer. Math. Soc., Providence, 1968.

BEREZANSKII, IU. M. and U. B. OROCKO

[1] A remark on the growth of eigenfunctions of selfadjoint operators, *Ukrain. Mat. Z.* 14 (1962) 180–184. MR 26 # 2880.

CALDERON, A. P. and A. ZYGMUND

[1] A note on the interpolation of linear operators, *Studia Math.* 12 (1951) 194–204. MR 13-754.

CARATHEODORY, C.

[1] *Vorlesungen über reelle Funktionen* (Reprint), Chelsea, New York, 1968.

CHANG, S. T.

[1] Continuity and complete continuity of the Uryson operator, *Acta Math. Sinica* 13 (1963) 204–215. MR 28 # 5357.

COURANT, R. and D. HILBERT

[1] *Methods of mathematical physics* (English translation) Interscience, New York, 1962.

DAY, M. M.

[1] The spaces L_p with $0 < p < 1$, *Bull. Amer. Math. Soc.* 46 (1940) 816–823. MR 2-102.

DUNFORD, N. and J. T. SCHWARTZ

[1] *Linear operators I*, Interscience, New York, 1958.

GAGLIARDO, E.

[1] Proprietá di alcune classi di funzioni in piu variabili, *Ricerche Mat.* 7 (1958) 102–137. MR 21 # 4290.

GELFAND, I. M.

[1] Abstrakte Funktionen und lineare Operatoren, *Math. Sb.* (*N.S.*) 4 (46) (1938) 235–286.

GLUSHKO, V. P.

[1] On operators of potential type and some imbedding theorems, *Dokl. Akad. Nauk SSSR* 126 (1959) 467–470. MR 21 # 7369.

GLUSHKO, V. P. and S. G. KREIN

[1] Fractional powers of differential operators and imbedding theorems, *Dokl. Akad. Nauk SSSR* 122 (1958) 963–966. MR 20 # 6578.

GOLOMB, M.

[1] Zur Theorie der nichtlinearen Integralgleichungen, Integralgleichungssysteme und allgemeinen Funktionalgleichungen, *Math. Zeit.* 39 (1934) 45–75.

[2] Über System von nichtlineare Integralgleichungen, *Publ. Math. Univ. Beograde* 5 (1936) 52–83.

HAAR, A.

[1] Zur Theorie der orthogonalen Funktionensysteme, *Math. Ann.* 69 (1910) 331–337.

HALMOS, P. R.

[1] *Measure theory*, Springer, Berlin, 1974.

HAMMERSTEIN, A.

[1] Nichtlineare Integralgleichungen nebst Anwendungen, *Acta Math.* 54 (1930) 117–176.

HARDY, G., D. LITTLEWOOD and G. POLYA

[1] *Inequalities*, Univ. Press, Cambridge, 1952.

HEINZ, E.

[1] Beitrage zur Störungstheorie der Spektralzerlegung, *Math. Ann.* 123 (1951) 415–438. MR 13–471.

HILLE, E. and R. S. PHILLIPS

[1] *Functional analysis and semi-groups*, rev. ed., Amer. Math. Soc. Providence, 1974.

ILIN V. A.

[1] Kernels of fractional order, *Mat. Sb.* (*N.S.*) 41 (83) (1957) 459–480. MR 19–661.

[2] On convergence of expansions in eigenfunctions of the Laplace operator, *Uspehi Mat. Nauk* (*N.S.*) 13 (1958) no. 1 (79), 87–180. MR 20 # 1828.

ILIN, V. P.

[1] On an imbedding theorem for a limiting exponent, *Dokl. Acad. Nauk SSSR* 96 (1954) 905–908. MR 16–121.

[2] Some functional inequalities of the type of imbedding theorems, *Dokl. Akad. Nauk SSSR* 123 (1958) 967–970. MR 21 # 103.

[3] Imbedding theorems, *Trudy Mat. Inst. Steklov* 53 (1959) 359–386. MR 22 # 3776.

KANTOROVIC, L. V.

[1] On integral operators, *Uspehi Mat. Nauk* (*N.S.*) 7 (1956) no. 2 (68), 3–29. MR 20 # 5432.

KANTOROVIC, L. V. and G. P. AKILOV

[1] *Functional analysis in normed spaces* (English translation), Pergamon, New York, 1964.

KANTOROVIC, L. V., B. Z. VULIKH and A. G. PINSKER

[1] *Functional analysis in partially ordered spaces*, Gosudarstv. Izdat. Tehn.-Teor. Lit., Moscow, 1950.

KATO, T.

[1] Note on some inequalities for linear operators, Math. Ann. 125 (1952) 208–212. MR 14-766.

[2] *Quadratic forms in Hilbert spaces and asymptotic perturbation*, Univ. of California, 1955. MR 17-514.

[3] Integration of the equation of evolution in a Banach space, *J. Math. Soc. Japan* 5 (1953) 208–234. MR 15-437.

[4] Note on fractional powers of dissipative operators, *Proc. Japan Acad.* 36 (1960) 94–96. MR 22 # 12400.

[5] Fractional powers of dissipative operators, I–II, *J. Math. Soc. Japan* 13 (1961) 246–274; 14 (1962) 242–248. MR 25 # 1453, 27 # 1851.

[6] Abstract evolution equations of parabolic type in Banach and Hilbert spaces, *Nagoya Math. J.* 19 (1961) 93–125. MR 26 # 631.

KOLESOV, Y. S.

[1] On some criteria for existence of stable periodic solutions of quasi-linear parabolic equations, *Dokl. Akad. Nauk SSSR* 157 (1964) 1288–1290. MR 29 # 3761.

KOLMOGOROV, A. N.

[1] Über Kompaktheit der Funktionenmengen bei der Konvergenz in Mittle, *Nachr. Ges. Göttingen Math.-Phys. Kl.* (1931) 60–63.

KOROTKOV, V. B.

[1] On integral operators with kernels of Carleman type, *Dokl. Akad. Nauk SSSR* 165 (1965) 748–751. MR 35 # 816.

KRASNOSELSKII, M. A.

[1] Criteria for continuity of some non-linear operators, *Ukrain. Mat. Z.* 2 (1950) 70–86. MR 13-954.

[2] Eigenfunctions of non-linear operators, asymptotically close to linear, *Dokl. Akad. Nauk SSSR* 74 (1950) 177–179. MR 12-187.

[3] Continuity of operator $\mathfrak{f}u(x) = f[x, u(x)]$, *Dokl. Akad. Nauk SSSR* 77 (1951) 185–188. MR 12 # 836.

[4] Splitting of operators acting from a space L_q to a space L_p, *Dokl. Akad. Nauk SSSR* 82 (1952) 333–336. MR 13–661.

[5] Properties of a root of a linear integral operator, *Dokl. Akad. Nauk SSSR* 88 (1953) 749–751. MR 14-1092.

[6] On a boundary value problem, *Izv. Akad. Nauk SSSR Ser. Mat.* 20 (1956) 241–252. MR 20 # 2506.

[7] *Topological methods in the theory of non-linear integral equations* (English translation), Pergamon, New York, 1964.

[8] On a theorem of M. Riesz, *Dokl. Akad. Nauk SSSR* 131 (1960) 246–248. MR 22 # 9852.

[9] *Positive solutions of operator equations* (English translation), Noordhoff, Groningen, 1964.

[10] *The operator of translation along the trajectories of differential equations* (English translation), Amer. Math. Soc. Providence, 1968.

KRASNOSELSKII, M. A. and S. G. KREIN

[1] Criteria for continuity and compactness of a linear operator in terms of properties of its square. *Voronez. Gos. Univ. Trudy Sem. Funkcional Anal.* 5 (1957) 98–101. MR 20 # 4784.

[2] *On differential equations in Banach spaces*, Proc. Third All-Union Math. Congress, 1958.

KRASNOSELSKII, M. A., S. G. KREIN and P. E. SOBOLEVSKII

[1] On differential equations with unbounded operators in Banach spaces, *Dokl. Akad. Nauk SSSR* 111 (1956) 19–22. MR 19-550.

[2] On differential equations with unbounded operators in Hilbert spaces, *Dokl. Akad. Nauk SSSR* 112 (1957) 990–993. MR 19-747.

KRASNOSELSKII, M. A. and L. A. LADYZENSKII

[1] Conditions for complete continuity of Uryson operators acting in a space L^p, *Trudy Moskov. Mat. Obsc.* 3 (1954) 307–320. MR 15–966.

[2] On the scope of notions of a u_0-concave operator, *Vyss. Ucebn. Zaved. Matematika* (1959) no. 5, 112–121. MR 24 # A1043.

KRASNOSELSKII, M. A., A. I. PEROV, A. I. PEVOLOCKII and P. P. ZABREIKO

[1] *Vector fields on the plane* (English translation), Academic, New York, 1966.

KRASNOSELSKII, M. A. and E. I. PUSTYLNIK

[1] Use of fractional powers of operators in study of Fourier expansions in eigenfunctions of differential operators, *Dokl. Akad. Nauk SSSR* 122 (1958) 978–981. MR 20 # 6579.

[2] On criteria for compactness of linear and non-linear integral operators, *Dokl. Akad. Nauk SSSR* 142 (1962) 25–28. MR 24 # A1615.

[3] Two theorems on integral operators, *Voronez. Gos. Univ. Trudy Sem. Funkcional Anal.* 7 (1963).

[4] On conditions of compactness of a non-linear integral operator, *Voronez. Gos. Univ. Trudy Sem. Funkcional Anal.* 7 (1963).

KRASNOSELSKII, M. A. and Y. B. RUTICKII

[1] Linear integral operators, acting in Orlicz spaces, *Dokl. Akad. Nauk SSSR* 85 (1952) 33–36. MR 14-57.

[2] Differentiability of non-linear integral operators in Orlicz spaces, *Dokl. Akad. Nauk SSSR* 89 (1953) 601–604. MR 15-137.

[3] Linear integral operators acting in Orlicz spaces, *Voronez. Gos. Univ. Trudy Sem. Funkcional Anal.* 2 (1956) 55–76. MR 18–811.

[4] On some non-linear operators in Orlicz spaces, *Dokl. Akad. Nauk SSSR* 117 (1957) 363–366. MR 20 # 1933.

[5] *Convex functions and Orlicz spaces* (English translation) Noordhoff, Groningen, 1962.

[6] Orlicz spaces and non-linear integral equations, *Trudy Moskow Mat. Obsc.* 7 (1958) 63–120. MR 21 # 7433.

KRASNOSELSKII, M. A., Y. B. RUTICKII and R. M. SULTANOV

[1] On a non-linear operator acting in a space of abstract functions, *Izv. Akad. Nauk Azerbaidyan SSR Ser. Fiz.-Tehn. Mat. Nauk.* (1959) no. 3, 15–21. MR 21 # 7458.

KRASNOSELSKII, M. A. and V. I. SOBOLEV

[1] On splitting of linear operators, *Uspehi Mat. Nauk* (*N.S.*) 12 (1957) no. 4 (68), 313–317. MR 19–666.

KRASNOSELSKII, M. A. and P. E. SOBOLEVSKII

[1] Fractional powers of operators, acting in Banach spaces, *Dokl. Akad. Nauk SSSR* 129 (1959) 499–502. MR 21 # 7447.

[2] Structure of the set of solutions of parabolic equations, *Dokl. Akad. Nauk SSSR* 146 (1962) 26–29. MR 27 # 2703.

[3] Structure of the set of solutions of parabolic equations, *Ukrain. Mat. Z.* 16 (1964) 319–333. MR 29 # 3763.

[4] *On some non-linear problems for partial differential equations*, Report at Soviet-America Symposium on Partial Differential Equations, Novosivisk, 1963.

KREIN, M. G.

[1] On compact linear operators in functional spaces with two norms, *Trudy Inst. Mat. Akad. Nauk Ukrain. SSRS* 9 (1947).

KREIN, M. G. and M. A. RUTMAN

[1] Linear operators leaving invariant a cone in a Banach space, *Uspehi Mat. Nauk* (*N.S.*) 3 (1948) no. 1 (23), 3–95. MR 10–256.

KREIN, S. G. and E. M. SEMENOV

[1] On a scale of spaces, *Dokl. Akad. Nauk SSSR* 138 (1961) 763–766. MR 25 # 4352.

KREIN, S. G. and P. E. SOBOLEVSKII

[1] Differential equations with abstract elliptic operators in Hilbert space, *Dokl. Akad. Nauk SSSR* 118 (1958) 233–236. MR 20 # 6043.

LADYZENSKAYA, O. A.

[1] *The mixed problems for a hyperbolic equation*, Gosdarst. Izdat. Tehn.-Teor. Lit., Moscow, 1953. MR 17-160.

LADYZENSKII, L. A.

[1] Conditions of complete continuity of an Uryson integral operator acting in space of continuous functions, *Dokl. Akad. Nauk SSSR* 96 (1954) 1105–1108. MR 16-48.

LJUBIC, YU. I.

[1] On the belonging of powers of an operator on a given vector to a certain linear class, *Dokl. Akad. Nauk SSSR* 102 (1955) 881–884. MR 17-176.

[2] On inequalities among powers of a linear operator, *Izv. Akad. Nauk SSSR Ser. Mat.* 24 (1960), 825–864. MR 24 # A436.

LORENTZ, G. G.

[1] Some new functional spaces, *Ann. of Math.* 51 (1950) 37–55. MR 11-442.

LUXEMBURG, W. A. J.

[1] *Banach function spaces*, Thesis, Delft, 1955. MR 17–285.

LUXEMBURG, W. A. J. and A. C. ZAANEN

[1] Some remarks on Banach function spaces, *Indag. Math.* 18 (1956) 110–119. MR 17-987.

[2] Notes on Banach function spaces, I–IV, *Indag. Math.* 25 (1963) 135–153; 239–263. MR 26 # 6723; 27 # 5119.

LJUSTERNIK, L. A. and V. I. SOBOLEV

[1] *Elements of functional analysis* (English translation) Ungar, New York, 1961.

MARCINKIEWICZ, Y.

[1] Sur l'interpolation d'opérateurs, *C. R. Acad. Sci. Paris* 208 (1939) 1272–1273.

MARKUSHEVIC, A. I.

[1] *Theory of analytic functions* (English translation), Prentice Hall, Englewood Cliffs, 1964.

MAZJA, V. G. and P. E. SOBOLEVSKII

[1] On generators of semi-groups, *Uspehi Mat. Nauk* (*N.S.*) 17 (1962) no. 6 (108), 151–154. MR 27 # 1838.

MELAMED, V. B.

[1] On computing the rotation of a completely continuous vector field in a critical case, *Sibirsk. Mat. Z.* 2 (1961) 414–427. MR 24 # A2227.

MIKHLIN, S. G.

[1] *Integral equations* (English translation), Gordon and Breach, New York, 1960.

[2] *The problem of the minimum of a quadratic functional* (English translation) Holden-Day, San-Francisco, 1965.

MIRANDA, C. Z.

[1] *Partial differential equations of elliptic type*, Springer, Berlin, 1955.

MIYADERA, I.

[1] On one-parameter semi-groups of operators, *J. Math. Tokyo* 1 (1951) 23–26. MR 14-564.

NAIMARK, M. A.

[1] *Linear differential operators* (English translation) Ungar, New York, 1967.

NATANSON, I. P.

[1] *Theory of functions of a real variable* (English translation) Ungar, New York, 1955–1960.

NEMYCKII, V. V.

[1] Théorèmes d'existence et d'unicité des solutions de quelques équations intégrales non-linéaires, *Mat. Sb.* 41 (1934) 438–452.

[2] On a class of non-linear integral equations, *Mat. Sb.* 41 (1934) 655–658.

NIRENBERG, L.

[1] Remarks on strongly elliptic partial differential equations, *Comm. Pure Appl. Math.* 8 (1955) 649–675. MR 17-742.

NUREKENOV, T.

[1] On a criterion for the complete continuity of an Uryson integral operator, *Izv. Akad. Nauk Kazav SSR* 15 (1965) no. 3, 13–16. MR 28 # 4325.

PETROVSKII, I. G.

[1] *Lectures on the theory of integral equations* (English translation) Graylock, Rochester N.Y., 1957.

PUSTYLNIK, E. I.

[1] On a representation of linear compact operators acting in Banach spaces, *Izv. Vyss. Ucebn. Zaved. Matematika* (1960) no. 2, 149–153. MR 24 # A1026.

[2] On fractional powers of unbounded operators, *Dokl. Akad. Nauk Ukrain. SSR* 10(1961) 1266–1270. MR 24 # A2249.

[3] On integral operators, acting in spaces L_p, *Dokl. Akad. Nauk SSSR* 146 (1962) 1271–1274. MR 26 # 6818.

[4] Interpolation theorems in spaces L_p with $p < 1$, *Sibirsk. Mat. Z.* 4 (1963) 318–324. MR 33 # 7862.

[5] On convergence of eigenfunction expansions of a compact operator in Banach spaces, *Sibirsk. Mat. Z.* 4 (1963) 705–708. MR 28 # 475.

[6] Operators of potential type for the case of sets with abstract measures, *Dokl. Akad. Nauk SSSR* 156 (1964) 519–520.

RIEMENSCHNEIDER, S. D.

[1] The L-characteristics of linear operators on $L^{1/\alpha}([0, 1])$, *J. Functional Analysis* 8 (1971) 405–421. MR 44 # 7329.

RIESZ, F.

[1] Untersuchungen über Systeme integrierbarer Funktionen, *Math. Ann.* 69 (1910) 449–497.

[2] Über lineare Funktionalgleichungen, *Acta. Math.* 41 (1917) 71–98.

RIESZ, F. and B. SZ.-NAGY

[1] *Functional analysis* (English translation) Ungar, New York, 1955.

RIESZ, M.

[1] Sur les maxima des formes bilinéaires et sur les fonctionnelles linéaires, *Acta Math.* 49 (1926) 465–479.

RUTICKII, Y. B.

[1] On a non-linear operator in Orlicz spaces, *Dokl. Akad. Nauk Ukrain. SSR* 3 (1952) 161–166. MR 15-719.

[2] On a property of a compact linear operator acting in Orlicz spaces, *Uspehi Mat. Nauk* (*N.S.*) 11 (1956) no. 2 (68), 201–208. MR 17–1113.

[3] *On integral equations with exponential non-linearity*, Proc. Third All-Union Math. Congress, 1959.

[4] Integral operators in Orlicz spaces, *Dokl. Akad. Nauk SSSR* 145 (1962) 1000–1003. MR 25 # 5393.

[5] New criteria for continuity and compactness of integral operators in Orlicz spaces, *Izv. Vyss. Ucebn. Zaved. Matematika* (1962) no. 5, 87–100. MR 28 # 1483.

[6] Scales of Orlicz spaces and interpolation theorems, *Dokl. Akad. Nauk SSSR* 149 (1963) 32–35. MR 28 # 3315.

[7] Scales of Orlicz spaces and interpolation theorems, *Voronez. Gos. Univ. Trudy Sem. Funkcional Anal.* 7 (1963).

SEMENOV, E. M.

[1] On a scale of spaces with interpolation properties, *Dokl. Akad. Nauk SSSR* 148 (1963) 1038–1041. MR 26 # 2870.

SLOBODECKII, L. N.

[1] Sobolev spaces and their applications to boundary value problems of partial differential equations, *Leningrad. Gos. Univ. Ucen. Zap. Ser. Mat. Nauk* 197 (1958).

[2] Estimates of solutions of elliptic systems in L_p, *Dokl. Akad. Nauk SSSR* 123 (1958) 616–619. MR 21 # 5061.

[3] Estimates of solutions of elliptic and parabolic systems in L_2, *Vestnik Leningrad Univ.* 15 (1960) 28–47. MR 22 # 2794.

SMIRNOV, V. I.

[1] *A course of higher mathematics* (English translation) Pergamon, New York, 1964.

SMOLICKII, K. L.

[1] On summability of potentials, *Uspehi Mat. Nauk* (*N.S.*) 12 (1957) no. 4 (76), 349–356. Mr 20 # 5433.

SOBOLEV, V. I.

[1] On splitting of linear operators, *Dokl. Akad. Nauk SSSR* 111 (1956) 951–954. MR 18–912.

SOBOLEV, S. L.

[1] On a theorem of functional analysis, *Mat. Sb.* (*N.S.*) 4 (46) (1938) 471–497.

[2] *Some applications of functional analysis in mathematical physics* (English translation), Amer. Math. Soc., Providence, 1963.

SOBOLEVSKII, P. E.

[1] On equations with operators forming an acute angle, *Dokl. Akad. Nauk SSSR* 116 (1957) 754–757. MR 20 # 4194.

[2] On first-order differential equations in Hilbert space with a variable positive definite selfadjoint operator, whose fractional powers have the same domain, *Dokl. Akad. Nauk SSSR* 123 (1958) 984–987. MR 23 # A3905.

[3] On an inequality for elliptic operators, *Voronez. Gos. Univ. Trudy Sem. Funkcional Anal.* 6 (1958).

[4] On non-stationary equations of viscous fluid dynamics, *Dokl. Akad. Nauk SSSR* (1959) 45–48. MR 22 # 1763.

[5] On logarithm of normally positive type operators, *Uspehi Mat. Nauk* (*N.S.*) 16 (1961) no. 4 (100), 206–208.

[6] On equations of parabolic type in Banach spaces with a unbounded variable operator, whose fractional powers have the same domain, *Dokl. Akad. Nauk SSSR* 138 (1961) 59–62. MR 28 # 2360.

[7] Estimates of Green functions of second order partial differential equations of parabolic type, *Dokl. Akad. Nauk SSSR* 138 (1961) 313–316. MR 23 # A3379.

[8] On Green functions of arbitrary fractional powers of elliptic operators, *Dokl. Akad. Nauk SSSR* 142 (1961) 804–807. MR 25 # 311.

[9] On local and non-local existence theorems for second order non-linear parabolic equations, *Dokl. Akad. Nauk SSSR* 136 (1961) 292–295. MR 25 # 5291.

[10] On equations of parabolic type in Banach spaces, *Trudy Moskov. Mat. Obsc.* 10 (1961) 297–350. MR 25 # 5297.

[11] *Investigations of partial differential equations with use of fractional powers of operators*, Proc. Fourth All-Union Math. Congress, Leningrad, 1961.

[12] *Theory of fractional powers of operators in Banach spaces and its applications to the study of equations of parabolic type*, Doctor Dissertation, 1962.

[13] On application of method of fractional powers of selfadjoint operators to investigation of Navier-Stokes equations, *Dokl. Akad. Nauk SSSR* 155 (1964) 50–53. MR 28 # 4258.

[14] On fractional powers of weakly positive type operators, *Dokl. Akad. Nauk SSSR* 166 (1965) 1296–1299. MR 33 # 3137.

SOLOMJAK, M. Z.

[1] Application of theory of semi-groups to the study of differential equations in Banach spaces, *Dokl. Akad. Nauk SSSR* 112 (1958) 766–769. MR 21 # 3775.

STEIN, E. M.

[1] Interpolation of linear operators, *Trans. Amer. Math. Soc.* 83 (1956) 482–492. MR 18-575.

STEIN, E. M. and G. WEISS

[1] An extension of a theorem of Marcinkiewicz and some of its applications, *J. Math. Mech.* 8 (1959) 263–284. MR 21 # 5888.

TAMARKIN, J. D.

[1] On the compactness of the space L_p, *Bull. Amer. Math. Soc.* 38 (1932) 79–84.

TAMARKIN, J. D. and A. ZYGMUND

[1] Proof of a theorem of Thorin, *Bull. Amer. Math. Soc.* 50 (1944) 219–282. MR 5-229.

THORIN, E. O.

[1] Convexity theorems generalizing those of M. Riesz and Hadamard, *Comm. Sem. Math. Univ. Lund* 9 (1948) 1–58. MR 10-21.

URYSON, P. S.

[1] On a type of non-linear integral equations, *Mat. Sb.* 31 (1924) 236–255.

VAINBERG, M. M.

[1] On the continuity of some operators of a special form, *Dokl. Akad. Nauk SSSR* 73 (1950) 253–255. MR 12–111.

[2] Some problems of the differential calculus in linear spaces, *Uspehi Mat. Nauk* (*N.S.*) 7 (1952) no. 4 (50), 55–102. MR 14–384.

[3] On the structure of an operator, *Dokl. Akad. Nauk SSSR* 92 (1953) 213–216. MR 15–439.

[4] On some properties of quadratic forms in spaces L_q, *Dokl. Akad. Nauk SSSR* 100 (1955) 845–848. MR 16–934.

[5] Nemyckii operators, *Ukrain. Mat. Z.* 7 (1955) 363–378. MR 16–934.

[6] *Variational methods for study of non-linear operators* (English translation) Holden-Day, San Francisco, 1965.

VALLÉE-POUSSIN, CH.

[1] *Cours d'analyse infinitesimal* (Reprint) Dover, New York, 1946.

WANG, S. W.

[1] Differentiability of Nemyckii operator, *Dokl. Akad. Nauk SSSR* 150 (1963) 1198–1201. MR 27 # 1794.

[2] The converse of a theorem of Ladyzenskii on the complete continuity of the Uryson operator, *Acta Math. Sinica* 13 (1963) 254–261. MR 30 # 3351.

WEISS, G.

[1] An interpolation theorem of sublinear operators on H_p spaces, *Proc. Amer. Math. Soc.* 8 (1957) 92–99. MR 19-869.

YOSIDA, K.

[1] Fractional powers of infinitesimal generators and the analyticity of the semi-groups generated by them, *Proc. Japan Acad.* 36 (1960) 86–89. MR 22 # 12399.

[2] *Semi-group theory and the integration problem of diffusion equations*, Proc. Intern. Congress Math., Amsterdam, 1954.

ZABREIKO, P. P.

[1] On continuity of a non-linear operator, *Sibirsk. Mat. Z.* 5 (1960) 958–960. MR 30 # 473.

[2] On some properties of linear operators acting in spaces L_p, *Dokl. Akad. Nauk SSSR* 159 (1964) 975–977. MR 30 # 472.

[3] On compactness of u_0-bounded linear operators in spaces L_p, Functional analysis and theory of function 2, *Ucen. Zap. Kazan Univ.* 124 (1965) 110–113. MR 32 # 6233.

[4] On continuity and complete continuity of non-linear Uryson operators, *Dokl. Akad. Nauk SSSR* 161 (1965) 1007–1010. MR 31 # 2610.

[5] On differentiability of non-linear operators in spaces L_p, *Dokl. Akad. Nauk SSSR* 166 (1966) 1039–1042. MR 33 # 1772.

ZABREIKO, P. P. and M. A. KRASNOSELSKII

[1] Calculation of the index of a fixed point of a completely continuous vector field, *Sibirsk. Mat. Z.* 5 (1964) 509–531. MR 29 # 3839.

[2] On the *L*-characteristics of operators, *Uspehi Mat. Nauk* (*N.S.*) 19 (1964) no. 5 (119), 187–189.

ZABREIKO, P. P., M. A. KRASNOSELSKII and E. I. PUSTYLNIK

[1] On fractional powers of elliptic operators, *Dokl. Akad. Nauk SSSR* 165 (1965). 990–993, MR 33 # 406.

[2] A problem of fractional powers of operators, *Uspehi Mat. Nauk* (*N.S.*) 20 (1965) no. 6 (126), 87–89. MR 33 # 7867.

[3] On a proof of Mercer's theorem, *Bul. Akad. Stiince RSS Moldoven* 7 (1965) 98–100. MR 33 # 1767.

ZABREIKO, P. P. and E. I. PUSTYLNIK

[1] On continuity and complete continuity of non-linear integral operators in spaces L_p, *Uspehi Mat. Nauk* (*N.S.*) 19 (1964) no. 2 (116), 204–205.

[2] On interpolation properties of compactness, Functional analysis and theory of function 2, *Ucen. Zap. Kazan Univ.* 124 (1965) 114–118. MR 32 # 6207.

ZAANEN, A. C.

[1] On a certain class of Banach spaces, *Ann. of Math.* 47 (1946) 654–666. MR 8-158.

[2] Note on a certain class of Banach spaces, *Indag. Math.* 11 (1949) 148–158. MR 11-38.

[3] *Linear analysis*, Noordhoff, Groningen, 1953.

[4] Integral transformations and their resolvents in Orlicz and Lebesgue spaces, *Compositio Math.* 10 (1952) 56–94. MR 14-767.

ZYGMUND, A.

[1] On a theorem of Marcinkiewicz concerning interpolation theorem, *J. Math. Pures Appl.* 35 (1956) 223–248. MR 18-321.

Index of terminologies

Index of notations

Author index